Audio Engineering and Technology

Audio Engineering and Technology

Edited by **Celia Lopez**

Willford Press

New York

Published by Willford Press,
118-35 Queens Blvd., Suite 400,
Forest Hills, NY 11375, USA
www.willfordpress.com

Audio Engineering and Technology
Edited by Celia Lopez

International Standard Book Number: 978-1-68285-232-3 (Hardback)

Printed in China.

Contents

Preface

Audio engineering is a growing field of science that integrates scientific principles and technologies in relation to sound production, recording, processing and mixing. Audio engineers and experts involved in this field adopt different innovative methods and technologies to modify, mix and recreate sound for a wide variety of commercial and creative activities. The field of audio engineering also encompasses developments and technological advancements in the discipline of acoustic engineering that aims to improve the understanding and use of various sound waves, vibrations and audible sounds. This book provides a comprehensive insight into the varied aspects of audio engineering like active sound and noise control, music and audio coding, room acoustics, analyses and syntheses of various sound waves and signals, etc. It is an excellent reference book for all the students, researchers, audio engineers and experts who are looking for an in-depth explanation of the significant concepts in audio science & technology and their applied aspects.

The researches compiled throughout the book are authentic and of high quality, combining several disciplines and from very diverse regions from around the world. Drawing on the contributions of many researchers from diverse countries, the book's objective is to provide the readers with the latest achievements in the area of research. This book will surely be a source of knowledge to all interested and researching the field.

In the end, I would like to express my deep sense of gratitude to all the authors for meeting the set deadlines in completing and submitting their research chapters. I would also like to thank the publisher for the support offered to us throughout the course of the book. Finally, I extend my sincere thanks to my family for being a constant source of inspiration and encouragement.

Editor

1

Bayesian group sparse learning for music source separation

Jen-Tzung Chien* and Hsin-Lung Hsieh

Abstract

Nonnegative matrix factorization (NMF) is developed for parts-based representation of nonnegative signals with the sparseness constraint. The signals are adequately represented by a set of basis vectors and the corresponding weight parameters. NMF has been successfully applied for blind source separation and many other signal processing systems. Typically, controlling the degree of sparseness and characterizing the uncertainty of model parameters are two critical issues for model regularization using NMF. This paper presents the *Bayesian group sparse learning* for NMF and applies it for single-channel music source separation. This method reconstructs the rhythmic or repetitive signal from a *common subspace* spanned by the shared bases for the whole signal and simultaneously decodes the harmonic or residual signal from an *individual subspace* consisting of separate bases for different signal segments. A *Laplacian scale mixture* distribution is introduced for sparse coding given a sparseness control parameter. The relevance of basis vectors for reconstructing two groups of music signals is automatically determined. A Markov chain Monte Carlo procedure is presented to infer two sets of model parameters and hyperparameters through a sampling procedure based on the conditional posterior distributions. Experiments on separating single-channel audio signals into rhythmic and harmonic source signals show that the proposed method outperforms baseline NMF, Bayesian NMF, and other group-based NMF in terms of signal-to-interference ratio.

Keywords: Bayesian sparse learning; Signal reconstruction; Subspace approach; Group sparsity; Nonnegative matrix factorization; Single-channel source separation

1 Introduction

Many problems in audio, speech and music processing can be tackled through matrix factorization. Different cost functions and constraints may lead to different factorized matrices. This procedure can identify underlying sources from the mixed signals through blind source separation [1]. Nonnegative matrix factorization (NMF) is designed to find an approximate factorization $X \approx AS$ for a data matrix X into a basis matrix A and a weight matrix S which are all nonnegative [2]. Some divergence measures have been proposed to derive solutions to NMF [3,4]. NMF provides a useful learning tool for clustering as well as for classification. When a portion of labeled data are available, the semi-supervised NMF was developed for an improved classification system [5]. Different from standard principal component analysis (PCA) and independent component analysis (ICA), NMF only allows additive combination due to the nonnegative constraints on matrices A and S. Nevertheless, nonnegative PCA and nonnegative ICA were proposed for blind source separation in the presence of nonnegative image and music sources [6].

On the other hand, NMF conducts a parts-based sparse representation where only a few components or bases are relevant for representation of input nonnegative matrix X. The sparseness constraint is imposed in objective function [2]. An automatic relevance determination (ARD) scheme [7-9] is developed to determine relevant bases for sparse representation. Such sparse coding is efficient and robust. However, controlling the sparseness or smoothness is influential for system performance. Bayesian learning is beneficial to deal with sparse representation [9] and model regularization [7]. In [10], Bayesian learning was performed for sparse representation of image data where Laplacian distribution was used as prior density. The ℓ_1-regularized optimization was comparably performed. In addition, the group-based NMF [11] was proposed to

*Correspondence: jtchien@nctu.edu.tw
Department of Electrical and Computer Engineering, National Chiao Tung University, Taiwan 30010, Republic of China

capture the intra-subject variations and the inter-subject variations in EEG signals. In [12], the group sparse NMF was proposed by minimizing the Itakura-Saito divergence between X and AS. In [13], NMF was applied for drum source separation where the factorized components were partitioned into rhythmic sources and harmonic sources. No Bayesian learning was performed in [11-13].

More recently, a Bayesian NMF approach [14] was proposed for model selection and image reconstruction. This approach inferred NMF model by a variational Bayes method and a Markov chain Monte Carlo (MCMC) algorithm. In [15], a Bayesian NMF with gamma priors for source signals and mixture weights was implemented through a MCMC algorithm. In [16], the Bayesian NMF with Gaussian likelihood and exponential prior was constructed for image feature extraction where the posterior distribution was approximated by Gibbs sampling procedure. In [17], a Bayesian approach for blind separation of linear mixtures of sources was developed. The Student t distribution for mixture weights was introduced to achieve sparse basis representation. The underdetermined noisy mixtures were separated. However, the case of nonnegative source was not applied. Besides, single-channel source separation is known as an underdetermined problem. In [18], the harmonic structure information was adopted to estimate the demixed instrumental sources. In [19], the NMF was applied for single-channel speech separation where the speech of target speaker over that of masking speaker was enhanced by using sparse dictionaries learned on a phoneme level for individual speakers.

This paper addresses the problem of underdetermined source separation based on NMF for an application to music source separation [20]. The uses of NMF and Bayesian theory to source separation are not new since they have been many papers [11-13,15]. But, to our best knowledge, the novelty of this paper is to propose *Bayesian group sparse* (*BGS*) *learning* using *Laplacian distribution* and *Laplacian scale mixture* (*LSM*) *distribution* and apply it for single-channel music signal separation. We present a group-based NMF where the groups of common bases and individual bases are estimated for blind separation of rhythmic sources and harmonic sources, respectively. Bayesian sparse learning is developed by introducing LSM distributions as the priors for two groups of reconstruction weights. Gamma priors are used to represent two groups of nonnegative basis components. The BGS-NMF algorithm is accordingly established. A MCMC algorithm is derived to infer BGS-NMF parameters and hyperparameters according to *full Bayesian* theory. The rhythmic sources and harmonic sources are reconstructed through the relevant bases in common subspace and individual subspace, respectively. In the experiments, the proposed BGS-NMF is evaluated and compared with the other NMF methods for single-channel separation of audio signals into rhythmic signals and harmonic signals. From comparative study, we find that the improvement of separation performance benefits from Bayesian modeling, group basis representation, and sparse signal reconstruction. Sparser priors identify fewer but more relevant bases and correspondingly lead to a better performance in terms of signal-to-interference ratio.

The remaining of this paper is organized as follows. In the next section, the related studies on NMF and group basis representation are surveyed. Some Bayesian learning approaches are addressed. Section 3 highlights on the construction of BGS-NMF model as well as the inference procedure based on MCMC algorithm. The conditional posterior distributions of different parameters and hyperparameters are derived in the sampling procedure. Section 4 reports a series of experiments on underdetermined music source separation with different music sources. The convergence condition in MCMC sampling is investigated. The evaluation of demixed signals in terms of signal-to-interference ratio is reported. Finally, the conclusions drawn by this study are provided in Section 5.

2 Background survey

In what follows, nonnegative matrix factorization (NMF) and its extensions to different regularization functions are introduced. Several approaches to group basis representation are addressed. Group sparse coding is surveyed. Then Bayesian learning methods for matrix factorization and other related tasks are introduced.

2.1 Nonnegative matrix factorization

NMF is a linear model where the observed signals, factorized signals, and source signals are all assumed to be nonnegative. Given a data matrix $X = \{X_{ik}\}$, NMF estimates two factorized matrices $A = \{A_{ij}\}$ and $S = \{S_{jk}\}$ by minimizing the reconstruction error between X and AS. In [2], the sparseness constraint was imposed on minimization of an objective function $\mathcal{F}$ which is based on a regularized error function

$$\|X - AS\|^2 + \eta_a \sum_i \sum_j f(A_{ij}) + \eta_s \sum_j \sum_k f(S_{jk}) \quad (1)$$

where $\eta_a \geq 0$ and $\eta_s \geq 0$ are regularization parameters and different sparseness measures could be used, e.g., $f(S_{jk}) = |S_{jk}|$, $f(S_{jk}) = S_{jk}$, $f(S_{jk}) = S_{jk} \ln(S_{jk})$, etc. Several extensions of NMF have been proposed. In [21], the nonnegative matrix partial co-factorization (NMPCF) was proposed for rhythmic source separation. Given the magnitude spectrogram as input data matrix X, NMPCF decomposes the music signal into a drum or rhythmic part and a residual or harmonic part $X \approx A_r S_r + A_h S_h$ with the factorized matrices including basis matrix and weight matrix for rhythmic source $\{A_r, S_r\}$ and for harmonic source $\{A_h, S_h\}$. The prior knowledge from drum-only signal $Y \approx A_r S_r$ given the same rhythmic bases A_r is incorporated in joint minimization of two Euclidean error functions

$$\|X - A_r S_r - A_h S_h\|^2 + \eta \, \|Y - A_r S_r\|^2 \tag{2}$$

where η is a trade-off between the first and the second reconstruction errors due to X and Y, respectively. In [22], the mixed signals were divided into L segments. Each segment $X^{(l)}$ is decomposed into common and individual parts which reflect the rhythmic and harmonic sources, respectively. The common bases A_r are shared for different segments due to high temporal repeatability in rhythmic sources. The individual bases $A_h^{(l)}$ are separate for individual segment l due to the changing frequency and low temporal repeatability. The resulting objective function consists of a weighted Euclidean error function and the regularization terms due to bases A_r and $A_h^{(l)}$ which are expressed by

$$\sum_{l=1}^{L} \omega^{(l)} \|X^{(l)} - A_r S_r^{(l)} - A_h^{(l)} S_h^{(l)}\|^2 + \eta L \|A_r\|^2 + \eta \sum_{l=1}^{L} \|A_h^{(l)}\|^2 \tag{3}$$

where $\{\omega^{(l)}, S_r^{(l)}, S_h^{(l)}\}$ denotes the segment-dependent weights and weight matrices for common basis and individual basis, respectively. This is a NMPCF for L segments. The solutions to these NMFs are derived and implemented by the multiplicative update rules so that nonnegative constraints are met for individual model parameters. For example, the terms in gradient of objective function $\mathcal{F}$ with respect to nonnegative parameter A are divided into positive terms and negative terms $\frac{\partial \mathcal{F}}{\partial A} = \left[\frac{\partial \mathcal{F}}{\partial A}\right]^+ - \left[\frac{\partial \mathcal{F}}{\partial A}\right]^-$ where $\left[\frac{\partial \mathcal{F}}{\partial A}\right]^+ > 0$ and $\left[\frac{\partial \mathcal{F}}{\partial A}\right]^- > 0$. The multiplicative update rule is yielded by

$$A \leftarrow A \otimes \left[\frac{\partial \mathcal{F}}{\partial A}\right]^- \oslash \left[\frac{\partial \mathcal{F}}{\partial A}\right]^+ \tag{4}$$

where $\otimes$ and $\oslash$ denote element-wise multiplication and division, respectively.

2.2 Group basis representation

The signal reconstruction methods in (2) and (3) correspond to the group basis representation where two groups of bases A_r and $A_h^{(l)}$ are applied. The separation of single-channel mixed signal into two source signals is achieved. The issue of underdetermined source separation is resolved. In [11], the group-based NMF (GNMF) was developed by conducting group analysis and constructing two groups of bases. The intra-subject variations for a subject in different trials and the inter-subject variations for different subjects could be compensated. Given the L subjects or segments, the lth segment is generated by $X^{(l)} \approx A_r^{(l)} S_r^{(l)} + A_h^{(l)} S_h^{(l)}$ where $A_r^{(l)}$ denotes the common bases which capture the intra and inter-subject variations and $A_h^{(l)}$ denotes the individual bases which reflect the residual information. In general, different common bases $A_r^{(l)}$ should be close together since these bases represent the shared information in mixed signal. Contrarily, individual bases $A_h^{(l)}$ characterize individual features which should be discriminated and mutually far apart [11]. The object function of GNMF is formed by

$$\begin{aligned}
&\sum_{l=1}^{L} \|X^{(l)} - A_r^{(l)} S_r^{(l)} - A_h^{(l)} S_h^{(l)}\|^2 + \eta_a \sum_{l=1}^{L} \|A_r^{(l)}\|^2 \\
&+ \eta_a \sum_{l=1}^{L} \|A_h^{(l)}\|^2 \\
&+ \eta_{a_r} \sum_{l=1}^{L} \sum_{m=1}^{L} \|A_r^{(l)} - A_r^{(m)}\|^2 \\
&- \eta_{a_h} \sum_{l=1}^{L} \sum_{m=1}^{L} \|A_h^{(l)} - A_h^{(m)}\|^2.
\end{aligned} \tag{5}$$

In (5), the second and third terms are seen as the ℓ_2 regularization functions, the fourth term enforces the distance between different common bases to be small, and the fifth term enforces the distance between different individual bases to be large. Regularization parameters $\{\eta_a, \eta_{a_r}, \eta_{a_h}\}$ are used. The NMPCFs in [21,22] and GNMF in [11] did not consider sparsity in group basis representation.

More generally, a group sparse coding algorithm [23] was proposed for basis representation of group instances $\{X_k, k \in \mathcal{G}\}$ where objective function is defined by

$$\sum_{k \in \mathcal{G}} \left\| X_k - \sum_{j=1}^{|D|} S_j^k A_j \right\|^2 + \eta \sum_{j=1}^{|D|} \|S_j\|. \tag{6}$$

All the instances within a group $\mathcal{G}$ share the same dictionary D with basis vectors $\{A_j\}_{j=1}^{|D|}$. The weight matrix $\{S_j\}_{j=1}^{|D|}$ consists of nonnegative vectors $S_j = [S_j^1, \ldots, S_j^{|\mathcal{G}|}]^T$. The weight parameters $\{S_j^k\}$ are estimated for different group instances $k \in \mathcal{G}$ using different bases $j \in \mathcal{D}$. In (6), ℓ_1 regularization term is incorporated to carry out group sparse coding. The group sparsity was further extended to structural sparsity for dictionary learning and basis representation. Nevertheless, nonnegative constraints were not imposed on bases $\{A_j\}$ and observed signals $\{X_k\}$. Basically, all the above-mentioned methods [2,11,21-24] did not apply probabilistic framework. No Bayesian learning was considered.

2.3 Bayesian learning approaches

Model regularization is critical for improving the generalization of a learning machine to new data [7]. Conducting Bayesian learning shall compensate the variations of the estimated parameters and accordingly improve model regularization. Typically, NMF and group basis representation are viewed as learning machine which is based on a set of bases. Following the perspective of relevance vector machines [8,9], Bayesian sparse learning is beneficial to identify relevant bases for regularized basis representation. To do so, sparse priors based on Student t distribution [17] and Laplacian distribution [10,25] could act as regularization functions and merged with likelihood function to come up with *a posteriori* probability. Maximizing the logarithm of *a posteriori* probability is equivalent to minimizing the ℓ_1-regularized error function if Laplacian prior is applied. Hyperparameters of sparse priors are then used as the regularization parameter which controls the trade-off between a reconstruction error function and a sparsity-favorable penalty function.

In the literature, a probabilistic matrix factorization (PMF) [26] for $X = A^{\mathrm{T}}S$ was proposed by assuming Gaussian noise for each independent entry of data matrix $X = \{X_{ik}\}$ by $p(X|A,S,\alpha) = \prod_{i=1}^{N}\prod_{k=1}^{M}\mathcal{N}(X_{ik}|A_i^T S_k, \alpha^{-1})$ and assuming Gaussian priors $p(A|\alpha_{\mathrm{a}}) = \prod_{i=1}^{N}\mathcal{N}(A_i|0,\alpha_{\mathrm{a}}^{-1}I)$ and $p(S|\alpha_{\mathrm{s}}) = \prod_{k=1}^{M}\mathcal{N}(S_k|0,\alpha_{\mathrm{s}}^{-1}I)$ where $\{\alpha,\alpha_{\mathrm{a}},\alpha_{\mathrm{s}}\}$ is a set of precision parameters of Gaussians. Here, A_i denotes the ith column of A and S_k denotes the kth column of S. Learning for PMF is equivalent to maximizing the log posterior likelihood

$$\begin{aligned}\ln p(A,S|X,\alpha,\alpha_{\mathrm{a}},\alpha_{\mathrm{s}}) &= \ln p(X|A,S,\alpha) \\ &\quad + \ln p(A|\alpha_{\mathrm{a}}) + \ln p(S|\alpha_{\mathrm{s}}) + C \end{aligned} \tag{7}$$

with respect to A and S. In (7), C is a constant. This optimization turns out to minimizing the sum-of-squares error function with quadratic regularization terms

$$\sum_{i=1}^{N}\sum_{k=1}^{M}(X_{ik} - A_i^T S_k)^2 + \eta_{\mathrm{a}}\sum_{i=1}^{N}\|A_i\|^2 + \eta_{\mathrm{s}}\sum_{k=1}^{M}\|S_k\|^2. \tag{8}$$

The regularization terms are determined from hyperparameters by $\eta_{\mathrm{a}} = \alpha_{\mathrm{a}}/\alpha$ and $\eta_{\mathrm{s}} = \alpha_{\mathrm{s}}/\alpha$. Bayesian learning of PMF was performed through MCMC algorithm where Gaussian-Wishart priors for Gaussian mean vectors and precision matrices were assumed. There was no constraint on nonnegative matrices by using PMF. No sparse learning was considered.

In [27], a full Bayesian NMF was implemented to determine the number of bases according to the marginal likelihood. Furthermore, Bayesian nonparametric approach to NMF was proposed in [28] where model structure was determined through Gamma process NMF. This method was applied to find both latent sources in spectrograms and their number. In [25], the group sparse coding [23] was upgraded with Bayesian interpretation. Bayesian sparse learning was only developed for single-sample basis representation. In [29], the group sparse priors were presented for maximum *a posteriori* estimation of covariance matrix which was used in Gaussian graphical model. More recently, the group sparse hidden Markov models (HMMs) [30] were proposed to represent a sequence of observations and have been successfully applied for speech recognition. A set of common bases were shared for representation of speech samples across HMM states, while a set of individual bases were employed to represent speech samples within individual HMM states. Bayesian group sparse learning was performed for speech recognition [30] and signal separation [20] by using Laplacian scale mixture distribution.

3 Bayesian group sparse matrix factorization

Previous NMF methods [11,13,21] were developed to extract task-specific nonnegative factors, but they did not simultaneously consider the *uncertainty* of model parameters and control the *sparsity* of weight parameters. In [23,25], the group sparse coding and its Bayesian extension did not impose *nonnegative* constraints in data matrix X and factorized matrices A and S. This paper presents a new Bayesian group sparse learning for NMF (denoted by BGS-NMF) and applied it for single-channel music source separation.

3.1 Model construction

In this study, magnitude spectrogram $X = \{X^{(l)}\}$ of a mixed audio signal is calculated and chopped into L segments for implementation of BGS-NMF algorithm. The audio signal is assumed to be mixed from two kinds of source signals. One is rhythmic or repetitive source signal and the other is harmonic or residual source signal. As illustrated in Figure 1, BGS-NMF aims to decompose a nonnegative matrix $X^{(l)} \in \mathcal{R}_+^{N\times M}$ of the lth segment into a product of two nonnegative matrices $A^{(l)}S^{(l)}$. A linear decomposition model is constructed in a form of

$$X^{(l)} = A_{\mathrm{r}}S_{\mathrm{r}}^{(l)} + A_{\mathrm{h}}^{(l)}S_{\mathrm{h}}^{(l)} + E^{(l)} \tag{9}$$

where $A_{\mathrm{r}} \in \mathcal{R}_+^{N\times D_{\mathrm{r}}}$ denotes the shared basis matrix for all segments $\{X^{(l)}, l = 1,\ldots,L\}$; $A_h^{(l)} \in \mathcal{R}_+^{N\times D_h}$ and $E^{(l)}$ denotes the individual matrix and the noise matrix for a given segment l, respectively. Typically, common bases capture the repetitive patterns which continuously happen in different segments of a whole signal. Individual bases are used to compensate the residual information that common bases could not handle. Without loss of generality, common bases and individual bases are applied

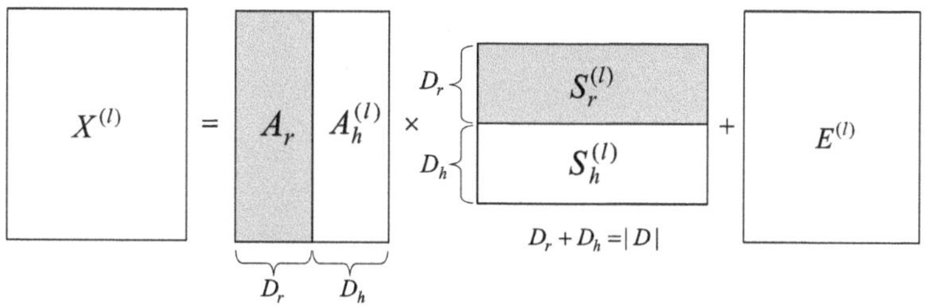

Figure 1 Illustration for group basis representation. There are $|D|$ bases in the dictionary.

to recover the rhythmic signal and the harmonic signal, respectively, from a mixed audio signal. Such a signal recovery problem could be interpreted from a perspective of subspace approach. Namely, an observed signal is demixed into one signal from *principal subspace* spanned by common bases and the other signal from *minor subspace* spanned by individual bases [31]. Moreover, the sparseness constraint is imposed on two groups of reconstruction weights $S_r^{(l)} \in \mathcal{R}_+^{D_r \times M}$ and $S_h^{(l)} \in \mathcal{R}_+^{D_h \times M}$. It is assumed that the reconstruction weights of rhythmic sources $S_r^{(l)}$ and harmonic sources $S_h^{(l)}$ are independent, but the dependencies between reconstruction weights within each group are allowed. Assuming that the kth noise vector $E_k^{(l)}$ is Gaussian distributed with zero mean and $N \times N$ diagonal covariance matrix $\Sigma^{(l)} = \text{diag}\{[\Sigma^{(l)}]_{ii}\}$ which is shared for all samples within a segment l, the likelihood function of an audio signal segment $X^{(l)}$ is expressed by

$$p(X^{(l)}|\Theta^{(l)}) = \prod_{i=1}^{N}\prod_{k=1}^{M} \mathcal{N}(X_{ik}^{(l)} \,|[A_r S_r^{(l)}]_{ik} + [A_h^{(l)} S_h^{(l)}]_{ik}, [\Sigma^{(l)}]_{ii}) . \tag{10}$$

BGS-NMF model is therefore constructed with parameters $\Theta^{(l)} = \{A_r, A_h^{(l)}, S_r^{(l)}, S_h^{(l)}, \Sigma^{(l)}\}$.

3.2 Priors for Bayesian group sparse learning

From Bayesian perspective, the uncertainties of BGS-NMF parameters, expressed by prior densities, are considered to assure *model regularization*. Using BGS-NMF model, the common bases A_r are constructed to represent the characteristics of repetitive patterns for different data segments, while the individual bases $A_h^{(l)}$ are estimated to reflect unique information in each segment l. Sparsity control is enforced in the corresponding reconstruction weights $S_r^{(l)}$ and $S_h^{(l)}$ so that relevant bases are retrieved for group basis representation. In accordance with [15], the nonnegative basis parameters are assumed to be gamma distributed by

$$p(A_r) = \prod_{i=1}^{N}\prod_{j=1}^{D_r} \mathcal{G}([A_r]_{ij} \,|\alpha_{rj}, \beta_{rj}) \tag{11}$$

$$p(A_h^{(l)}) = \prod_{i=1}^{N}\prod_{j=1}^{D_h} \mathcal{G}([A_h^{(l)}]_{ij} \,|\alpha_{hj}^{(l)}, \beta_{hj}^{(l)}) \tag{12}$$

where $\Phi_a^{(l)} = \{\{\alpha_{rj}, \beta_{rj}\}, \{\alpha_{hj}^{(l)}, \beta_{hj}^{(l)}\}\}$ denotes the hyperparameters of gamma distributions and $\{D_r, D_h\}$ denote the numbers of common bases and individual bases, respectively. Gamma distribution is an exponential family distribution for *nonnegative data*. Its two parameters $\{\alpha, \beta\}$ can be adjusted to fit different shapes of distributions. In (11) and (12), all entries in matrices A_r and $A_h^{(l)}$ are assumed to be independent.

Importantly, we control the sparsity of reconstruction weights by using prior density based on the *Laplacian scale mixture* (LSM) distribution [25]. The LSM of a reconstruction weight of common basis is constructed by $[S_r^{(l)}]_{jk} = (\lambda_{rj}^{(l)})^{-1} u_{rj}^{(l)}$ where $u_{rj}^{(l)}$ is a Laplacian distribution $p(u_{rj}^{(l)}) = \frac{1}{2}\exp\{-|u_{rj}^{(l)}|\}$ with scale 1 and $\lambda_{rj}^{(l)}$ is an inverse scale parameter. Accordingly, the parameter $[S_r^{(l)}]_{jk}$ has a Laplacian distribution

$$p([S_r^{(l)}]_{jk} \,|\lambda_{rj}^{(l)}) = \frac{\lambda_{rj}^{(l)}}{2}\exp\{-\lambda_{rj}^{(l)}[S_r^{(l)}]_{jk}\} \tag{13}$$

which is controlled by a positive continuous mixture parameter $\lambda_{rj}^{(l)} \geq 0$. Considering a gamma distribution for inverse scale parameter, i.e., $p(\lambda_{rj}^{(l)}) = \mathcal{G}(\lambda_{rj}^{(l)} \,|\gamma_{rj}^{(l)}, \delta_{rj}^{(l)})$, the marginal distribution of a reconstruction weight can be calculated by [25]

$$p([S_r^{(l)}]_{jk}) = \int_0^{\infty} p([S_r^{(l)}]_{jk} \,|\lambda_{rj}^{(l)}) p(\lambda_{rj}^{(l)}) d\lambda_{rj}^{(l)} = \frac{\gamma_{rj}^{(l)} (\delta_{rj}^{(l)})^{\gamma_{rj}^{(l)}}}{2(\delta_{rj}^{(l)} + [S_r^{(l)}]_{jk})^{\gamma_{rj}^{(l)}+1}} . \tag{14}$$

In (13) and (14), the constraint $[S_r^{(l)}]_{jk} \geq 0$ has been considered. This LSM distribution is obtained by adopting the property that gamma distribution is the *conjugate prior* for Laplacian distribution. In application of image coding, LSM distribution was estimated and measured to be sparser than Laplacian distribution by approximately a factor of 2 [25]. Figure 2 compares Gaussian, Laplacian, and LSM distributions with specific parameters. In this example, LSM is the sharpest distribution among these distributions. In addition, a truncated LSM prior for nonnegative parameter $[S_r^{(l)}]_{jk} \in \mathcal{R}_+$ is adopted, namely, the distribution of negative parameter is forced to be zero. The sparse prior for reconstruction weight for individual basis $[S_h^{(l)}]_{jk}$ is also expressed by LSM distribution with hyperparameter $\{\gamma_{hj}^{(l)}, \delta_{hj}^{(l)}\}$. The hyperparameters of BGS-NMF is formed by $\Phi^{(l)} = \{\Phi_a^{(l)}, \Phi_s^{(l)} = \{\gamma_{rj}^{(l)}, \delta_{rj}^{(l)}, \gamma_{hj}^{(l)}, \delta_{hj}^{(l)}\}\}$. Figure 3 displays a graphical representation for construction of BGS-NMF with different parameters $\Theta^{(l)}$ and hyperparameters $\Phi^{(l)}$.

By combining the likelihood function in (10) and the prior densities in (11) to (13), the negative logarithm of posterior distribution $-\ln p(A_r, A_h^{(l)}, S_r^{(l)}, S_h^{(l)}|X)$ can be calculated and arranged as a new objective function expressed by

$$\begin{aligned}
&\sum_{l=1}^{L}\sum_{i=1}^{N}\sum_{k=1}^{M}(X_{ik}^{(l)} - [A_r S_r^{(l)}]_{ik} - [A_h^{(l)} S_h^{(l)}]_{ik})^2 \\
&\quad + \eta_a L \sum_{i=1}^{N}\sum_{j=1}^{D_r}((1-\alpha_{rj})\ln[A_r]_{ij} \\
&\quad + \beta_{rj}[A_r]_{ij}) \\
&\quad + \eta_a \sum_{l=1}^{L}\sum_{i=1}^{N}\sum_{j=1}^{D_h}((1-\alpha_{hj}^{(l)})\ln[A_h^{(l)}]_{ij} \\
&\quad + \beta_{hj}^{(l)}[A_h^{(l)}]_{ij}) + \eta_{s_r}\sum_{l=1}^{L}\sum_{j=1}^{D_r}\sum_{k=1}^{M}[S_r^{(l)}]_{jk} \\
&\quad + \eta_{s_h}\sum_{l=1}^{L}\sum_{j=1}^{D_h}\sum_{k=1}^{M}[S_h^{(l)}]_{jk}
\end{aligned} \tag{15}$$

where $\{\eta_a, \eta_{s_r}, \eta_{s_h}\}$ denote the regularization parameters for two groups of bases and reconstruction weights. Some BGS-NMF parameters or hyperparameters have been absorbed in these regularization parameters. Comparing with the objective functions (3) for NMPCF, (5) for GNMF, and (8) for PMF, the optimization of (15) for BGS-NMF shall lead to two groups of signals which are reconstructed from the sparse common bases A_r and sparse individual bases $A_h^{(l)}$. The regularization terms due to two gamma bases are additionally considered. Different from the Bayesian NMF (BNMF) [15], BGS-NMF conducts group sparse learning which does not only characterize the within-segment harmonic information but also represent the across-segment rhythmic regularity. Sparse sets of basis vectors are further determined for sparse representation. Basically, BGS-NMF follows a general objective function. By applying different hyperparameter values $\{\alpha_{rj}, \beta_{rj}, \alpha_{hj}^{(l)}, \beta_{hj}^{(l)}\}$, probability structures, and prior distributions for $\{A_r, A_h^{(l)}, S_r^{(l)}, S_h^{(l)}\}$, BGS-NMF can be realized to find solutions to NMF [2], NMPCF [21], GNMF [11], PMF [26], and BNMF [15]. Notably, the objective function in (15) is written for comparative study among different methods. This function only considers BGS-NMF

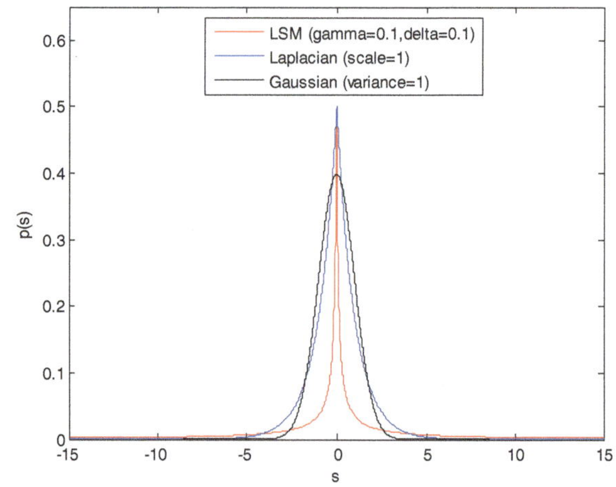

Figure 2 Comparison of Gaussian, Laplacian, and LSM distributions.

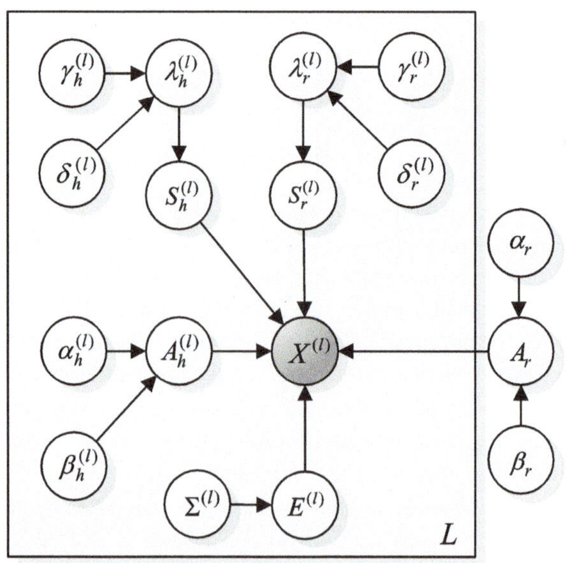

Figure 3 A graphical representation for BGS-NMF.

based on Laplacian prior. BGS-NMF algorithms with Laplacian prior and LSM prior shall be both implemented in the experiments. Nevertheless, in what follows, we address the model inference procedure for *BGS-NMF with LSM prior*.

3.3 Model inference

The full Bayesian framework for BGS-NMF model based on the posterior distribution of parameters and hyperparameters $p(\Theta, \Phi|X)$ is not analytically tractable. A stochastic optimization scheme is adopted. We develop a MCMC sampling algorithm for approximate inference through iteratively generating samples of parameters Θ and hyperparameters Φ according to the posterior distribution. This algorithm converges by those samples. The key idea of MCMC sampling is to simulate a stationary ergodic Markov chain whose samples asymptotically follow the posterior distribution $p(\Theta, \Phi|X)$. The estimates of parameters Θ and hyperparameters Φ are then computed via Monte Carlo integrations on the simulated Markov chains. For simplicity, the segment index l is neglected in derivation of MCMC algorithm for BGS-NMF. At each new iteration $t+1$, the BGS-NMF parameters $\Theta^{(t+1)}$ and hyperparameters $\Phi^{(t+1)}$ are sequentially sampled in an order of $\{A_\mathrm{r}, S_\mathrm{r}, A_\mathrm{h}, S_\mathrm{h}, \Sigma, \alpha_\mathrm{r}, \beta_\mathrm{r}, \alpha_\mathrm{h}, \beta_\mathrm{h}, \lambda_\mathrm{r}, \lambda_\mathrm{h}, \gamma_\mathrm{r}, \delta_\mathrm{r}, \gamma_\mathrm{h}, \delta_\mathrm{h}\}$ according to their corresponding conditional posterior distributions. In this subsection, we describe the calculation of conditional posterior distributions under BGS-NMF parameters $\{A_\mathrm{r}, S_\mathrm{r}, A_\mathrm{h}, S_\mathrm{h}, \Sigma\}$. The conditional posterior distributions for hyperparameters $\{\alpha_\mathrm{r}, \beta_\mathrm{r}, \alpha_\mathrm{h}, \beta_\mathrm{h}, \lambda_\mathrm{r}, \lambda_\mathrm{h}, \gamma_\mathrm{r}, \delta_\mathrm{r}, \gamma_\mathrm{h}, \delta_\mathrm{h}\}$ are derived in the Appendix.

1. Sampling of $[A_\mathrm{r}]_{ij}$. First of all, the common basis parameter $[A_\mathrm{r}^{(t+1)}]_{ij}$ is sampled by the conditional posterior distribution

$$p([A_\mathrm{r}]_{ij}|X_i^T, \Theta_{A_{rij}}^{(t)}, \Phi_{A_{rij}}^{(t)}) \propto p(X_i^T|\Theta_{A_{rij}}^{(t)})p([A_\mathrm{r}]_{ij}|\Phi_{A_{rij}}^{(t)}) \quad (16)$$

where $\Theta_{A_{rij}}^{(t)} = \{[A_\mathrm{r}^{(t+1)}]_{i(1:j-1)}, [A_\mathrm{r}^{(t)}]_{i(j+1:D_\mathrm{r})}, S_\mathrm{r}^{(t)}, A_\mathrm{h}^{(t)}, S_\mathrm{h}^{(t)}, \Sigma^{(t)}\}$ and $\Phi_{A_{rij}}^{(t)} = \{\alpha_{rj}^{(t)}, \beta_{rj}^{(t)}\}$. Here, X_i denotes the ith row vector of X. Notably, for each sampling, we use the preceding bases $[A_\mathrm{r}^{(t+1)}]_{i(1:j-1)}$ at new iteration $t+1$ and subsequent bases $[A_\mathrm{r}]_{i(j+1:D_\mathrm{r})}^{(t)}$ at current iteration t. The likelihood function can be arranged as a Gaussian distribution of $[A_\mathrm{r}]_{ij}$

$$p(X_i^T|\Theta_{A_{rij}}^{(t)}) \propto \exp\left\{-\frac{([A_\mathrm{r}]_{ij} - \mu_{A_{rij}}^{\mathrm{likel}})^2}{2[\sigma_{A_{rij}}^{\mathrm{likel}}]^2}\right\} \quad (17)$$

where $\mu_{A_{rij}}^{\mathrm{likel}} = [\sigma_{A_{rij}}^{\mathrm{likel}}]^{-2}\sum_{k=1}^{M}([S_\mathrm{r}^{(t)}]_{jk}\,\varepsilon_{ik}^{(-j)})$, $\varepsilon_{ik}^{(-j)} = X_{ik} - (\sum_{m=1}^{j-1}[A_\mathrm{r}^{(t+1)}]_{im}[S_\mathrm{r}^{(t)}]_{mk} + \sum_{m=j+1}^{D_\mathrm{r}}[A_\mathrm{r}^{(t)}]_{im}[S_\mathrm{r}^{(t)}]_{mk}) - \sum_{m=1}^{D_\mathrm{h}}[A_\mathrm{h}^{(t)}]_{im}[S_\mathrm{h}^{(t)}]_{mk}$ and $[\sigma_{A_{rij}}^{\mathrm{likel}}]^2 = [\Sigma^{(t)}]_{ii}\,(\sum_{k=1}^{M}[S_\mathrm{r}^{(t)}]_{jk})^{-1}$. By combining likelihood function of (17) and gamma prior $p([A_\mathrm{r}]_{ij}|\Phi_{A_{rij}}^{(t)})$ of (11), the conditional posterior distribution in (16) is derived in a form of

$$[A_r]_{ij}^{\alpha_{rj}^{(t)}-1}\ \exp\left\{-\frac{([A_r]_{ij} - \mu_{A_{rij}}^{\mathrm{post}})^2}{2[\sigma_{A_{rij}}^{\mathrm{post}}]^2}\right\}\mathbb{I}_{[0,+\infty[}([A_r]_{ij}) \quad (18)$$

where $\mu_{A_{rij}}^{\mathrm{post}} = \mu_{A_{rij}}^{\mathrm{likel}} - \beta_{rj}^{(t)}[\sigma_{A_{rij}}^{\mathrm{likel}}]^2$, $[\sigma_{A_{rij}}^{\mathrm{post}}]^2 = [\sigma_{A_{rij}}^{\mathrm{likel}}]^2$, and $\mathbb{I}_{[0,+\infty[}(z)$ denotes an indicator function which has value either 1 if $z \in [0, +\infty[$ or 0 for the other case. In (18), the posterior distribution for negative $[A_\mathrm{r}]_{ij}$ is forced to be zero. Derivations of (17) and (18) are detailed in the Appendix. However, (18) is not an usual distribution, therefore its sampling requires the use of a *rejection sampling* method, such as the Metropolis-Hastings algorithm [32]. Using this algorithm, an *instrumental distribution* $q([A_\mathrm{r}]_{ij})$ is chosen to fit at best the target distribution (18) so that high rejection condition is avoided or equivalently rapid convergence toward true parameter could be achieved. In case of rejection, the previous parameter sample is used, namely, $[A_\mathrm{r}^{(t+1)}]_{ij} \leftarrow [A_\mathrm{r}^{(t)}]_{ij}$. Generally, the shape of target distribution is characterized by its mode and width. The instrumental distribution is constructed as a truncated Gaussian distribution which is calculated by

$$q([A_\mathrm{r}]_{ij}) = \mathcal{N}_+([A_\mathrm{r}]_{ij}|\mu_{A_{rij}}^{\mathrm{inst}}, [\sigma_{A_{rij}}^{\mathrm{inst}}]^2). \quad (19)$$

In (19), the mode $\mu_{A_{rij}}^{\mathrm{inst}}$ is obtained by finding the roots of a quadratic equation of $[A_\mathrm{r}]_{ij}$ which appears in the exponent of the posterior distribution in (18). Derivation for the mode $\mu_{A_{rij}}^{\mathrm{inst}}$ is detailed in the Appendix. In case of complex-valued root or negative-valued root, the mode is forced by $\mu_{A_{rij}}^{\mathrm{inst}} = 0$. The width of instrumental distribution is controlled by $[\sigma_{A_{rij}}^{\mathrm{inst}}]^2 = [\sigma_{A_{rij}}^{\mathrm{post}}]^2$.

2. Sampling of $[S_\mathrm{r}]_{jk}$. The sampling of reconstruction weight of common basis $[S_\mathrm{r}^{(t+1)}]_{jk}$ depends on the conditional posterior distribution

$$p([S_\mathrm{r}]_{jk}|X_k, \Theta_{S_{rjk}}^{(t)}, \Phi_{S_{rjk}}^{(t)}) \propto p(X_k|[S_\mathrm{r}]_{jk}, \Theta_{S_{rjk}}^{(t)})\ p([S_\mathrm{r}]_{jk}|\Phi_{S_{rjk}}^{(t)}) \quad (20)$$

where $\Theta_{S_{rjk}}^{(t)} = \{A_\mathrm{r}^{(t+1)}, [S_\mathrm{r}^{(t+1)}]_{(1:j-1)k}, [S_\mathrm{r}^{(t)}]_{(j+1:D_\mathrm{r})k}, A_\mathrm{h}^{(t)}, S_\mathrm{h}^{(t)}, \Sigma^{(t)}\}$ and $\Phi_{S_{rjk}}^{(t)} = \lambda_{rj}^{(t)}$. X_k is the kth column of X. Again, the preceding weights $[S_\mathrm{r}^{(t+1)}]_{(1:j-1)k}$ at new iteration $t+1$ and subsequent weights $[S_\mathrm{r}^{(t)}]_{(j+1:D_\mathrm{r})k}$ at current iteration tx are used. The likelihood function is rewritten as a Gaussian distribution of $[S_{\mathrm{r}}]_{jk}$ given by

$$p(X_k \mid [S_{\mathrm{r}}]_{jk}, \Theta_{S_{rjk}}^{(t)}) \propto \exp\left\{-\frac{([S_{\mathrm{r}}]_{jk} - \mu_{S_{rjk}}^{\mathrm{likel}})^2}{2[\sigma_{S_{rjk}}^{\mathrm{likel}}]^2}\right\}. \quad (21)$$

The Gaussian parameters are obtained by $\mu_{S_{rjk}}^{\mathrm{likel}} = [\sigma_{S_{rjk}}^{\mathrm{likel}}]^{-2}\sum_{i=1}^{N}([\Sigma^{(t)}]_{ii}^{-1}[A_{\mathrm{r}}^{(t+1)}]_{ij}\varepsilon_{ik}^{(-j)})$, $\varepsilon_{ik}^{(-j)} = X_{ik} - (\sum_{m=1}^{j-1}[A_{\mathrm{r}}^{(t+1)}]_{im}[S_{\mathrm{r}}^{(t+1)}]_{mk} + \sum_{m=j+1}^{D_{\mathrm{r}}}[A_{\mathrm{r}}^{(t+1)}]_{im}[S_{\mathrm{r}}^{(t)}]_{mk}) - \sum_{m=1}^{D_{\mathrm{h}}}[A_h^{(t)}]_{im}[S_{\mathrm{h}}^{(t)}]_{mk}$ and $[\sigma_{S_{rjk}}^{\mathrm{likel}}]^2 = (\sum_{i=1}^{N}[\Sigma^{(t)}]_{ii}^{-1}([A_{\mathrm{r}}^{(t+1)}]_{ij})^2)^{-1}$. Given the Gaussian likelihood and Laplacian prior, the conditional posterior distribution is calculated by

$$\lambda_{rj}^{(t)} \exp\left\{-\frac{([S_{\mathrm{r}}]_{jk} - \mu_{S_{rjk}}^{\mathrm{post}})^2}{2[\sigma_{S_{rjk}}^{\mathrm{post}}]^2}\right\} \mathbb{I}_{[0,+\infty[}([S_{\mathrm{r}}]_{jk}) \quad (22)$$

where $\mu_{S_{rjk}}^{\mathrm{post}} = \mu_{S_{rjk}}^{\mathrm{likel}} - \lambda_{rj}^{(t)}[\sigma_{S_{rjk}}^{\mathrm{likel}}]^2$ and $[\sigma_{S_{rjk}}^{\mathrm{post}}]^2 = [\sigma_{S_{rjk}}^{\mathrm{likel}}]^2$. Notably, the hyperparameters $\{\gamma_{rj}^{(t+1)}, \delta_{rj}^{(t+1)}\}$ in LSM prior are also sampled and used to sample LSM parameter $\lambda_{rj}^{(t+1)}$ based on a gamma distribution. Here, Metropolis-Hastings algorithm is applied again. The best instrumental distribution $q([S_r]_{jk})$ is selected to fit (22). This distribution is derived as a truncated Gaussian distribution $\mathcal{N}_+([S_r]_{jk} \mid \mu_{S_{rjk}}^{\mathrm{inst}}, [\sigma_{S_{rjk}}^{\mathrm{inst}}]^2)$ where the mode $\mu_{S_{rjk}}^{\mathrm{inst}}$ is derived by finding the root of a quadratic equation of $[S_{\mathrm{r}}]_{jk}$ and the width is obtained by $[\sigma_{S_{rjk}}^{\mathrm{inst}}]^2 = [\sigma_{S_{rjk}}^{\mathrm{post}}]^2$. In addition, the conditional posterior distributions for sampling the individual basis parameter $[A_{\mathrm{h}}^{(t+1)}]_{ij}$ and its reconstruction weight $[S_{\mathrm{h}}^{(t+1)}]_{jk}$ are similar to those for sampling $[A_{\mathrm{r}}^{(t+1)}]_{ij}$ and $[S_{\mathrm{r}}^{(t+1)}]_{jk}$, respectively. We do not address these two distributions.

3. *Sampling of* $[\Sigma]_{ii}^{-1}$. The sampling of the inverse of noise variance $([\Sigma]_{ii}^{(t+1)})^{-1}$ is performed according to the conditional posterior distribution

$$p([\Sigma]_{ii}^{-1} \mid X_i^T, \Theta_{\Sigma_{ii}}^{(t)}, \Phi_{\Sigma_{ii}}^{(t)}) \propto p(X_i^T \mid [\Sigma]_{ii}^{-1}, \Theta_{\Sigma_{ii}}^{(t)}) p([\Sigma]_{ii}^{-1} \mid \Phi_{\Sigma_{ii}}^{(t)}) \quad (23)$$

where $\Theta_{\Sigma_{ii}}^{(t)} = \{A_{\mathrm{r}}^{(t+1)}, S_{\mathrm{r}}^{(t+1)}, A_{\mathrm{h}}^{(t+1)}, S_{\mathrm{h}}^{(t+1)}\}$ and $p([\Sigma]_{ii}^{-1} \mid \Phi_{\Sigma_{ii}}^{(t)}) = \mathcal{G}([\Sigma]_{ii}^{-1} \mid \alpha_{\Sigma_{ii}}, \beta_{\Sigma_{ii}})$. The resulting posterior distribution can be derived as a new gamma distribution with updated hyperparameters $\alpha_{\Sigma_{ii}}^{\mathrm{post}} = \frac{M}{2} + \alpha_{\Sigma_{ii}}$ and $\beta_{\Sigma_{ii}}^{\mathrm{post}} = \frac{1}{2}\sum_{k=1}^{M}(X_{ik} - \sum_{m=1}^{D_r}[A_r^{(t+1)}]_{im}[S_r^{(t+1)}]_{mk} - \sum_{m=1}^{D_h}[A_h^{(t+1)}]_{im}[S_h^{(t+1)}]_{mk})^2 + \beta_{\Sigma_{ii}}$. In the experiments, we conduct MCMC sampling procedure for $t_{\max}$ iterations. However, the first $t_{\min}$ iterations are not stable. These burn-in samples are abandoned. The marginal posterior estimates of common basis $[\hat{A}_{\mathrm{r}}]_{ij}$, individual basis $[\hat{A}_{\mathrm{h}}]_{ij}$ and their reconstruction weights $[\hat{S}_{\mathrm{r}}]_{jk}$ and $[\hat{S}_{\mathrm{h}}]_{jk}$ are calculated by finding the following sample means, e.g.,

$$[\hat{A}_{\mathrm{r}}]_{ij} = \frac{1}{t_{\max} - t_{\min}} \sum_{t=t_{\min}+1}^{t_{\max}} [A_{\mathrm{r}}]_{ij}^{(t)}. \quad (24)$$

With these posterior estimates, the rhythmic source and the harmonic source are calculated by $\hat{A}_{\mathrm{r}}\hat{S}_{\mathrm{r}}$ and $\hat{A}_{\mathrm{h}}\hat{S}_{\mathrm{h}}$, respectively. The BGS-NMF algorithm is completed. Different from BNMF [15], the proposed BGS-NMF conducts a *group sparse learning* based on *LSM distribution*. Common bases A_{r} are shared for different data segments l. The group sparse learning performs well in our experiments.

4 Experiments

In this study, BGS-NMF is implemented to estimate two audio source signals from a single-channel mixed signal. One source signal contains rhythmic pattern which is constructed by the bases shared for all audio segments while the other source contains harmonic information which is represented via bases from individual segments. Bayesian sparse learning is performed to conduct probabilistic reconstruction based on the relevant group bases. Some experiments are reported to evaluate the performance of model inference and signal reconstruction.

4.1 Experimental setup

In the experiments, we sampled six rhythmic signals and six harmonic signals from http://www.free-scores.com/index_uk.php3 and http://www.freesound.org/. Six mixed music signals were collected as follows: 'music 1,' bass+piano; 'music 2,' drum+guitar; 'music 3,' drum+violin; 'music 4,' cymbal+organ; 'music 5,' drum+saxophone; and 'music 6,' cymbal+singing, which contained combinations of different rhythmic and harmonic source signals. Three different drum signals and two different cymbal signals were included. For each set of experimental data, we applied a different mixing matrix music 1 (1.2667 − 1.9136), music 2 (1.1667 − 1.9136), music 3 (−1.2667 1.6136), music 4 (1.8667 1.1136), music 5 (−1.1667 2.8136), and music 6 (1.9617 1.1510) to simulate the corresponding single-channel mixed signal. Each audio signal was 21 s long. Readers may access http://chien.cm.nctu.edu.tw/bgs-nmf to listen to the twelve source signals and the corresponding six mixed signals.

The specification of 44,100-Hz sampling rate and 16-bit resolution was used in the collected audio signals. In our implementation, the magnitude of fast Fourier transform of audio signal was extracted every 1,024 samples with 512 samples in frame overlapping. Each mixed signal was equally chopped into L segments for music source separation. Each segment had a length of 3 s. Sufficient rhythmic signal existed within a segment. The numbers of common bases and individual bases were empirically set to be 15 and 10, respectively, i.e., $D_{\rm r} = 15$ and $D_{\rm h} = 10$. The common bases were sufficiently allocated so as to capture the shared base information from different segments. The initial common bases $A_{\rm r}^{(0)}$ and individual bases $A_{\rm h}^{(0)}$ were estimated by applying k-means clustering using the automatically detected rhythmic and harmonic segments, respectively. The detection was based on a classifier using Gaussian mixture model. We performed 1,000 Gibbs sampling iterations ($t_{\max} = 1,000$). The separation performance was evaluated according to the signal-to-interference ratio (SIR) in decibels

$$\mathrm{SIR\ (dB)} = 10\log_{10}\left[\frac{\sum_{l=1}^{L}\sum_{k=1}^{M}\|X_k^{(l)}\|^2}{\sum_{l=1}^{L}\sum_{k=1}^{M}\|\hat{X}_k^{(l)} - X_k^{(l)}\|^2}\right]. \quad (25)$$

The interference was measured by the Euclidean distance between original signal $\{X_k^{(l)}\}$ and reconstructed signal $\{\hat{X}_k^{(l)}\}$ for different samples k in different segments l. These signals include rhythmic signals $\{[\hat{A}_{\rm r}\hat{S}_{\rm r}^{(l)}]_k\}$ and harmonic signals $\{[\hat{A}_{\rm h}^{(l)}\hat{S}_{\rm h}^{(l)}]_k\}$.

For system initialization at $t = 0$, we detected two short segments with only rhythmic signal and harmonic signal and applied them for finding rhythmic parameters $\{A_{\rm r}^{(0)}, S_{\rm r}^{(0)}\}$ and harmonic parameters $\{A_{\rm h}^{(0)}, S_{\rm h}^{(0)}\}$, respectively. This prior information was used to implement five NMF methods for single-channel source separation. We carried out baseline NMF [2], Bayesian NMF (BNMF) [15], group-based NMF (GNMF) [11] (or NMPCF [22]), and the proposed BGS-NMF under consistent experimental conditions. To evaluate the effect of sparse priors in BGS-NMF for music source separation, we additionally realized BGS-NMF by applying Laplacian distribution. For this realization, the sampling steps of LSM parameters $\{\gamma_{r_j}, \delta_{r_j}, \gamma_{h_j}, \delta_{h_j}\}$ were ignored. The BGS-NMFs with Laplacian distribution (denoted by BGS-NMF-LP) and LSM distribution (BGS-NMF-LSM) were compared. All these NMFs were implemented for different segments l. Basically, the NMF model [2] was realized by using multiplicative updating algorithm in (4). The BNMF [15] conducted Bayesian learning of NMF model where MCMC sampling was performed, and gamma distributions were assumed for bases and reconstruction weights. No group sparse learning was considered in NMF and BNMF. Using NMPCF [22] or GNMF [11], the common bases and individual bases were constructed by applying multiplicative updating algorithm. No probabilistic framework was involved. The ℓ_2-norm regularization for basis parameters $A_{\rm r}$ and $A_{\rm h}^{(l)}$ was considered. There was no sparseness constraint imposed on reconstruction weight parameters $S_{\rm r}^{(l)}$ and $S_{\rm h}^{(l)}$. Only the result of GNMF method was reported. Using GNMF, the regularization parameters in (5) were empirically determined as $\{\eta_{\rm a} = 0.35, \eta_{\rm a_r} = 0.2, \eta_{\rm a_h} = 0.2\}$. Nevertheless, the Bayesian group sparse learning is presented in BGS-NMF-LP and BGS-NMF-LSM algorithms. Using this algorithm, the uncertainties of bases and reconstruction weights are represented by gamma distributions and LSM distributions, respectively. MCMC algorithm is developed to sample BGS-NMF parameters $\Theta^{(t+1)}$ and hyperparameters $\Phi^{(t+1)}$. The groups of common bases $A_{\rm r}$ and individual bases $A_{\rm h}$ are estimated to capture between-segment repetitive patterns and within-segment residual information, respectively. The relevant bases are detected via sparse priors in accordance with Laplacian or LSM distributions. Using BGS-NMF-LP, we sampled the parameters and hyperparameters by using different frames from six music signals and automatically calculated the averaged values of regularization parameters in (15) as $\{\eta_{\rm a} = 0.41, \eta_{\rm s_r} = 0.31, \eta_{\rm s_h} = 0.26\}$. The regularization parameters in (5) and (15) reflect different physical meanings in objective function. The computational cost and the model size are also examined. The computation times of running MATLAB codes were measured by a personal computer with Intel Core 2 Duo 2.4-GHz CPU and 4-GB RAM. In our investigation, the computation times of demixing an audio signal with 21 s long were measured as 3.1, 12.1, 16.2, 20.9, and 21.2 min by using NMF, BNMF, GNMF, and the proposed BGS-NMF-LP and BGS-NMF-LSM respectively. In addition, BNMF, GNMF, BGS-NMF-LP, and BGS-NMF-LSM were measured to be 2.5, 4.5, 5.2, and 5.3 times the model size of the baseline NMF respectively.

4.2 Evaluation for MCMC iterative procedure

In this set of experiments, the sampling process of BGS-NMF algorithm is evaluated. The control parameter of sparsity λ_{rj} and its hyperparameters γ_{rj} and δ_{rj} for common basis are investigated. Figure 4 displays an example of MCMC iterative sampling process for LSM parameter $\lambda_{rj}^{(t+1)}$. The value of samples converges after 200 iterations. Also, Figure 5 shows an example of iterative sampling process for LSM hyperparameters $\gamma_{rj}^{(t+1)}$ and $\delta_{rj}^{(t+1)}$. Convergence condition is good in these examples. MCMC samples converge after 200 iterations. Empirically, the parameter t_{min} is specified as 200 when

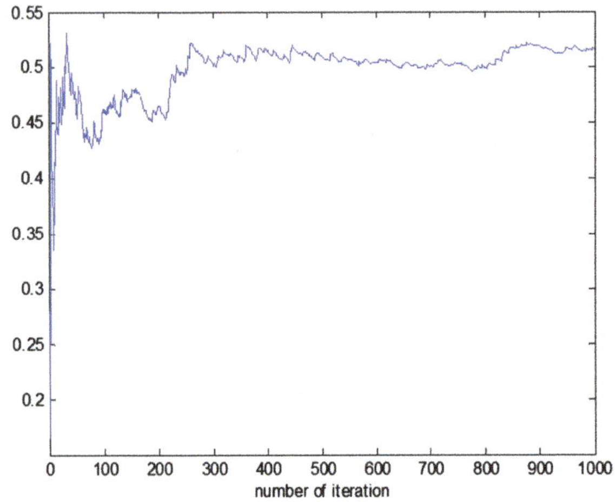

Figure 4 An example of iterative sampling process for LSM parameter $\lambda_{rj}^{(t+1)}$.

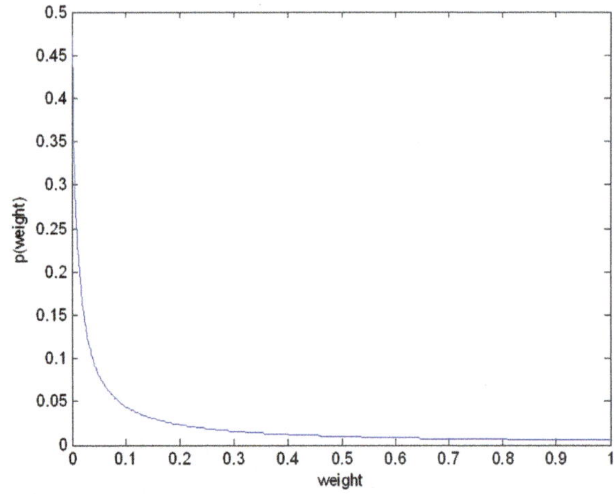

Figure 6 An estimated distribution of reconstruction weight of common basis $p([S_r]_{jk}|\gamma_{rj}, \delta_{rj})$.

calculating posterior estimates of BGS-NMF parameters as given in (24). In addition, Figure 6 shows an estimated distribution of reconstruction weight of common basis $p([S_r]_{jk}|\gamma_{rj}, \delta_{rj})$ where only nonnegative $[S_r]_{jk}$ is valid in the distribution. This distribution is shaped as a LSM distribution which is estimated from the 2nd segment of "music 2".

4.3 Evaluation for single-channel music source separation

A quantitative comparison over different NMFs is conducted by measuring SIRs of reconstructed rhythmic signal and reconstructed harmonic signal. Table 1 shows the experimental results on six mixed music signals. These six signals come from twelve different source signals.

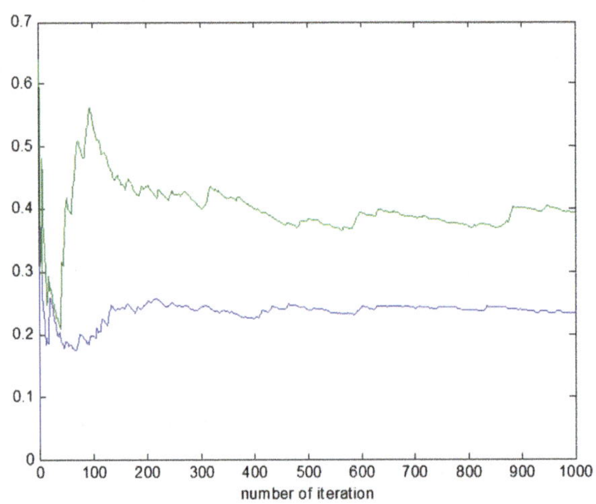

Figure 5 An example of iterative sampling process for LSM hyperparameters $\gamma_{rj}^{(t+1)}$ (green curve) and $\delta_{rj}^{(t+1)}$ (blue curve).

The averaged SIRs are reported in the last row. Comparing NMF and BNMF, we find that BNMF obtains higher SIRs on the reconstructed signals. Further, BNMF is more robust to different combination of rhythmic signals and harmonic signals. The variation of SIRs using NMF is relatively high. Bayesian learning provides model regularization for NMF. On the other hand, GNMF (or NMPCF) performs better than BNMF in terms of averaged SIR of the reconstructed signals. The key difference between BNMF and GNMF is the reconstruction of rhythmic signal. BNMF estimates the rhythmic bases for individual segments while GNMF (or NMPCF) calculates the shared rhythmic bases for different segments. Prior information $\{A_r^{(0)}, S_r^{(0)}, A_h^{(0)}, S_h^{(0)}\}$ is applied for these methods. From these results, we confirm the importance of basis grouping in signal reconstruction based on NMF. In particular, BGS-NMF-LP and BGS-NMF-LSM perform better than other NMF methods. BGS-NMF-LSM even outperforms BGS-NMF-LP in terms of SIRs. Reconstruction weights modeled by LSM distributions are better than those by Laplacian distributions. Sparser reconstruction weights identify fewer but more relevant basis vectors for signal separation. Nevertheless, among these five related NMFs, the highest SIRs of reconstructed signals are achieved by using BGS-NMF-LSM. The SIRs of reconstructed rhythmic and harmonic signals are measured as 8.13 dB and 8.40 dB which are higher than 3.71 dB and 3.38 dB by using NMF, 4.87 dB and 4.61 dB by using BNMF, 5.63 dB and 5.71 dB by using GNMF and 7.91 dB and 8.11 dB by using BGS-NMF-LP, respectively. Basically, the superiority of BGS-NMF-LSM to other NMFs is three-fold, i.e. *Bayesian probabilistic modeling, group basis representation* and *sparse reconstruction weight.* Again, compared to GNMF, the proposed BGS-NMF-LP and BGS-NMF-LSM

Table 1 Comparison of SIR (in dB) of the reconstructed rhythmic signal and harmonic signal based on NMF, BNMF, GNMF, BGS-NMF-LP and BGS-NMF-LSM

	NMF		BNMF		GNMF		BGS-NMF-LP		BGS-NMF-LSM	
	Rhythmic	Harmonic	Rhythmic	Harmonic	Rhythmic	Harmonic	Rhythmic	Harmonic	Rhythmic	Harmonic
Music 1	6.47	4.17	6.33	4.29	9.19	6.10	9.61	8.32	9.86	8.63
Music 2	6.30	1.10	8.08	5.18	8.22	3.03	8.33	7.13	8.55	7.45
Music 3	3.89	-1.11	5.16	3.80	6.01	3.22	8.44	8.52	8.63	8.79
Music 4	2.66	6.03	3.28	6.28	3.59	8.36	7.97	9.52	8.20	9.78
Music 5	1.85	3.71	3.03	2.55	3.97	6.44	8.11	8.22	8.35	8.50
Music 6	1.06	6.37	3.34	5.56	2.78	7.10	5.00	6.93	5.19	7.23
Average	*3.71*	*3.38*	*4.87*	*4.61*	*5.63*	*5.71*	*7.91*	*8.11*	*8.13*	*8.40*

Six mixed music signals are investigated.

obtain a more robust performance in SIRs against different music source signals. Figure 7 shows the waveforms of a drum signal, a saxophone signal and the resulting mixed signal in "music 5". Figure 8 displays the spectrograms of these three signals. Figure 9 demonstrates the spectrograms of the reconstructed drum signal and saxophone signal using BGS-NMF-LSM. For the other five mixed signals, the performance of reconstructed signals in single-channel music source separation is shown at http://chien.cm.nctu.edu.tw/bgs-nmf.

5 Conclusions

This paper has presented the Bayesian group sparse learning and applied it for single-channel nonnegative source separation. The basis vectors in NMF were grouped into two partitions. The first group was the common bases which were used to explore the inter-segment repetitive characteristics, while the second was the individual bases which were applied to represent the intra-segment harmonic information. The LSM distribution was introduced to express sparse reconstruction weights for two groups of basis vectors. Bayesian learning was incorporated into group basis representation with model regularization. The MCMC algorithm or the Metropolis-Hastings algorithm was developed to conduct approximate inference of model parameters and hyperparameters. Model parameters were used to find the decomposed rhythmic signals and harmonic signals. Hyperparameters were used to control the sparsity of reconstructed weights and the generation of basis parameters. In the experiments, we implemented the

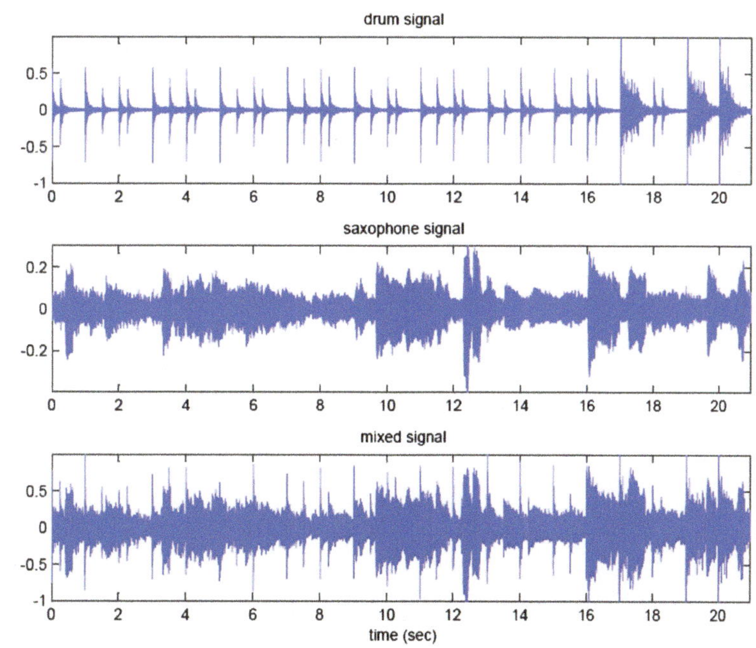

Figure 7 Waveforms of music 5 containing a drum signal, a saxophone signal, and their mixed signal.

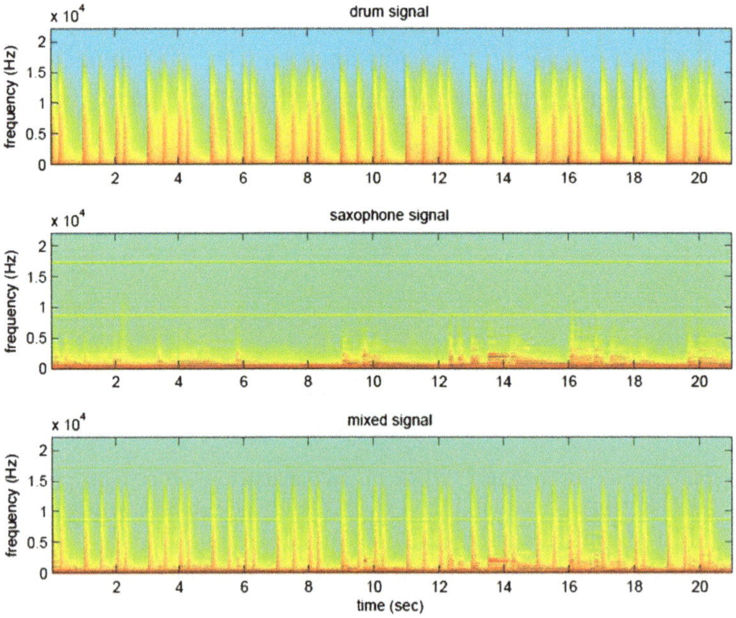

Figure 8 **Spectrograms of music 5 containing a drum signal, a saxophone signal, and their mixed signal.**

proposed BGS-NMFs for underdetermined source separation. The convergence condition of sampling procedure for approximate inference was investigated. The performance of BGS-NMF-LP and BGS-NMF-LSM was shown to be robust to the different kinds of rhythmic and harmonic sources and mixing conditions. BGS-NMF-LSM outperformed the other NMFs in terms of SIRs. The BGS-NMF controlled by LSM distribution performed better than that controlled by Laplacian distribution. In the future, the system performance of BGS-NMF may be further improved by some other considerations. For example, the numbers of common bases and individual bases could be automatically selected according to Bayesian framework by using marginal likelihood. The group sparse learning could be extended for constructing hierarchical NMF where hierarchical grouping of basis vectors is examined. The underdetermined separation under different number of sources and sensors could be tackled. Also, the online learning could be involved to update segment-based parameters and hyperparameters [33,34]. The evolutionary BGS-NMFs shall work for nonstationary single-channel blind source separation. In addition, more evaluations shall be conducted by using realistic data with larger amount of mixed speech signals from different application domains, such as meetings and call centers.

Appendix

Derivations for inference of BGS-NMF parameters and hyperparameters

We address some derivations for model inference of BGS-NMF parameters and hyperparameters. First, the

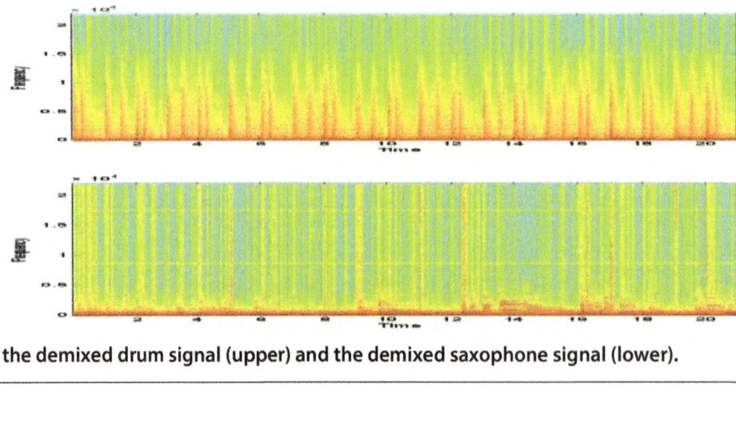

Figure 9 **Spectrograms of the demixed drum signal (upper) and the demixed saxophone signal (lower).**

exponent of the likelihood function $p(X_i^T|[A_r^{(t+1)}]_{i(1:j-1)}, [A_r^{(t)}]_{i(j+1:D_r)}, S_r^{(t)}, A_h^{(t)}, S_h^{(t)}, \Sigma^{(t)})$ in (16) is expressed by

$$-\frac{1}{2[\Sigma^{(t)}]_{ii}}\sum_{k=1}^{M}\left[X_{ik}-\sum_{m=1}^{j-1}[A_r^{(t+1)}]_{im}[S_r^{(t)}]_{mk}\right.$$
$$-[A_r]_{ij}[S_r^{(t)}]_{jk}-\sum_{m=j+1}^{D_r}[A_r^{(t)}]_{im}[S_r^{(t)}]_{mk}$$
$$\left.-\sum_{m=1}^{D_h}[A_h^{(t)}]_{im}[S_h^{(t)}]_{mk}\right]^2 \tag{26}$$

which can be manipulated as a quadratic function of parameter $[A_r]_{ij}$ and leads to (17). The conditional posterior distribution $p([A_r]_{ij}|X_i^T, \Theta_{A_{rij}}^{(t)}, \Phi_{A_{rij}}^{(t)})$ is then derived by combining (17) and (11) and turns out to be

$$[A_r]_{ij}^{\alpha_{rj}^{(t)}-1}$$
$$\exp\left\{-\frac{[A_r]_{ij}^2-2(\mu_{A_{rij}}^{\text{likel}}-\beta_{rj}^{(t)}[\sigma_{A_{rij}}^{\text{likel}}]^2)[A_r]_{ij}+[\mu_{A_{rij}}^{\text{likel}}]^2)}{2[\sigma_{A_{rij}}^{\text{likel}}]^2}\right\}$$
$$\mathbb{I}_{[0,+\infty[}([A_r]_{ij}) \tag{27}$$

which is proportional to (18). In addition, when finding the mode of (18), we take logarithm of (18) and solve a corresponding quadratic equation of $[A_r]_{ij}$ as

$$\frac{\partial}{\partial[A_r]_{ij}}\left\{(\alpha_{rj}^{(t)}-1)\ln[A_r]_{ij}-\frac{([A_r]_{ij}-\mu_{A_{rij}}^{\text{post}})^2}{2[\sigma_{A_{rij}}^{\text{post}}]^2}\right\}=0$$
$$\Rightarrow[A_r]_{ij}^2-\mu_{A_{rij}}^{\text{post}}[A_r]_{ij}-(\alpha_{rj}^{(t)}-1)[\sigma_{A_{rij}}^{\text{post}}]^2=0. \tag{28}$$

By defining $\triangle=(\mu_{A_{rij}}^{\text{post}})^2+4(\alpha_{rj}^{(t)}-1)[\sigma_{A_{rij}}^{\text{post}}]^2$, the mode is determined by

$$\mu_{A_{rij}}^{\text{inst}}=\begin{cases}0, & \text{if } \triangle<0\\ \max\{\frac{1}{2}(\mu_{A_{rij}}^{\text{post}}+\sqrt{\triangle}),0\}, & \text{else.}\end{cases} \tag{29}$$

On the other hand, following the model inference in Section 3.3, we continue to describe the MCMC sampling algorithm and the calculation of conditional posterior distributions for the remaining BGS-NMF hyperparameters $\{\alpha_r, \beta_r, \alpha_h, \beta_h, \lambda_r, \lambda_h, \gamma_r, \delta_r, \gamma_h, \delta_h\}$.

4. Sampling of α_{rj}. The hyperparameter $\alpha_{rj}^{(t+1)}$ is sampled according to a conditional posterior distribution which is obtained by combining a likelihood function of $[A_r]_{ij}$ and an exponential prior density of α_{rj} with parameter $\lambda_{\alpha_{rj}}$. The resulting distribution is written by

$$p(\alpha_{rj}|[A_r^{(t+1)}]_{ij}, \beta_{rj}^{(t)}) \propto \left(\frac{1}{\Gamma(\alpha_{rj})}\exp\{\lambda_{\alpha_{rj}}^{\text{post}}\alpha_{rj}\}\right)^{D_r}\mathbb{I}_{[0,+\infty[}(\alpha_{rj}) \tag{30}$$

where $\lambda_{\alpha_{rj}}^{\text{post}}=\ln\beta_{rj}^{(t)}+(1/D_r)\sum_{j=1}^{D_r}\ln[A_r^{(t+1)}]_{ij}-(1/D_r)\lambda_{\alpha_{rj}}$. This distribution does not belong to a known family, so the Metropolis-Hastings algorithm is applied. An instrumental distribution $q(\alpha_{rj})$ is obtained by fitting the term within the brackets of (30) through a gamma distribution as detailed in [15].

5. Sampling of β_{rj}. The hyperparameter $\beta_{rj}^{(t+1)}$ is sampled according to a conditional posterior distribution which is obtained by combining a likelihood function of $[A_r]_{ij}$ and a gamma prior density of β_{rj} with parameters $\{\alpha_{\beta_{rj}}, \beta_{\beta_{rj}}\}$, i.e.,

$$p(\beta_{rj}|[A_r^{(t+1)}]_{ij}, \alpha_{rj}^{(t+1)}) \propto (\beta_{rj})^{D_r\alpha_{rj}^{(t+1)}}$$
$$\times\exp\left\{-\beta_{rj}\sum_{j=1}^{D_r}[A_r^{(t+1)}]_{ij}\right\}\mathcal{G}(\beta_{rj}|\alpha_{\beta_{rj}}, \beta_{\beta_{rj}}). \tag{31}$$

The resulting distribution is arranged as a new gamma distribution $\mathcal{G}(\beta_{rj}|\alpha_{\beta_{rj}}^{\text{post}}, \beta_{\beta_{rj}}^{\text{post}})$ where $\alpha_{\beta_{rj}}^{\text{post}}=1+D_r\alpha_{rj}^{(t+1)}+\alpha_{\beta_{rj}}$ and $\beta_{\beta_{rj}}^{\text{post}}=\sum_{j=1}^{D_r}[A_r^{(t+1)}]_{ij}+\beta_{\beta_{rj}}$. Here, we do not describe the sampling of $\alpha_{hj}^{(t+1)}$ and $\beta_{hj}^{(t+1)}$ since the conditional posterior distributions for sampling these two hyperparameters are similar to those for sampling of $\alpha_{rj}^{(t+1)}$ and $\beta_{rj}^{(t+1)}$.

6. Sampling of λ_{rj} or λ_{hj}. For sampling of scaling parameter $\lambda_{rj}^{(t+1)}$, the conditional posterior distribution is obtained by

$$p(\lambda_{rj}|[S_r^{(t+1)}]_{j(k=1:M)}, \gamma_{rj}^{(t)}, \delta_{rj}^{(t)}) \propto \prod_{k=1}^{M}p([S_r^{(t+1)}]_{jk}|\lambda_{rj})p(\lambda_{rj}|\gamma_{rj}^{(t)}, \delta_{rj}^{(t)})$$
$$\propto(\lambda_{rj})^{M\gamma_{rj}^{(t)}}\exp\left\{-M\lambda_{rj}\left(\delta_{rj}^{(t)}+\sum_{k=1}^{M}[S_r^{(t+1)}]_{jk}\right)\right\}. \tag{32}$$

7. Sampling of γ_{rj}. The sampling of LSM parameter $\gamma_{rj}^{(t+1)}$ is performed by using the conditional posterior distribution which is derived by combining a likelihood function of λ_{rj} and an exponential prior density of γ_{rj} with parameter $\lambda_{\gamma_{rj}}$. The resulting distribution is expressed as

$$p(\gamma_{rj}|\lambda_{rj}^{(t+1)}, \delta_{rj}^{(t)}) \propto \frac{1}{\Gamma(\gamma_{rj})} \exp\{\lambda_{\gamma_{rj}}^{\text{post}} \gamma_{rj}\} \mathbb{I}_{[0,+\infty[}(\gamma_{rj}), \tag{33}$$

where $\lambda_{\gamma_{rj}}^{\text{post}} = \ln \delta_{rj}^{(t)} + \frac{\gamma_{rj}-1}{\gamma_{rj}} \ln \lambda_{rj}^{(t+1)} - \lambda_{\gamma_{rj}}$. Again, we need to find an instrumental distribution $q(\gamma_{rj})$ which optimally fits the conditional posterior distribution $p(\gamma_{rj}|\lambda_{rj}^{(t+1)}, \delta_{rj}^{(t)})$. An approximate gamma distribution is found accordingly. The Metropolis-Hastings algorithm is then applied.

8. Sampling of δ_{rj}. The sampling of the other LSM parameter $\delta_{rj}^{(t+1)}$ is performed by using the conditional posterior distribution which is derived from a likelihood function of λ_{rj} and a gamma prior density of δ_{rj} with parameters $\{\alpha_{\delta_{rj}}, \beta_{\delta_{rj}}\}$

$$\begin{aligned} p(\delta_{rj}|\lambda_{rj}^{(t+1)}, \gamma_{rj}^{(t+1)}) &\propto (\delta_{rj})^{\gamma_{rj}^{(t+1)}} \\ &\exp\{-\delta_{rj}\lambda_{rj}^{(t+1)}\} \mathcal{G}(\delta_{rj}|\alpha_{\delta_{rj}}, \beta_{\delta_{rj}}). \end{aligned} \tag{34}$$

This distribution can be arranged as a new gamma distribution $\mathcal{G}(\delta_{rj}|\alpha_{\delta_{rj}}^{\text{post}}, \beta_{\delta_{rj}}^{\text{post}})$ where $\alpha_{\delta_{rj}}^{\text{post}} = D_r \gamma_{rj}^{(t+1)} + \alpha_{\delta_{rj}}$ and $\beta_{\delta_{rj}}^{\text{post}} = \lambda_{rj}^{(t+1)} + \beta_{\delta_{rj}}$. Similarly, the conditional posterior distributions for sampling $\gamma_{hj}^{(t+1)}$ and $\delta_{hj}^{(t+1)}$ could be formulated by referring those for sampling $\gamma_{rj}^{(t+1)}$ and $\delta_{rj}^{(t+1)}$, respectively.

Competing interests

Both authors declare that they have no competing interests.

Acknowledgments

The authors acknowledge anonymous reviewers for their constructive feedback and helpful suggestions. This work has been partially supported by the National Science Council, Taiwan, Republic of China, under contract NSC 100-2628-E-009-028-MY3.

References

1. A Cichocki, R Zdunek, S Amari, in *Proceedings of International Conference on Acoustic, Speech and Signal Processing (ICASSP)*. New algorithms for non-negative matrix factorization in applications to blind source separation (IEEE, Piscataway, 2006), pp. 621–624
2. PO Hoyer, Non-negative matrix factorization with sparseness constraints. J. Mach. Lear. Res. **5**, 1457–1469 (2004)
3. J-T Chien, H-L Hsieh, Convex divergence ICA for blind source separation. IEEE Trans. Audio, Speech, Language Process. **20**(1), 290–301 (2012)
4. R Kompass, A generalized divergence measure for nonnegative matrix factorization. Neural Comput. **19**, 780–791 (2007)
5. H Lee, J Yoo, S Choi, Semi-supervised nonnegative matrix factorization. IEEE Signal Process. Lett. **17**(1), 4–7 (2010)
6. MD Plumbley, Algorithms for nonnegative independent component analysis. IEEE Trans. Neural Netw. **14**(3), 534–543 (2003)
7. CM Bishop, *Pattern Recognition and Machine Learning* (Springer Science, New York, 2006)
8. G Saon, J-T Chien, Bayesian sensing hidden Markov models. IEEE Trans. Audio, Speech Language, Process. **20**(1), 43–54 (2012)
9. ME Tipping, Sparse Bayesian learning and the relevance vector machine. J Mach. Learn. Res. **1**, 211–244 (2001)
10. SD Babacan, R Molina, AK Katsaggelos, Bayesian compressive sensing using Laplace priors. IEEE Trans. Image Process. **19**(1), 53–63 (2010)
11. H Lee, S Choi, in *Proceedings of the International Conference on Artificial Intelligence and Statistics (AISTATS)*, vol. 15. Group nonnegative matrix factorization for EEG classification (JMLR, 2009), pp. 320–327
12. A Lefevre, F Bach, C Fevotte, Itakura-Saito, in *Proceedings of the International Conference on Acoustic, Speech and Signal Processing (ICASSP)*. Nonnegative matrix factorization with group sparsity (Prague Congress Center, 22–27 May 2011), pp. 21–24
13. M Kim, J Yoo, K Kang, S Choi, in *Proceedings of the International Conference on Acoustic, Speech and Signal Processing (ICASSP)*. Blind rhythmic source separation: nonnegativity and repeatability (IEEE, Piscataway, 2010), pp. 2006–2009
14. AT, Cemgil, Bayesian inference for nonnegative matrix factorization models. University of Cambridge, Technical Report CUED/F-INFENG/TR.609, 2008
15. S Moussaoui, D Brie, Mohammad-A Djafari, C Carteret, Separation of non-negative mixture of non-negative sources using a Bayesian approach and MCMC sampling. IEEE Trans. Signal Process. **54**(11), 4133–4145 (2006)
16. MN Schmidt, O Winther, LK Hansen, in *Proceedings of the International Conference on Independent Component Analysis and Signal Separation*, Paraty, March 2009. Lecture Notes in Computer Science, vol. 5441. Bayesian non-negative matrix factorization (Springer, Heidelberg, 2009), pp. 540–547
17. C Fevotte, SJ Godsill, A Bayesian approach for blind separation of sparse sources. IEEE Trans. Audio, Speech, Language Process. **14**(6), 2174–2188 (2006)
18. Z Duan, Y Zhang, C Zhang, Z Shi, Unsupervised single-channel music source separation by average harmonic structure modeling. IEEE Trans. on Audio, Speech, Language Process. **16**(4), 766–778 (2008)
19. MN Schmidt, RK Olsson, in *Proceedings of the Annual Conference of International Speech Communication Association (INTERSPEECH)* Single-channel speech separation using sparse non-negative matrix factorization (Pittsburgh, 17–21 September 2006), pp. 2614–2617
20. J-T Chien, H-L Hsieh, in *Proceedings of the Annual Conference of International Speech Communication Association (INTERSPEECH)*. Bayesian group sparse learning for nonnegative matrix factorization (Portland, 9–13 September 2012), pp. 1552–1555
21. J Yoo, M Kim, K Kang, Choi S, in *Proceedings of the International Conference on Acoustic, Speech and Signal Processing (ICASSP)* Nonnegative matrix partial co-factorization for drum source separation (IEEE, Piscataway, 2010), pp. 1942–1945
22. M Kim, J Yoo, K Kang, S Choi, Nonnegative matrix partial co-factorization for spectral and temporal drum source separation. IEEE J. Sel. Top. Signal Process. **5**(6), 1192-1204 (2011)
23. S Bengio, F Pereira, Y Singer, D Strelow, in *Advances in Neural Information Processing Systems (NIPS)*, vol. 22. Group sparse coding (NIPS La Jolla, 2009), pp. 82–89
24. R Jenatton, J Mairal, G Obozinski, F Bach, in *Proceedings of the International Conference on Machine Learning (ICML)*. Proximal methods for sparse hierarchical dictionary learning (Haifa, 21–25 June 2010)
25. PJ Garrigues, BA Olshausen, in *Advances in Neural Information Processing Systems (NIPS)*, vol. 23. Group sparse coding with a Laplacian scale mixture prior (NIPS La Jolla, 2010), pp. 676–684
26. R Salakhutdinov, A Mnih, in *Proceedings of the International Conference on Machine Learning (ICML)*. Bayesian probabilistic matrix factorization using Markov chain Monte Carlo (Helsinki, 5–9 July 2008), pp. 880–887
27. M Zhong, M Girolami, in *Proceedings of the International Conference on Artificial Intelligence and Statistics (AISTATS)*. Reversible jump MCMC for non-negative matrix factorization (Clearwater Beach, 16–18 April 2009), pp. 663–670
28. MD Hoffman, DM Blei, PR Cook, in *Proceedings of the International Conference on Machine Learning (ICML)*. Bayesian nonparametric matrix factorization for recorded music (Haifa, 21–24 June 2010)

29. M Marlin, BM, Schmidt, KP Murphy, in *Proceedings of the Conference on Uncertainty in Artificial Intelligence (UAI)*. Group sparse priors for covariance estimation (Montreal, 18–21 June 2009), pp. 383–392
30. J-T Chien, C-C Chiang, in *Proceedings of the Annual Conference of International Speech Communication Association (INTERSPEECH)*. Group sparse hidden Markov models for speech recognition (Portland, 9–13 September 2012), pp. 2646–2649
31. J-T Chien, C-W Ting, Factor analyzed subspace modeling and selection. IEEE Trans. Audio, Speech Language Process. **16**(1), 239–248 (2008)
32. S Chib, E Greenberg, Understanding the Metropolis-Hastings algorithm. Am. Statistician. **49**(4), 327–335 (1995)
33. J-T Chien, H.-L Hsieh, Nonstationary source separation using sequential and variational Bayesian learning. IEEE Trans. Neural Netw. Learn. Syst. **24**(5), 681–694 (2013)
34. H-L Hsieh, J-T Chien, in *Proceedings of the International Conference on Acoustic, Speech and Signal Processing (ICASSP)*. Nonstationary and temporally-correlated source separation using Gaussian process (Prague Congress Center, 22–27 May 2011), pp. 2120–2123

2

N-dimensional N-microphone sound source localization

Ali Pourmohammad* and Seyed Mohammad Ahadi

Abstract

This paper investigates real-time N-dimensional wideband sound source localization in outdoor (far-field) and low-degree reverberation cases, using a simple N-microphone arrangement. Outdoor sound source localization in different climates needs highly sensitive and high-performance microphones, which are very expensive. Reduction of the microphone count is our goal. Time delay estimation (TDE)-based methods are common for N-dimensional wideband sound source localization in outdoor cases using at least $N + 1$ microphones. These methods need numerical analysis to solve closed-form non-linear equations leading to large computational overheads and a good initial guess to avoid local minima. Combined TDE and intensity level difference or interaural level difference (ILD) methods can reduce microphone counts to two for indoor two-dimensional cases. However, ILD-based methods need only one dominant source for accurate localization. Also, using a linear array, two mirror points are produced simultaneously (half-plane localization). We apply this method to outdoor cases and propose a novel approach for N-dimensional entire-space outdoor far-field and low reverberation localization of a dominant wideband sound source using TDE, ILD, and head-related transfer function (HRTF) simultaneously and only N microphones. Our proposed TDE-ILD-HRTF method tries to solve the mentioned problems using source counting, noise reduction using spectral subtraction, and HRTF. A special reflector is designed to avoid mirror points and source counting used to make sure that only one dominant source is active in the localization area. The simple microphone arrangement used leads to linearization of the non-linear closed-form equations as well as no need for initial guess. Experimental results indicate that our implemented method features less than 0.2 degree error for angle of arrival and less than 10% error for three-dimensional location finding as well as less than 150-ms processing time for localization of a typical wideband sound source such as a flying object (helicopter).

Keywords: Sound source localization; ITD; TDE; TDOA; PHAT; ILD; HRTF

1. Introduction

Source localization has been one of the fundamental problems in sonar [1], radar [2], teleconferencing or videoconferencing [3], mobile phone location [4], navigation and global positioning systems (GPS) [5], localization of earthquake epicenters and underground explosions [6], microphone arrays [7], robots [8], microseismic events in mines [9], sensor networks [10,11], tactile interaction in novel tangible human-computer interfaces [12], speaker tracking [13], surveillance [14], and sound source tracking [15]. Our goal is real-time sound source localization in outdoor environments, which necessitates a few points to be considered. For localizing such sound sources, a far-field assumption is usual. Furthermore, our experiments confirm that placing localization system in suitable higher heights often reduces the reverberation degree, especially for flying objects. Also, many such sound source signals are wideband signals. Moreover, outdoor high-accuracy sound source localization in different climates needs highly sensitive and high-performance microphones which are very expensive. Therefore, reduction of the number of microphones is very important, which in turn leads to reduced localization accuracies using conventional methods. Here, we intend to introduce a real-time accurate wideband sound source localization system in low degree reverberation far-field outdoor cases using fewer microphones.

The structure of this paper is as follows. After a literature review, we explain HRTF-, ILD-, and TDE-based methods and discuss TDE-based phase transform (PHAT). In Section 4, we explain sound source angle of arrival and location calculations using ILD and PHAT. Section 5

* Correspondence: pourmohammad@aut.ac.ir
Electrical Engineering Department, Amirkabir University of Technology, 424 Hafez Ave., Tehran 15914, Iran

covers the introduction of TDE-ILD-based method to two-dimensional (2D) half-plane sound source localization using only two microphones. Section 6 includes simulation of this method for 2D cases, where according to simulation results and due to the use of ILD, we introduce source counting. In Section 7, we propose, and in Section 8, we implement our TDE-ILD-HRTF-based method for 2D whole-plane and three-dimensional (3D) entire-space sound source localization. Section 9 includes the implementations. Finally conclusions will be made in Section 10.

2. Literature review

Passive sound source localization methods, in general, can be divided into direction of arrival (DOA) [16], time difference of arrival (TDOA) or TDE or interaural time difference (ITD)- [17-20], ILD- [21-24], and HRTF-based methods [25-30]. DOA-based beamforming and sub-space methods typically need a large number of microphones for estimation of narrowband source locations in far-field cases and wideband source locations in near-field cases. Also, they have higher processing needs in comparison to other methods. Many localization methods for near-field cases have been proposed in the literature, such as maximum likelihood (ML), covariance approximation, multiple signal estimation (MUSIC), and estimation of signal parameters via rotational invariance techniques (ESPRIT) [16]. However, these methods are not applicable to the localization of wideband signal sources in far-field cases with small number of microphones. On the other hand, ILD-based methods are mostly applicable to the case of a single dominant sound source (high signal-to-noise ratio (SNR)) [21-24]. TDE-based methods with high sampling rates are commonly used for 2D and 3D high-accuracy wideband near-field and far-field sound source localization. In the case of ILD or TDE-based methods, minimum number of microphones required is three for 2D positioning and four for the 3D case [17-20,31-39]. Finally, HRTF-based methods are applicable only to the case of calculating arrival angle in azimuth or elevation [30].

TDE- and ILD-based outdoor far-field accurate wideband sound source localization in different climates needs highly sensitive and high-performance microphones which are very expensive. In the last decade, some papers were published which introduce 2D sound source localization methods using just two microphones in indoor cases using TDE- and ILD-based methods simultaneously. However, it is noticeable that using ILD-based methods requires only one dominant source to be active, and it is known that, by using a linear array in the proposed TDE-ILD-based method, two mirror points will be produced simultaneously (half-plane localization) [40]. In this paper, we apply this method in outdoor (low-degree reverberation) cases for a dominant sound source. We also propose a novel method to have 2D whole-plane (without producing two mirror points) and 3D entire-space dominant sound source localization using TDE-, ILD-, and HRTF-based methods simultaneously (TDE-ILD-HRTF method). Based on the proposed method, a special reflector for the implemented simple microphone arrangement is designed, and source counting method is used to find that only one dominant sound source is active in the localization area.

In TDE and ILD localization approaches, calculations are carried out in two stages: estimation of time delay or intensity level differences and location calculation. Correlation is most widely used for time delay estimation [17-20,31-39]. The most important issue in this approach is high-accuracy time delay estimation between microphone pairs. Meanwhile, the most important issue in ILD-based approach is high accuracy level difference measurement between microphone pairs [21-24]. Also, numerous results were published in the last decades for the second stage, i.e., location calculation. Equation complexities and large processing times are the important obstacles faced at this stage. In this paper, we propose a simple microphone arrangement that solves both these problems simultaneously.

In the above-mentioned first stage, the classic methods of source localization from time delay estimates by detecting radio waves were Loran and Decca [31]. However, generalized cross-correlation (GCC) using a ML estimator, proposed by Knapp and Carter [32], is the most widely used method for TDE. Later, a number of techniques were proposed to improve GCC in the presence of noise [3,33-36]. As GCC is based on an ideal signal propagation model, it is believed to have a fundamental weakness of inability to cope well with reverberant environments. Some improvements were obtained by cepstral prefiltering by Stephenne and Champagne [37]. Even though more sophisticated techniques exist, they tend to be computationally expensive and are thus not well suited for real-time applications. Later, PHAT was proposed by Omologo and Svaizer [38]. More recently, a new PHAT-based method was proposed for high-accuracy robust speaker localization, known as steered response pattern-phase or power-phase transform (SRP-PHAT) [39]. Its disadvantage is higher processing time in comparison to PHAT, as it requires a search of a large number of candidate locations. According to the fact that in the application of this research, the number of candidate locations is much higher due to direction and distance estimation; this disadvantage does not allow us to use it in real-time applications.

In the last decade, according to the fact that the received signal energy is inversely proportional to the squared distance between the source and the receiving sensor, there has been some interest in using the received signal level at different sensors for source localization. Due to the spatial separation of the sensors, the source signal will arrive at the sensors with different levels so that the level differences can be utilized for source localization. Sheng and Hu [41] followed by Blatt and Hero [42] have proposed different algorithms for locating sources using a sensor network based on energy measurements. Birchfield and Gangishetty [22] applied ILD to sound source localization. While these works used only ILD-based methods to locate a source, Cui et al. [40] tried 2D sound source localization by a pair of microphones using TDE- and ILD-based methods simultaneously. When the source signal is captured at the sensors, both time delay and signal level information are used for source localization. This technique is applicable for 2D localization with two sensors only. Also, due to the use of a linear array, it generates two mirror points simultaneously (half-plane localization). Ho and Sun [24] addressed a more common scenario of 3D localization using more than four sensors to improve the source location accuracy.

Human hearing system allows finding sound sources direction of arrival in 3D with just two ears. Pinnas, shoulders, and head diffract the incoming sound waves [43]. These propagation effects collectively are termed the HRTF. Batteau reported that the external ears play an important role in estimating the elevation angle of arrival [44]. Roffler and Butler [45], Oldfield and Parker [46], and Hofman et al. [47] have tried to find experimental evidence for this claim by using a Plexiglas headband to flatten the pinna against the head [45]. Based on HRTF measuring, Brown and Duda have made an extensive experimental study and provided empirical formulas for the multipath delays produced by pinna [48]. Although more sophisticated HRTF models have been proposed [49], the Brown-Duda model has the advantage that it provides an analytical relationship between the multipath delays and the azimuth and elevation angles. Recently, Sen and Nehorai considered the Brown-Duda HRTF model as an example to model the frequency-dependent head-shadow effects and the multipath delays close to the sensors for analyzing a 3D direction finding system with only two sensors inspired by the human auditory system [43]. However, they did not consider white noise gain error or spatial aliasing error in their model. They computed the asymptotic frequency domain Cramer-Rao bound (CRB) on the error of the 3D direction estimate for zero-mean wide-sense stationary Gaussian source signals. It should be noted that HRTF-based works are just able to estimate the azimuth and elevation angles [50]. In the last decades, some papers were published which tried to apply HRTF along with TDE for azimuth and elevation angle of arrival estimation [30]. According to this ability, we apply HRTF in our TDE-ILD-based localization system for solving the ambiguity in the generation of two mirror location points. We named it TDE-ILD-HRTF method.

Given a set of TDEs and ILDs from a small set of microphones using PHAT- and ILD-based methods, respectively, the second stage of a two-stage algorithm determines the best point for the source location. The measurement equations are non-linear. The most straightforward way is to perform an exhaustive search in the solution space. However, this is computationally expensive and inefficient. If the sensor array is known to be linear, the position measurement equations are simplified. Carter focused on a simple beamforming technique [1]. However, it requires a search in the range and bearing space. Also, beamforming methods need many more microphones for high-accuracy source localization. The linearization solution based on Taylor-series expansion by Foy [51] involves iterative processing, typically incurs high computational complexity, and for convergence, requires a tolerable initial estimate of the position. Hahn proposed an approach [20] that assumes a distant source. Abel and Smith proposed an explicit solution that can achieve the Cramer-Rao Lower Bound (CRLB) in the small error region [52]. The situation is more complex when sensors are distributed arbitrarily. In this case, emitter position is determined from the intersection of a set of hyperbolic curves defined by the TDOA estimates using non-Euclidean geometry [53,54]. Finding the solution is not easy as the equations are non-linear. Schmidt has proposed a formulation [18] in which the source location is found as the focus of a conic passing through three sensors. This method can be extended to an optimal closed-form localization technique [55]. Delosme [56] proposed a gradient method for search in a localization procedure leading to computation of optimal source locations from noisy TDOA's. Fang [57] gave an exact solution when the number of TDOA measurements is equal to the number of unknowns (coordinates of transmitter). This solution, however, cannot make use of extra measurements, available when there are extra sensors, to improve position accuracy. The more general situation with extra measurements was considered by Friedlander [58], Schau and Robinson [59], and Smith and Abel [55]. These methods are not optimum in the least-squares sense and perform worse in comparison to the Taylor-series method.

Although closed-form solutions have been developed, their estimators are not optimum. The divide and conquer (DAC) method [60] from Abel can achieve optimum performance, but it requires that the Fisher information is sufficiently large. To obtain a precise position estimate at reasonable noise levels, the Taylor-series method [51] is commonly employed. It is an iterative method that starts with an initial guess and improves the estimate at each step by determining the local linear least-squares (LS) solution. An initial guess close to the true solution is needed to avoid local minima. Selection of such a starting point is not simple in practice. Moreover, convergence of the iterative process is not assured. It also suffers from convergence problem and large LS computational burden as the method is iterative. Within the last few years, some papers were published on improving LS and closed-form methods [16,23,61-65]. Based on closed-form hyperbolic-intersection method, we will explain our proposed method, which using a simple arrangement of two microphones for 2D cases and three microphones for 3D cases, can simplify non-linear equations of this method to have a linear equation. Although there have been attempts to linearize closed-form non-linear equations through algebraic means, such as [7,16,56,63], our proposed method with simple pure geometrical linearization needs less microphones and features accurate localization and less processing time.

3. Basic methods

3.1. HRTF

Human beings are able to locate sound sources in three dimensions in range and direction, using only two ears. The location of a source is estimated by comparing difference cues received at both ears, among which are time differences of arrival and intensity differences. The sound source specifications are modified before entering the ear canal. The head-related impulse response (HRIR) is the name given to the impulse response relating the source location and the ear location. Therefore, convolution of an arbitrary sound source with the HRIR leads to what would have been received at the ear canal. The HRTF is the Fourier transform of HRIR and thus represents the filtering properties due to diffractions and reflections at the head, pinna, and torso. Hence, HRTF can be computed through comparing the original and modified signals [25-29,43-50]. Consider $X_c(k)$ being the Fourier transform of sound source signal at one ear (modified signal at either left or right ear) and $Y_c(k)$ to be that of the original signal. HRTF of that source can be found as [30]:

$$H_c(k) = \frac{Y_c(k)}{X_c(k)} \tag{1}$$

In detail:

$$|H_c(k)| = \frac{|Y_c(k)|}{|X_c(k)|} \tag{2}$$

$$\arg H_c(k) = \arg Y_c(k) - \arg X_c(k) \tag{3}$$

$$H_c(k) = |H_c(k)| e^{j \arg H_c(k)}. \tag{4}$$

In fact, $H_c(k)$ contains all the direction-dependent and direction-independent components. Therefore, in order to have pure HRTF, the direction-independent elements have to be removed from $H_c(k)$. If $C_c(k)$ is the known CTF (common transfer function), then DTF (directional transfer function), $D_c(k)$, can be found as:

$$D_c(k) = \frac{Yc(k)}{C_c(k)X_c(k)}. \tag{5}$$

3.2. ILD-based localization

We consider two microphones for localizing a sound source. Signal $s(t)$ propagates through a generic free space with noise and no (or low degree of) reverberation. According to the so-called inverse-square law, the signal received by the two microphones can be modeled as [21-24,40-42]:

$$s_1(t) = \frac{s(t-T_1)}{d_1} + n_1(t) \tag{6}$$

and

$$s_2(t) = \frac{s(t-T_2)}{d_2} + n_2(t), \tag{7}$$

where d_1 and d_2 are the distances and T_1 and T_2 are time delays from source to the first and second microphones, respectively. Also $n_1(t)$ and $n_2(t)$ are additive white Gaussian noises. The relative time shift between the signals is important for TDE but can be ignored in ILD. Therefore, if we find the delay between the two signals and shift the delayed signal in respect to the other one, the signal received by the two microphones can be modeled as:

$$s_1(t) = \frac{s(t)}{d_1} + n_1(t) \tag{8}$$

and

$$s_2(t) = \frac{s(t)}{d_2} + n_2(t). \tag{9}$$

Now, we assume that the sound source is audible and in a fixed location. Also, it is available during the time interval [0, W] where W is the window size. The energy received by the two microphones can be obtained by integrating the square of the signal over this time interval [21-24,40-42]:

$$E_1 = \int_o^w s_1^2(t)\,dt = \frac{1}{d_1^2}\int_o^w s^2(t)\,dt + \int_o^w n_1^2(t)\,dt \tag{10}$$

$$E_1 = \int_o^w s_2^2(t)\,dt = \frac{1}{d_2^2}\int_o^w s^2(t)\,dt + \int_o^w n_2^2(t)\,dt. \tag{11}$$

According to (10) and (11), the received energy decreases in relation to the inverse of the square of the distance to the source. These equations lead us to a simple relationship between the energies and distances:

$$E_1 \cdot d_1^2 = E_2 \cdot d_2^2 + \eta, \tag{12}$$

where $\eta = \int_o^w \left[d_1^2 n_1^2(t) + d_2^2 n_2^2(t)\right] dt$ is the error term. If (x_1, y_1) is the coordinates of the first microphone, (x_2, y_2) is the coordinates of the second microphone and (x_s, y_s) is the coordinates of the sound source with respect to the origin located at array center, then:

$$d_1 = \sqrt{(x_1 - x_s)^2 + (y_1 - y_s)^2} \tag{13}$$

$$d_2 = \sqrt{(x_2 - x_s)^2 + (y_2 - y_s)^2}. \tag{14}$$

Now using (12), (13), and (14), we can localize the sound source (Section 4.1.).

3.3. TDE-based localization

Correlation-based methods are the most widely used time delay estimation approaches. These methods use the following simple reasoning for the estimation of time delay. The autocorrelation function of $s_1(t)$ can be written in time domain as [17-20,31-39]:

$$R_{s_1 s_1}(\tau) = \int_{-\infty}^{+\infty} s_1(t) \cdot s_1(t - \tau)\, dt. \tag{15}$$

Dualities between time and frequency domains for autocorrelation function of $s_1(t)$ with the Fourier transform $S_1(f)$, results in frequency domain presentation as:

$$R_{s_1 s_1}(\tau) = \int_{-\infty}^{+\infty} S_1(f) \cdot S_2^*(f) e^{j2\pi f\tau} df. \tag{16}$$

According to (15) and (16), if the time delay τ is zero, this function's value will be maximized and will be equal to the energy of $s_1(t)$. The cross-correlation of two signals $s_1(t)$ and $s_2(t)$ is defined as:

$$R_{s_1 s_2}(\tau) = \int_{-\infty}^{+\infty} S_1(f) \cdot S_2^*(f) e^{j2\pi f\tau} df. \tag{17}$$

If $s_2(t)$ is considered to be the delayed version of $s_1(t)$, this function features a peak at the point equal to the time delay. This delay can be expressed as:

$$\tau_{12} = \mathrm{argmax}_\tau R_{s_1 s_2}(\tau). \tag{18}$$

In an overall view, the time delay estimation methods are as follows [17-20,31-39]:

- Correlation-Based Methods: Cross-Correlation (CC), ML, PHAT, Average Square Difference Function (ASDF)
- Adaptive Filter-Based Methods: Sync Filter, LMS.

Advantages of PHAT are accurate delay estimation in the case of wideband and quasi-periodic/periodic signals, good performance in noisy and reflective environments, sharper spectrum due to the use of better weighting function and higher recognition rate. Therefore, PHAT is used in cases where signals are detected using arrays of microphones and additive environmental and reflective noises are observed. In such cases, the signal delays cannot be accurately found using typical correlation-based methods as the correlation peaks cannot be precisely extracted. PHAT is a cross-correlation-based method used for finding the time delay between the signals. In PHAT, similar to ML, weighting functions are used along with the correlation function, i.e.,

$$\phi_{\mathrm{PHAT}}(f) = \frac{1}{|G_{s_1 s_2}(f)|}, \tag{19}$$

where $G_{s_1 s_2}(f)$ is the cross-correlation-based power spectrum. The overall function used in PHAT for the estimation of delay between two signals is defined as:

$$R_{s_1 s_2}(\tau) = \int_{-\infty}^{+\infty} \phi_{\mathrm{PHAT}}(f) \cdot G_{s_1 s_2}(f) e^{j2\pi f\tau} df \tag{20}$$

$$D_{\mathrm{PHAT}}(f) = \mathrm{argmax}_\tau R_{s_1 s_2}(\tau), \tag{21}$$

where D_{PHAT} is the delay calculated using PHAT. $G_{s_1 s_2}(f)$ is found as:

$$G_{s_1 s_2}(f) = \int_{-\infty}^{+\infty} r_{s_1 s_2}(\tau) e^{j2\pi f\tau} df \tag{22}$$

where

$$r_{s_1 s_2}(\tau) = E[s_1(t) s_2(t + \tau)]. \tag{23}$$

4. ILD- and PHAT-based angle of arrival and location calculation methods

4.1. Using ILD method

Assuming two microphones are on x-axis and have a distance of R ($R = D/2$) from origin (Figure 1), we can rewrite (13) and (14) as:

$$d_1 = \sqrt{(R - x_s)^2 + {y_s}^2} \tag{24}$$

$$d_2 = \sqrt{(-R - x_s)^2 + {y_s}^2}. \tag{25}$$

Therefore, we can rewrite (12) as:

$$\left(\frac{E_1}{E_2}\right) \cdot d_1^2 = d_2^2 + (\eta / E_2). \tag{26}$$

Assuming $m = \frac{E_1}{E_2}$ and $n = \eta / E_2$, (26) is written as:

$$m\left[(R - x_s)^2 + y_s\right]^2 = \left[(-R - x_s) + {y_s}^2\right] + n. \tag{27}$$

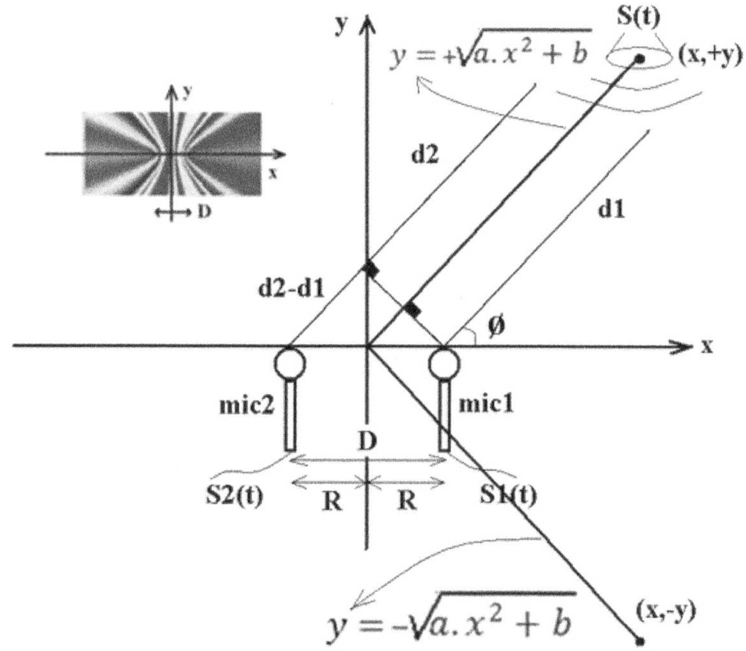

Figure 1 Hyperbolic geometrical location of 2D sound source localization using two microphones.

Using x and y instead of x_s and y_s, (27) will become:

$$x^2 - \left[\frac{2R(m+1)}{m-1}\right]x + y^2 = \frac{n}{m-1} - R^2 \tag{28}$$

$$\left(x - \frac{R(m+1)}{m-1}\right)^2 + y^2 = \frac{1}{m-1}\left(n + \frac{4m}{m-1}R^2\right) \tag{29}$$

$$\begin{cases} (x-k)^2 + y^2 = l \\ k = \dfrac{R(m+1)}{m-1} \\ l = \dfrac{1}{m-1}\left(n + \dfrac{4m}{m-1}R^2\right) \end{cases}. \tag{30}$$

Therefore, source location is on a circle with center coordinates (k, 0) and radius $(\sqrt{l})$. Now, using a new microphone to find a new equation, in combination with one of the first or second microphones, helps us to have another circle which leads to source location with different center coordinates and different radii relative to the first circle. Intersection of the first and second circles gives us source location x and y [22,40].

4.2. Using PHAT method

Assuming a single frequency sound source with a wavelength equal to λ to have a distance from the center of two microphones equal to r, this source will be in far-field if [51-65]:

$$r > \frac{2D^2}{\lambda}, \tag{31}$$

where D is the distance between two microphones. In the far-field case, the sound can be considered as having the same angle of arrival to all microphones, as shown in Figure 1. If $s_1(t)$ is the output signal of the first microphone and $s_1(t)$ is that of the second microphone (Figure 1), taking into account the environmental noise, and according to the so-called inverse-square law, the signal received by the two microphones can be modeled as (6) and (7). The relative time shift between the signals is important for TDOA but can be ignored in ILD. Also, the attenuation coefficients ($1/d_1$ and $1/d_2$) are important for ILD but can be ignored in TDOA. Therefore, assuming T_D is the time delay between the two received signals, the cross-correlation between $s_1(t)$ and $s_2(t)$ is [51-65]:

$$R_{s_1s_2}(\tau) = \int_{-\infty}^{+\infty} s_1(t)s_2(t+\tau)dt. \tag{32}$$

Since $n_1(t)$ and $n_2(t)$ are independent, we can write

$$R_{s_1s_2}(\tau) = \int_{-\infty}^{+\infty} s(t)s(t-T_D+\tau)dt. \tag{33}$$

Now, the time delay between these two signals can be measured as:

$$\tau = \mathrm{argmax}_{TD} R_{s_1s_2}(\tau). \tag{34}$$

Correct measurement of the time delay needs the distance between the two microphones to be:

$$D \le \frac{\lambda}{2}, \tag{35}$$

since when D is greater than $\frac{\lambda}{2}$, T_D is greater than π, and therefore, time delay is measured as $\tau = -(T_D - \pi)$. According to Figure 1, the cosine of the angle of arrival is

$$\cos(\phi) = \frac{d_2 - d_1}{D} = \frac{(t_2 - t_1)\, v_{sound}}{D} = \frac{T_D . v_{sound}}{D} = \frac{\tau_{21} . v_{sound}}{D}. \tag{36}$$

Here, v_{sound} is sound velocity in air. The delay time τ_{21} is measurable using the cross-correlation function between the two signals. However, the location of source cannot be measured this way. We can measure the distance between source and each of the microphones as (13) and (14). The difference between these two distances will be

$$d_2 - d_1 = \tau_{21} \cdot v_{sound}. \tag{37}$$

Using x and y instead of x_s and y_s, τ_{21} will be

$$\tau_{21} = \frac{\sqrt{(x-x_2)^2 + (y-y_2)^2} - \sqrt{(x-x_1)^2 + (y-y_1)^2}}{v_{sound}}. \tag{38}$$

This is an equation with two unknowns, x and y. Assuming the distances of both microphones from the origin to be R ($D = 2R$) and both located on x axis,

$$\tau_{21} = \frac{\sqrt{(x+R)^2 + y^2} - \sqrt{(x-R)^2 + y^2}}{v_{sound}}. \tag{39}$$

Simplifying the above equation will result in:

$$\begin{cases} y^2 = a.x^2 + b \\ a = \dfrac{4R^2}{v_{sound} \cdot \tau_{21}{}^2} - 1 \\ b = \dfrac{v_{sound}{}^2 \cdot \tau_{21}{}^2}{4} - R^2 \end{cases}, \tag{40}$$

where y has hyperbolic geometrical location relative to x, as shown in Figure 1. In order to find x and y, we need to add another equation to (38) for the first and a new (third) microphone so that:

$$\begin{cases} \tau_{21} = \dfrac{\sqrt{(x-x_2)^2 + (y-y_2)^2} - \sqrt{(x-x_1)^2 + (y-y_1)^2}}{v_{sound}} \\ \tau_{31} = \dfrac{\sqrt{(x-x_2)^2 + (y-y_3)^2} - \sqrt{(x-x_1)^2 + (y-y_1)^2}}{v_{sound}} \end{cases}. \tag{41}$$

It is noticeable that these are non-linear equations (Hyperbolic-intersection Closed-Form method) and numerical analysis should be used to calculate x and y, which will increase localization processing times. Also in this case, the solution may not converge.

5. TDE-ILD-based 2D sound source localization

Using either TDE or ILD to calculate source location (x and y) in 2D cases needs at least three microphones. Using TDE and ILD simultaneously helps us calculate source location using only two microphones. According to (26) and (37) and this fact that in a high SNR environment, the noise term $\eta/E2$ can be neglected, after some algebraic manipulations, we derive [40]

$$(x_s - x_1)^2 + (y_s - y_1)^2 = \left(\frac{\tau_{21} \cdot v_{sound}}{1 - \sqrt{m}}\right)^2 = r\frac{2}{1} \tag{42}$$

and

$$(x_s - x_2)^2 + (y_s - y_2)^2 = \left(\frac{\tau_{21} \cdot v_{sound} \cdot \sqrt{m}}{1 - \sqrt{m}}\right)^2 = r_2^2 \tag{43}$$

Intersection of two circles determined by (42) and (43), with center $(x_1, y_1)and$ (x_1, y_1), and radius r_1 and r_2, respectively, gives the exact source position. In $E_1 = E_2$ ($m - 1$) case, both the hyperbola and the circle determined by (26) and (37) degenerate a line perpendicular bisector of microphone pair. Consequently, there will be no intersection to determine source position. Trying to obtain a closed-form solution to this problem, we transform the expression by [40]:

$$x_1 x_s + y_1 y_s = \frac{1}{2}\left(k_1^2 - r_1^2 + R_s^2\right) \tag{44}$$

and

$$x_2 x_s + y_2 y_s = \frac{1}{2}\left(k_2^2 - r_s^2 + R_s^2\right), \tag{45}$$

where

$$k_1^2 = x_1^2 + y_1^2, k_2^2 = x_2^2 + y_2^2 \text{ and } R_s^2 = x_s^2 + y_s^2. \tag{46}$$

Rewriting (44) and (45) into matrix form results in:

$$\begin{bmatrix} x_1 & y_1 \\ x_2 & y_2 \end{bmatrix} \begin{bmatrix} x_s \\ y_s \end{bmatrix} = \frac{1}{2}\left\{\begin{bmatrix} k_1^2 - r_1^2 \\ k_2{}^2 - r_2{}^2 \end{bmatrix} + R_s^2 \begin{bmatrix} 1 \\ 1 \end{bmatrix}\right\} \tag{47}$$

and

$$\begin{bmatrix} x_s \\ y_s \end{bmatrix} = \begin{bmatrix} x_1 & y_1 \\ x_2 & y_2 \end{bmatrix}^{-1} \left(\frac{1}{2}\left\{\begin{bmatrix} k_1^2 - r_1^2 \\ k_2^2 - r_2^2 \end{bmatrix} + R_s^2 \begin{bmatrix} 1 \\ 1 \end{bmatrix}\right\}\right). \tag{48}$$

If we define

$$p = \begin{bmatrix} p1 \\ p2 \end{bmatrix} = \frac{1}{2}\begin{bmatrix} x1 & y1 \\ x2 & y2 \end{bmatrix}^{-1} \begin{bmatrix} 1 \\ 1 \end{bmatrix} \tag{49}$$

and

$$q = \begin{bmatrix} q1 \\ q2 \end{bmatrix} = \frac{1}{2}\begin{bmatrix} x_1 & y_1 \\ x_2 & y_2 \end{bmatrix}^{-1} \begin{bmatrix} k_1^2 - r_2^1 \\ k_2^2 - r_2^2 \end{bmatrix}, \tag{50}$$

then the source coordinates can be expressed with respect to Rs:

$$X = \begin{bmatrix} x_s \\ y_s \end{bmatrix} = \begin{bmatrix} q_1 + p_1 R_s^2 \\ q_2 + p_2 R_s^2 \end{bmatrix}. \quad (51)$$

Inserting (46) into (51), the solution to Rs is obtained as:

$$R_s^2 = \frac{0_1 \pm 0_2}{0_3}, \quad (52)$$

where

$$0_1 = 1 - p_1 q_1 + p_2 q_2,$$

$$0_2 = \sqrt{(1 - p_1 q_1 + p_2 q_2)^2 + (p_1^2 + p_2^2)(q_1^2 + q_2^2)},$$

$$0_3 = p_1^2 + p_2^2.$$

The positive root gives the square of distance from source to origin. Substituting Rs into (51), the final source coordinate will be obtained [40].

However, a rational solution requires prior information of evaluation regions. It is known to us that, by using a linear array, two mirror points will be produced simultaneously. Assuming two microphones are on x-axis ($y_1 = y_2 = 0$) and have distance R from origin (Figure 1), According to (49) and (50), we cannot find p and q. Therefore, we cannot consider such a microphone arrangement. However, using this microphone arrangement simplifies equations. According to (26) and (37), we can intersect circle and hyperbola to find source location, x and y. For intersection of circle and hyperbola, firstly we rewrite (42) and (43), respectively, as

$$(x_s - x_1)^2 + (y_s - y_1)^2 = r_1^2 \quad (53)$$

and

$$(x_s - x_2)^2 + (y_s - y_2)^2 = r_2^2. \quad (54)$$

Using microphones coordinate values x and y instead of x_s and y_s, we will have

$$(x - R)^2 + (y - 0)^2 = r_1^2 \quad (55)$$

and

$$(x + R)^2 + (y - 0)^2 = r_2^2; \quad (56)$$

therefore,

$$r_1^2 - (x - R)^2 = r_2^2 - (x + R)^2 \quad (57)$$

which results in

$$r_2^2 - r_1^2 = 4Rx. \quad (58)$$

Hence, the sound source location can be calculated as:

$$x = (r_2^2 - r_1^2)/4R \quad (59)$$

and

$$y = \pm\sqrt{r_1^2 - (x - R)^2} \quad (60)$$

We remember again that by using a linear array, two mirror points will be produced simultaneously. This means that we can localize 2D sound source only in half-plane.

6. Simulations of TDE-ILD-based method and discussion

In order to use the introduced method for sound source localization in low reverberant outdoor cases, firstly we performed simulations. We tried to evaluate the accuracy of this method in a variety of SNRs for some environmental noises. We considered two microphones on x-axis ($y_1 = y_2 = 0$) with 1-m distance from the origin ($x_1 = 1\ m$, $x_2 = -1\ m$ (D = 2R = 2)) (Figure 1). In order to use PHAT for the calculation of time delay between the signals of the two microphones, we used a helicopter sound (wideband and quasi-periodic signal) with a length of approximately 4 s, downloaded from the internet (www.freesound.com) as our sound source. For different source locations and for an ambient temperature of 15°C, first we calculated sound speed in air using:

$$v_{\text{sound}} = 20.05\sqrt{273.15 + \text{Temperature(Centigrade)}}. \quad (61)$$

Then, we calculated d_1 and d_2 using (24) and (25), and using (37), we calculated time delay between the received signals of the two microphones. For time delay positive values, i.e., sound source nearer to the first microphone (mic1 in Figure 1), we delayed second microphone signal with respect to the first microphone signal, and for negative values, i.e., sound source nearer to the second microphone (mic2 in Figure 1), did the opposite. Then using (6) and (7), we divided the first microphone signal by d_1 and the second microphone signal by d_2 to have correct attenuation in signals according to the source distances from microphones. Finally, we tried to calculate source location using the proposed TDE-ILD method (Section 5) in a variety of SNRs for some environmental noises.

For a number of signal-to-noise ratios (SNRs) for white Gaussian, pink and babble noises from 'NATO RSG-10 Noise Data,' 16-bit Quantization, and 96,000-Hz sampling frequency (hence, we upsampled NATO RSG-10 from its original 190,00 to 96,000 Hz), simulation results are shown in Figure 2 for source location of ($x = 10$, $y = 10$).

Assuming a high audio sampling rate, fractional delays are negligible and delays are rounded to nearest sampling point. Simulation results show larger localization error for SNRs lower than 10 dB. This issue occurs due to the use of ILD. Hence, we have to use this method for the case of only a dominant source to have accurate localization. Using spectral subtraction and source counting are useful for reducing localization error in SNRs lower than zero.

7. Our proposed TDE-ILD-HRTF method

Using TDE-ILD-based method, dual microphone 2D sound source localization is applicable. However, it is known that, by using a linear array in TDE-ILD-based method, two mirror points will be produced simultaneously (half-plane localization) [40]. Also, according to TDE-ILD-based simulation results (Section 6), it is noticeable that using ILD-based method needs only one dominant

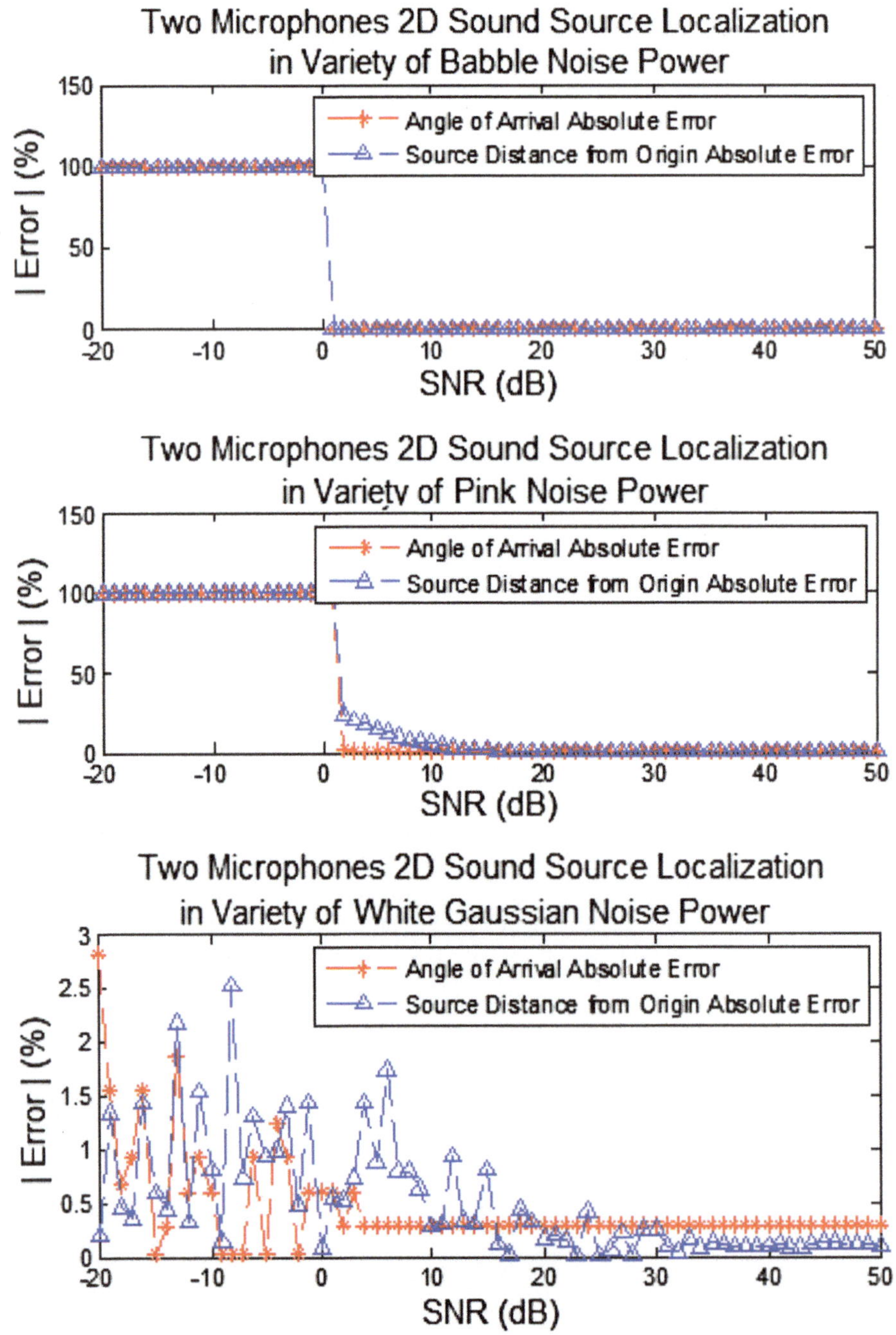

Figure 2 Simulation results for a variety of SNRs.

high SNR source to be active in localization area. Our proposed TDE-ILD-HRTF method tries to solve these problems using source counting, noise reduction using spectral subtraction, and HRTF.

7.1. Source counting method

According to previous discussions, ILD-based method needs to use source counting to find that one dominant source is active for high-resolution localization. If more than one source is active in localization area, it cannot calculate $m = E_1/E_2$ correctly. Therefore, we would need to count active and dominant sound sources and decide on localization of one sound source if only one source is dominant enough. PHAT gives us the cross-correlation vector of two microphone output signals. The number of dominant peaks of the cross-correlation vector gives us the number of dominant sound sources [66]. We consider only one source signal to be a periodic signal as:

$$s(t) = s(t + T). \tag{62}$$

If the signals window is greater than T, calculating cross-correlation between the output signals of the two microphones gives us one dominant peak and some weak peaks with multiples of T distances. However, using signals window of approximately equal to T or using non-periodic source signal would lead to only one dominant peak when calculating cross-correlation between the output signals of the two microphones. This peak value is delayed equal to the number of samples between the two microphones' output signals. Therefore, if one sound source is dominant in the localization area, only one dominant peak value will be in cross-correlation vector. Now, we consider having two sound sources $s(t)$ and $s'(t)$ in high SNR localization area. According to (6) and (7), we have

$$s_1(t) = s(t - T_1) + s'(t - T_1') \tag{63}$$

$$s_2(t) = s(t - T_2) + s'(t - T_2'). \tag{64}$$

According to (32), we have

$$R_{s_1 s_2}(\tau) = \int_{-\infty}^{+\infty} \left(s(t - T_1) + s'(t - T_1')\right)\left(s(t - T_2 + \tau) + s'(t - T_2' + \tau)\right)dt \tag{65}$$

$$R_{s_1 s_2}(\tau) = R1_{s_1 s_2}(\tau) + R2_{s_1 s_2}(\tau) + R3_{s_1 s_2}(\tau) + R4_{s_1 s_2}(\tau) \tag{66}$$

where

$$R1_{s_1 s_2}(\tau) = \int_{-\infty}^{+\infty} (t - T_1) \cdot s(t - T_2 + \tau)dt \tag{67}$$

$$R2_{s_1 s_2}(\tau) = \int_{-\infty}^{+\infty} (t - T_1) \cdot s'(t - T_2' + \tau)dt \tag{68}$$

$$R3_{s_1 s_2}(\tau) = \int_{-\infty}^{+\infty} s'(t - T_1') \cdot s(t - T_2 + \tau)dt \tag{69}$$

$$R4_{s_1 s_2}(\tau) = \int_{-\infty}^{+\infty} s'(t - T_1') \cdot s'(t - T_2' + \tau)dt. \tag{70}$$

Using (34), $\tau_1 = T_2 - T_1$ gives us a maximum value for $R1_{s_1 s_2}(\tau), \tau_2 = T_2' - T_1$, gives us a maximum value for $R2_{s_1 s_2}(\tau), \tau_3 = T_2 - T_1'$, gives us a maximum value for $R3_{s_1 s_2}(\tau)$ *and* $\tau_4 = T_2' - T_1'$, and gives us a maximum value for $R4_{s_1 s_2}(\tau)$. Therefore, we will have four peak values in cross-correlation vector. However, according to this fact that (67) and (70) are cross-correlation functions of a signal with delayed version of itself, and (68) and (69) are cross-correlation functions of two different signals, τ_1 and τ_4 maximum values are dominant with respect to τ_2 and τ_3 values. Now, we conclude in two dominant sound sources area, cross-correlation vector will have two dominant values and therefore equal count dominant values for more than two dominant sound sources signals as multiple power spectrum peaks in DOA-based multiple sound source beamforming methods [16]. Therefore, counting dominant cross-correlation vector values, we can find the number of active and dominant sound sources in localization area.

7.2. Noise reduction using spectral subtraction

In order to apply ILD in TDE-ILD-based dual microphone 2D sound source localization, source counting is used to find that one dominant high SNR source is active in localization area. Source counting was proposed to calculate the number of active sources in localization area. Furthermore, spectral subtraction can be used for noise reduction and therefore increasing active source's SNR. Also, according to the background noise, such as wind, rain, and babble, we can consider a background spectrum estimator. In the most practical cases, we can assume that the noisy signal can be modeled as the sum of the clean signal and the noise [67]:

$$s_n(t) = s(t) + n(t). \tag{71}$$

Also according to this fact that the signal and noise are generated by independent sources, they are considered uncorrelated. Therefore, the noisy spectrum can be written as:

$$S_n(w) = S(w) + N(w). \tag{72}$$

During the silent periods, i.e., periods without target sound, it can be estimated background noise spectrum, considering the noise to be stationary. Then, the noise magnitude spectrum can be subtracted from the noisy input magnitude spectrum. In non-stationary noise cases, there can be used an adaptive background noise spectrum estimation procedure [67].

7.3. Using HRTF method

Using a linear array in TDE-ILD-based dual microphone 2D sound source localization method leads to two mirror points produced simultaneously. Adding an HRTF-inspired idea, whole-plane dual microphone 2D sound source localization would be possible. This idea was published in [68], and it is reviewed here again. Researchers have used an artificial ear with a spiral shape. This is a special type of pinna with a varying distance from a microphone placed in the center of its spiral [30]. However, we consider a half-cylinder instead of artificial ear [68]. Due to the use of such a reflector, a constant notch position is created for all variations (0 to180°) of sound signal angle of arrival (Figure 3). However, clearly, and as shown in the figure, a complete half-cylinder scatters the sound waves from sources behind it. In order to overcome this problem we consider slits on the surface of the half-cylinder (Figure 3). If d is the distance between the reflector (half-cylinder) and the microphone (placed at the center), a notch will be created when it is equal to quarter of the wavelength of sound, λ, plus any multiples of λ/2. For such wavelengths, the incident waves are reduced by reflected waves [30]:

$$n\cdot\left(\frac{\lambda}{2}\right)+\left(\frac{\lambda}{4}\right)=d\, n=0,1,2,3\ldots \tag{73}$$

These notches will appear at the following frequencies:

$$f=\frac{c}{\lambda}=\frac{(2n+1).c}{4d}. \tag{74}$$

Covering only microphone 2 in Figure 1 by reflector, calculating the interaural spectral difference results in

$$\begin{aligned}|\Delta \mathrm{H}(f)| &= \left|10\log_{10}H_1(f)-10\log_{10}H_2(f)\right| \\ &= \left|10\log_{10}\frac{H_1(f)}{H_2(f)}\right|.\end{aligned} \tag{75}$$

High $|\Delta H(f)|$ values indicate that the sound source is in front, while negligible values indicate that sound source is at the back. One important point is that in order to have the same spectrum in both microphones when the sound source is at the back, careful design of the slits is necessary.

7.4. Extension of dimensions to three

According to the use of half-cylinder reflector in the proposed localization system, this approach is only applicable in 2D cases. Using a half-sphere reflector instead of the half-cylinder makes it usable in 3D sound source localization (Figure 4). Adding the half-sphere reflector only to microphone 2 in Figure 5 allows the localization system to localize sound sources in 3D cases.

For 3D case, we can consider a plane that passes through source position and x-axis and which makes an angle of θ with the x-y plane, and we name it source-plane (Figure 5). This plane consists of the microphones 1 and 2. According to these assumptions, using (36), we can calculate ϕ which is the angle of arrival in source-plane and is equal to angle of arrival in x-y plane. Then, using (59) and (60), we can also calculate x and y in 2D source-plane (not in x-y plane in 3D case) and therefore calculate sound source distance $r\sqrt{x^2+y^2}$. Introducing a new microphone (mic3 in Figure 5) with $y = 0$ and $x =$

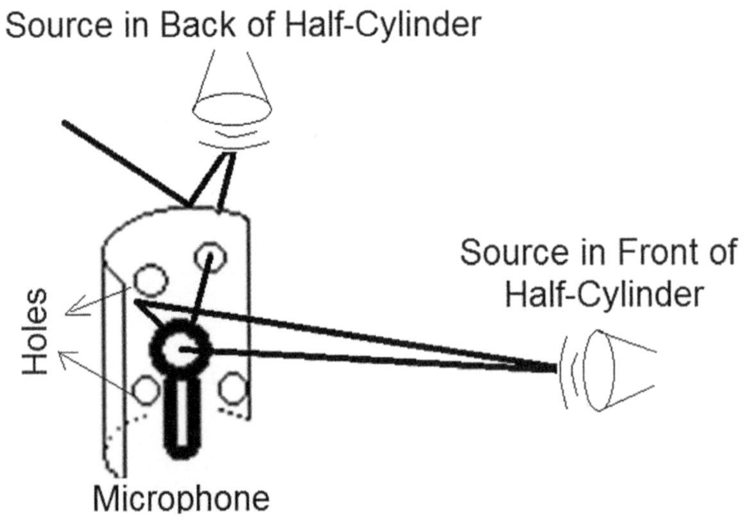

Figure 3 The half-cylinder used instead of artificial ear for 2D cases.

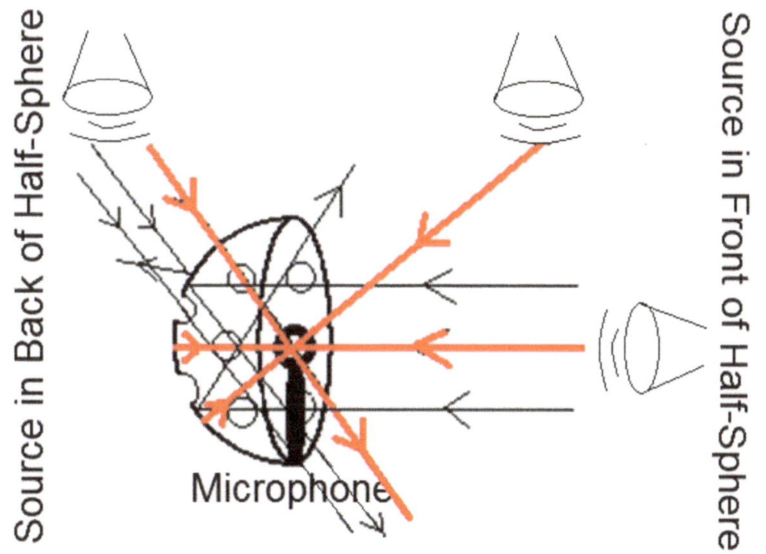

Figure 4 **Half-sphere used instead of artificial ear for 3D cases.**

$z = R$ helps us calculate the angle θ. Therefore, using (36), we can calculate θ as:

$$\theta = \cos^{-1}\left(\frac{v_{\text{sound}} \cdot \tau_{31}}{R}\right). \tag{76}$$

Using half-sphere reflector decreases accuracy of time delay and intensity level deference estimation between microphones 1 and 2 due to the change in the spectrum of second microphone's signal. However, covering the third microphone instead of the second microphone by half-sphere reflector only decreases the accuracy of time delay estimation between the first and third microphones (Figure 5). Multiplying the inverse function of notch-filter in third microphone's spectrum leads to increase in accuracy. Now using r, ϕ, and θ, we can calculate source location x, y and z in 3D case:

$$\begin{cases} x = r.\cos(\phi).\sin(\theta) \\ y = r.\sin(\phi), \sin(\theta) \\ z = r.\cos(\theta) \end{cases}. \tag{77}$$

The reasons for choosing the shape of sphere for the reflector are as follows [69]. The simplest type of

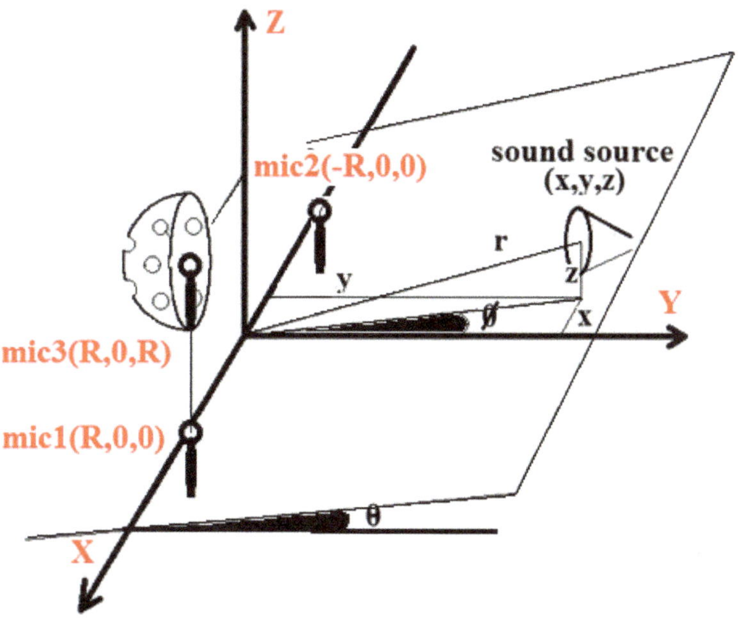

Figure 5 **3D sound source localization using TDE-ILD-HRTF method.**

reflector is a plane reflector introduced to direct signal in a desired direction. Clearly, using this type of reflector, the distance between reflector and microphone, *d* in (73), varies with respect to source position in 3D cases leading to a change in notch position within spectrum. The change in notch position may not be suitable as it might occur out of the spectral band of interest. To better adjust the energy in the forward direction, the geometrical shape of the plane reflector must be changed so as not to allow radiation in the back and side directions.

A possible arrangement for this purpose consists of two plane reflectors joined on one side to form a corner. This type of reflector returns the signal exactly in the same direction as it is received. Because of this unique feature, reflected wave acquired by the microphone is unique, which is our aim. Whereas having more than one reflected wave causes a higher energy value with respect to a single reflected wave at the microphone, which causes a deep notch. But using this type of reflector, *d* is varied with respect to the source position in 3D cases which is related to notch position in spectrum. It has been shown that if a beam of parallel waves is incident upon a parabola reflector, the radiation will focus at a spot which is known as the focal point. This point in spherical reflector does not match to the center in accordance with this fact that the center has equal distance (*d*) to all points of the reflector surface. Because of this feature, reflected wave which is received by the microphone is unique, whatever the position of the sound source, in 3D cases and belongs to the wave passing the center. The center of the spherical reflector with radius *R* is located at *O* (Figure 6). The sound wave axis strikes the reflector at *B*. From the Law of Reflection and angle geometry of parallel lines, the marked angles are equal. Hence, *BFO* is an equilateral triangle. Dropping the median from *F* which is stapled to *BO* and using trigonometry, the focal point is calculated as:

$$\frac{R}{2b} = \cos(\theta) \rightarrow b = \frac{R}{2\cos\theta} \tag{78}$$

$$\rightarrow F = R - \frac{R}{2\cos(\theta)}. \tag{79}$$

F is not equal to *R* for all θ values. The half-spherical reflector scatters the sound source waves hitting it from back. Therefore, we consider some slits on its surface (Figure 4). Obviously, the slits need to be designed in a fashion that leads to the same spectrum in both microphones when sound source is in back. When a plane wave hits a barrier with a single circular slit narrower than signal wavelength (λ), the wave bends and emerges from the slit as a circular wave [70]. If *L* is the distance

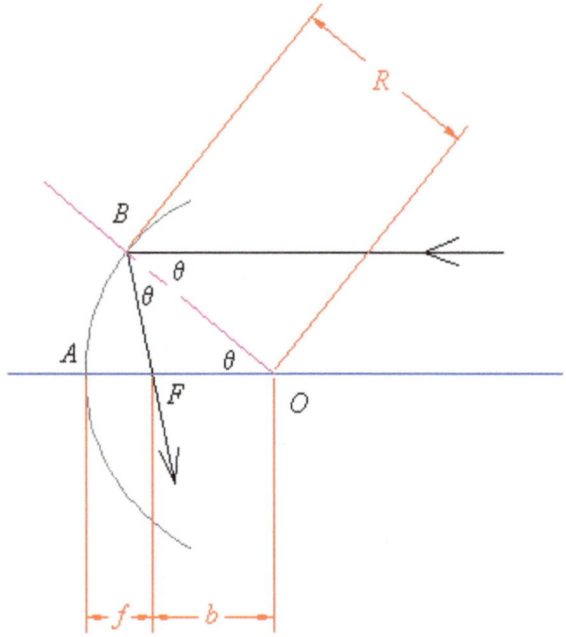

Figure 6 Focal point in spherical type sound wave reflectors.

between the slit and the viewing screen and *d* is the slit width, based on the assumption that $L >> d$ (Fraunhofer scattering), the wave intensity distribution observed (on a screen) at an angle θ with respect to the incident direction is given by:

$$\text{if } k = \frac{\pi . d}{\lambda} \sin(\theta) \rightarrow \frac{I(\theta)}{I(\theta)} = \frac{\sin^2(k)}{k^2}, \tag{80}$$

where $I(\theta)$ is wave intensity in direction of observation and $I(0)$ is maximum wave intensity of diffraction pattern (central fringe). This relationship will have a value of zero each time $\sin^2(k) = 0$. This occurs when

$$k = \pm m\pi \text{ or } \frac{\pi . d}{\lambda} \sin(\theta) = \pm m\pi, \tag{81}$$

yielding the following condition for observing minimum wave intensity from a single circular slit:

$$\sin(\theta) = \frac{m\lambda}{d} m = 0, \pm 2, \ldots. \tag{82}$$

This relationship is satisfied for integer values of *m*. Increasing values of *m* gives minima at correspondingly larger angles. The first minimum will be found for $m = 1$, the second for $m = 2$, and so forth. If $\frac{d}{\lambda}\sin(\theta)$ is less than one for all values of θ, i.e., when the size of the aperture is smaller than a wavelength ($d < \lambda$), there will be no minima. As we need more than one circular slit on half-sphere reflector's surface, we consider a parallel wave incident on a barrier that consists of two closely spaced narrow circular slits S1 and S2. The narrow circular slits split the incident wave into two coherent waves. After passing through the slits, the two waves spread out due to diffraction interfere with one another. If this transmitted wave is made to fall on a screen some distance away, an interference pattern of bright and dark fringes are observed on the screen, with the bright ones corresponding to regions of maximum wave intensity while the dark ones corresponding to those of minimum wave intensity. As discussed before, for all points on the screen where the path difference is some integer multiple of the wavelength, the two waves from the slits S1 and S2 arrive in phase and bright fringes are observed. Thus, the condition for producing bright fringes is as in (82). Similarly, dark fringes are produced on the screen if the two waves arriving on the screen from slits S1 and S2 are exactly out of phase. This happens if the path difference between the two waves is an odd integer multiple of half-wavelengths:

$$\sin(\theta) = \frac{\left(m+\frac{1}{2}\right)\lambda}{d} m = 0, \pm 1, \pm 2, \ldots \quad (83)$$

Of course, this condition is changed using a half-sphere instead of a plane (Figure 7), where according to the same distance R to the center, the two waves from the slits S1 and S2 arrive in phase at the center and bright fringes are observed there. This signal intensity magnification is not suitable for the case of estimating intensity level difference between two microphones where one of them is covered by this reflector. However, covering the third microphone by half-sphere reflector is suitable where there is no need to estimate the intensity level difference between two microphones 1 and 3.

8. TDE-ILD-HRTF approach algorithm

According to previous discussions, the following steps can be considered for our proposed method:

1. Setup microphones and hardware.
2. Calculate the sound recording hardware set (microphone, preamplifier, and sound card) amplification normalizing factor.
3. Apply voice activity detection. Is there valuable signal? yes → go to 4 no → go to 3.
4. Obtain $s_1(t)$, $s_2(t)$, and $s_3(t) \rightarrow m = E_1/E_2$.
5. Remove DC from the signals. Then normalize them regarding the sound intensity.
6. Hamming window signals regarding their stationary parts (at least about 100 ms for wideband quasi-periodic helicopter sound or twice that).
7. Apply FFT to signals.
8. Cancel noise using spectral subtraction.
9. Apply PHAT to the signals in order to calculate τ_{21} and τ_{31} in frequency domain (index of first maximum value). Then find the second maximum value of cross-correlation vector.
10. If the first maximum value is not dominant enough with respect to the second maximum value, go to the next windows of signals and do not calculate sound source location, otherwise:

a.

$$\Phi = \cos^{-1}\left(\frac{v_{\text{sound}} \cdot \tau_{21}}{2R}\right) \text{and} \, \theta = \cos^{-1}\left(\frac{v_{\text{sound}} \cdot \tau_{31}}{R}\right)$$

b.

$$v_{\text{sound}} = 20.05\sqrt{273.15 + \text{Temperature(Centigrade)}}$$

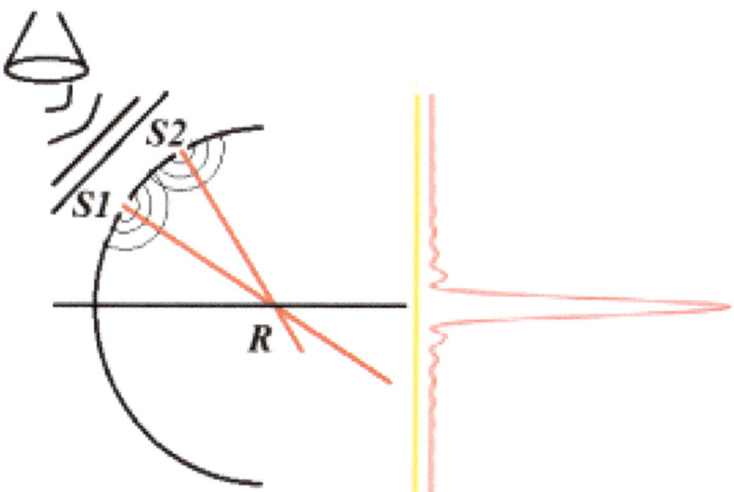

Figure 7 Intensity profile of the diffraction pattern resulting from a plane wave passing through two narrow circular slits in a half-sphere.

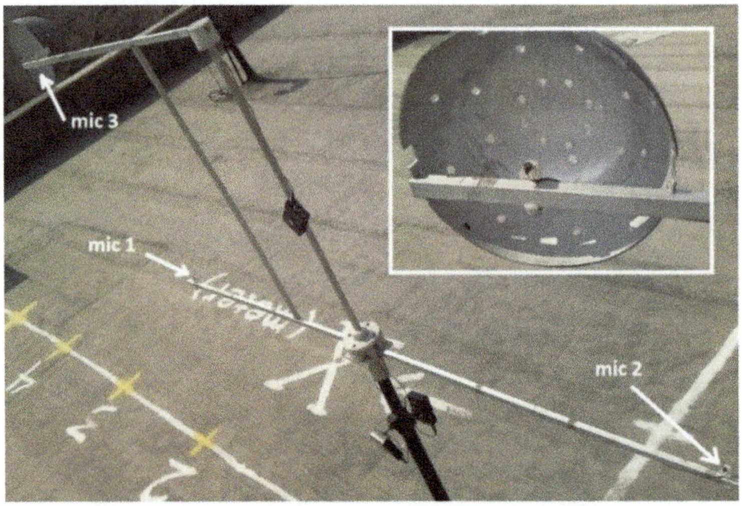

Figure 8 Setup of the proposed 3D sound source localization method using four microphones.

c.

$$r_1 = \frac{\tau_{21}.v_{\text{sound}}}{1-\sqrt{m}} \text{ and } r_2 = \frac{\tau_{21}.v_{\text{sound}}.\sqrt{m}}{1-\sqrt{m}}$$

d.

$$x = \left(r_2^2 - r_1^2\right)/4R \text{ and } y = \pm\sqrt{r_1^2-(x-R)^2}$$

e.

$$|\Delta H(f)| = \left|10\log_{10}\frac{H_1(f)}{H_3(f)}\right|$$

f.

$$\text{if}(|\Delta H(f)|\approx 0)\ y = -\sqrt{r_1^2-(x-R)^2}$$

$$\text{else } y = \sqrt{r_1^2-(x-R)^2}$$

$$\rightarrow \begin{cases} x_s = r.\cos(\Phi).\sin(\theta) \\ y_s = r.\sin(\Phi).\sin(\theta) \\ z_s = r.\cos(\theta) \end{cases}$$

g. Go to 3.

9. Hardware and software implementations and results

We implemented our proposed hardware using 'MB800H' motherboard and DELTA IOIO LT sound card from M-Audio featuring eight analogue sound inputs, 96-kHz sampling frequency, and up to 16-bit resolution. Three MicroMic microphones of type 'C 417' with B29L pre-amplifier were used for the implementation (Figure 8). The reason for using this type of microphone is its very flat spectral response up to 5 kHz. Visual C++ used for software implementation. According to the very low sensitivity of the used microphones (10 mV/Pa), a loudspeaker was used for the sound source. Initially, we used microphones one and two to evaluate the accuracy of the angle of arrival ϕ as well as microphones one and three to evaluate the accuracy of the angle of arrival θ. We considered 1 m($R = D/2 = 1$) distance between every microphone and the origin. Sound localization results showed that sound velocity changed in different temperatures. Since the sound velocity was used in most of the steps of the proposed methods, sound source location calculations were carried out inaccurately. Therefore, (61) was used based on a thermometer reading in order to more

Table 1 Results of the azimuth angle of arrival based on hardware implementation

ϕ Real angle (degrees)	ϕ Proposed method results for angle of arrival (degrees)	Absolute value of error (degrees)
0	0.18	0.18
15	15.13	0.13
30	29.85	0.15
45	45.18	0.18
60	60.14	0.14
75	74.91	0.09
90	89.83	0.17
105	104.82	0.18
120	120.14	0.14
135	135.11	0.11
150	150.14	0.14
165	165.09	0.09
180	179.93	0.07

Table 2 Results of the elevation angle of arrival based on hardware implementation

θ Real angle (degrees)	θ Proposed method results for angle of arrival (degrees)	Absolute value of error (degrees)
−10	−10.18	0.18
−5	−5.16	0.16
0	0.05	0.05
15	14.15	0.15
30	29.91	0.09
45	45.19	0.19
60	60.13	0.13
75	75.11	0.11
90	89.82	0.18

accurately find the sound velocity in different temperatures. Angle of arrival calculations using downloaded wideband quasi-periodic helicopter sound resulted in Tables 1 and 2. Acquisition hardware was initialized to 96-kHz sampling frequency and 16-bit resolution. Also signal acquisition window length was set to 100 ms. A custom 2-cm-diameter reflector (Figure 8) was placed behind the third microphone. Using the aforementioned sound source, the sound spectra shown in Figure 9 were measured using the third microphone. Figure 9a shows the spectrum when the sound source has been placed behind the reflector and Figure 9b when placed in front of it. Notches can be spotted in the second spectrum according to the reflector diameter and (74). Finally, using our proposed TDE-ILD-HRTF approach, first we tried to find that a dominant source is active in localization area. Then, using (72), we tried to reduce background noise and localize sound source in 3D space. Table 3 shows the improved results after the application of noise reduction method. Although PHAT features high performance in noisy and reflective environments, due to the rather constant and high valued signal-to-reverberation ratio (SRR) in outdoor cases, this feature could not be evaluated. We tried to find the effect of probable reverberation on our localization system results. In order to alleviate reverberation, our experiments indicated that at least a distance of 1.2 m with surrounding buildings would be necessary. Although the loudspeaker's power, the hardware set (microphone, preamplifier, and sound card) amplification factor and the distance of the microphones with the surrounding buildings indicate the SNR value for reverberate signals in a real environment, this distance can be calculated, but it can also be indicated experimentally, which is easier. Tables 1 and 2 indicate that our implemented 3D sound source localization method features less than 0.2 degree error for angle of arrival. Table 3 indicates less than 10% error for 3D

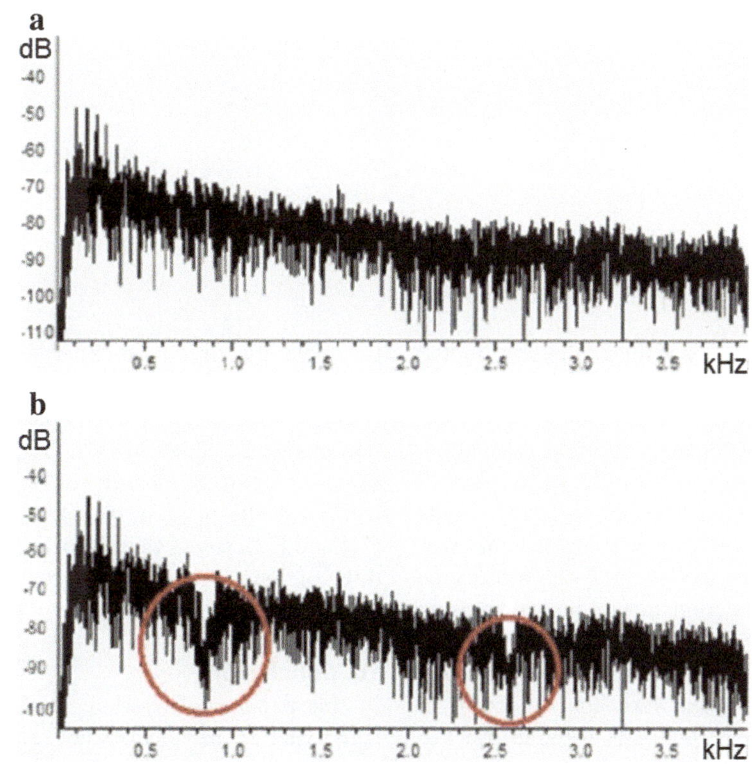

Figure 9 Recorded sound spectrum using the third microphone. **(a)** When the source has been placed behind the reflector; **(b)** when the source has been placed in front of the reflector. The logarithmic (dB) graphs indicate spectral level in comparison to 16-bit full scale.

Table 3 Results for proposed 3D sound source localization method (Section 7) using noise reduction procedure

Real location (*m*)			Proposed method results (*m*)			Absolute error (*m*)		
x	*y*	*z*	*x*	*y*	*z*	*x*	*y*	*z*
1	10	5	1.9	9.2	4.1	0.9	0.8	0.9
2	10	5	2.9	9.2	4.3	0.9	0.8	0.7
3	10	5	2.2	10.7	5.6	0.8	0.7	0.6
4	10	5	3.2	10.7	5.8	0.8	0.7	0.8
5	10	5	5.5	9.5	5.8	0.5	0.5	0.8
6	10	5	6.5	10.5	4.5	0.5	0.5	0.5
7	10	5	7.7	10.8	4.4	0.7	0.8	0.6
8	10	5	8.6	10.8	4.4	0.6	0.8	0.6
9	10	5	9.8	9.1	4.2	0.8	0.9	0.8
10	10	5	9.5	9.4	5.9	0.5	0.6	0.9
10	15	5	10.6	15.7	5.5	0.6	0.7	0.5
10	20	5	10.7	19.2	5.6	0.7	0.8	0.6
10	25	5	9.2	23.9	5.7	0.8	1.1	0.7
10	30	5	10.5	31.7	4.3	0.5	1.7	0.7
10	9	5	10.8	9.7	5.8	0.8	0.7	0.8
10	8	5	9.3	8.7	5.8	0.7	0.7	0.8
10	1	5	10.6	6.5	5.9	0.6	0.5	0.9
10	6	5	9.5	6.6	4.1	0.5	0.6	0.9
10	5	5	10.5	4.5	4.2	0.5	0.5	0.8
10	4	5	10.6	3.4	5.7	0.6	0.6	0.7
10	3	5	9.3	3.6	5.5	0.7	0.6	0.5
10	2	5	10.8	±2.8	5.8	0.8	0.8	0.8
10	1	5	10.1	±0.2	4.3	0.1	0.8	0.7
10	−1	5	10.6	±0.4	4.1	0.6	0.6	0.9
10	−2	5	10.5	±2.7	5.6	0.5	0.7	0.6
10	−3	5	10.4	−2.8	5.5	0.4	0.2	0.5
10	−4	5	9.7	−3.7	5.5	0.3	0.3	0.5
10	−5	5	9.4	−5.2	5.9	0.6	0.2	0.9

location finding. It also indicates ambiguity in sign finding between 0.5° and −0.5° as well as 175° and 185° due to the implemented reflector edges. We measured processing times of less than 150 ms for every location calculation. Comparing angle of arrival results with the results reported in various cited papers, the best reported results (on approximately similar tasks) were 1 degree limit error in [71], 2 degree limit error in [72], and within 8 degree error in [73]. Note that the fundamental limit of bearing estimates determined by Cramer-Rao Lower Bound (CRLB) for $D = 1$ cm spacing of the microphone pair, sampling frequency of 25 kHz and estimation duration of 100 ms, is computed to be 1° [71]. This is found equal to 1.4° for our system. However, only few respectable references have calculated 2D or 3D source locations accurately. Most of such research reports, besides calculating the angle of arrival, used either least-square error estimation or Kalman filtering techniques to predict motion path of sound sources, in order to overcome their local calculation errors. Examples are [2], [13], [15], and [19]. Meanwhile, comparison of our results with these and other works (results on approximately similar tasks), such as [72] and [73], shows the approximately same accuracy nature of our proposed source location measurement approach.

10. Conclusion

In this paper, we reported on the simulation of TDE-ILD-based 2D half-plane sound source localization using only two microphones. Reduction of the microphone count was our goal. Therefore, we also proposed and implemented TDE-ILD-HRTF-based 3D entire-space sound source localization using only three microphones. Also, we used spectral subtraction and source counting methods in low-degree reverberation outdoor cases to increase localization accuracy. According to Table 3, implementation results show that the proposed method has led to less than 10% error for 3D location finding. This is a higher accuracy in source location measurement in comparison with similar researches which did not use spectral subtraction and source counting. Also, we indicated that partly covering one of the microphones by a half-sphere reflector leads to entire-space N-dimensional sound source localization using only N-microphones.

Competing interests

The authors declare that they have no competing interests.

Authors' information

AP was born in Azerbaijan. He has a Ph.D. in Electrical Engineering (Signal Processing) from the Electrical Engineering Department, Amirkabir University of Technology, Tehran, Iran. He is now an invited lecturer at the Electrical Engineering Department, Amirkabir University of Technology and has been teaching several courses (C++ programming, multimedia systems, digital signal processing, digital audio processing, and digital image processing). His research interests include statistical signal processing and applications, digital signal processing and applications, digital audio processing and applications (acoustic modeling, speech coding, text to speech, audio watermarking and steganography, sound source localization, determined and underdetermined blind source separation and scene analyzing), digital image processing and applications (image and video coding, image watermarking and steganography, video watermarking and steganography, object tracking and scene matching) as well as BCI.

SMA received his B.S. and M.S. degrees in Electronics from the Electrical Engineering Department, Amirkabir University of Technology, Tehran, Iran, in 1984 and 1987, respectively, and his Ph.D. in Engineering from the University of Cambridge, Cambridge, U.K., in 1996, in the field of speech processing. Since 1988, he has been a Faculty Member at the Electrical Engineering Department, Amirkabir University of Technology, where he is currently an associate professor and teaches several courses and conducts research in electronics and communications. His research interests include speech processing, acoustic modeling, robust speech recognition, speaker adaptation, speech enhancement as well as audio and speech watermarking.

References

1. GC Carter, Time delay estimation for passive sonar signal processing. IEEE T Acoust S. **ASSP-29**, 462–470 (1981)
2. E Weinstein, Optimal source localization and tracking from passive array measurements. IEEE T. Acoust. S. **ASSP-30**, 69–76 (1982)
3. H Wang, P Chu, *Voice source localization for automatic camera pointing system in videoconferencing, in Proceedings of the ICASSP, New Paltz, NY, 19–22*, 1997
4. J Caffery Jr, G Stüber, Subscriber location in CDMA cellular networks. IEEE Trans. Veh. Technol. **47**, 406–416 (1998)
5. JB Tsui, *Fundamentals of Global Positioning System Receivers* (Wiley, New York, 2000)
6. F Klein, *Finding an earthquake's location with modern seismic networks* (Earthquake Hazards Program, Northern California, 2000)
7. Y Huang, J Benesty, GW Elko, RM Mersereau, Real-time passive source localization: A practical linear-correction least-squares approach. IEEE T. Audio P. **9**, 943–956 (2001)
8. VJM Michaud, F Rouat, J Letourneau, *Robust Sound Source Localization Using a Microphone Array on a Mobile Robot* (in Proceedings of the IEEE/RSJ International Conference on Intelligent Robots and Systems, vol. 2, Las Vegas), pp. 1228–1233
9. BLF Daku, JE Salt, L Sha, AF Prugger, An algorithm for locating microseismic events, in *Proceedings of the IEEE Canadian Conference on Electrical and Computer Engineering*. Niagara Falls. **2–5**, 2311–2314 (May 2004)
10. N Patwari, JN Ash, S Kyperountas, AO Hero III, RL Moses, NS Correal, Locating the nodes. IEEE Signal Process. Mag. **22**, 54–69 (2005)
11. S Gezici, T Zhi, G Giannakis, H Kobayashi, A Molisch, H Poor, Z Sahinoglu, Localization via ultra-wideband radios: a look at positioning aspects for future sensor networks. IEEE Signal Process. Mag. **22**, 70–84 (2005)
12. JY Lee, SY Ji, M Hahn, C Young-Jo, *Real-Time Sound Localization Using Time Difference for Human-Robot Interaction* (in Proceedings of the 16th IFAC World Congress, Prague, 2005)
13. WK Ma, BN Vo, SS Singh, A Baddeley, Tracking an unknown time-varying number of speakers using TDOA measurements: a random finite set approach. IEEE Trans. Signal. Process. **54**(9), 3291–3304 (2006)
14. KC Ho, X Lu, L Kovavisaruch, Source localization using TDOA and FDOA measurements in the presence of receiver location errors: analysis and solution. IEEE Trans Signal. Process. **55**(2), 684–696 (2007)
15. V Cevher, AC Sankaranarayanan, JH McClellan, R Chellappa, Target tracking using a joint acoustic video system. IEEE Trans. Multimedia. **9**(4), 715–727 (2007)
16. X Zhengyuan, N Liu, BM Sadler, *A Simple Closed-Form Linear Source Localization Algorithm* (in IEEE MILCOM, Orlando, FL, 2007), pp. 1–7
17. JPV Etten, Navigation systems: fundamentals of low and very-low frequency hyperbolic techniques. Elec. Comm. **45**(3), 192–212 (1970)
18. RO Schmidt, A new approach to geometry of range difference location. IEEE Trans. Aerosp. Electron Syst. **AES-8**(6), 821–835 (1972)
19. HB Lee, A novel procedure for assessing the accuracy of hyperbolic multilateration systems. IEEE Trans. Aerosp. Electron Syst. **110**(1), 2–15 (1975)
20. WR Hahn, Optimum signal processing for passive sonar range and bearing estimation. J. Acoust. Soc. Am. **58**, 201–207 (1975)
21. P Julian, A Comparative, Study of Sound Localization Algorithms for Energy Aware Sensor Network Nodes. IEEE Trans. Circuits and Systems - I. **51**(4) (2004)
22. ST Birchfield, R Gangishetty, *Acoustic Localization by Interaural Level Difference* (in Proceedings of the ICASSP, Philadelphia, PA, 2005), pp. 1109–1112
23. KC Ho, M Sun, An accurate algebraic Closed-Form solution for energy-based source localization. IEEE Trans. Audio, Speech. Lang. Process **15**, 2542–2550 (2007)
24. KC Ho, M Sun, Passive Source Localization Using Time Differences of Arrival and Gain Ratios of Arrival. IEEE Trans. Signal. Process **56**, 2 (2008)
25. JH Hebrank, D Wright, Spectral cues used in the localization of sound sources on the median plane. J. Acoust. Soc. Am. **56**, 1829–1834 (1974)
26. JC Middlebrooks, JC Makous, DM Green, Directional sensitivity of sound-pressure level in the human ear canal. J Acoust Soc Am. **1**, 86 (1989)
27. F Asano, Y Suzuki, T Sone, Role of spectral cues in median plane localization. J. Acoust. Soc. Am. **88**, 159–168 (1990)
28. RO Duda, WL Martens, Range dependence of the response of a spherical head model. J. Acoust. Soc. Am. **104**(5), 3048–3058 (1998)
29. CI Cheng, GH Wakefield, Introduction to head-related transfer functions (HRTFS): representations of HRTFs in time, frequency, and space. J. of the Audio Engin. Soc. **49**(4), 231–248 (2001)
30. A Kulaib, M Al-Mualla, D Vernon, *2D Binaural Sound Localization for Urban Search and Rescue Robotic*. in *Proceedings of the 12th Int* (Conference on Climbing and Walking Robots, Istanbul, 2009), pp. 9–11
31. S Razin, Explicit (non-iterative) Loran solution. J. Inst. Navigation **14**(3) (1967)
32. C Knapp, G Carter, The generalized correlation method for estimation of time delay. IEEE Trans. Acoust., Speech. Signal Process. **ASSP-24**(4), 320–327 (1976)
33. MS Brandstein, H Silverman, A practical methodology for speech localization with microphone arrays. Comput Speech Lang. **11**(2), 91–126 (1997)
34. MS Brandstein, JE Adcock, HF Silverman, A Closed-Form location estimator for use with room environment microphone arrays. IEEE Trans. Speech Audio Process. **5**(1), 45–50 (1997)
35. P Svnizer, M Matnssoni, M Omologo, *Acoustic Source Location in a THREE-Dimensional Space Using Crosspower Spectrum Phase* (in Proceedings of the ICASSP, Munich, 1997), p. 231
36. E Lleida, J Fernandez, E Masgrau, *Robust continuous speech recog. sys. based on a microphone array* (in Proceedings of the ICASSP, Seattle, WA, 1998), pp. 241–244
37. A Stephenne, B Champagne, *Cepstral prefiltering for time delay estimation in reverberant environments* (in Proceedings of the ICASSP, Detroit, MI, 1995), pp. 3055–3058

38. M Omologo, P Svaizer, *Acoustic Event Localization using a Crosspower-Spectrum Phase based Technique* (in Proceedings of the ICASSP, vol. 2, Adelaide, 1994), pp. 273–276
39. C Zhang, D Florencio, Z Zhang, *Why Does PHAT Work Well in Low Noise, reverberative Environments* (in Proceedings of the ICASSP, Las Vegas, NV, 2008), pp. 2565–2568
40. W Cui, Z Cao, J Wei, *DUAL-Microphone Source Location Method in 2-D Space* (in Proceedings of the ICASSP, Toulouse, 2006), pp. 845–848
41. X Sheng, YH Hu, Maximum likelihood multiple-source localization using acoustic energy measurements with wireless sensor networks. IEEE Trans Signal Process **53**(1), 44–53 (2005)
42. D Blatt, AD Hero III, Energy-based sensor network source localization via projection onto convex sets. IEEE Trans Signal Process **54**(9), 3614–3619 (2006)
43. S Sen, A Nehorai, Performance analysis of 3-D direction estimation based on head-related transfer function. IEEE Trans. Audio Speech Lang. Process **17**(4) (2009)
44. DW Batteau, The role of the pinna in human localization. Series B, Biol. Sci **168**, 158–180 (1967)
45. SK Roffler, RA Butler, Factors that influence the localization of sound in the vertical plane. J. Acoust. Soc. Am. **43**(6), 1255–1259 (1968)
46. SR Oldfield, SPA Parker, Acuity of sound localization: a topography of auditory space. Perception **13**(5), 601–617 (1984)
47. PM Hofman, JGAV Riswick, AJV Opstal, Relearing sound localization with new ears. Nature Neurosci. **1**(5), 417–421 (1998)
48. CP Brown, *Modeling the Elevation Characteristics of the Head-Related Impulse Response* (San Jose State University, San Jose, CA, M.S. Thesis, 1996)
49. P Satarzadeh, VR Algazi, RO Duda, *Physical and filter pinna models based on anthropometry* (in Proceedings of the 122nd Conv, Audio Eng. Soc, Vienna, 2007), p. 7098
50. S Hwang, Y Park, Y Park, *Sound Source Localization using HRTF database* (in ICCAS2005, Gyeonggi-Do, 2005)
51. WH Foy, Position-location solution by Taylor-series estimation. IEEE Trans. Aerosp. Electron Syst. **AES-12**, 187–194 (1976)
52. JS Abel, JO Smith, Source range and depth estimation from multipath range difference measurements. IEEE Trans. Acoust. Speech. Signal. Process. **37**, 1157–1165 (1989)
53. DMY Sommerville, *Elements of Non-Euclidean Geometry* (Dover, New York, 1958). pp. 260–1958
54. LP Eisenhart, *A Treatise on the Differential Geometry of Curves and Surfaces* (New York, Dover, 1960), pp. 270–271. originally Ginn, 1909
55. JO Smith, JS Abel, Closed-Form least-squares source location estimation from range-difference measurements. IEEE Trans. Acoust. Speech Signal Process. **ASSP-35**, 1661–1669 (1987)
56. JM Delosme, M Morf, B Friedlander, *A linear equation approach to locating sources from time-difference-of-arrival measurements* (in Proceedings of the ICASSP, Denver, CO, 1980), pp. 09–11
57. BT Fang, Simple solutions for hyperbolic and related position fixes. IEEE Trans. Aerosp. Electron Syst. **26**, 748–753 (1990)
58. B Friedlander, A passive localization algorithm and its accuracy analysis. IEEE J. Oceanic Eng. **OE-12**, 234–245 (1987)
59. HC Schau, AZ Robinson, Passive source localization employing intersecting spherical surfaces from time-of-arrival differences. IEEE Trans. Acoust. Speech Signal Process. **ASSP-35**, 1223–1225 (1987)
60. JS Abel, A divide and conquer approach to least-squares estimation. IEEE Trans. Aerosp. Electron. Syst. **26**, 423–427 (1990)
61. A Beck, P Stoica, J Li, Exact and approximate solutions of source localization problems. IEEE Trans. Signal. Process. **56**(5), 1770–1778 (2008)
62. HC So, YT Chan, FKW Chan, Closed-Form Formulae for Time-Difference-of-Arrival Estimation. IEEE Trans Signal Process **56**, 6 (2008)
63. MD Gillette, HF Silverman, A linear closed-form algorithm for source localization from time-differences of arrival. IEEE Signal Process. Lett. **15**, 1–4 (2008)
64. EG Larsson, D Danev, Accuracy comparison of LS and squared-range ls for source localization. IEEE Trans. Signal. Process **58**, 2 (2010)
65. N Ono, S Sagayama, R-means localization: a simple iterative algorithm for range-difference-based source localization. in Proceedings of the ICASSP. **14–19**, 2718–2721 (March 2010)
66. ZE Chami, A Guerin, A Pham, C Servière, *A PHASE-based Dual Microphone Method to Count and Locate Audio Sources in Reverberant Rooms* (in IEEE Workshop on Applications of Signal Processing to Audio and Acoustics, New Paltz, NY, 2000), pp. 18–21
67. SV Vaseghi, *Advanced Digital Signal Processing and Noise Reduction* (3rd edn, John Wiley & Sons, Ltd, Hoboken, NJ, 2006)
68. A Pourmohammad, SM Ahadi, TDE-ILD-HRTF-based 2D whole-plane sound source localization using only two microphones and source counting. Int. J. Info. Eng. **2**(3), 307–313 (2012)
69. CA Balanis, *Antenna Theory: Analysis and Design (John Wiley & Sons Inc* (NJ, Hoboken, 2005)
70. DC Giancoli, *Physics for Scientists and Engineers*, vol. I (Upper Saddle River, Prentice Hall, 2000)
71. B Friedlander, On the Cramer-Rao bound for time delay and Doppler estimation. IEEE Trans. Inf. Theory. **30**(3), 575–580 (1984)
72. A Gore, A Fazel, S Chakrabartty, Far-Field Acoustic Source Localization and Bearing Estimation Using ΣΔ Learners. IEEE Trans. circuits and systems-I: regular papers **57**(4) (2010)
73. SH Chen, JF Wang, MH Chen, ZW Sun, MJ Liao, SC Lin, SJ Chang, *A Design of Far-Field Speaker Localization System using Independent Component Analysis with Subspace Speech Enhancement* (in Proc. 11th IEEE Int. Symp. Mult., San Diego, CA, 2009), pp. 14–16

3

A memory efficient finite-state source coding algorithm for audio MDCT coefficients

Sumxin Jiang*, Rendong Yin and Peilin Liu

Abstract

To achieve a better trade-off between the vector dimension and the memory requirements of a vector quantizer (VQ), an entropy-constrained VQ (ECVQ) scheme with finite memory, called finite-state ECVQ (FS-ECVQ), is presented in this paper. The scheme consists of a finite-state VQ (FSVQ) and multiple component ECVQs. By utilizing the FSVQ, the inter-frame dependencies within source sequence can be effectively exploited and no side information needs to be transmitted. By employing the ECVQs, the total memory requirements of the FS-ECVQ can be efficiently decreased while the coding performance is improved. An FS-ECVQ, designed for the modified discrete cosine transform (MDCT) coefficients coding, was implemented and evaluated based on the Unified Speech and Audio Coding (USAC) scheme. Results showed that the FS-ECVQ achieved a reduction of the total memory requirements by about 11.3%, compared with the encoder in USAC final version (FINAL), while maintaining a similar coding performance.

1 Introduction

It is well known that a memoryless vector quantizer (VQ) can achieve performance arbitrarily close to the rate-distortion (R/D) function of the source, if the codevector dimension is large enough [1]. However, with the increase of the codevector dimension, the memory requirements and the computational complexity of the VQ will also increase exponentially. Furthermore, it will be difficult to design a practical VQ with high performance in a high-dimensional space. Consequently, various product code-vector quantization methods [2-5] have been proposed as alternative solutions. These methods cut down the memory requirements and reduce the computational complexity with a moderate loss of quantization performance. Among the widely reported product code techniques, split vector quantizer (SVQ), which was first proposed by Paliwal and Atal [6] for linear predictive coding (LPC) parameters quantization, receives extensive attention. In a SVQ, the input vector is first split into multiple subvectors [7], and then the resulting subvectors are quantized independently [8,9]. Although the SVQ cuts down the memory requirements and reduces the computational complexity of a memoryless VQ, it ignores the correlations between the subvectors and, hence, leads to a coding loss, referred to as 'split loss' [10].

In order to recover the split loss, many techniques have been developed. So and Paliwal [2,11] have proposed a switched SVQ (SSVQ) method, which adds multiple different SVQs to the input vector space so as to exploit the global dependencies. Based on SSVQ, a Gaussian mixture model (GMM)-based SSVQ (GMM-SSVQ) was proposed by Chatterjee et al. [12], where the distribution of the source is modeled by a GMM. Furthermore, a GMM-based Karhunen-Loève transform (KLT) domain SSVQ was proposed by Lee et al. [13], which was constructed by adding a region-clustering algorithm to the GMM-SSVQ. To better exploit the probability density function (pdf) of the source, Chatterjee and Sreenivas [14] developed a switched conditional pdf-based SVQ where the vector space is partitioned into non-overlapping Voronoi regions, and the source pdf of each switching Voronoi region is modeled by a multivariate Gaussian. Although these methods efficiently recover the split loss, most of them simply focus on removing intra-frame redundancies and fail to exploit inter-frame redundancies.

In addition, ordinary VQs can generally be divided into two groups: entropy-constrained VQ (ECVQ) [15] and resolution-constrained VQ (RCVQ) [16], and the above-mentioned methods are mainly proposed for the RCVQ and can hardly be applied on the ECVQ [17]. In the other

*Correspondence: microsum2005@sjtu.edu.cn
Department of Electronic Engineering, Shanghai Jiao Tong University, Shanghai 200241, China

side, an ECVQ usually achieves better R/D performance than a RCVQ does [18]. This is mainly owing to the length function contained in the ECVQ which allocates a different number of bits to different vector indices according to the probability of their appearance. Therefore, an ECVQ with recovered split loss would achieve a higher R/D performance than a RCVQ does.

To better recover the split loss of a SVQ, the finite-state VQ (FSVQ) can usually be resorted to, which is able to efficiently take advantage of the inter-frame dependencies. FSVQ [19,20], which incorporates memory into a memoryless VQ, is intrinsically a prediction-based technique. An FSVQ can be regarded as a finite-state machine [21], which contains multiple states, each corresponding to a certain state codebook. The state transition is determined by a next-state function based on the information obtained from the previously encoded vectors. Thus, the FSVQ utilizes the previous encoded vectors to predict the current input [22] and, therefore, efficiently exploits the redundancies among the input vectors and achieves a considerable increase in the R/D performance over a memoryless VQ.

In this paper, a composite quantizer, called FS-ECVQ, is introduced, in which multiple ECVQs are combined with a FSVQ. In FS-ECVQ [23], this FSVQ serves as a classifier which splits the source sequence into multiple clusters. To achieve better classification performance, the FSVQ draws the current decision based on information obtained from a number of previous adjacent vectors, even from those in previous frames, and thus better exploits the inter-frame redundancies than an ordinary SVQ does. After that, a specially designed ECVQ is applied on each cluster derived from the FSVQ. Among the resulting clusters, the more frequently a cluster occurred, the higher vector dimension it will be assigned. Through this method, the total memory requirements can be significantly reduced and the coding performance can be obviously improved. Moreover, within each component ECVQ, multiple length functions are devised for coding the indices of input vectors, each corresponding to a certain pdf model. To select the optimal length function for each vector index, another FSVQ is introduced. This FSVQ predicts the source pdf of the current vector index based on the information obtained from its previous adjacent ones, and then the length function with the highest matching probability is chosen. Through this method, the 'mismatch' between the designed pdf and the source pdf can be efficiently decreased. Thus, the FS-ECVQ will be more robust than an ordinary SVQ.

The organization of this paper is as follows. In Section 2, some fundamentals about VQ, FSVQ, and ECVQ are introduced. Section 3 deals with the design of the FS-ECVQ. Then, in Section 4, a practical FS-ECVQ aimed at coding the audio-modified discrete cosine transform (MDCT) coefficients in the MPEG Unified Speech and Audio Coding (USAC) [24] is implemented and tested. Finally, conclusions are presented in Section 5.

2 Preliminaries

Since FS-ECVQ is based on FSVQ and ECVQ, in this section we will review the classical results of these vector quantization theories under the high rate assumption.

2.1 Vector quantization

Generally, a VQ, q, consists of four elements: encoder ϕ, decoder ψ, index coder ζ, and codebook $\mathcal{C}$. Suppose that random vector, $\mathbf{x}$, with pdf, f, is quantized by quantizer q and the corresponding reconstructed vector is $\hat{\mathbf{x}}$. Then, for a given measurable space $(\Omega, \mathcal{F})$ consisting of a k-dimensional Euclidean space Ω and its Borel subset, the mappings of quantizer q can be described as follows:

- Encoder ϕ: $\Omega \rightarrow \mathcal{I}$, where $\mathcal{I}$ is a countable index set. Each element in $\mathcal{I}$ corresponds to a different codevector contained in codebook $\mathcal{C}$. The aim of encoder ϕ is to find the index of the best matching vector in codebook $\mathcal{C}$ for input vector $\mathbf{x}$ according to a given distortion criterion
- Decoder ψ: $\mathcal{I} \rightarrow \Omega$, which is used to reconstruct the vector in space Ω according to the received vector index
- Index coder ζ: $\mathcal{I} \rightarrow$ {bitstream}, which transforms the index sequence generated from encoder ϕ to a bitstream
- Codebook $\mathcal{C}$, which is used by both encoder ϕ and decoder ψ to generate the optimal codevector indices or to find the corresponding codevectors

The average rate and the entropy of quantizer q are

$$R_f(q) = E(\zeta(\phi(\mathbf{x}))) = \sum_i p_i \zeta(i) \tag{1}$$

$$H_f(q) = -\sum_i p_i \ln p(i) \tag{2}$$

respectively, where p_i denotes the occurrence probability of index i. According to the result in [25], it implies that

$$R_f(q) \geq H_f(q) \tag{3}$$

with equality if and only if $\zeta(i) = -\ln p_i$. Therefore, the optimal length function of quantizer q for pdf f is

$$\zeta(i) = -\ln p_i. \tag{4}$$

The performance of quantizer q can be measured by an average distortion

$$D_f(q) = E(d(\mathbf{x}_n, \hat{\mathbf{x}}_n)). \tag{5}$$

In our work, Euclidean distance, $d(\mathbf{x}, \hat{\mathbf{x}}) = \|\mathbf{x} - \hat{\mathbf{x}}\|^2$, is used as the distortion measure, where $\|\cdot\|$ denotes the l_2 norm.

2.2 Finite-state vector quantization

FSVQ is a VQ with a time-varying encoder and decoder pair [21], which is realized by means of a finite-state machine. Assume that a FSVQ contains M distinct states, $S_1, \ldots, S_M$, whose corresponding state codebooks are, $\mathcal{C}_1, \ldots, \mathcal{C}_M$, respectively. Suppose that $\mathbf{x}_n$ is the input vector, whose current state is $s_n \in \{S_1, \ldots, S_M\}$. Then, by searching the codebook $\mathcal{C}_m$, corresponding to the current state s_n, for the best matching codevector $\hat{\mathbf{x}}_n$, the input vector $\mathbf{x}_n$ can be quantized, whose vector index is denoted as i_n.

In FSVQ, the current state s_n is achieved using a next-state function [26], γ, which can be written as

$$s_n = \gamma(i_{n-1}, s_{n-1}) \tag{6}$$

where i_{n-1} and s_{n-1} are the index and state of the last vector $\mathbf{x}_{n-1}$, respectively. Thus, the state transition is determined by the next-state function, and the current state s_n can be considered as a prediction to the input vector $\mathbf{x}_n$ based on the previously encoded vectors. Once the current state s_n is obtained, the encoding procedure of the FSVQ [20] can be written as

$$i_n = \phi(\mathbf{x}_n, s_n) \tag{7}$$

which implies that the input vector $\mathbf{x}_n$ is quantized in the codebook $\mathcal{C}_m$ corresponding to the current state s_n.

Similarly, the decoding procedure of the FSVQ is also based on the current state s_n. In this procedure, the received vector index i_n and its current state s_n are combined to reconstruct the input vector $\mathbf{x}_n$. The decoding procedure can be shown as

$$\hat{\mathbf{x}}_n = \psi(s_n, i_n) \tag{8}$$

which implies that the received vector index, i_n, is decoded in the codebook $\mathcal{C}_m$ corresponding to the current state s_n.

In FSVQ, encoder ϕ and decoder ψ are synchronized using the following coding rule:

$$i_n = \operatorname*{argmin}_{i \in \mathcal{C}_m} d\left(\mathbf{x}_n, \psi(i, s_n)\right). \tag{9}$$

2.3 Entropy-constrained vector quantization

The design of an ECVQ is to find a set of reconstruction vectors which minimizes the average distortion between the source and its reconstruction, subject to a constraint on the index entropy [15]. To obtain a common conclusion, Gray et al. [25,27] investigated the variable-rate ECVQ using a Lagrangian formulation in which a Lagrangian multiplier $\lambda > 0$ is defined for each rate.

Assume that the pdf, f, of random vector $\mathbf{x}$ is absolutely continuous with respect to Lebesgue measure, that $h(f) = -\int f(x) \ln f(x) dx$ exists and is finite and that $H_f(q_1) < \infty$, where q_1 is a cubic lattice quantizer with unit volume cells, the Lagrangian distortion of ECVQ, q, can be given by

$$J_f(\lambda, q) = D_f(q) + \lambda R_f(q) \tag{10}$$

and the optimal performance can be written as

$$J_f^*(\lambda) \triangleq \inf_q J_f(\lambda, q) = \inf_q \{D_f(q) + \lambda R_f(q)\} \tag{11}$$

where $D_f(q)$ and $R_f(q)$, obtained from (5) and (1), are the average rate and average distortion of quantizer q, respectively.

In order to demonstrate the variable-rate results of the research done by Gray et al. in a simplified form, we introduce the following notations:

$$\xi(f, \lambda, q) \triangleq \frac{J_f(\lambda, q)}{\lambda} + \frac{k}{2} \ln \lambda - h(f) \tag{12}$$

$$\xi(f, \lambda) = \frac{J_f^*(\lambda)}{\lambda} + \frac{k}{2} \ln \lambda - h(f) \tag{13}$$

$$\xi_k \triangleq \inf_{\lambda > 0} \left(\frac{J_{\mu_1}(\lambda)}{\lambda} + \frac{k}{2} \ln \lambda \right) \tag{14}$$

where ξ_k is a finite constant and μ_1 is the uniform pdf on the k-dimensional unit cube $C_1 = [0, 1)^k$. Then, the main result of the researches done by Gray et al. [25] is the following:

$$\lim_{\lambda \to 0} \xi(f, \lambda) = \xi_k. \tag{15}$$

This result guarantees that if a pdf f satisfies the conditions of (15), then there exists an optimal quantizer q for f in the sense that for any decreasing λ converging to 0, its optimal performance is ξ_k.

Mismatch appears if there exists any difference between the designed pdf and the source pdf. Suppose that the designed pdf is g and the source pdf is f, then according to the mismatch theorem proposed in [28], the minimal distortion of quantizer q can be given as

$$\lim_{\lambda \to 0} \xi(f, \lambda, q) = \xi_k + I(f \| g) \tag{16}$$

where $I(f \| g)$ is the relative entropy and can be given as

$$I(f \| g) = \int f(x) \ln \frac{f(x)}{g(x)} dx. \tag{17}$$

Compared with (15), it can be seen that the mismatch resulted from applying an asymptotically optimal quantizer for pdf g to a source sequence with pdf f is exactly the relative entropy of the source pdf f to the design pdf g, $I(f \| g)$.

3 Quantizer design

Compared with a conventional ECVQ, such as the quantization methods of USAC, whose architecture is to be described in details in Section 4, the FS-ECVQ can be taken as a super-ECVQ, in which multiple component ECVQs are contained and all of them are combined to a FSVQ. Thus, the FS-ECVQ is composed of two steps. The first step is to split the source sequence into multiple clusters using the FSVQ (main FSVQ), and the second step is to apply a dedicated conventional ECVQ to each cluster. Suppose that the largest available vector dimension is 8, the whole coding scheme of a FS-ECVQ can be demonstrated as Figure 1.

3.1 Main FSVQ

The major function of the main FSVQ is to partition the input space into four non-overlapped clusters according to the four states contained. For each resulting cluster, a component ECVQ is constructed holding a different vector dimension and different memory requirements. By this means, the total memory requirements could be efficiently decreased. The state transition is determined by a next-state function, which is the key component of the main FSVQ. In the following part of this section, we will mainly discuss the construction of the next-state function.

The next-state function of the main FSVQ is built on the dependencies among audio MDCT coefficients. In practice, audio signals are usually divided into a series of time intervals (often referred to as 'frames') due to their long-term time-varying property, and over a specified frame they are assumed to be stationary. Thus, as an expression of the audio signal in MDCT domain, there are high dependencies among the MDCT coefficients of the adjacent frames as well as among the coefficients within one frame. As a result, based on the inter- and intra-frame correlations, the MDCT coefficients of current frame could be estimated through prediction methodology. In our work, the audio MDCT coefficient frames are further divided into small blocks, and then, by estimating the shape properties of these blocks among multiple sequential frames, the next-state function is constructed in order to exploit both the intra- and inter-frame dependencies. In fact, within an audio MDCT coefficient sequence, the occurrence frequency of a block is highly related to its shape features. Suppose the block size is 4, then the relationship of the shape feature and the occurrence of a block are demonstrated in Figure 2.

To characterize the shape of a block, three statistical parameters block energy, e, block deviation, σ, and block skewness, g, are employed in our work. Let μ be the mean value of block $\mathbf{x}$. Then the parameters e, σ, and g can be written as

$$e = \frac{1}{N}\sum_{i=1}^{N} x_i^2 \tag{18}$$

$$\sigma = \frac{1}{N}\sum_{i=1}^{N} (x_i - \mu)^2 \tag{19}$$

$$g = \frac{\frac{1}{N}\sum_{i=1}^{N}(x_i - \mu)^3}{\left(\frac{1}{N}\sum_{i=1}^{N}(x_i - \mu)^2\right)^{3/2}} \tag{20}$$

where x_i and N are the ith element and the length of block $\mathbf{x}$, respectively. To describe the shape feature of block $\mathbf{x}$ in a simplified form, a new statistical parameter, $V_{\mathbf{x}}$, is defined, which is given as

$$V_{\mathbf{x}} = (\sigma + e) \cdot (1 + \log(1 + |g|)). \tag{21}$$

Once a source sequence is split into a series of blocks, the value of $V_{\mathbf{x}}$ will be calculated for each block. Thus, a mapping can be established between the $V_{\mathbf{x}}$ set, composed of all the possible values of $V_{\mathbf{x}}$, and the input space Ω. Then, by splitting the possible values of $V_{\mathbf{x}}$ into two segments, we can partition the input space Ω into two clusters, Ω_k and Ω_k^C. Here, k denotes the dimension of

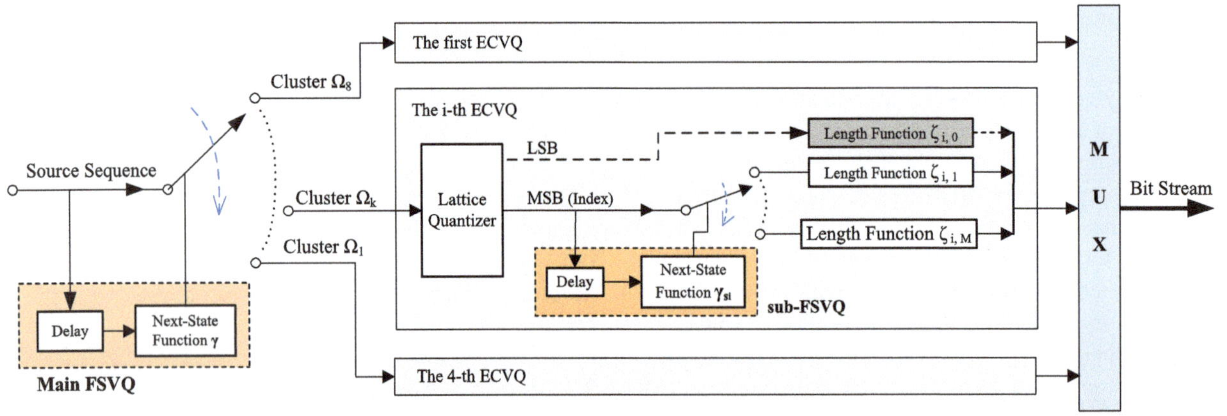

Figure 1 Block diagram of the proposed FS-ECVQ scheme, which contains a main FSVQ and four component ECVQs.

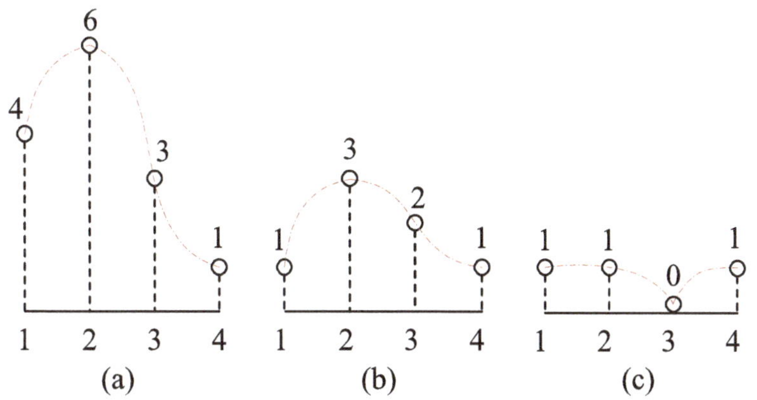

Figure 2 Three typical block shapes within an audio MDCT coefficient sequence. The blocks appeared **(a)** scarcely, **(b)** moderately, and **(c)** frequently.

cluster Ω_k. To implement the split, a threshold V_T is employed, whose value is obtained by maximizing the coding gain of the FS-ECVQ under the constraint of the total memory requirements using the training data. As for the two resulting clusters, Ω_k is supposed to contain the blocks occurring relatively frequently, whereas Ω_k^C is assumed to hold those occurring relatively scarcely.

To construct the next-state function, four previous blocks, A, B, C, and D, which are adjacent to the current block, $\mathbf{x}$, can be employed [29]. For simplicity, we assume that the current block and its previous neighbors form a Markov chain [26]. The relative positions of all these blocks are demonstrated in Figure 3. Assume that the shape parameters of the four adjacent blocks are independent measurements, then according to the research done by Nasrabadi et al. [30], the conditional joint posterior probability, which the next-state function is built on, can be given as

$$P(V_{\mathbf{x}}|(V_A, V_B, V_C, V_D)) = \frac{P(V_{\mathbf{x}}) \prod_{i=A}^{D} P(V_i|V_{\mathbf{x}})}{\prod_{i=A}^{D} P(V_i)} \quad (22)$$

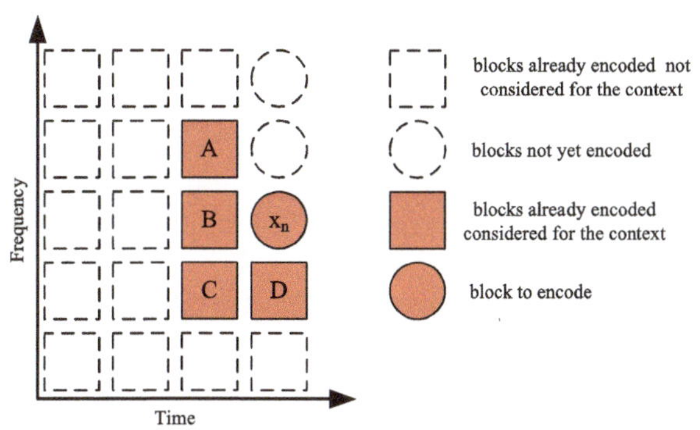

Figure 3 The input block *x* and the previously encoded adjacent blocks *A*, *B*, *C*, and *D*. These blocks are used to obtain the current state *s* by the next-state function of the main FSVQ.

where $V_{\mathbf{x}}$ and V_i are the shape parameters of block $\mathbf{x}$ and its four neighbors A, B, C, and D, respectively. Suppose that $P(V_{\mathbf{x}})$ and $P(V_i)$ are measured independently and considered to be equal, then probability $P(V_i|V_{\mathbf{x}})$ will be equal to probability $P(V_{\mathbf{x}}|V_i)$, which represents a conditional probability of the parameter $V_{\mathbf{x}}$ given one of its neighbors V_i, for $i = A, B, C$, and D, and can be obtained through recording all the possible cluster pairs occurring together using the training data. Assume that all the shape parameters obey the same probability distribution, then the conditional joint probability, $P(V_{\mathbf{x}}|(V_A, V_B, V_C, V_D))$, will only depend on the four conditional probabilities $P(V_{\mathbf{x}}|V_i)$, for $i = A, B, C$, and D, and the other parameters in (22) will be constant for any input block. Therefore, we can build the next-state function (6) on these four conditional probabilities, and the current state of the main FSVQ, s, can be given by

$$s = \gamma(V_A, V_B, V_C, V_D) = \max_{\mathbf{x} \in \{\Omega_k, \Omega_k^C\}} \prod_{i=A}^{D} P(V_{\mathbf{x}}|V_i) \quad (23)$$

which denotes an estimation of the cluster to which the current block is most likely to be classified.

To split the source sequence into smaller clusters, a pyramidal decomposing algorithm is employed, as demonstrated in Figure 4. In this algorithm, a block, $\mathbf{x}$, is first separated from the source sequence, whose length is set to be the largest available vector dimension, supposed to be 8. Then, the current state s of the obtained block $\mathbf{x}$, which is calculated through (23), is compared with a given threshold, T_8. If current state s is lower than T_8, block $\mathbf{x}$ will be taken as an element belonging to cluster Ω_8. Else, it would be equally decomposed into two smaller blocks, $\mathbf{x}_4^{(1)}$ and $\mathbf{x}_4^{(2)}$, whose vector dimensions are both 4, and then the block $\mathbf{x}_4^{(1)}$ will be taken as the new current block. Once again, the current state s is calculated and is compared with another threshold, T_4. If the obtained state s is

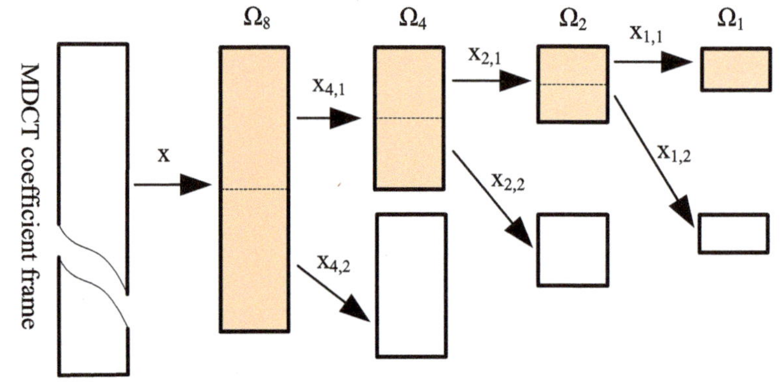

Figure 4 Pyramidal decomposing procedure of input vector (block) x. It is split into subvectors with an optimal vector dimension.

lower than T_4, the block $\mathbf{x}_4^{(1)}$ will be taken as an element belonging to cluster Ω_4. Else, it will be decomposed once more. This procedure continues iteratively until the lowest available vector dimension, supposed to be 1, is reached. Since each threshold can be regarded as the occurrence frequency of a block, then the current blocks considered to be with low-occurrence frequency, will be split iteratively, until a suitable vector dimension is found. The whole procedure is summarized in Algorithm 1.

Algorithm 1 The method to split the source sequence into 4 non-overlapped clusters.

Require:

The source sequence, the largest available vector dimension 8, and the thresholds T_k, for $k \in \{8, 4, 2, 1\}$;

Ensure:

1: Set the vector dimensions of current block, $\mathbf{x}$, and its four previous neighbors, A, B, C, and D, to be 8; And set $k = 8$.
2: Calculate current state, s, according to (23);
3: If $s \geq T_k$, then $k \leftarrow k/2$; else goto step 5;
4: Equally split current block $\mathbf{x}$ into $\mathbf{x}_k^{(1)}$ and $\mathbf{x}_k^{(2)}$, and let $\mathbf{x}_k^{(1)}$ be the new current block $\mathbf{x}$. And then, set the dimension of its four previous neighbors, A, B, C, and D all to be k, and go back to step 2;
5: Cluster Ω_k is selected and the current block $\mathbf{x}$ belongs to Ω_k;
6: If the source sequence is unfinished, go back to step 1;
7: **return** Cluster Ω_k, for $k = 8, 4, 2, 1$;

At beginning, there is no previous block, and therefore, an original state, s_0, ought to be initialized by the main FSVQ.

3.2 ECVQ

Based on the research done by Gray et al. [25], in our work, Z_n lattice quantizer and arithmetic coder are selected as the lattice quantizer and the length function of each component ECVQ, respectively. Unlike conventional ECVQ [15,17], where all the vector indices generated from the lattice quantizer share a same length function regardless of their possible differences, in our work multiple length functions are available and the optimal one is selected by another FSVQ (sub-FSVQ) for each generated vector index. Moreover, to improve the robustness and, at the same time, decrease the memory requirements of each component ECVQ, the design of sub-FSVQ is optimized and an iterative method to merge the similar length functions is proposed.

The length functions are implemented by an arithmetic coder, which are based on the pdf model of the input index. Hence, the main work of the sub-FSVQ is to search for the optimal one among a predesigned collection of pdf models based on the information obtained from previous indices.

3.2.1 Lattice quantizer

The issue whether an optimal ECVQ has a finite or infinite number of codevectors has been in-depth investigated by Gyärgy and Linder [31]. They found that ECVQ has a finite number of codevectors only if the tail of the source distribution is lighter than the tail of a Gaussian distribution. With respect to the probability distribution of an audio MDCT coefficient sequence, Yu et al. [32] show that the generalized Gaussian function with distribution parameter $r = 0.5$ provides a good approximation. Moreover, in practice, the possible values of the audio MDCT coefficients are always finite and concentrated in a finite range. Therefore, in our work, all codevectors of the lattice quantizer are simply constrained in the range

$$P\{||X|| \leq t_0\} \geq p_0 \tag{24}$$

where X denotes an input vector, and t_0 and p_0 are two thresholds that constrain the norm and the probability of input vector X, respectively.

Since all the codevectors are constructed within the range (24), the input vectors outside the range will suffer a larger quantization loss than those inside the range. Such circumstances are usually required to be avoided for audio MDCT coefficients quantization. To keep the possible quantization error constant, the input vector which falls outside the range (24) will be split into two parts, the least significant bits (LSB) and the most significant bits (MSB), and then the two parts are encoded separately. Let $\mathbf{x} = (x_1, \ldots, x_k)$ be a candidate vector, whose vector dimension is k. Assume that after each split the generated MSB and LSB are denoted by $\mathbf{x}^*$ and $B_i = (b^i_0, b^i_1, \ldots, b^i_k)$, respectively, where i denotes the i-th split. To indicate an overflow happens, a symbol, *escape symbol*, is employed. The whole procedure is demonstrated in Algorithm 2.

Algorithm 2 The split of an overflowed vector $\mathbf{x}$ into the LSB and MSB parts.

Require:

The overflowed vector $\mathbf{x} = (x_1, \ldots, x_k)$ and an array, B_i, for $i = 0, 1, 2, \ldots$, to preserve the resulting LSBs.

Ensure:

1: Set $B_0 = 0$ and $i = 0$;
2: If $||\mathbf{x}|| \leq t_0$, then exit; {not overflowed, no split is needed}
3: Split vector $\mathbf{x}$ into the LSB and MSB parts:
$B_i = (b^i_0 = x_0 \,\&\, 1, \ldots, b^i_k = x_k \,\&\, 1)$,
$\mathbf{x}^* = (x_0 \gg 1, \ldots, x_k \gg 1)$;
And then, $i \leftarrow i + 1$ and set $B_i = 0$;
4: If $||\mathbf{x}^*|| > t_0$, $\mathbf{x} \leftarrow \mathbf{x}^*$, go back to step 3; {still overflowed, split it again}
5: MSB is the resulting vector $\mathbf{x}^*$ and LSBs are B_i, for $i = 0, \ldots, I$, where I is the number of split;
6: **return** The LSB sequence $B_0, \ldots, B_I$, MSB $\mathbf{x}^*$, and I, which denotes the number of *escape symbol*;

3.2.2 Sub-FSVQ

This FSVQ is used to search for the optimum in a predesigned collection of length functions, which are used to encode the current vector index generated from the lattice quantizer. The next-state function of the sub-FSVQ, γ_{s_i}, is built on the four previous indexes I_A, I_B, I_C, and I_D, adjacent to the current input, $I_{\mathbf{x}}$. Since the ECVQ holds a finite number of codevectors, the simplest way to construct the next-state function is to enumerate all the possible combinations of the four neighbors, each denoting a certain state. But with the increase of the number of codevectors, the possible number of current states will be extremely large, and thus, the memory requirements and the computation cost skyrocket.

To reduce the number of possible current states, the different dependencies between the current index and its four previous neighbors must be taken into account. In practice, less emphasis is placed on indices I_A and I_C than on indices I_B and I_D. This is due to the fact that among the four neighbors, current vector $\mathbf{x}$ is less relevant to vectors A and C than to vectors B and D. Thus, we apply the operation $||\cdot||_2$ to vectors A and C, so as to reduce the number of their possible values.

The location of the current vector should also be considered. The frame, current vector located, can be generally classified into two types: the normal frame and the reset frame. In addition, within a frame the current vector can be located at the normal position or the starting position. Thus, there exist four cases, as demonstrated in Figure 5. Specially, if the current vector is located at the starting position of a reset frame, there will be no adjacent vector to build the next-state function, then a special state, s_{s_0}, should be assigned.

As a result, the next-state function of the sub-FSVQ can be written as

$$s_{s_i} = \gamma_{s_i}(I_B, I_D, I_{||A||_2}, I_{||C||_2}) = \begin{cases} t_0 I_B + t_1 I_D + t_2(I_{||C||_2} + I_{||A||_2}) & \text{Case: } (a) \\ t_0 I_B + t_2 I_{||A||_2} & \text{Case: } (b) \\ t_1 I_D & \text{Case: } (c) \\ s_{s_0} & \text{Case: } (d) \end{cases} \tag{25}$$

where i denotes that the sub-FSVQ belongs to the i-th ECVQ and t_0, t_1, and t_2 are three constants making each combination of the four indices corresponding to a different current state. This is feasible since for an audio MDCT coefficient sequence, the values of the four variables, I_B, I_D, $I_{||A||_2}$, and $I_{||C||_2}$, are all finite, and then according to their maximum possible values, it is easy to find the possible values of the three constants.

3.2.3 Length function

The length functions are realized by an arithmetic coder holding multiple pdf models. There are two difficulties in building an optimal arithmetic coder for an optimal ECVQ. First, the memory requirements for saving the predesigned pdf models will become infeasible as the number of states derived from (25) increases. Second, as the volumes of the partitions split by the sub-FSVQ shrink, the available data may not provide credible pdf estimation. Popat and Picard [33] proposed a solution to the second problem using a Gaussian mixture model (GMM) for describing the source pdf. Thus, this work mainly focuses on reducing the memory requirements for saving the pdf models necessary for the arithmetic coder.

The memory requirements can be reduced by merging the similar pdf models. However, according to (16), if one pdf model is replaced by another, mismatch will inevitably

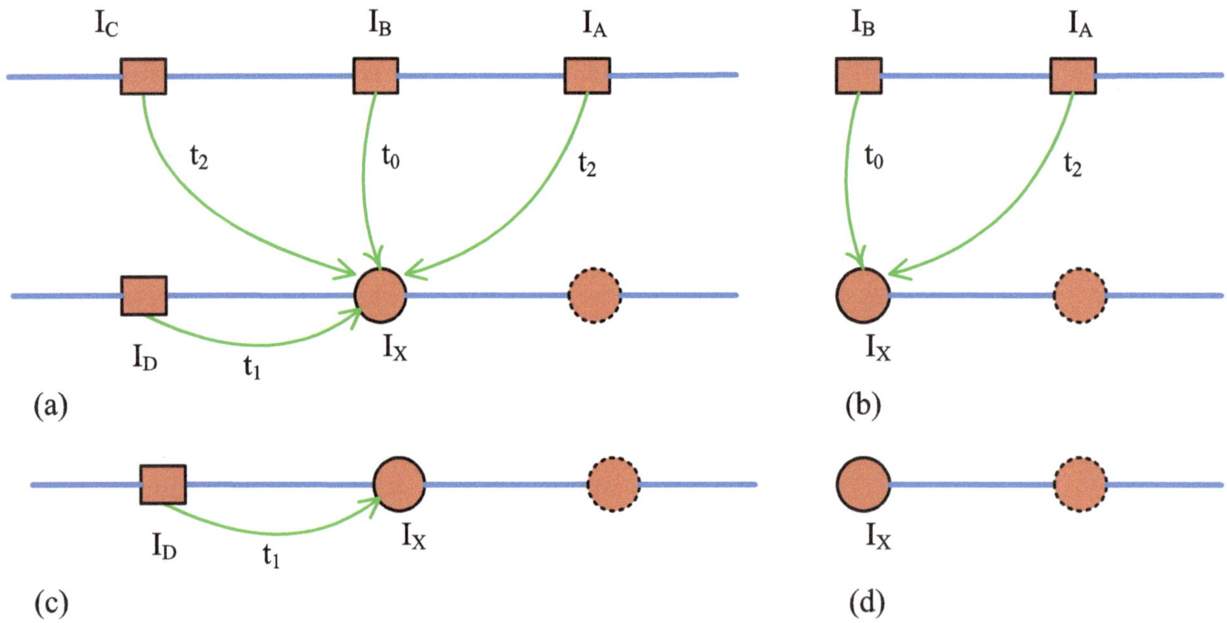

Figure 5 Possible previous adjacent indexes used to calculate the current state s_{s_i} by sub-FSVQ in four situations. (a) Normal frame and normal position. **(b)** Normal frame but starting position. **(c)** Reset frame and normal position. **(d)** Reset frame but starting position.

take place. Let g be the true pdf of the input signal and suppose that Ω_g is its support. Assume that $\{S_m; m \in \mathcal{U}\}$, whose corresponding pdf model is $\{g_m; m \in \mathcal{U}\}$ for $\mathcal{U} = \{1, \ldots, M\}$, is a finite partition of Ω_g and that $P_g(S_m) \leq 0$ for all m. Assume also that model g_m is replaced by another model, g_n, then according to (17) the mismatch of the pdf model pair, g_m and g_n, denoted by $d_{\text{mis}}(m, n)$, can be given as

$$d_{\text{mis}}(m,n) = \int_{S_m} \rho_m g_m(x) \ln \frac{g_m(x)}{g_n(x)} dx = \rho_m I(g_m || g_n) \qquad (26)$$

where ρ_m, which equals to the probability $P_g(S_m)$, is the weight of model g_m. Thus, the mismatch d_{mis} can be seen as a distance measure of a pdf model pair. The more similar the two models are, the smaller is the mismatch. Therefore, we can efficiently decrease the memory requirements for saving the pdf models by merging the model pairs, which hold small enough mismatches, into a new pdf model with a negligible loss of the coding performance.

For a pdf model collection, once we have obtained the d_{mis} values of each model pair, we can merge the ones with minimal d_{mis} values into a new pdf model so as to reduce the memory requirements. If the memory size is still above the requirements, the mergence of the similar pdf models should be continued. But once a new pdf model is generated, the mismatches among pdf models should be updated first. And then, a new merge can be executed. The whole procedure will be carried out iteratively, until the memory size reaches the requirements. Once the final pdf models are obtained, a remapping between these models and their corresponding states is needed.

4 Results

In USAC [34], an up-to-date MPEG standardization, MDCT plays an important role [35]. In the USAC encoder, the MDCT coefficients are firstly companded with a power low function before scalar quantization, achieving in effect a non-uniform scalar quantization. And then, the residuals are further entropy coded. To improve the performance of MDCT coefficients quantization and coding, a novel scheme [29], which combined a scalar quantization with a context-based entropy coding, was developed in the USAC. In this new scheme, the input tuples (blocks) were first quantized by a scalar quantizer (SQ), and then the generated tuple indices were further encoded through a context-based arithmetic encoder. In the USAC final version (FINAL), the tuple length of this scheme was selected to be 2, in order to decrease the total memory requirements.

To further reduce the memory requirements and improve the R/D performance of the MDCT coefficients quantization and coding, a FS-ECVQ was implemented and tested based on the USAC final version. The implemented FS-ECVQ consisted of three component ECVQs, ECVQ_CB4, ECVQ_CB2, and ECVQ_CB1, of which the vector dimensions were 4, 2, and 1, respectively.

To make an easy comparison with the FINAL, the FS-ECVQ was divided into two parts, SQ, which was formed by merging the scaling steps contained in the three component ECVQs and constructed just the same as the one in the FINAL, and the core module of FS-ECVQ, which was referred to FS-ECVQ for simplicity. Thus, the FS-ECVQ and the FINAL would share the same source sequence and the same quantization error and only differ in their coding performance. Therefore, the remainder of this section was mainly focused on evaluating the coding performance of the FINAL and the FS-ECVQ.

4.1 Memory requirements

The total memory requirements of the FINAL and the FS-ECVQ were demonstrated in Table 1. From Table 2, it could be seen that the number of codevectors in FS-ECVQ and FINAL were 85 and 17, respectively. This implied that the equivalent vector dimension of FS-ECVQ would be slightly higher than 2, the dimension of FINAL. Generally, fewer codevectors would lead to a smaller number of vector indices and a smaller memory requirements of each cumulative distribution function (cdf) model. Thus, compared with the FINAL, the FS-ECVQ held a much higher memory requirements for preserving the cdf models.

Compared with FINAL, the FS-ECVQ was less memory exhausting in cdf model decision. This was mainly due to the two FSVQs (main FSVQ and sub-FSVQ), which adaptively reshaped the input blocks and merged the states with similar cdf models to be a new one, while at the same time no side information was needed to be transmitted. Thus, the number of states needed to be conserved contained in sub-FSVQ would be much fewer than those contained in the context-model of the FINAL. As a result, the FS-ECVQ further reduced the total memory requirements of the FINAL by up to 11.3%.

The number of codevectors (codebook size) and the memory requirements for saving the cdf models of FINAL and FS-ECVQ were demonstrated in Table 2. It could be seen that the FS-ECVQ employed three different codebooks, whose dimensions were 4, 2, and 1, respectively.

Table 1 Memory requirements for the two methods: FINAL and FS-ECVQ

Table name	Description	Words of 32 b	
		FINAL	FS-ECVQ
Model decision	For selecting the optimal cdf model	927.5	192
Cdf models	Required for saving the cdf models	512	1,075
Others	Other requirements	1.5	11.5
Total	-	1,441.0	1,278.5

Table 2 Number of codevectors, models, and memory requirements for FINAL and FS-ECVQ

Codebook	FINAL			FS-ECVQ		
	Vector	Model	ROM	Vector	Model	ROM
Dimension 1	-	-	-	10	27	108
Dimension 2	17	64	512	26	31	264
Dimension 4	-	-	-	49	27	703
Total	17	64	512	85	85	1,075

Among these codebooks, the 4-dimensional codebook was assigned the largest number of codevectors, whereas the 1-dimensional one was assigned the least. Through this means, the equivalent vector dimension of the FS-ECVQ would be reduced, and therefore, its memory requirements would be efficiently decreased.

4.2 Average computational complexity

The average computational complexities of the FINAL and the FS-ECVQ, whose units were the weighted million operations per second (WMOPS), were shown in Table 3. From this table, it could be seen that the FS-ECVQ and FINAL held a similar average complexity. The average complexity of FS-ECVQ was mainly due to its main FSVQ. In FS-ECVQ, the main FSVQ was used to estimate which cluster the current block would be classified into according to the shape parameters of its four previous adjacent blocks. To obtain these shape parameters, cubic terms were introduced which obviously increased the total computational complexity.

As the cubic terms usually led to a large computation, to reduce the computational complexity, a look-up table was employed in the FS-ECVQ so that the FS-ECVQ held a similar computational complexity as the FINAL. In practice, the size of the look-up table was dependent on the selection of the threshold of the main FSVQ. In our work, to calculate the threshold of current block, four previous neighbors were employed. Since the current block and its four neighbors were highly correlated and usually hold a similar envelope shape, the largest element of all the codevectors could be constrained to a small value, such as 8. Thus, the size of the look-up table for storing the cubic terms would be very small, about two words.

4.3 Rate performance

Nine audio items, covering speech, music, and mixed speech/music signals, were used for the training of the

Table 3 Average complexity numbers for decoding 32 kbps stereo reference quality bitstreams for quantizers FINAL and FS-ECVQ

Operating mode	USAC-FINAL	FS-ECVQ
PCU (MHz)	0.607	0.605

main-FSVQ, sub-FSVQ, and cdf models, of which the bitrates ranged from 12 to 64 kbps, and the length of every item was about 2 h. And among them, four were mono while the others were stereo items. Another nine audio items, also covering speech, music, and mixed speech/music signals, were chosen as the testing set for the FINAL and the FS-ECVQ, of which the bitrates ranged from 12 to 64 kbps and the length of every item was about 3 min. Among them, four were mono while the others were stereo. The testing results were shown in Table 4, where the percentage column represented the increment of the coding gain of the current method over the FINAL.

The table demonstrated that the FINAL and the FS-ECVQ achieved a similar coding performance in all the nine items. This denoted that the FINAL and the FS-ECVQ both could efficiently remove the redundancies within audio MDCT coefficient sequences. Moreover, both FINAL and FS-ECVQ obtained more coding gains in the low bitrate items than in the high bitrate items. These phenomena were mainly due to the fact that the nine items have different pdf of MDCT coefficients. In FS-ECVQ, a different source distribution would lead to a different calling ratio of its three component ECVQs.

4.4 Main FSVQ

The main FSVQ split the input vectors into subvectors according to a pyramidal decomposing method, by which the MDCT coefficient sequence was partitioned into three clusters, Ω_4, Ω_2, and Ω_1. The component ECVQs applied on these resulting clusters were ECVQ_D4, ECVQ_D2, and ECVQ_D1, respectively. To decompose an input vector, in cluster Ω_4 and Ω_2, the main FSVQ would first calculate two shape parameters from the two pairs of previous adjacent blocks B, D and A, C, respectively, via their corresponding block energies and block skewness, and then, compare them with the two thresholds, T_{bd} and T_{ac}, respectively. Thus, a different combination of the thresholds would lead to a different distribution of the MDCT coefficients among the three component ECVQs, and consequently a different coding gain of the FS-ECVQ. The different combinations of T_{bd} and T_{ac} in the two clusters and their corresponding results were all demonstrated in Table 5. From the table, at least two points could be derived.

Table 4 Bitrates of quantizers FINAL and FS-ECVQ for nine audio items

Operating mode	FINAL	FS-ECVQ	
	(kbps)	(kbps)	(%)
Test 1, 64 kbps stereo	48.59	48.77	−0.33
Test 2, 32 kbps stereo	24.56	24.55	0.04
Test 3, 24 kbps stereo	17.59	17.57	0.11
Test 4, 20 kbps stereo	14.87	14.86	0.09
Test 5, 16 kbps stereo	11.71	11.69	0.17
Test 6, 24 kbps mono	18.97	18.98	−0.05
Test 7, 20 kbps mono	15.41	15.42	−0.06
Test 8, 16 kbps mono	12.18	12.17	0.08
Test 9, 12 kbps mono	8.78	8.76	0.23
Average	19.18	19.18	0.03

First, the thresholds of cluster Ω_4 had a larger impact on the coding gain than cluster Ω_2 did, which could be explained by the fact that the variation range of the coding gains on Ω_4 was much wider than that on Ω_2. Furthermore, within a level threshold T_{bd} had a larger impact on the coding gain than threshold T_{ac} did. Since T_{bd} and T_{ac} were obtained from adjacent blocks B, D and A, C, respectively, this proved the assumption that B, D were more significant than A, C.

Second, the component ECVQ, ECVQ_D4, gains than the two others. From Table 5, it could be observed that most of the MDCT coefficients were encoded by ECVQ_D4. Therefore, to obtain the optimal performance, the promotion of performance of ECVQ_D4 should be of the highest priority.

The calling ratios of the three component ECVQs in the nine testing items were demonstrated in Figure 6. It could also be observed that among all the nine items, the calling frequency of ECVQ_D4 was the highest, whereas the frequency of ECVQ_D1 was the lowest. As the vector dimensions ECVQ_D4, ECVQ_D2, and ECVQ_D1, were 4, 2, and 1, respectively, the calling rations of them implied that most of the MDCT coefficients in each testing item were encoded by the 4-dimensional ECVQ and only a very small amount of them were encoded by the 1-dimensional one. Through this way, FS-ECVQ achieved a relatively high coding performance. Furthermore, among all the nine items, the more frequently ECVQ_D4 was called, the larger coding gains the FS-ECVQ obtained. This explained why FS-ECVQ was more efficient in coding low bitrate items than the high bitrate ones.

4.5 ECVQ

As each component ECVQ contained two stages, lattice quantization and entropy coding, we would first assess the quantization stage and then, the entropy coding stage.

4.5.1 *Quantization stage*

To assess the quantization stage, we took LSB as a major indicator. There were at least three reasons. First, LSB appeared if and only if an input vector fell outside the range constrained by the lattice quantizer, and thus, LSB could be seen as the sign of the appearance of error in the quantization stage. Therefore, the lower occurrence frequency of LSB would usually denote fewer quantization errors in the quantization stage, and as a result, a

Table 5 The effects on the three component ECVQs and coding gains

Pyramidal decomposition	Threshold		ECVQ_D4		ECVQ_D2		ECVQ_D1		Gains (%)
	T_{bd}	T_{ac}	Ratio (%)	LSB (%)	Ratio (%)	LSB (%)	Ratio (%)	LSB (%)	
Ω_4[a]	9		77.395	1.403	16.769	1.873	5.837	9.883	3.0626
	10		79.276	1.623	14.895	2.075	5.830	9.895	2.9917
	12	30	82.300	2.104	11.884	2.501	5.816	9.914	2.7091
	17		82.517	2.159	11.669	2.529	5.814	9.916	2.6779
	24		83.410	2.403	10.785	2.673	5.805	9.925	2.5094
	∞		87.733	4.947	7.728	2.672	4.539	10.129	0.9096
		24	78.733	1.579	15.436	2.014	5.832	9.892	3.0123
		27	78.836	1.588	15.333	2.025	5.831	9.893	3.0094
	10	33	79.520	1.645	14.651	2.105	5.829	9.896	2.9834
		36	79.606	1.654	14.566	2.117	5.829	9.896	2.9812
		∞	81.651	1.917	12.624	2.363	5.725	10.062	2.7537
Ω_2[b]	32				13.849	1.647	6.875	8.455	2.9771
	45				14.542	1.880	6.183	9.363	3.0022
	67	243	79.276	1.623	15.219	2.241	5.505	10.443	2.9449
	75				15.496	2.436	5.228	10.950	2.8997
	∞				17.364	6.231	3.360	12.663	1.4433
		189			14.727	2.021	5.997	9.634	2.9563
		216			14.776	2.036	5.948	9.707	2.9535
	65	249	79.276	1.623	14.922	2.085	5.802	9.941	2.9929
		257			14.942	2.098	5.782	9.972	2.9916
		∞			15.537	2.548	5.817	10.975	2.8963

[a]Set the thresholds of Ω_2 to be $T_{bd} = 65$, $T_{ac} = 243$; [b]set the thresholds of Ω_4 to be $T_{bd} = 10$, $T_{ac} = 30$. Meanwhile, the main FSVQ is of different thresholds in 24 kbps stereo.

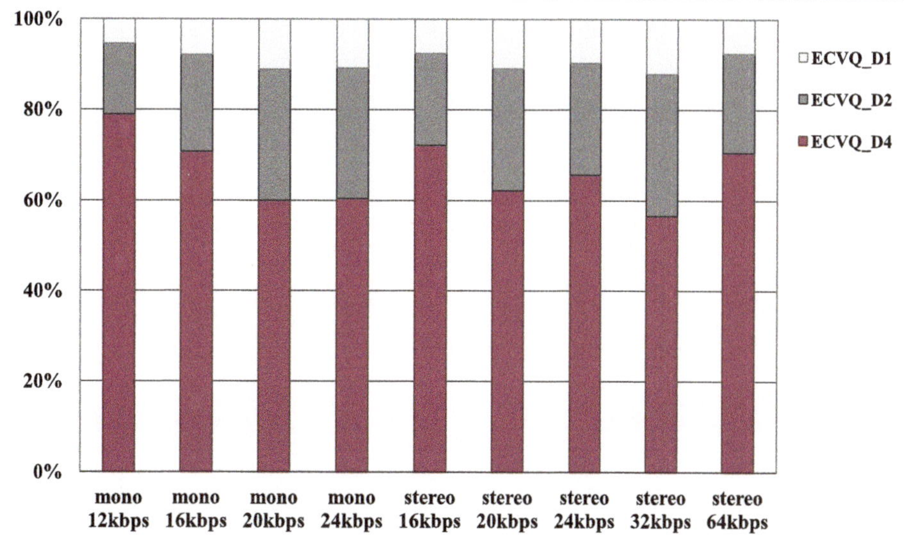

Figure 6 The calling ratios of the three component quantizers. The three components ECVQ_D4, ECVQ_D2, and ECVQ_D1 have vector dimensions of 4, 2, and 1, respectively.

higher coding gain achieved by the component ECVQ. Second, by adjusting the threshold T_{bd} and T_{ac}, we could achieve different occurrence frequency of LSB and thus make different trade-off between the coding gain and the memory requirements. At last, the ratio among the three LSB occurrence frequencies is correlated with the distribution of quantization errors among the three component ECVQs. A higher LSB occurrence frequency denoted more quantization errors distributed to the corresponding component ECVQ.

The LSB occurrence in each component ECVQ significantly influenced the final coding gain of the FS-ECVQ, which could be seen from the Table 5. For an input vector, if the LSB appeared, the ECVQ would consume much more bits than that for encoding it directly. There were two methods for reducing the appearance of LSB: to enlarge the range of the corresponding codebook or to shrink the range constrained by the threshold. However, the first method would lead to an increase in the memory requirements, while the second would degrade the coding gain. Therefore, a trade-off must be made between the memory requirements and the coding gain. Among the three ECVQs, ECVQ_D4 had the least percentage of LSBs while ECVQ_D1 had the largest. By this means, the FS-ECVQ could save the memory requirements while keeping the coding gain as high as possible.

4.5.2 The length functions

In each component ECVQ, the length function was realized by an arithmetic coder, which employed the

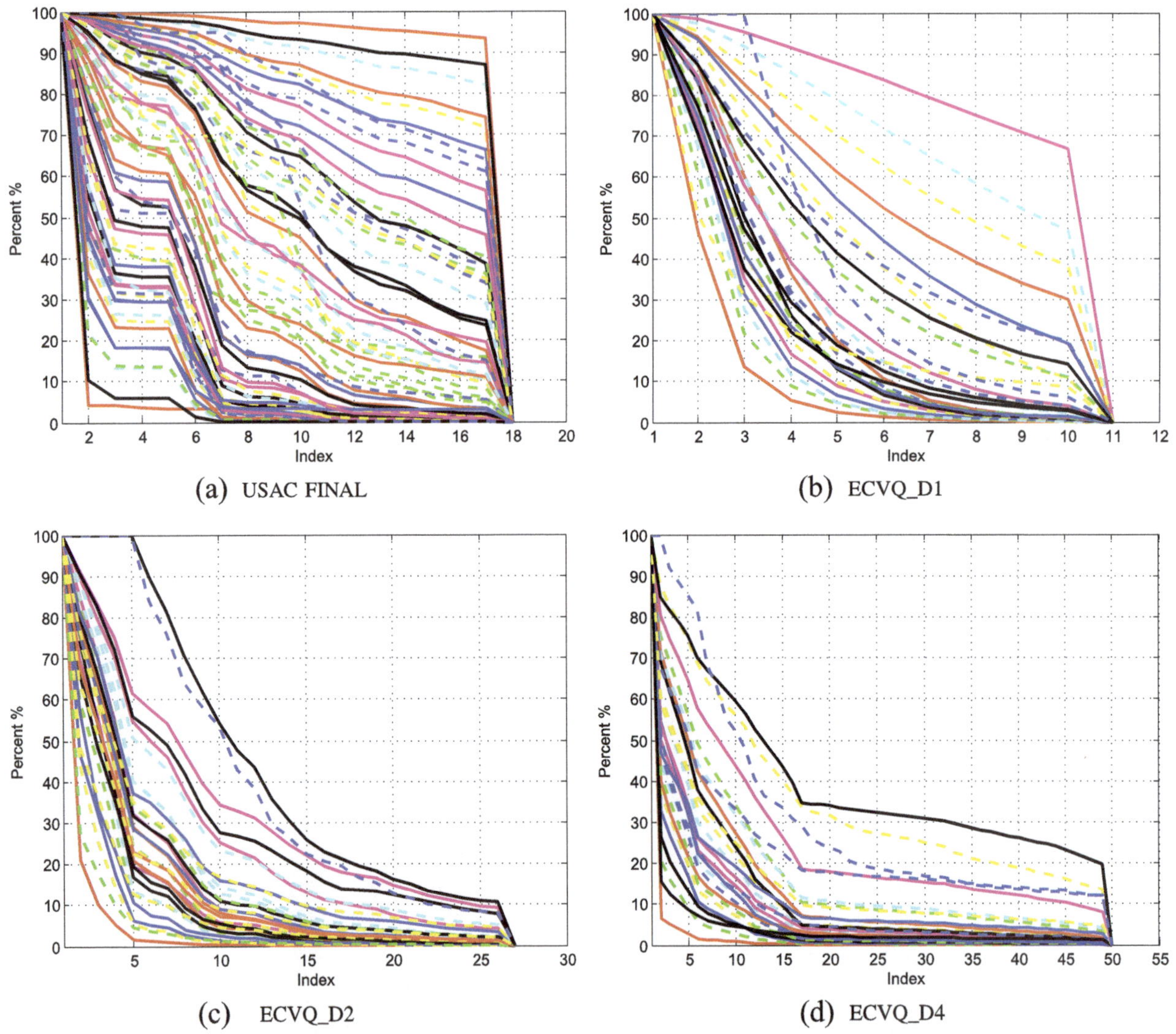

Figure 7 The cdf models contained in the FINAL and the FS-ECVQ. (a) The 64 cdf models which are contained in the USAC FINAL; **(b)** 27 cdf models contained in the ECVQ_D1 of the FS-ECVQ; **(c)** 31 cdf models contained in the ECVQ_D2 of the FS-ECVQ; **(d)** 27 cdf models contained in the ECVQ_D4 of the FS-ECVQ.

sub-FSVQ to search for the optimum in a predesigned cdf model collection. The cdf models of FINAL and FS-ECVQ were demonstrated in Figure 7. From the figure, it could be seen that the cdf model numbers of the FINAL and FS-ECVQ were 64 and 85, respectively. Essentially, the cdf models contained were used to fit the pdf of the MDCT coefficient sequence. A larger number of cdf models generally would provide a higher accuracy fitting of the source pdf. Therefore, the FS-ECVQ could obtain a higher performance than the FINAL, theoretically.

Although the FINAL contained less cdf models than the FS-ECVQ did, it obtained similar coding performance to the FS-ECVQ. This was mainly owing to the cdf model selection method used in FINAL, which accurately selected the optimal cdf model for each input vector index. However, it was more complicated than that used in FS-ECVQ. This could be seen from the fact that the memory requirements for the cdf model selection in FINAL was much larger than those in FS-ECVQ, as demonstrated in Table 1.

5 Conclusions

In this paper, an ECVQ with finite memory, called FS-ECVQ, is proposed. In the FS-ECVQ, a FSVQ, namely the main FSVQ, is used to partition the source sequence into multiple non-overlapped clusters. Then to each cluster, an ECVQ is applied. Within each ECVQ, its length function is taken by an arithmetic coder holding multiple predesigned cdf models. To select the optimal cdf model for each input vector, another FSVQ, namely the sub-FSVQ, is employed.

Owing to the main FSVQ which effectively exploits the inter-frame dependencies, the source sequence is split into multiple clusters and no side information is needed to be transmitted. Moreover, the main FSVQ assigned different vector dimensions to the resulting clusters. The more frequently a cluster appears, the higher vector dimension is allocated. This helps the FS-ECVQ to efficiently reduce its total memory requirements while, at the same time, maintaining a relatively high coding performance. Finally, for each input vector, the sub-FSVQ selects the best matching cdf model, which adds robustness to the FS-ECVQ.

There are multiple ways to realize the proposed FS-ECVQ. First of all, if the quantizing errors generated from the lattice quantizer are directly discarded, then the FS-ECVQ is equivalent to an ordinary ECVQ. However, if the quantizing errors are taken as the LSBs and encoded by an additional length function, the FS-ECVQ will be equal to an uniform quantizer. In addition, if the quantization steps of all the component ECVQs are separated from the FS-ECVQ, then the FS-ECVQ becomes an entropy encoder. The FS-ECVQ can also be used in coding the speech, image, and video signals, and even any other source sequence with non-uniform distribution.

Competing interests

The authors declare that they have no competing interests.

Acknowledgements

The authors wish to thank the anonymous reviewers for their detailed comments and suggestions, which have been extremely helpful in improving the clarity and quality of this paper. This work was supported by the National Natural Science Foundation of China under Grant No. 61171171.

References

1. A Gersho, RM Gray, *Vector Quantization and Signal Compression.* (Wiley, New York, 1994)
2. S So, KK Paliwal, Efficient product code vector quantisation using the switched split vector quantiser. Digit Signal Process. **17**, 138–171 (2007)
3. R Gray, D Neuhoff, Quantization. Inform. Theory, IEEE Trans. **44**(6), 2325–2383 (1998)
4. A Subramaniam, B Rao, PDF optimized parametric vector quantization of speech line spectral frequencies. Speech Audio Process IEEE Trans. **11**(2), 130–142 (2003)
5. S So, K Paliwal, Multi-frame GMM-based block quantisation of line spectral frequencies for wideband speech coding, in *Proceedings in IEEE International Conference on Acoustics, Speech, and Signal Processing, (ICASSP '05), vol. 1* (Philadelphia, March 2005), pp. 121–124
6. K Paliwal, B Atal, Efficient vector quantization of LPC parameters at 24 bits/frame. Speech Audio Process IEEE Trans. **1**, 3–14 (1993)
7. M Bouzid, S Cheraitia, M Hireche, Switched split vector quantizer applied for encoding the LPC parameters of the 2.4 Kbits/s MELP speech coder, in *7th International Multi-Conference on Systems Signals and Devices* (Amman, Jordan, June 2010), pp. 1–5
8. J Leis, S Sridharan, Adaptive vector quantization for speech spectrum coding. Digit Signal Process. **9**(2), 89–106 (1999)
9. S Chatterjee, T Sreenivas. Optimum switched split vector quantization of LSF parameters.Signal Process. **88**(6), 1528–1538 (2008)
10. F Nordin, T Eriksson, On split quantization of LSF parameters, in *Proceedings on IEEE International Conference on Acoustics, Speech, and Signal Processing (ICASSP '04), vol. 1* (Montreal, May 2004), pp. I–157–60
11. S So, KK Paliwal, A comparative study of LPC parameter representations and quantisation schemes for wideband speech coding. Digit Signal Process. **17**, 114–137 (2007)
12. S Chatterjee, T Sreenivas, Gaussian mixture model based switched split vector quantization of LSF parameters, in *IEEE International Symposium on Signal Processing and Information Technology* (Giza, December 2007), pp. 1054–1059
13. Y Lee, W Jung, MY Kim, GMM-based KLT-domain switched-split vector quantization for LSF coding. Signal Process Lett. IEEE. **18**(7), 415–418 (2011)
14. S Chatterjee, T Sreenivas, Switched conditional PDF-based split VQ using gaussian mixture model. Signal Process Lett. IEEE. **15**, 91–94 (2008)
15. P Chou, T Lookabaugh, R Gray, Entropy-constrained vector quantization. Acoustics Speech Signal Process. IEEE Trans. **37**, 31 –42 (1989)
16. T Lookabaugh, R Gray, High-resolution quantization theory and the vector quantizer advantage. Inform Theory IEEE Trans. **35**(5), 1020–1033 (1989)
17. D Zhao, J Samuelsson, M Nilsson, On entropy-constrained vector quantization using gaussian mixture models. Commun IEEE Trans. **56**(12), 2094–2104 (2008)
18. A Vasilache, Rate-distortion models for entropy constrained lattice quantization, in *IEEE International Conference on Acoustics Speech and Signal Processing, (ICASSP '10)* (Dallas, March 2010), pp. 4698–4701
19. J Foster, R Gray, M Dunham, Finite-state vector quantization for waveform coding. Inform Theory IEEE Trans. **31**(3), 348–359 (1985)
20. BS Andras Cziho, IL ETC, An optimization of finite-state vector quantization for image compression. Signal Process Image Commun. **15**(6), 545–558 (2000)
21. P Yahampath, M Pawlak, On finite-state vector quantization for noisy channels.Commun. IEEE Trans. **52**(12), 2125–2133 (2004)

22. RF Chang, YL Huang, Finite-state vector quantization by exploiting interband and intraband correlations for subband image coding. Image Process IEEE Trans. **5**(2), 374–378 (1996)
23. S Jiang, R Yin, P Liu, A finite-state entropy-constrained vector quantizer for audio MDCT coefficients coding, in *International Conference on Audio, Language and Image Processing, (ICALIP 2012)* (Shanghai, July 2012), pp. 218–223
24. ISO/IEC JTC1/SC29/WG11, Call for proposals on unified speech and audio coding (2007). [http://mpeg.chiariglione.org/standards/mpeg-d/unified-speech-and-audio-coding]
25. R Gray, T Linder, J Li, A Lagrangian formulation of Zador's entropy-constrained quantization theorem. Inform Theory IEEE Trans. **48**(3), 695–707 (2002)
26. N Nasrabadi, S Rizvi, Next-state functions for finite-state vector quantization. Image Process IEEE Trans. **4**(12), 1592–1601 (1995)
27. R Gray, J Li, On Zador's entropy-constrained quantization theorem, in *Proceedings on Data Compression Conference, (DCC 2001)* (Snowbird, March 2001), pp. 3–12
28. R Gray, T Linder, Mismatch in high-rate entropy-constrained vector quantization.Inform. Theory IEEE Trans. **49**(5), 1204–1217 (2003)
29. G Fuchs, V Subbaraman, M Multrus, Efficient context adaptive entropy coding for real-time applications, in *IEEE International Conference on Acoustics, Speech and Signal Processing, (ICASSP '11)* (Prague, May 2011), pp. 493–496
30. N Nasrabadi, C Choo, Y Feng, Dynamic finite-state vector quantization of digital images. Commun IEEE Trans. **42**(5), 2145–2154 (1994)
31. A Gyorgy, T Linder, P Chou, B Betts, Do optimal entropy-constrained quantizers have a finite or infinite number of codewords Inform Theory IEEE Trans. **49**(11), 3031–3037 (2003)
32. R Yu, X Lin, S Rahardja, C Ko, A statistics study of the MDCT coefficient distribution for audio, in *IEEE International Conference on Multimedia and Expo, (ICME '04) vol. 2* (Taipei, June 2004), pp. 1483–1486
33. K Popat, R Picard, Cluster-based probability model and its application to image and texture processing. Image Process IEEE Trans. **6**(2), 268–284 (1997)
34. ISO/IEC JTC 1/SC 29N11510, Information technology - MPEG audio technologies Part 3: unified speech and audio coding (2010). [http://mpeg.chiariglione.org/standards/mpeg-d/unified-speech-and-audio-coding]
35. M Neuendorf, P Gournay, M Multrus, J Lecomte, B Bessette, R Geiger, S Bayer, G Fuchs, J Hilpert, N Rettelbach, R Salami, G Schuller, R Lefebvre, B Grill, Unified speech and audio coding scheme for high quality at low bitrates, in *IEEE International Conference on Acoustics, Speech and Signal Processing, (ICASSP '09)* (Taipei, April 2009), pp. 1–4

Efficient voice activity detection algorithm using long-term spectral flatness measure

Yanna Ma* and Akinori Nishihara

Abstract

This paper proposes a novel and robust voice activity detection (VAD) algorithm utilizing long-term spectral flatness measure (LSFM) which is capable of working at 10 dB and lower signal-to-noise ratios(SNRs). This new LSFM-based VAD improves speech detection robustness in various noisy environments by employing a low-variance spectrum estimate and an adaptive threshold. The discriminative power of the new LSFM feature is shown by conducting an analysis of the speech/non-speech LSFM distributions. The proposed algorithm was evaluated under 12 types of noises (11 from NOISEX-92 and speech-shaped noise) and five types of SNR in core TIMIT test corpus. Comparisons with three modern standardized algorithms (ETSI adaptive multi-rate (AMR) options AMR1 and AMR2 and ITU-T G.729) demonstrate that our proposed LSFM-based VAD scheme achieved the best average accuracy rate. A long-term signal variability (LTSV)-based VAD scheme is also compared with our proposed method. The results show that our proposed algorithm outperforms the LTSV-based VAD scheme for most of the noises considered including difficult noises like machine gun noise and speech babble noise.

1 Introduction

Voice activity detection (VAD) is a method to discriminate speech segments from input noisy speech. It is an integral part to many speech and audio processing applications and is widely used within the field of speech communication for achieving high coding efficiency and low bit rate transmission. Examples include noise reduction for digital hearing aid devices [1], mobile communication services [2], voice recognition systems [3], compression [4], and speech coding [5].

A typical VAD system consists of two core parts: feature extraction and speech/non-speech decision mechanism. Researchers have proposed a variety of features exploiting different properties of speech and noise to achieve better VAD performance. In early VAD algorithms, short-term energy [6] and zero-crossing rate [7] were widely used features because of their simplicity. However, the performance degrades easily when faced with low signal-to-noise ratio (SNR) or non-stationary background noise. To solve this problem, robust acoustic features such as spectrum [8], autocorrelation [9], power in the band-limited region [10], and higher-order statistics [11] have been proposed for VAD. Most of those methods assume the background noise to be stationary during a certain period; thus, they are sensitive to changes in SNR of the observed signal. Some works [12,13] proposed noise estimation and adaptation for improving VAD robustness, but those methods are computationally expensive. Most of those features mentioned work sufficiently well in stationary noise and higher than 10-dB SNR cases. When facing with lower SNR cases or when the background noise contains complex audible events appearing occasionally, such as babble noise in a cafeteria and machinery noise in a factory, there will be cases when most of the speech spectrum is corrupted, which destroys the overall statistical as well as structural properties of the speech signal [14]. In general, VAD algorithms based on a particular feature or specific set of features are still far from efficient especially when they are operating in adverse acoustic conditions. Therefore, the VAD algorithm in low SNRs and some specific noises such as speech babble noise and machine gun noise still remains challenging and requires the design of further robust features and algorithms.

All VAD features mentioned are extracted from the short-term analysis frames (usually 20 to 40 ms), and decisions are made at each frame. In contrast with the use of frame level features, Ramirez et al. [12] proposed

*Correspondence: mayanna@nh.cradle.titech.ac.jp
Department of Communications and Integrated Systems, Tokyo Institute of Technology, Tokyo 152-8552, Japan

the use of a long-term spectral divergence feature to discriminate speech from noise. It requires average noise spectrum magnitude information which is not accurately available in practice. Moreover, Ghosh et al. [15] proposed a long-term signal variability (LTSV)-based VAD which uses a very long window to estimate the averaged spectrogram as well as for computing long-term entropies of each frequency band. This LTSV-based VAD yields a great improvement for SNRs smaller than 5 dB but becomes saturated when SNRs are higher than 5 dB.

Spectral flatness is a measure of the width, uniformity, and noisiness of the power spectrum. A high spectral flatness indicates that the spectrum has a similar amount of power in all spectral bands, and the graph of the spectrum would appear relatively flat and smooth; this would sound similar to white noise. A low spectral flatness indicates that the spectral power is less uniform in frequency structure, and this would typically sound like speech. Therefore, the analysis over a long window for exploiting the spectral flatness of the signal will be beneficial for distinguishing speech from noise. In this paper, we propose a novel VAD algorithm based on long-term spectral flatness measure (LSFM). The discriminative power of the proposed LSFM feature will be verified by researching the distribution of LSFM measure for speech and non-speech in terms of their misclassification rate for various noises. We have experimentally evaluated its performance under a variety of noise types and SNR conditions.

The structure of the rest of this article is arranged as follows. Section 2 discusses the LSFM feature and its discriminative ability. Section 3 presents the proposed LSFM-based VAD algorithm including the choice for proper parameters and the design of an adaptive threshold. Section 4 contains the speech and noise database and metrics used in the evaluation. Section 5 provides the experimental results. Finally, a conclusion of this work and the discussion are given in Section 6.

2 Long-term spectral flatness measure and its discriminative power

Speech is a highly non-stationary signal, while background noise can be considered to be stationary over relatively long periods. The rationale behind the LSFM

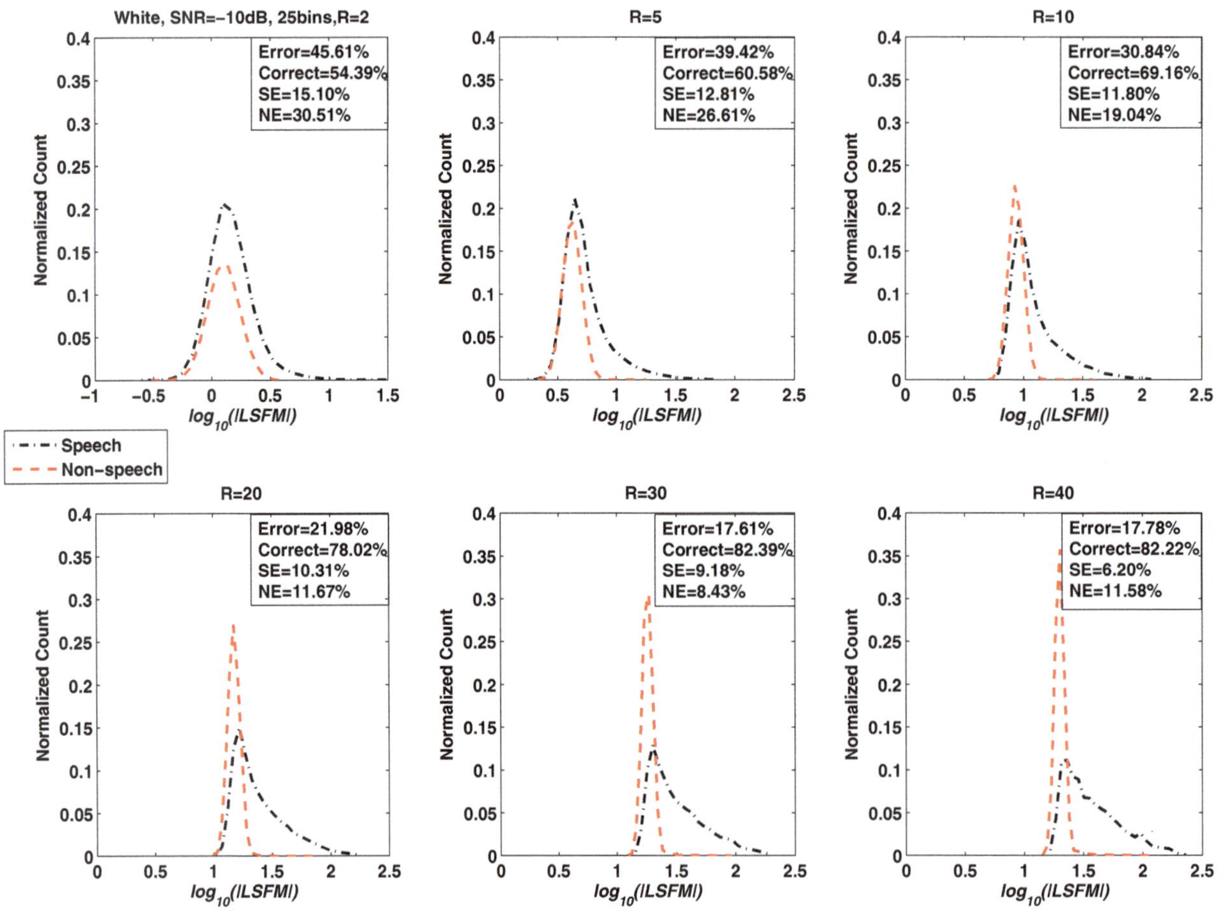

Figure 1 LSFM measure as a function of long-term window length (*R*) in additive white noise (SNR = −10 dB).

feature is that the observed signal spectrum evinces more structure when the signal of interests is present compared to when it is absent. This increase in the structure of the signal may be characterized by a reduction in the flatness of the magnitude spectrum of the short-time Fourier representation of the signal [16].

2.1 Long-term spectral flatness measure

The LSFM feature is computed using the spectra of the last R frames of the input signal $x(n)$. The LSFM feature, $L_x(m)$, at the mth frame and across all the chosen frequency is then calculated by dividing the geometric mean of the power spectrum by the arithmetic mean of the power spectrum. To expand the dynamic range, it is measured on a logarithmic scale, ranging from zero to minus infinity as:

$$L_x(m) = \sum_k \log_{10} \frac{\mathrm{GM}(m, \omega_k)}{\mathrm{AM}(m, \omega_k)}, \tag{1}$$

where the geometric mean $\mathrm{GM}(m, \omega_k)$ and arithmetic mean $\mathrm{AM}(m, \omega_k)$ of the power spectrum is calculated as:

$$\mathrm{GM}(m, \omega_k) = \sqrt[R]{\prod_{n=m-R+1}^{m} S(n, \omega_k)}, \tag{2}$$

$$\mathrm{AM}(m, \omega_k) = \frac{1}{R} \sum_{n=m-R+1}^{m} S(n, \omega_k). \tag{3}$$

The short-time spectrum $S(n, \omega_k)$ used in this research is estimated using the Welch-Bartlett method which averages the spectral estimates of M consecutive frames. The expressions are

$$S(n, \omega_k) = \frac{1}{M} \sum_{p=n-M+1}^{n} |X(p, \omega_k)|^2, \tag{4}$$

$$X(p, \omega_k) = \sum_{l=(p-1)N_{\mathrm{sh}}+1}^{N_w+(p-1)N_{\mathrm{sh}}} w(l-(p-1)N_{\mathrm{sh}}-1)x(l)e^{-j\omega_k l}, \tag{5}$$

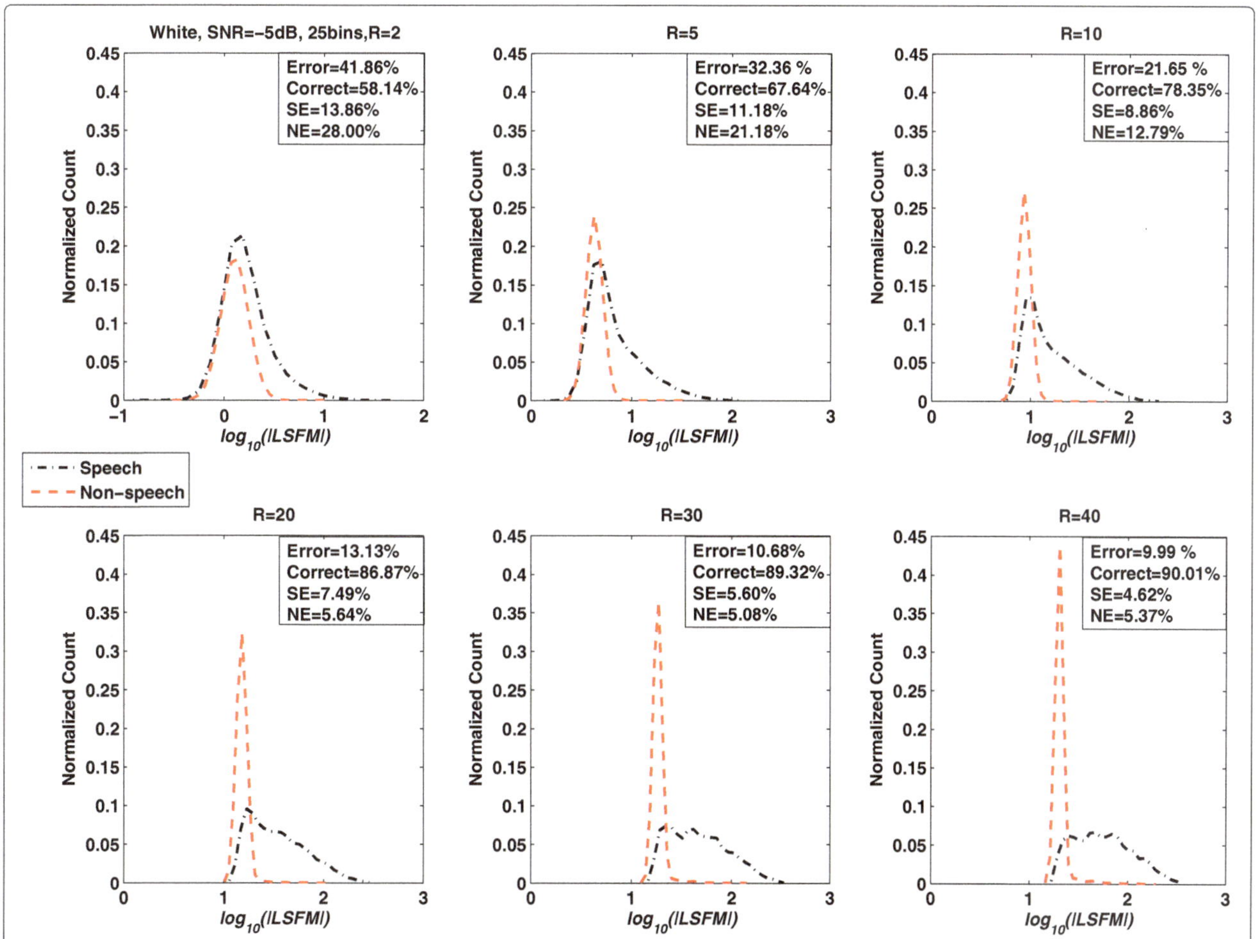

Figure 2 LSFM measure as a function of long-term window length (*R*) in additive white noise (SNR = −5 dB).

where $X(p, \omega_k)$ is the short-time Fourier transform coefficient at frequency ω_k of the pth frame. $w(i)$ is the short-time Hann window, where $i \in [0, N_w)$. N_w is the frame length and N_{sh} is the frame shift duration in terms of samples.

According to AM-GM inequality, the geometric mean, $GM(m, \omega_k)$, is smaller than or equal to the arithmetic mean, $AM(m, \omega_k)$, with equality being achieved if and only if all $S(n, \omega_k)$ are the same. Therefore, from Equation 1, we can conclude that the LSFM feature, $L_x(m)$, is in the range $(-\infty, 0]$ with the maximum value acquired when the geometric mean is equal to the arithmetic mean.

2.2 The LSFM feature distributions of speech and non-speech

In this subsection, the distributions of the LSFM feature are investigated in order to clarify the motivation for utilizing the proposed LSFM feature as a VAD algorithm and demonstrate the discriminative power of this feature.

The test set consisting of 16 individual speakers (8 male, 8 female), each speaking 10 phonetically balanced English sentences, is randomly chosen from the TIMIT training corpus [17]. The LSFM feature values were computed at every frame from noisy speech. The LSFM measure, $L_x(m)$, is considered to be $L_{S+N}(m)$ if there are speech samples between $(m - R + 1)$th and mth frame. Otherwise, it is decided to be $L_N(m)$. The overlap area of the two distributions (L_{S+N} and L_N) is considered to be the error caused by misclassification. The lower the misclassification rate is, the better the separation. The sampling frequency of the test signal is 16 kHz, and the Hann window has a length of 20 ms and a 10-ms shift. M is fixed to be 10, and $\{\omega_k\}$ is uniformly distributed between the frequency range 500 Hz to 4 kHz. The total misclassification error among these realizations of L_{S+N} and L_N was computed by comparing with the phonetic level transcription [17] of the TIMIT training corpus.

First, the distributions of the LSFM feature as a function of the long-term window length (R = 2, 5, 10, 20, 30, and 40) for white noise at five SNR levels (−10, − 5, 0, 5, and 10 dB) were studied. The results are shown in Figures 1, 2, 3, 4, and 5. The total misclassification

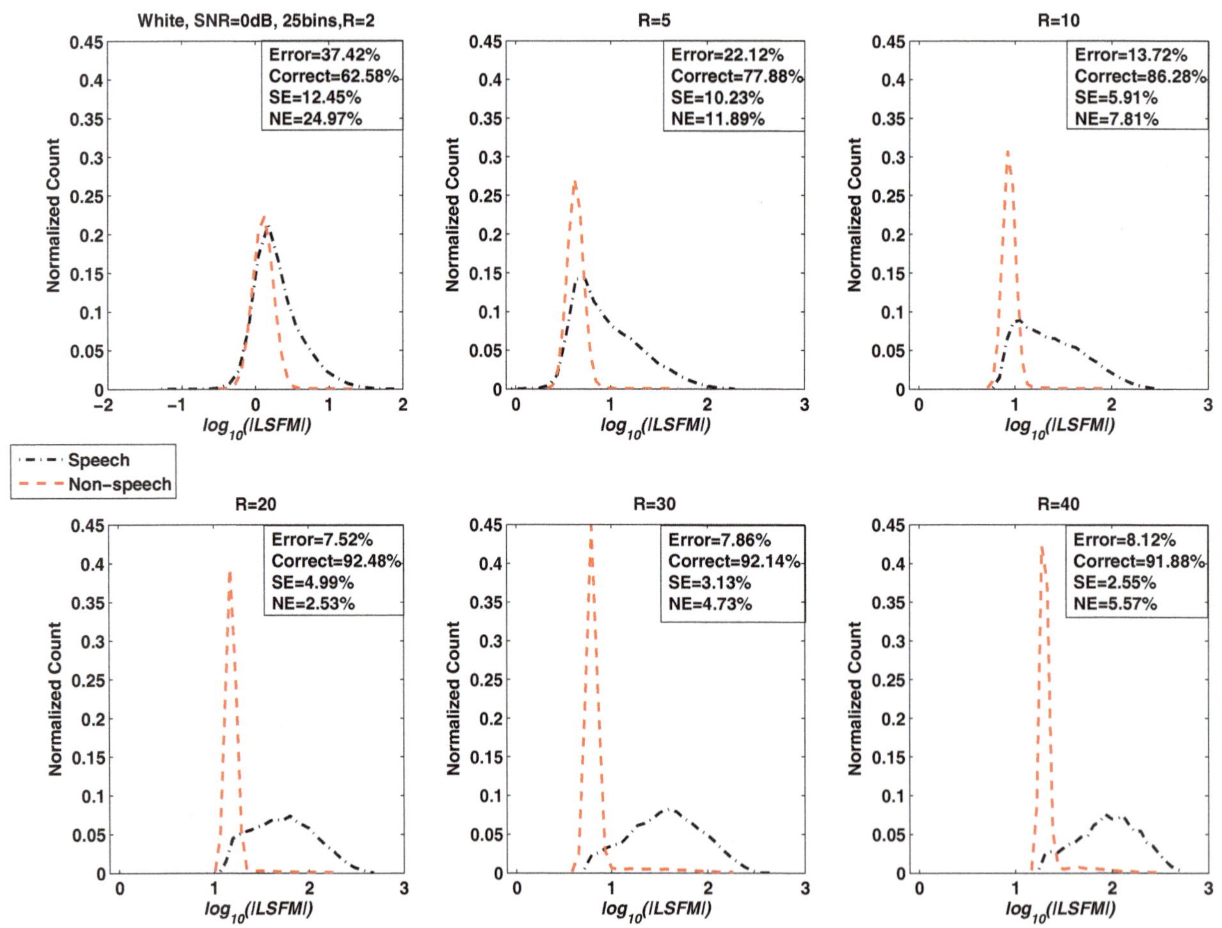

Figure 3 LSFM measure as a function of long-term window length (R) in additive white noise (SNR = 0 dB).

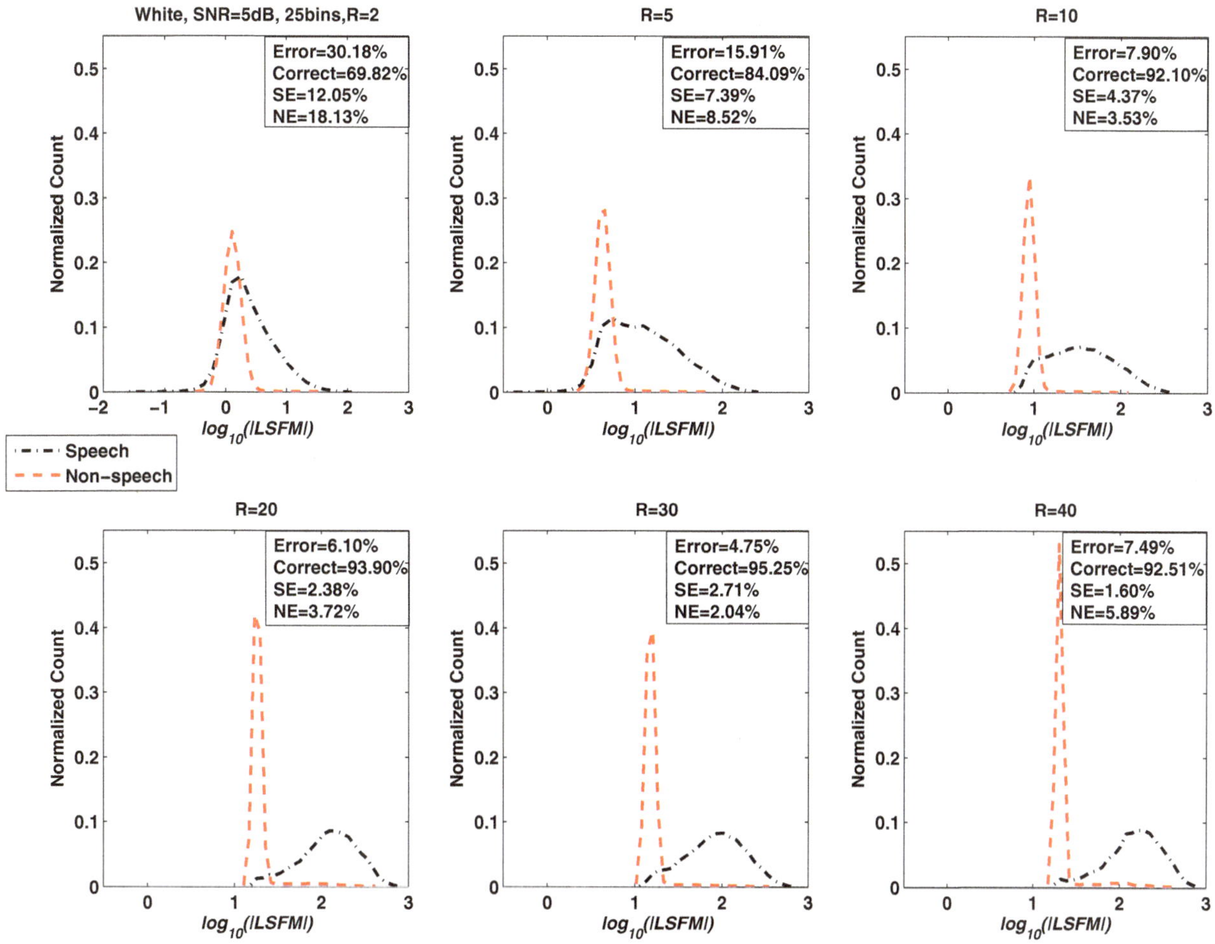

Figure 4 **LSFM measure as a function of long-term window length (*R*) in additive white noise (SNR = 5 dB).**

error (Error), accuracy rate (Correct), speech detection error (SE), and non-speech detection error (NE) are displayed on the upper or lower right of each subfigure. The total misclassification error was reduced by 61.4% (−10 dB), 74.5% (−5 dB), 79.0% (0 dB), 84.3% (5 dB), and 82.9% (10 dB) when the window length R was increased from 2 to 30 frames. The percentage is the ratio between the reduced misclassification error (when R was changed from 2 to 30 frames) and the misclassification error when R was 2.

The distributions of the LSFM feature for all 12 kinds of noises at 0-dB SNR were investigated. M is fixed to be 10, and R is chosen to be 30. The discriminative power of this LSFM feature can be measured by the separateness of its distribution for speech and non-speech. As shown in Figures 6, 7, and 8, there is overlap between the histograms of $\log_{10}(L_{S+N})$ and $\log_{10}(L_N)$. We calculated the total misclassification error which is the sum of the speech detection error and non-speech detection error. From the figures, we can conclude that for most noises considered (9 out of 12 kinds of noises), the proposed LSFM feature resulted in a misclassification error smaller than 10%: white (7.86%), pink (7.75%), tank (7.47%), military vehicle (7.75%), jet cockpit (9.32%), HF channel (8.30%), F-16 cockpit (8.89%), car interior (8.14%), and speech shaped (7.84%). For factory floor (25.86%), machine gun (45.42%), and speech babble (24.08%), the misclassification errors were comparatively high. The factory floor is that of cutting noise, that of the machine gun is impulsive in nature, and that of speech babble is speechlike. One possible reason for the poor performance is the mismatch of M and R.

3 The proposed LSFM-based VAD algorithm

The proposed VAD algorithm assumes that the signal spectrum is more organized during speech segments than during noise segments [18]. It adopts the average spectrum over a long-term window instead of instantaneous values of the spectrum. Typically, a periodogram is

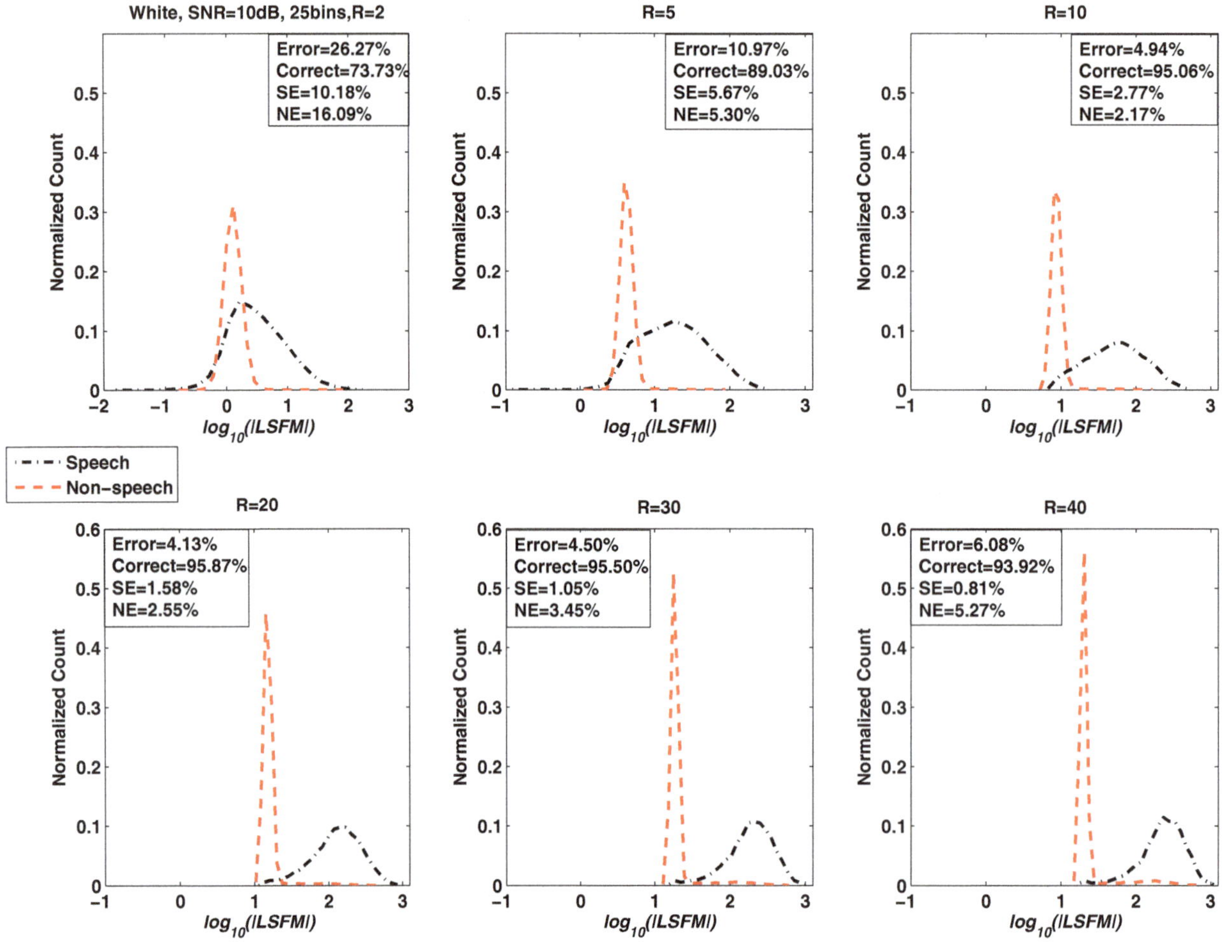

Figure 5 LSFM measure as a function of long-term window length (R) in additive white noise (SNR = 10 dB).

commonly employed for spectrum estimation, but it is well known that the periodogram is an inconsistent spectral estimator. According to [8], the Welch-Bartlett method [19] was found to give a good trade-off between variance reduction and spectral resolution reduction. Therefore, in our proposed algorithm, the signal spectrum is estimated using the Welch-Bartlett method.

A block diagram of the proposed LSFM-based VAD algorithm is shown in Figure 9. The algorithm can be described as follows. The input speech signal is decomposed into frames of 20 ms in length with an overlap of 10 ms by the Hann window. The spectrum of the segmented signal is estimated using the Welch-Bartlett method. At the mth frame, the LSFM feature $L_x(m)$ is computed using the previous R frames. The initial decision about whether there contains speech in the last R frames is made through the comparison with an adaptive threshold. The initial decision is denoted by V_{INL}. If there is a speech frame existing over the previous R frames ending at the mth frame, $V_{\text{INL}}(m) = 1$; otherwise, $V_{\text{INL}}(m) = 0$ and there are only non-speech frames over the previous R frames. We adopt the voting scheme proposed by Ghosh et al. [15] to make the final VAD decisions on a 10-ms interval. First, the initial decisions, $V_{\text{INL}}(m)$, $V_{\text{INL}}(m+1), \ldots, V_{\text{INL}}(m+R-1)$, are collected for those long windows which overlap with the target 10-ms interval. Then, the target 10-ms interval is marked to be speech if there is 80% or above of those initial decisions that contain speech; otherwise, it is marked as non-speech. The 80% was gotten empirically, which provided the maximum VAD accuracy for most noises tested over five SNR levels.

In general, speech is a low-pass signal, and the frequency range of 500 Hz to 4 kHz is crucial for speech intelligibility [20]. Hence, for a better discrimination, the start bin, k_s, and the end frequency bin, k_e, are calculated by:

$$k_s = N_{\text{DFT}}(\frac{500}{f_s}), \tag{6}$$

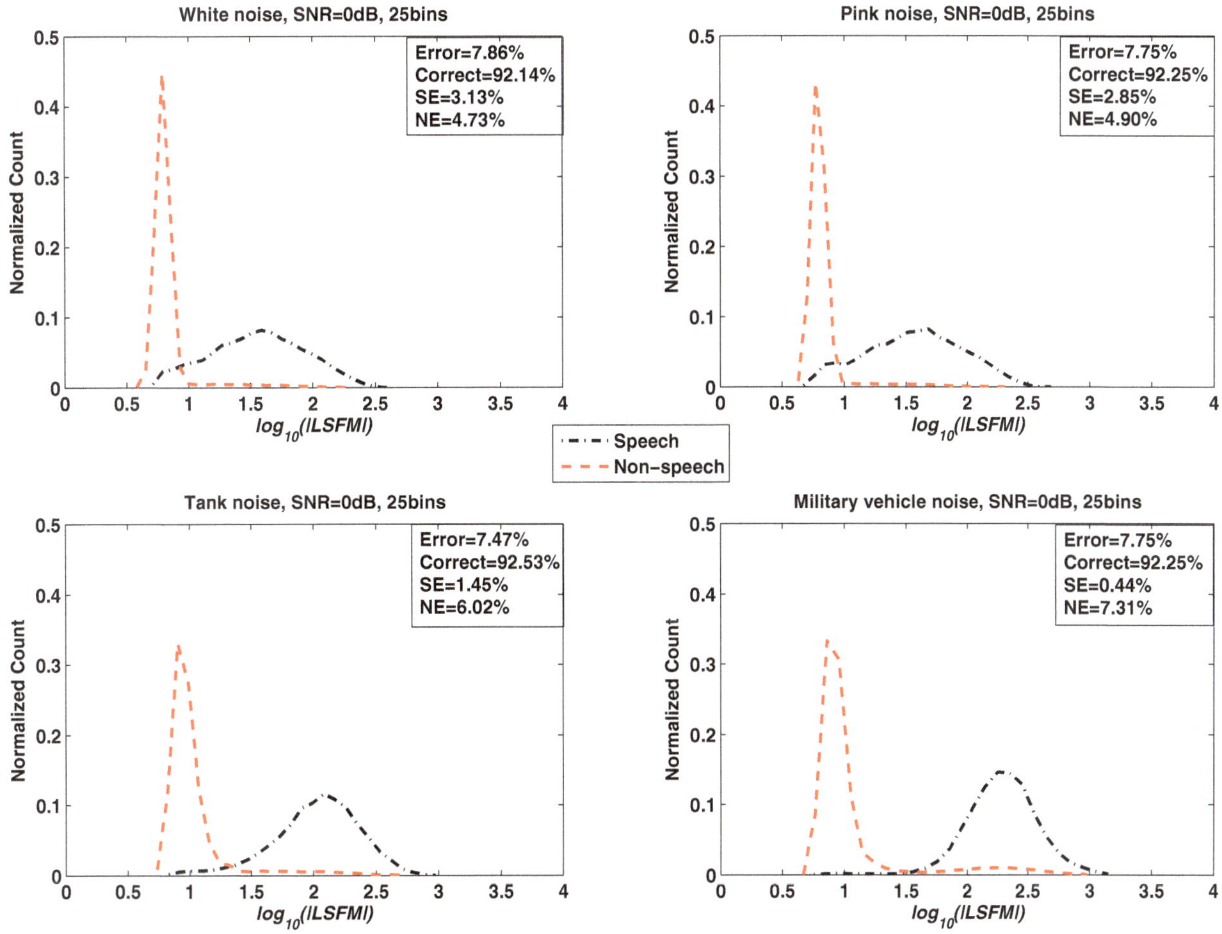

Figure 6 Histogram of the logarithmic LSFM measure for white, pink, tank, and military vehicle noises (SNR = 0 dB). Upper left: white noise, upper right: pink noise, lower left: tank noise, and lower right: military vehicle noise.

$$k_e = N_{\text{DFT}}\left(\frac{4,000}{f_s}\right), \tag{7}$$

in which f_s is the sampling frequency and N_{DFT} is the order of Discrete Fourier Transform (DFT) which is used to calculate the spectral estimate of the observed signal. In our experiment, $f_s = 16$ kHz and $N_{\text{DFT}} = 512$. The frequencies, ω_k , are uniformly distributed between 500 Hz and 4 kHz.

An illustrative example of the VAD output is shown in Figure 10. A high spectral flatness indicates that the spectrum has a similar amount of power in all spectral bands, which would sound similar to white noise, and the graph of the spectrum would appear relatively flat and smooth. A low spectral flatness indicates that the spectral power is concentrated in a relatively small number of bands; this means that the spectrum is more organized, and the graph of the spectrum would appear 'spiky'. Hence, the spectral flatness measure is a good feature for VAD.

3.1 Selection of M and R

M and R are parameters used for computing the LSFM feature L_x. We want to choose the appropriate M and R so that the separateness of the distribution for noise and speech is maximized since the more it is separated, the better the final VAD decision is. The total misclassification errors (sum of speech detection error and non-speech detection error) for all combinations of M (1, 5, 10, 20, 30, and 40) and R (5, 10, 20, 30, and 40) are computed over 12 types of noise for five SNR levels (−10, −5, 0, 5, and 10 dB). The test speech set is the same with the one we used for the demonstration of the discriminative power of the proposed LSFM feature in Section 2.2.

The total misclassification error as a function of different combinations of M and R is shown in Figures 11, 12, 13, and 14. The best combination of M and R for each noise at each SNR level is written on the upper or lower right of each subfigure. After the summed up frequency of each M and R that appeared in the subfigures, we conclude that (10, 30) is the optimal combination

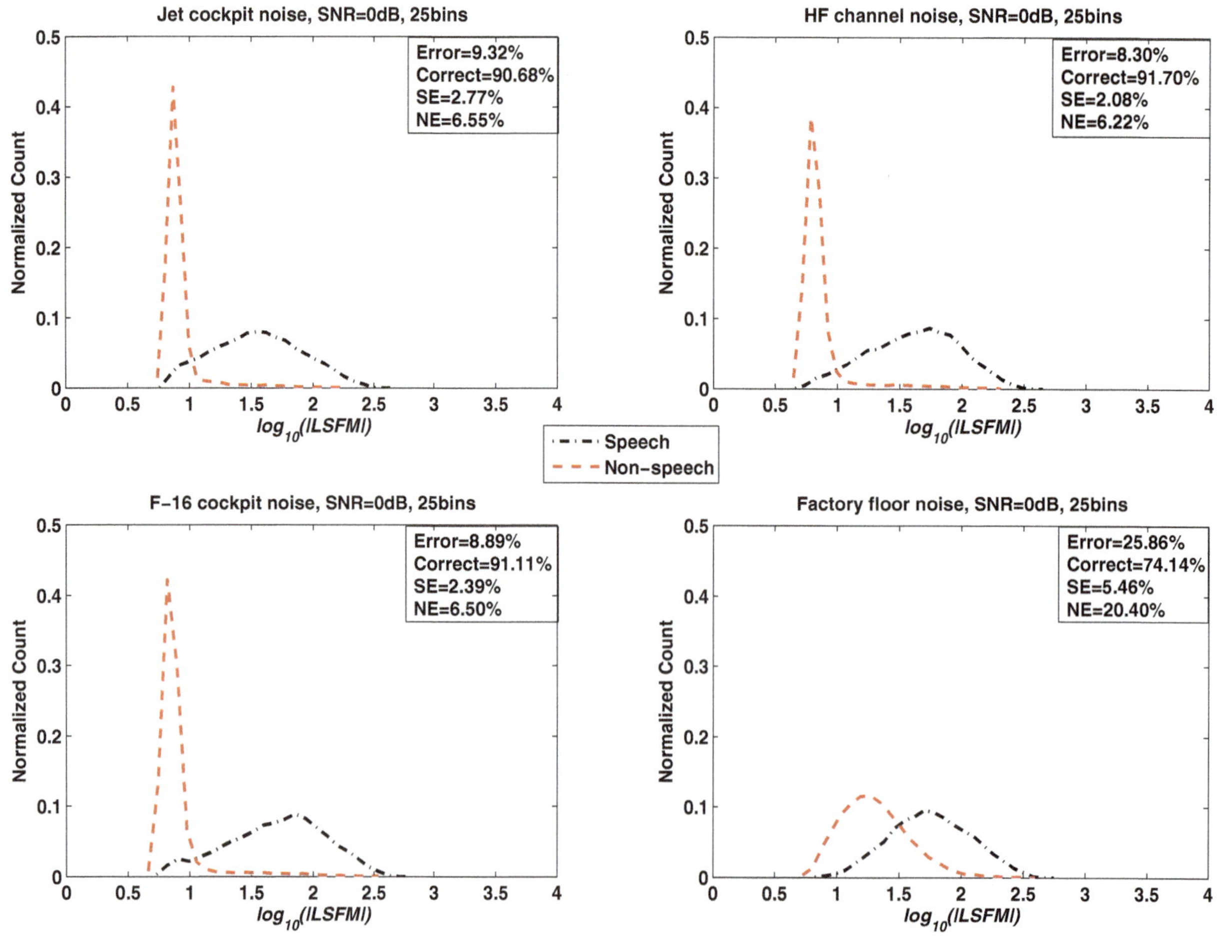

Figure 7 Histogram of the logarithmic LSFM measure for jet cockpit, HF channel, F-16 cockpit, and factory floor noises (SNR = 0 dB). Upper left: jet cockpit noise, upper right: HF channel noise, lower left: F-16 cockpit noise, and lower right: factory floor noise.

which appeared most frequently ($M = 10$ appeared 25 times out of the 60 subfigures in total; $R = 30$ also appeared 25 times out of 60 subfigures in total). This fixed combination (10, 30) is then adopted for all the following tests.

Furthermore, from Figures 11, 12, 13, and 14, we observe that for the same R value, if M is increased from 1 to 10, the total misclassification error is decreased for most cases tested. However, when M is larger than 10, even if M is increased, the total misclassification error stops decreasing any further. This observation verified the choice of 10 to be the optimal value of M.

It is also worth mentioning that for those noise types and SNR levels whose optimal M and R combination is not (10, 30), the fixed combination (10, 30) still works well. Table 1 shows the points that the total misclassification error of adopting the fixed combination (10, 30) is worse (higher error value) than utilizing the best combination of M and R for each noise type and SNR level shown in the subfigures. Except for cutting factory floor noise and impulsive machine gun noise, the differences are all less than five points.

For machine gun noise, the optimal choice of R is 5 for all SNR levels. Machine gun noise is an impulsive noise which consists of two types of sounds, namely gunshot and silence between gunshots [15]. When R is 30, the long analysis window would include both types of sounds. Therefore, the spectral power over these 30 frames will be less uniform; the LSFM feature value will then be small, and there will be more classification errors compared to the case when R is 5. Similarly, for factory floor noise, the optimal choice of M is 1 for all SNR levels. Factory floor noise [21] was recorded near plate-cutting and electrical welding equipment which shows a repetitive pattern. According to [19], the variance of estimated power spectrum will not be obviously reduced if the overlapped frames are highly correlated with each other. Therefore, averaging over M overlapped frames will cause

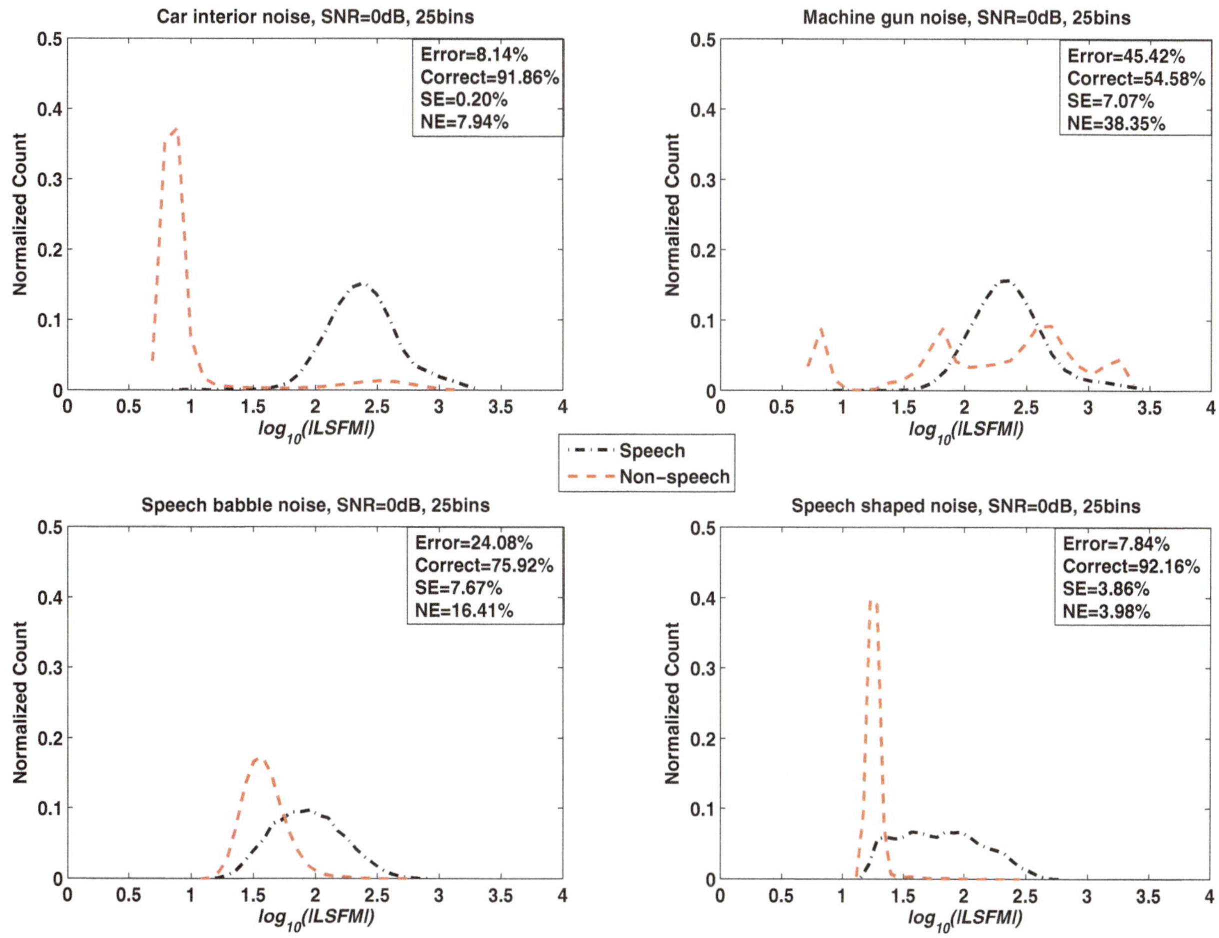

Figure 8 Histogram of the logarithmic LSFM measure for car interior, machine gun, speech babble, and speech-shaped noises (SNR = 0 dB). Upper left: car interior noise, upper right: machine gun noise, lower left: speech babble noise, and lower right: speech-shaped noise.

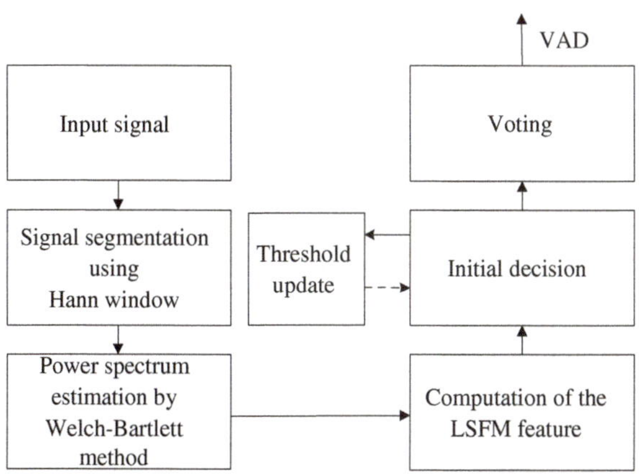

Figure 9 Block diagram of the proposed LSFM-based VAD algorithm.

more misclassification errors compared to the case when M is 1.

3.2 Adaptive threshold

Unlike M and R, a fixed threshold would lose its efficiency when facing varying acoustic environments. Therefore, it is more suitable to design an adaptive threshold [22]. From Equations 2 to 4, we can conclude that $(R + M - 1)$ frame (0.39 s for fixed $R = 30$ and $M = 10$) information is needed to acquire the first LSFM feature value. In our implementations, the initial 1.39 s of the test signal $x(n)$ is always assumed to be non-speech. From this 1.39 s of $x(n)$, 100 realizations of L_N can be collected and saved to ψ_{INL}.

The threshold is initialized to be

$$\text{THR}_{\text{INL}} = \min(\psi_{\text{INL}}). \tag{8}$$

To update the threshold at the mth frame, we used two buffers ψ_{S+N} and ψ_N. ψ_{S+N} stores the LSFM measures of the last 100 long window ending at the mth frame which was decided as containing speech; similarly, ψ_N stores

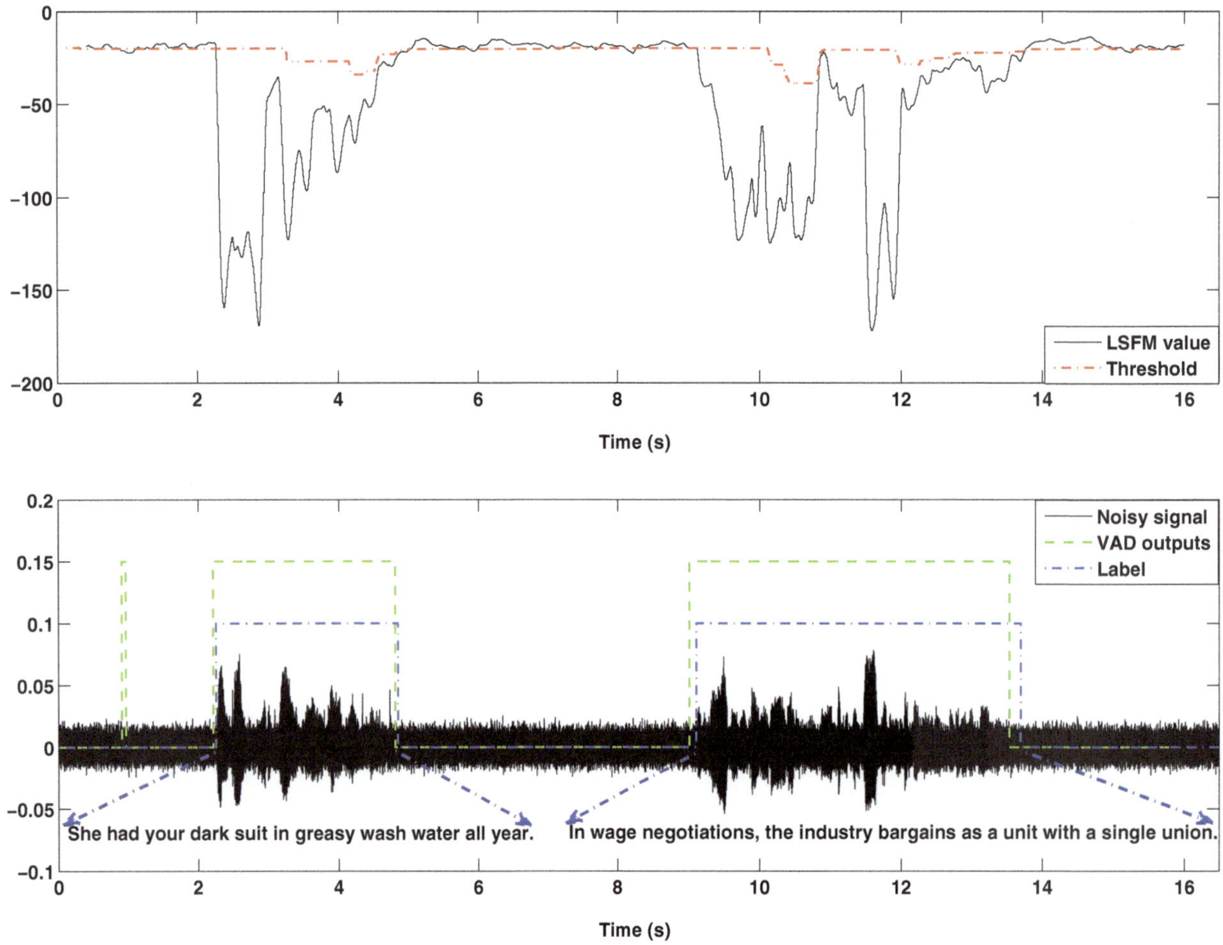

Figure 10 Illustrative example of the adaptive threshold and the VAD output, white noise, SNR = 0 dB. The upper figure shows the LSFM value and the corresponding adaptive threshold for each frame. The lower figure shows the VAD output decisions and the ground truth, namely 'Label'. The two sentences are as follows: (1) She had your dark suit in greasy wash water all year; (2) in wage negotiations, the industry bargains as a unit with a single union.

the LSFM measures of the last 100 long window ending at the mth frame which was decided as including non-speech information only. The adaptive threshold for the mth frame is then updated as:

$$\mathrm{THR}(m) = \lambda \times \min(\psi_{S+N}) + (1 - \lambda) \times \max(\psi_N), \quad (9)$$

where λ is the parameter of the convex combination. We experimentally found that $\lambda = 0.55$ results in the maximum accuracy rate in VAD decisions over the TIMIT training set.

4 Evaluation setup

The proposed VAD algorithm was trained and tested using a speech database that is phonetically balanced. The system was evaluated using the error rate and accuracy rate metrics.

4.1 Database description

For the evaluation of VAD algorithms, TIMIT corpus is preferred since it provides manual transcription down to word and phoneme levels. The reference labels are computed using the start and end times of the utterance obtained from the TIMIT transcription (.phn files). Some experiments are carried out on the core TIMIT test set consisting of 24 individual speakers (16 males, 8 females) of eight different dialects, each speaking 10 phonetically balanced English sentences. The utterances of TIMIT corpus are short (about 3.5 s), and around 90% of which are speech; this may introduce a bias when comparing the distributions of speech and non-speech. To reduce this effect and make it closer to real-world scenarios, 2-s silence was added before and after each utterance to simulate a typical telephone conversation [8,15,23] in which the ratio of speech to non-speech is almost 40% to 60%.

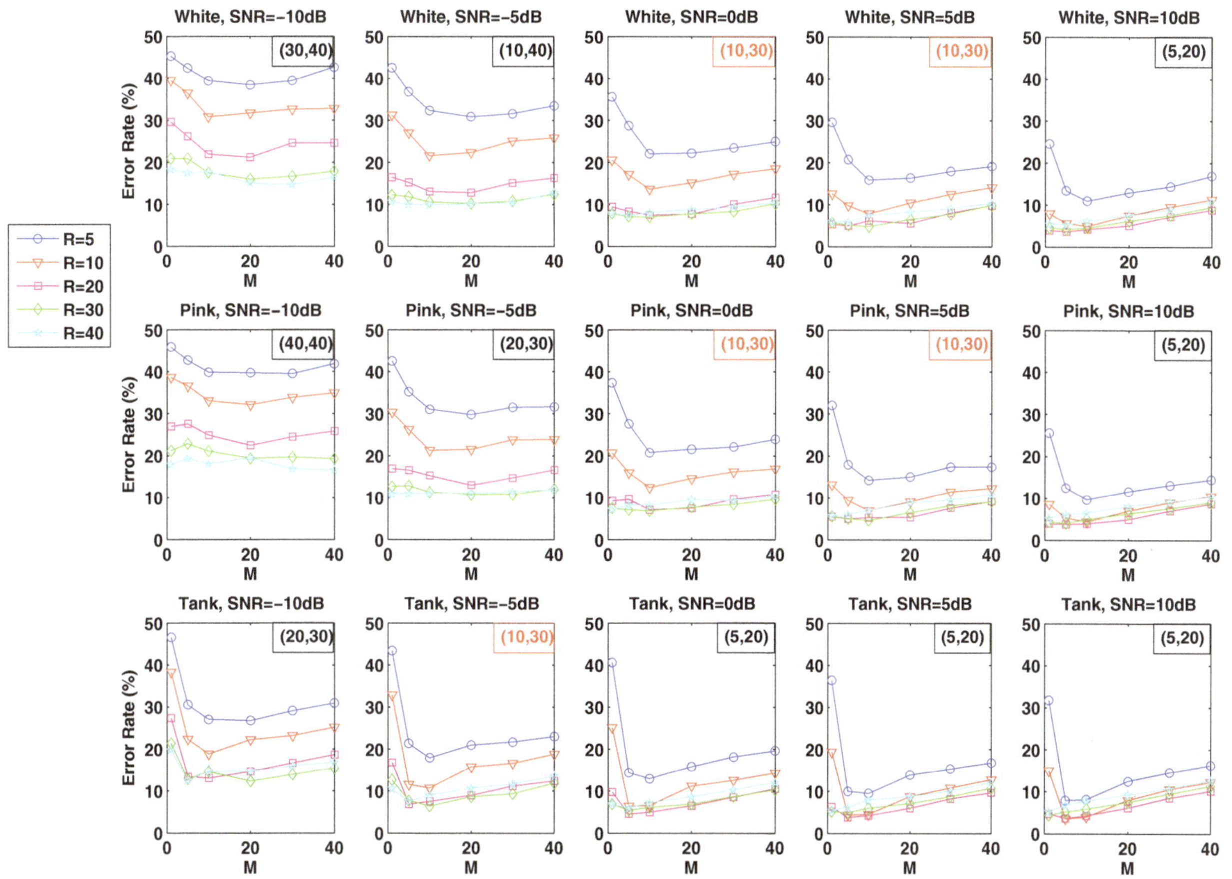

Figure 11 Total misclassification error as a function of M and R combination for white, pink, and tank noises. Upper row: white noise, middle row: pink noise, and lower row: tank noise. SNR= −10, −5, 0, 5, and 10 dB.

The noise of 11 categories taken from the NOISEX-92 database [21] and speech-shaped noise are added at five different SNR levels (−10, −5, 0, 5, and 10 dB) to the signal concatenated by all 240 sentences. The noise samples from the NOISEX-92 database are resampled to 16 kHZ according to the experiment requirement. Among the 12 kinds of noises, white noise and pink noise are stationary noises while others are all non-stationary noises, namely tank, military vehicle, jet cockpit, HF channel, F-16 cockpit, factory floor, car interior, machine gun, speech babble, and speech-shaped noises. The test set for each noise and SNR thus consisted of 28.10 min of noisy speech of which 62.51% was only noise.

4.2 Performance evaluation

Performance of a VAD algorithm can be evaluated both subjectively and objectively. In general, subjective evaluation is done through a listening test, and VAD decision errors are detected based on human perception [24]. On the other hand, objective evaluation relies on a mathematical criterion for judging. However, subjective listening tests like ABC [24] fail to consider the effect of the false alarm which is inappropriate for a thorough evaluation of a VAD algorithm [8]. Therefore, the objective evaluation scheme proposed by Freeman et al. [2] was adopted to evaluate the performance of the proposed VAD algorithm.

4.2.1 Error rate

The four traditional parameters that describe the error rate are as follows:

- *Front-end clipping (FEC).* Clipping introduced in passing from noise to speech activity.
- *Mid-speech clipping (MSC).* Clipping due to speech misclassified as noise in an utterance.
- *Noise detected as speech (NDS).* Noise detected as speech within a silence period.
- *Carry over (OVER).* Noise interpreted as speech due to the VAD flag remaining active in passing from speech activity to noise.

These four parameters are illustrated in Figure 15. Among them, FEC and MSC are indicators of true

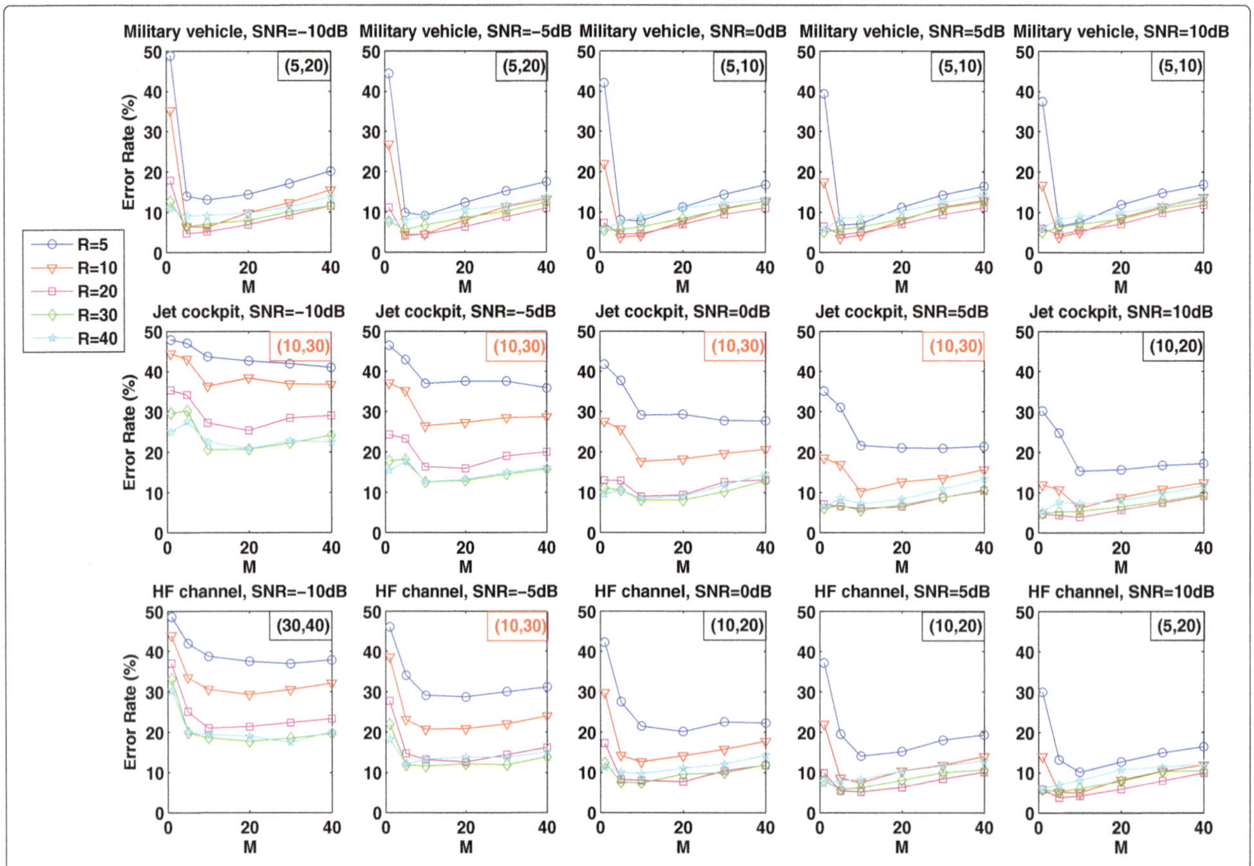

Figure 12 Total misclassification error as a function of M and R combination for military, jet cockpit, and HF channel noises. Upper row: military noise, middle row: jet cockpit noise, and lower row: HF channel noise. SNR= −10, −5, 0, 5, and 10 dB.

rejection, while NDS and OVER are indicators of false acceptance. Thus, in order to obtain the best overall system performance, all four parameters should be minimized.

4.2.2 *Accuracy rate*

Although the method described above provides useful objective information concerning the performance of a VAD algorithm, it only gives the error rate of the system. Parameters which describe the accuracy rate are needed for a thorough analysis of the detection results. Three parameters concerning the accuracy rate are described as follows:

- *CORRECT*. They are correct decisions made by VAD algorithm.
- *Speech hit rate (HR1).* Speech frames that are correctly detected among all speech frames.
- *Non-speech hit rate (HR0).* Non-speech frames that are correctly detected among all non-speech frames.

Among the three parameters, HR1 and HR0 define the fraction of all actual speech frames or non-speech frames that are correctly detected as speech frames or non-speech frames, respectively [12]. The speech hit rate and non-speech hit rate are calculated as follows:

$$\mathrm{HR1} = \frac{N_{1,1}}{N_1^{\mathrm{ref}}} \qquad \mathrm{HR0} = \frac{N_{0,0}}{N_0^{\mathrm{ref}}}, \tag{10}$$

where N_1^{ref} and N_0^{ref} are the numbers of real speech and non-speech frames in the whole database, respectively, while $N_{1,1}$ and $N_{0,0}$ are the numbers of speech and non-speech frames correctly classified. The overall accuracy rate (CORRECT) is then defined as:

$$\mathrm{CORRECT} = \frac{N_{1,1} + N_{0,0}}{N_1^{\mathrm{ref}} + N_0^{\mathrm{ref}}}. \tag{11}$$

All three parameters should be maximized to get the best performance.

5 Simulation results

In order to gain a comparative analysis of the proposed LSFM-based VAD performance, three modern standardized VAD schemes and one recent long-term algorithm, namely ETSI adaptive multi-rate (AMR) VAD options 1 and 2 (AMR1 and AMR2) [25], the G.729B VAD [26], and LTSV, were also evaluated. The implementations of these

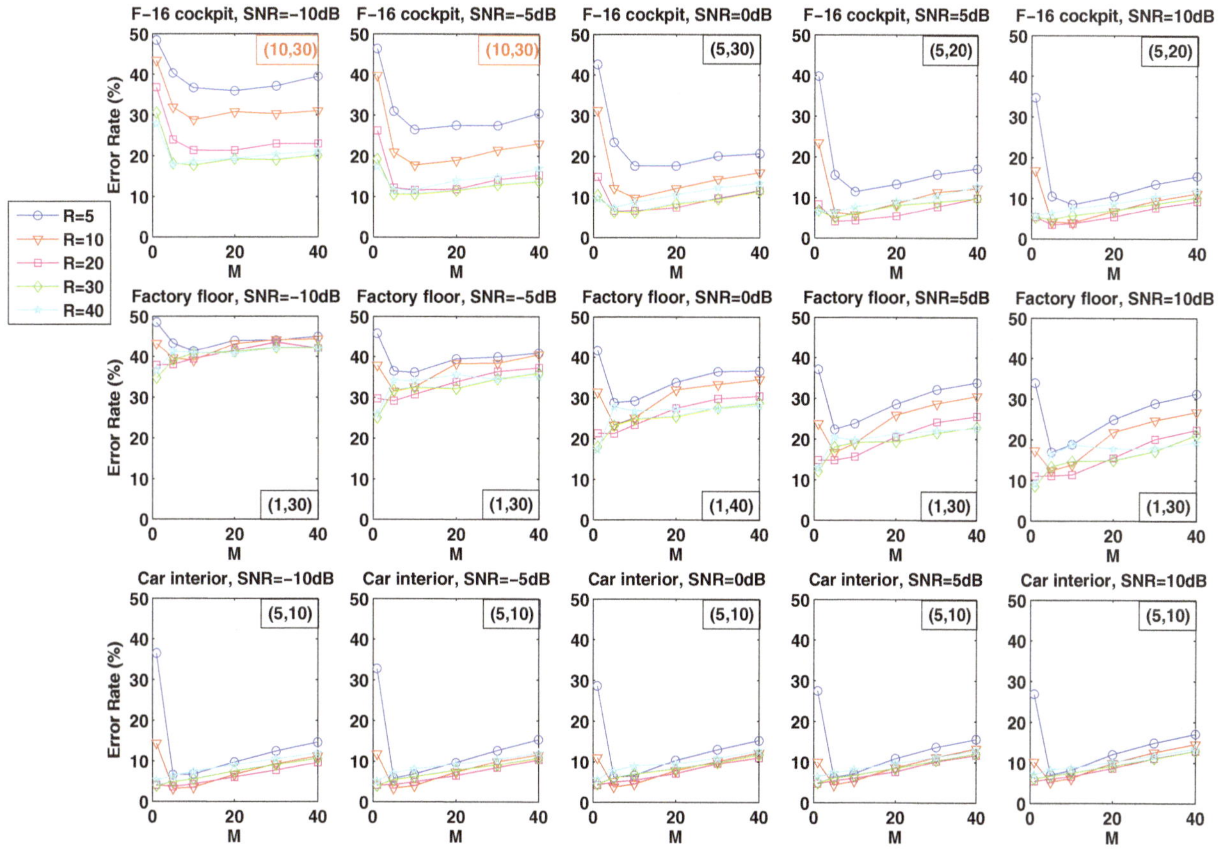

Figure 13 Total misclassification error as a function of M and R combination for F-16 cockpit, factory floor, and car interior noises. Upper row: F-16 cockpit noise, middle row: factory floor noise, and lower row: car interior noise. SNR= −10, −5, 0, 5, and 10 dB.

three schemes were taken from the authors' C implementations [27,28], respectively.

One important aspect of the comparison is the different frame lengths used. The proposed schemes, the G.729B VAD and LTSV-based VAD, produce a decision every 10 ms, while the AMR VADs need 20 ms. In order to be comparable, the frame-wise VAD decisions produced by the AMR VADs were compared to a set of reference labels generated every 20 ms from the TIMIT phonetic level transcription. Meanwhile, the proposed schemes, the G.729B VAD and LTSV-based VAD, were compared to a set of reference labels generated every 10 ms from the TIMIT phonetic level transcription. The TIMIT utterances were down-sampled to 8 kHz for the software implementations of the G.729B VAD and AMR VADs. The final VAD decisions were made, and the accuracy rate and error rate were computed for 12 noises and five SNRs.

5.1 Performance average over all twelve kinds of noises

In Figure 16, the proposed LSFM-based VAD is compared with three standards and LTSV-based VAD in terms of accuracy rate and error rate for SNR levels ranging from −10 to 10 dB. Note that the results in Figure 16 are averaged values for all 12 noises. The first row of the figure shows the accuracy rates which include CORRECT, HR1, and HR0. The behavior of the different VADs is analyzed. G.729B suffers poor CORRECT(62.74% at −10 dB) and HR1 (33.62% at −10 dB) with the increasing noise level, while it keeps a steady and relatively high HR0 for the whole range of SNRs (80.33% on average). AMR1 performs much better than G.729B for both CORRECT and HR1 while suffering degradation of HR0 when the SNR level is increased. AMR2 improves considerably over AMR1 in CORRECT mainly because of the high HR0 over all SNRs (88.96% on average) while yielding similar HR1 with AMR1. LTSV performs very well under low SNR conditions (80.73% CORRECT at −10 dB) but becomes saturated (around 91% since 5 dB) at higher SNRs. Our proposed LSFM-based VAD yields the best CORRECT for all SNRs and shows a steady improvement with the increased SNR.

Similarly, the second row of Figure 16 shows the error rates which include FEC, MSC, OVER, and NDS. G.729B performs the worst on average in terms of true rejection rate (FEC and MSC). However, the OVER of G.729B is the lowest among the five VADs tested. AMR1 and

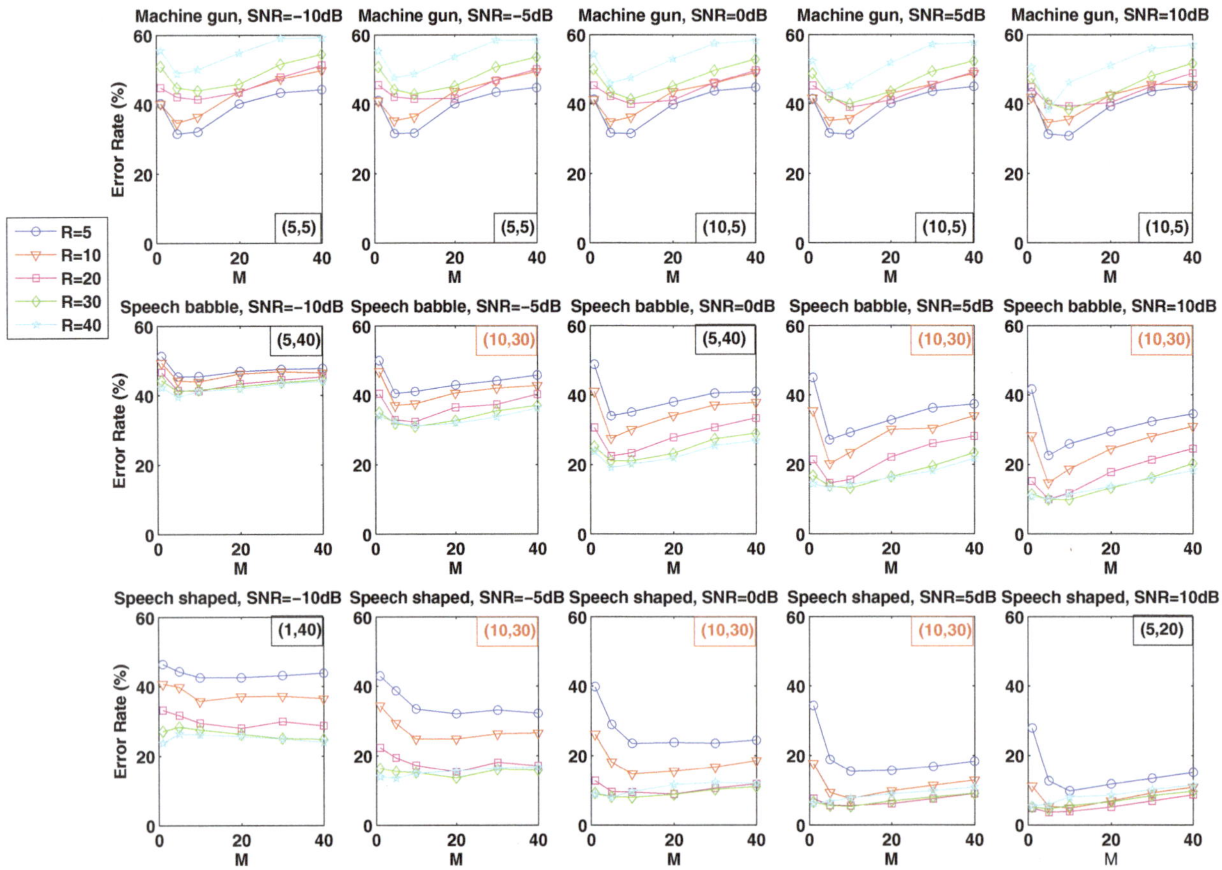

Figure 14 Total misclassification error as a function of M and R combination for machine gun, speech babble, and speech-shaped noises. Upper row: machine gun noise, middle row: speech babble noise, and lower row: speech-shaped noise. SNR= −10, −5, 0, 5, and 10 dB.

Table 1 Total misclassification error difference between adopting the fixed combination (10, 30) and utilizing the best (M, R) combination

Noise type	−10 dB	−5 dB	0 dB	5 dB	10 dB
White	2.85	0.69	0	0	0.87
Pink	4.49	0.64	0	0	1.19
Tank	2.32	0	1.59	2.20	2.17
Military vehicle	2.36	2.59	2.62	2.82	3.31
Jet cockpit	0	0	0	0	1.36
HF channel	1.15	0	0.57	1.02	2.27
F-16 cockpit	0	0	0.10	1.94	2.30
Factory floor	6.04	7.35	7.59	7.04	6.25
Car interior	2.33	2.82	3.16	2.33	2.21
Machine gun	12.45	11.33	9.83	8.74	7.56
Speech babble	2.11	0	2.068	0	0
Speech shaped	3.80	0	0	0	1.84

The numbers are the points that worse (higher error value) than utilizing the best (M, R) combination for each noise type and SNR level shown in the subfigures.

AMR2 yield similar true rejection rates for all tested SNRs, while AMR2 gives smaller false acceptance rate (NDS and OVER) especially for NDS (around four points less than AMR1 for all SNRs). LTSV leads to the lowest true rejection rate, while LSFM achieved the best performance in terms of NDS. The proposed LSFM-based VAD acquires a comparatively higher FEC in low SNRs (smaller than −5 dB) because of the averaging property of this algorithm shown in Equations 2, 3, and 4.

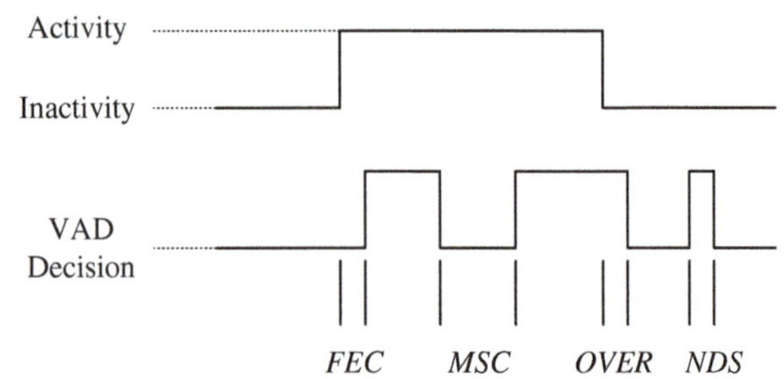

Figure 15 Objective parameters for performance evaluation.

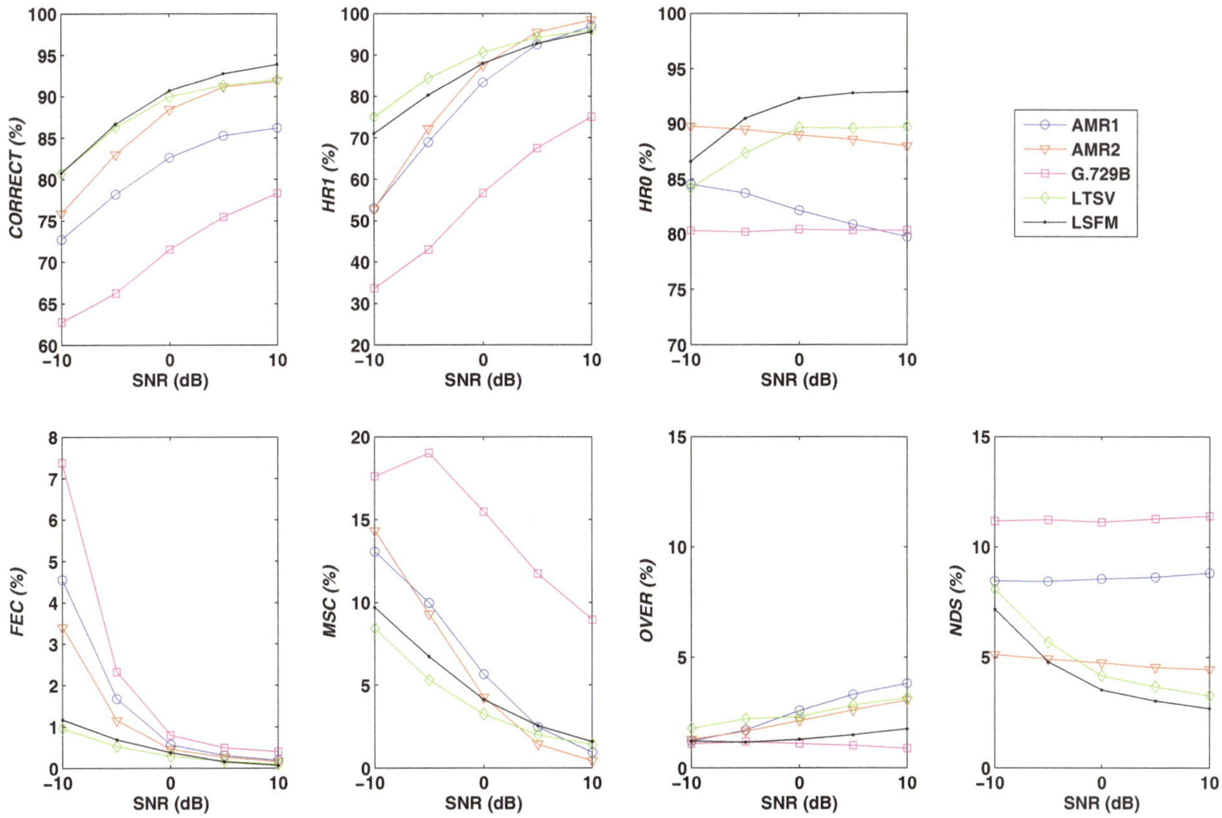

Figure 16 Accuracy and error rate comparisons for five VAD schemes averaged over 12 noises for five SNRs. Accuracy rate: CORRECT, HR1, and HR0; error rate: FEC, MSC, OVER, and NDS. Five VAD schemes: AMR1, AMR2, G.729B, LTSV, and LSFM. Five SNRs (−10, −5, 0, 5, and 10 dB).

Table 2 summarizes the results provided by LSFM-based VAD over the different VAD methods being evaluated by comparing them in terms of the average accuracy rate and error rate for all 12 noises over five SNR levels ranging from −10 to 10 dB. LSFM achieves the best CORRECT (88.95%) and HR0 (91.00%), while LTSV yields the best HR1 (88.04%).

5.2 Performance average over five SNRs

Figure 17 shows the three different accuracy rate evaluation metrics averaged over five SNRs for 12 kinds of noises computed for AMR1, AMR2, G.729B, LTSV, and LSFM-based VAD algorithms. From Figure 17, it is clear that in terms of CORRECT, LTSV is the best among all four reference VAD algorithms considered here. Hence, the proposed LSFM-based VAD is compared with the LTSV-based VAD. We observe that on average, the LSFM-based VAD is better than the LTSV-based VAD in terms of CORRECT for tank (0.52%), military vehicle (1.40%), F-16 cockpit (0.34%), car interior (2.12%), machine gun (1.88%), and speech babble (7.63%) noises, and it is worse for white (1.10%), pink (0.79%), jet cockpit (0.64%), HF channel (0.15%), factory floor (0.15%), and speech-shaped (0.58%) noises. The number in the bracket indicates the absolute CORRECT by which the proposed LSFM-based VAD is better or worse than the LTSV-based VAD. The mean CORRECT over all 12 noise types of our proposed LSFM-based VAD is 0.87% higher than that of the LTSV-based VAD. Furthermore, the proposed LSFM-based VAD outperforms LTSV-based VAD in terms of HR0 over most noises (11 out of 12) that were considered.

Table 2 Average performance comparison for all 12 noises over five SNR levels ranging from −10 to 10 dB

VAD	AMR1	AMR2	G.729B	LTSV	LSFM
CORRECT	81.00	86.07	70.87	88.08	88.95
HR1	78.96	81.25	55.16	88.04	85.53
HR0	82.22	88.96	80.33	88.10	91.00
FEC	1.46	1.09	2.28	0.41	0.49
MSC	6.42	5.94	14.56	4.07	4.93
OVER	2.53	2.15	1.06	2.46	1.39
NDS	8.59	4.75	11.24	4.98	4.24

Figure 18 shows the four different error rate evaluation metrics (FEC, MSC, OVER, and NDS), averaged over five SNRs for 12 kinds of noises, computed for AMR1, AMR2, G.729B, LTSV, and LSFM algorithms. From Figure 18, it

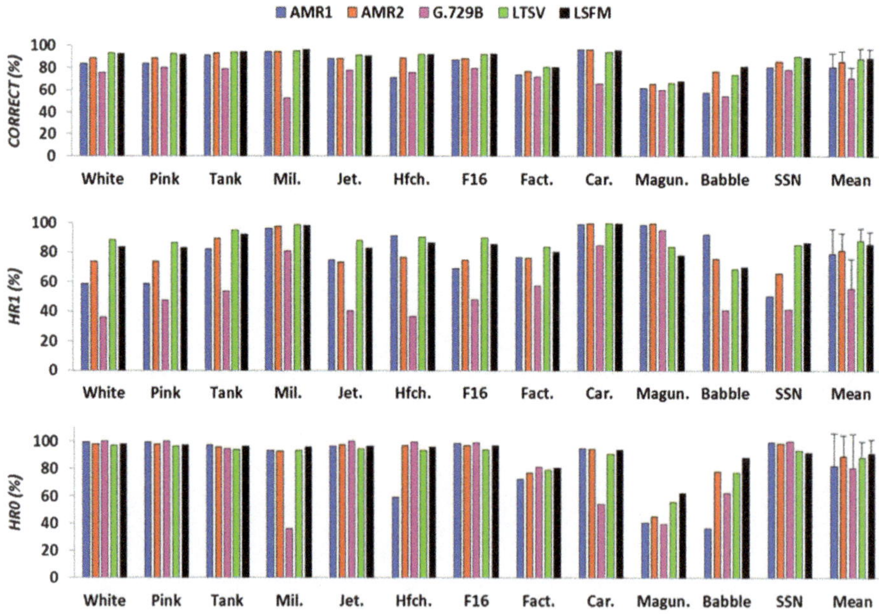

Figure 17 Accuracy rate comparisons for five VAD schemes averaged over five SNRs for 12 kinds of noises. Accuracy rate: CORRECT, HR1, and HR0. Five VAD schemes: AMR1, AMR2, G.729B, LTSV, and LSFM.

is clear that the performance of LSFM-based VAD outperforms the LTSV-based VAD in terms of OVER (all 12 noises) and NDS (9 out of 12). For example, The proposed LSFM-based VAD has a smaller NDS score for tank (0.26%), military vehicle (0.10%), jet cockpit (0.35%), HF channel(0.44%), F16 cockpit (0.80%), factory floor (0.42%), car interior (0.31%), machine gun (2.15%), and babble (5.83%) noises. The number in the bracket indicates the absolute NDS by which the proposed LSFM-based VAD is smaller than the LTSV-based VAD. Moreover, values of

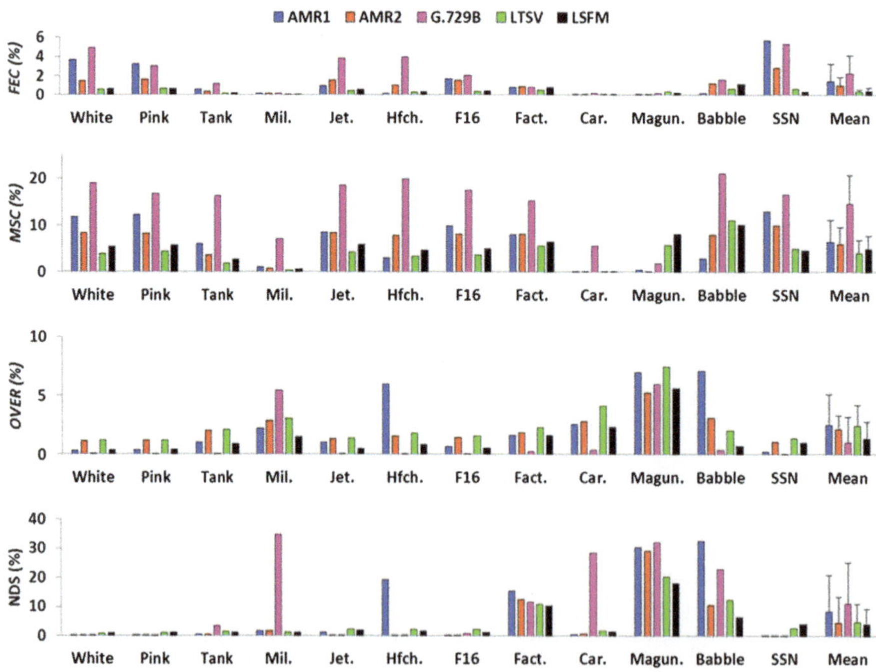

Figure 18 Error rate comparison of five VAD schemes averaged over five SNRs for 12 kinds of noises. Error rate: FEC, MSC, OVER, and NDS. Five VAD schemes: AMR1, AMR2, G.729B, LTSV, and LSFM.

standard deviation of our proposed LSFM-based VAD in terms of MSC, OVER, and NDS are all smaller than that of the LTSV-based VAD.

Thus, in consideration of both accuracy rate and error rate, the proposed VAD algorithm achieved the best compromise when compared with the four representative VADs analyzed.

6 Conclusions

The main contribution of this article was the introduction of an efficient long-term spectral flatness measure-based VAD algorithm. The motivation of exploring flatness measure along time frames using a long window was clarified by the LSFM feature distributions as a function of the long-term window length R. The discriminative power of the LSFM feature was verified in terms of the separateness of its distribution for noisy speech and non-speech signals. The decision threshold was adapted according to the previous 100 LSFM measures of speech and non-speech. Experiments were done on core TIMIT test set for 12 kinds of noises (11 from NOISEX-92 database and speech-shaped noise) across five different SNRs ranging from -10 to 10 dB. No *a priori* knowledge of noise characteristics was needed for training purposes. The performance of our proposed method was compared with the three standards (namely G.729B, AMR1, and AMR2) and with an emerging LTSV-based VAD algorithm. The results were analyzed by accuracy rate and error rate. Through extensive experiments, we showed that our proposed LSFM-based VAD achieved the best CORRECT, HR0, and NDS and among all tested schemes that averaged all 12 kinds of noises. Furthermore, we investigated the individual performance on each noise type. Our proposed LSFM-based VAD outperformed LTSV-based VAD for 6 out of 12 noise types tested especially for non-stationary impulsive machine gun noise and speechlike babble noise.

The test database used in the implementations was created to simulate typical conversational speech by inserting 2-s silence before and after each utterance from core TIMIT test corpus so that the ratio of speech to non-speech was almost 40% to 60%. While this simulates a conversational speech statistically, this is not very realistic in terms of randomness of pauses, hesitations, etc. Furthermore, depending on the choice of the long-term window length (R and M combination), the LSFM-based VAD application is expected to suffer a delay equal to the duration of the window ($R + M - 1$ frames). Therefore, a trade-off between the delay and robustness of VAD should be carefully considered before utilizing the proposed LSFM-based VAD algorithm.

Moreover, it is worth mentioning that there is a trade-off between HR1 and HR0. The increase of one may lead to a decrease of the other. Therefore, it should be noted that according to different applications, different (R, M) combinations and thresholds for voting scheme can be chosen to meet the variant requirement for HR1 and HR0. For example, HR1 is a crucial factor for speech coding, while high HR0 rate is necessary for most speech recognition-oriented systems.

Competing interests

Both authors declare that they have no competing interests.

References

1. K Itoh, M Mizushima, in *IEEE International Conference on Acoustics Speech and Signal Processing (ICASSP)*, vol. 1. Environmental noise reduction based on speech/non-speech identification for hearing aids (IEEE Piscataway, 1997), pp. 419–422
2. D Freeman, G Cosier, C Southcott, I Boyd, in *IEEE International Conference on Acoustics Speech and Signal Processing (ICASSP)*, vol. 1. The voice activity detector for the Pan-European digital cellular mobile telephone service (IEEE Piscataway, 1989), pp. 369–372
3. F Faubel, M Georges, K Kumatani, A Bruhn, D Klakow, in *2011 Joint Workshop on Hands-free Speech Communication and Microphone Arrays (HSCMA), Edinburgh,30 May–1 June 2011*. Improving hands-free speech recognition in a car through audio-visual voice activity detection (IEEE Piscataway, 2011), pp. 70–75
4. W Syed, HC Wu, in *Global Telecommunications Conference (GLOBECOM '07), Washington DC, 26–30 November 2007*. Speech waveform compression using robust adaptive voice activity detection for nonstationary noise in multimedia communications (IEEE Piscataway, 2007), pp. 3096–3101
5. A Kondoz, B Evans, in *3rd European Conference on Satellite Communications (ECSC-3), Manchester, 2–4 Nov 1993*. A high quality voice coder with integrated echo canceller and voice activity detector for VSAT systems (IEEE Piscataway, 1993), pp. 196–200
6. A Benyassine, E Shlomot, HY Su, E Yuen, in *IEEE Workshop on Speech Coding For Telecommunications Proceeding*. A robust low complexity voice activity detection algorithm for speech communication systems (IEEE Piscataway, 1997), pp. 97–98
7. LR Rabiner, MR Sambur, An algorithm for determining the endpoints of isolated utterances. Bell Syst. Techn. J. **54**(2), 297–315 (1975)
8. A Davis, S Nordholm, R Togneri, Statistical voice activity detection using low-variance spectrum estimation and an adaptive threshold. Audio, Speech, Lang. Proc. IEEE Trans. **14**(2), 412–424 (2006)
9. Z Shuyin, G Ying, W Buhong, in *First International Workshop on Education Technology and Computer Science (ETCS '09)*, vol. 3. Auto-correlation property of speech and its application in voice activity detection (IEEE Piscataway, 2009), pp. 265–268
10. M Marzinzik, B Kollmeier, Speech pause detection for noise spectrum estimation by tracking power envelope dynamics. Speech and Audio Proc. IEEE Trans. **10**(2), 109–118 (2002)
11. E Nemer, R Goubran, S Mahmoud, Robust voice activity detection using higher-order statistics in the LPC residual domain. Speech and Audio Proc. IEEE Trans. **9**(3), 217–231 (2001)
12. J Ramirez, J Segura, C Benitez, A Torre, A Rubio, Efficient voice activity detection algorithms using long-term speech information. Speech Commun. **42**(3–4), 271–287 (2004)
13. B Lee, M Hasegawa-Johnson, in *Proceedings of Biennial on DSP for In-Vehicle and Mobile Systems*, Minimum mean squared error a posteriori estimation of high variance vehicular noise. Istanbul, 17–19 June 2007
14. PC Khoa, Noise robust voice activity detection. Master's thesis, NangYang Technological University, 2012
15. P Ghosh, A Tsiartas, S Narayanan, Robust voice activity detection using long-term signal variability. Audio, Speech, and Lang. Proc. IEEE Trans. **19**(3), 600–613 (2011)
16. N Madhu, Note on measures for spectral flatness. Electron. Lett. **45**(23), 1195–1196 (2009)
17. JS Garofolo, LF Lamel, WM Fisher, JG Fiscus, DS Pallett, NL Dahlgren, V Zue, *TIMIT Acoustic-Phonetic Continuous Speech Corpus*. (Linguistic Data Consortium, Philadelphia, 1993)

18. P Renevey, A Drygajlo, in *IEEE Proceedings of 7th European Conference on Speech Communication and Technology (EUROSPEECH'2001)*. Entropy based voice activity detection in very noisy conditions (IEEE Piscataway, 2001), pp. 1887–1890
19. D Manolakis, V Ingle, S Kogon, *Statistical and Adaptive Signal Processing: Spectral Estimation, Signal Modeling, Adaptive Filtering, and Array Processing*. (Artech House, Norwood, 2005)
20. D Bies, *Engineering Noise Control: Theory and Practice*. (Taylor & Francis, New York, 2003)
21. A Varga, HJ Steeneken, Assessment for automatic speech recognition: II. NOISEX-92: a database and an experiment to study the effect of additive noise on speech recognition systems. Speech Commun. **12**(3), 247–251 (1993)
22. RV Prasad, A Sangwan, HS Jamadagni, MC Chiranth, R Sah, V Gaurav, in *Proceedings of the Seventh International Symposium on Computers and Communications (ISCC'02), Washington, DC*. Comparison of voice activity detection algorithms for VoIP (IEEE Piscataway, 2002), pp. 530–535
23. F Beritelli, S Casale, G Ruggeri, in *Proceedings of Acoustics, Speech, and Signal Processing (ICASSP '01)*, vol. 3. Performance evaluation and comparison of ITU-T/ETSI voice activity detectors (IEEE Piscataway, 2001), pp. 1425–1428
24. F Beritelli, S Casale, G Ruggeri, in *5th International Conference on Signal Processing Proceedings (WCCC-ICSP 2000)*, vol. 2. A psychoacoustic auditory model to evaluate the performance of a voice activity detector (IEEE Piscataway, 2000), pp. 807–810
25. ETSI, *Digital Cellular Telecommunications System (Phase 2+); Voice Activity Detector (VAD) for Adaptive Multi-Rate (AMR) Speech Traffic Channels; General Description*. (ETSI, Valbonne, 1999)
26. ITU, *Coding of Speech at 8 kbit/s Using Conjugate Structure Algebraic Code - Excited Linear Prediction Annex B: A Silence Compression Scheme for G729 Optimized for Terminals Conforming to Recommend V70*. (International Telecommunication Union, Geneva, 1996)
27. ETSI, *Digital Cellular Telecommunications System (Phase 2+); Adaptive Multi-Rate (AMR) Speech; ANSI-C Code for AMR Speech Codec*. (ETSI, Valbonne, 1998)
28. ITU, *Coding of Speech at 8 kbit/s Using Conjugate Structure Algebraic Code - Excited Linear Prediction Annex I: Reference Fixed-Point Implementation for Integrating G729 CS-ACELP Speech Coding Main Body with Annexes B, D and E*. (International Telecommunication Union, Geneva, 2000)

5

Identifying underlying articulatory targets of Thai vowels from acoustic data based on an analysis-by-synthesis approach

Santitham Prom-on[1,3*], Peter Birkholz[2] and Yi Xu[3]

Abstract

This paper investigates the estimation of underlying articulatory targets of Thai vowels as invariant representation of vocal tract shapes by means of analysis-by-synthesis based on acoustic data. The basic idea is to simulate the process of learning speech production as a distal learning task, with acoustic signals of natural utterances in the form of Mel-frequency cepstral coefficients (MFCCs) as input, VocalTractLab - a 3D articulatory synthesizer controlled by target approximation models as the learner, and stochastic gradient descent as the target training method. To test the effectiveness of this approach, a speech corpus was designed to contain contextual variations of Thai vowels by juxtaposing nine Thai long vowels in two-syllable sequences. A speech corpus consisting of 81 disyllabic utterances was recorded from a native Thai speaker. Nine vocal tract shapes, each corresponding to a vowel, were estimated by optimizing the vocal tract shape parameters of each vowel to minimize the sum of square error of MFCCs between original and synthesized speech. The stochastic gradient descent algorithm was used to iteratively optimize the shape parameters. The optimized vocal tract shapes were then used to synthesize Thai vowels both in monosyllables and in disyllabic sequences. The results, both numerically and perceptually, indicate that this model-based analysis strategy allows us to effectively and economically estimate the vocal tract shapes to synthesize accurate Thai vowels as well as smooth formant transitions between adjacent vowels.

Keywords: Articulatory target; Articulatory synthesis; Target approximation; Acoustic-to-articulatory inversion; Thai vowels

1. Introduction

Speaking requires an accurate control of highly variable successive articulatory movements, each involving simultaneous actions of multiple articulators [1,2], and all of them coordinated in such a way that many layers of meanings are simultaneously encoded [3]. Even more intriguingly, a human child seems to be able to acquire such highly intricate motor skills without specific articulatory instructions and without direct observation of the articulators of the mature speakers other than the visible ones such as the lips. The only definitive input the child receives that is articulatorily (as opposed to meaning-wise) informative is the acoustics of the speech utterances. Understanding how proper articulatory skills can be learned from acoustic data, a task known as acoustic-to-articulatory inversion, is therefore the key to our understanding of the nature of human speech acquisition and production. Such knowledge is also beneficial to both speech recognition [4] and speech synthesis [5].

Different approaches have been proposed to achieve acoustic-to-articulatory inversion [6-16]. These methods rely on either explicit mapping between acoustic and articulatory data [6-14] or optimization of articulatory synthesis model parameters [15,16]. Different methods of mapping between articulatory and acoustic data have been tested using probabilistic models such as hidden Markov models (HMM) [6,7], neural networks [8,9], codebooks [10-13], or filters [14]. Except the methods that are based on the task dynamic (TD) model, most of them, however, share the common drawback of the

* Correspondence: santitham@cpe.kmutt.ac.th
[1]Department of Computer Engineering, Faculty of Engineering, King Mongkut's University of Technology Thonburi, Bangkok 10140, Thailand
[3]Department of Speech, Hearing and Phonetic Sciences, University College London, London WC1N 1PF, UK
Full list of author information is available at the end of the article

mapping paradigm, i.e., the lack of the inclusion of speech production mechanism in the modeling process, in particular, the dynamic movement of speech gestures [1,2] that results in smooth spectral transitions observed in the natural acoustic data. An alternative approach is to use an analysis-by-synthesis strategy [15,16] in which parameters of a synthesis model are iteratively adjusted to minimize a cost function. The cost function can be the error from acoustic comparison between the original speech and speech synthesized with optimized parameters. This strategy, when implemented with an articulatory synthesizer with sufficient capacity to generate acoustic data from model parameters, may have the potential to achieve the closest simulation of speech learning behavior.

This paper reports the results of a study based on this alternative approach. The study attempts to identify underlying articulatory targets of Thai vowels by means of model-based optimization. Using the analysis-by-synthesis strategy, the underlying articulatory targets representing the vocal tract shapes are estimated from coarticulated disyllabic vowels. This modeling process iteratively adjusts the articulatory parameters to the optimal condition by minimizing the acoustical error between original and synthesized sounds generated with the tentative vocal tract parameters. The accuracy of these estimated vocal tract shapes are then evaluated by comparing the formants of the synthetic vowels to the formant trajectories of the natural utterances and to those of previous studies and by a listening experiment that compared the perceptual accuracy and naturalness of synthetic to those of natural speech.

2. Methods

2.1 Corpus

The corpus was designed to have full contextual vowel variations in Thai to facilitate the modeling process. Thai has nine vowels, in short and long minimal pairs, which are evenly spread across the vowel space [17]. To estimate the articulatory targets of all the Thai vowels, each utterance was designed to have two syllables consisting of only vowels, in the form of /V1 V2/, where both V1 and V2 are one of the nine long vowels (/aː/, /iː/, /uː/, /eː/, /ɛː/, /ɯː/, /ɤː/, /oː/, /ɔː/). Thus, there are 81 combinations in total. These disyllabic vowel sequences do not have any meanings. This design allows us to fully study the spectral changes resulting from transitions between vowels and to simulate their dynamics through computational modeling.

Speech data were recorded by a native Thai male speaker who had been living in the Greater Bangkok region in the past 20 years and had no self-reported speech or hearing disabilities. Recordings were done in a sound-treated room at the King Mongkut's University of Technology Thonburi, Bangkok, Thailand. The speaker was instructed to produce the disyllabic vowel sequences in a continuous manner with the mid tones for both syllables and without pauses between vowels as if they were in a simple noun-verb sentence. This makes the stress placed on the second syllable according to the general rule of Thai pronunciation. No particular normalization was done to remove the effect of stress. Nevertheless, the full factorial design of the corpus balances the occurrence of tones in both positions. The utterances were recorded at a sampling rate of 22.05 kHz and 16-b resolution.

The corpus was annotated using Praat [18]. Syllable boundaries were manually marked according to the concept of target approximation to be detailed later in Section 2.4. Briefly, the articulation of a segment is defined as a unidirectional movement toward its underlying target [19]. As a result, the moment a movement starts to turn away from the segmental target is viewed as the offset of one segment and onset of the next. Therefore, the boundary between two syllables was marked at the point where the spectrogram starts to change. This strategy, which was also used in our previous studies [20-23], differs from the conventional marking of syllable boundaries (*cf* [24] for evidence from production). Since the syllables contained only vowels, no consonantal boundaries were marked.

2.2 Overview of analysis-by-synthesis strategy for target estimation

Figure 1 shows a workflow diagram for the method used in this study. The basic idea, as mentioned in the 'Introduction,' is to estimate the underlying articulatory targets based on a distal learning strategy [25], in such a way that the learner (optimization system) is able to utilize the speech production system (articulatory synthesizer) to generate acoustic data that can be compared to the original speech. Vocal tract shapes as articulatory vowel targets are estimated for each utterance, starting from neutral positions and then iteratively adjusted until the overall acoustic error converges or the maximum number of steps is reached.

2.3 VocalTractLab, the articulatory synthesizer

The core of the analysis-by-synthesis strategy in this paper is VocalTractLab [26] - an articulatory synthesizer that can generate acoustic output based on articulatory parameters. VocalTractLab is capable of generating a full range of speech sounds by controlling vocal tract shapes, aerodynamics, and voice quality [27-29]. It consists of a detailed 3D model of the vocal tract that can be configured to fit the anatomy of any specific speaker, an advanced self-oscillating model of the vocal folds, and an efficient method for aeroacoustic simulation of the speech signal.

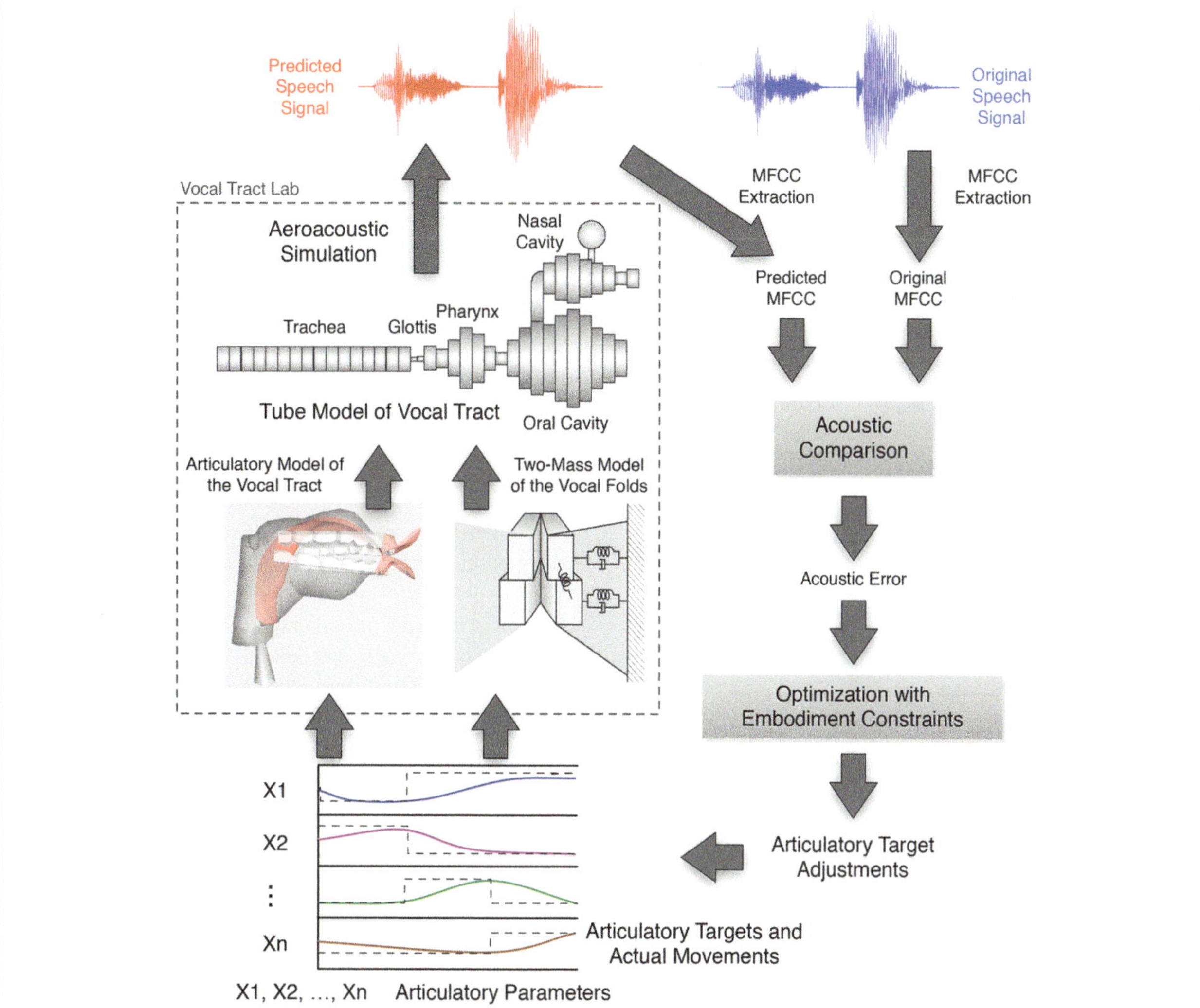

Figure 1 A schematic diagram of the articulatory target estimation.

The acoustic simulation of VocalTractLab approximates the trachea, the glottis, and the vocal tract as a series of cylindrical tube sections with variable lengths as shown in Figure 1. The aeroacoustic simulation is based on a transmission-line circuit representation of the tube system (see [30] for a detailed derivation). The simulation considers fluid dynamic losses at constrictions, as well as losses due to radiation, soft walls, and viscous friction.

2.4 Target approximation model

The control of the dynamics of VocalTractLab is based on the concept of sequential target approximation (TA), which has previously been implemented in various forms, as reviewed in [28]. The TA model implemented in VocalTractLab is illustrated in Figure 2. Compared to other articulatory models, TA is most similar to the task dynamic model [2], but with critical differences in several respects. Like the task dynamic model, TA simulates continuous articulatory trajectories as asymptotic movements toward underlying targets. Unlike the task dynamic model, however, TA does not assume that the target is always reached at the end of each target approximation interval. The unfinished target approximation movement thus often results in a highly dynamic articulatory trajectory that needs to be transferred to the next target approximation interval as its initial state. Such a transfer, as can be seen around the vertical dashed line in Figure 2, guarantees smooth and continuous articulatory movements across the TA boundaries.

Mathematically, VocalTractLab models articulatory trajectories as responses of target-driven high-order critically damped linear dynamic system [28]. The input to the system is a sequence of articulatory target positions. The dynamic system has another parameter, the time constant τ that controls the rate of target approximation. The basic

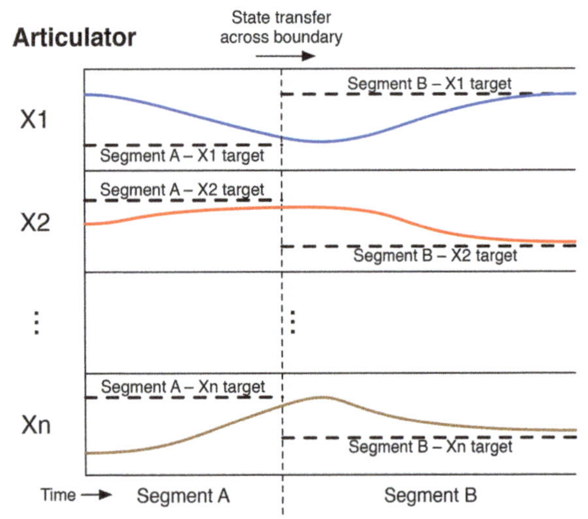

Figure 2 Target approximation model for controlling articulatory movements.

idea is to describe the dynamics of an articulator by a cascade of several identical first-order linear systems. The transfer function of such a system is

$$H(s) = \frac{Y(s)}{X(s)} = \frac{1}{(1+s\tau)^N} \tag{1}$$

where N is the order of the system and s is the complex frequency. The higher N, the more bell-shaped is the velocity profile associated with a step response of the system, a shape which is typically found in human target-directed movements [28]. However, with increasing order, the delay between input (target) and output (action) also increases. In VocalTractLab, sixth-order systems are used as a compromise.

The time-domain representation of Equation 1 can be derived using the inverse Laplace transform [28], which results in

$$y(t) = \left(c_0 + c_1 t + \cdots + c_{N-1} t^{N-1}\right) e^{-t/\tau} + x(t) \tag{2}$$

where for a static target, $x(t) = b$ is the position of the articulatory target. The time t is relative to the onset of the target interval. The coefficients c_i are calculated based on the initial conditions at the onset of the target [28], as can be shown by the following equation:

$$c_i = \begin{cases} y(0) - b & n = 0 \\ \left(y^{(n)}(0) - \sum_{i=0}^{n-1} c_i a^{n-i} \binom{n}{i} i!\right) / n! & 0 < n < N \end{cases} \tag{3}$$

Also, as a result of this purely sequential target approximation, no gestural overlap is assumed as far as any particular articulator is concerned. A target approximation movement does not start until the previous one is over. Any seeming overlap between adjacent movements is simulated by the combined effect of cross-boundary state transfer and the articulation strength of the later movement that determines how quickly it can eliminate the carryover effect exerted by the transferred state.

A key advantage of TA is that it allows the mapping of variant surface trajectories due to phonetic context, stress, speech rate, etc. to a single invariant target as demonstrated in our previous studies in prosody modeling [3,20-23,31]. TA simplifies the problem of inverse mapping from acoustics to underlying articulatory targets. This can be achieved because of the clear separation of the transient and the steady-state responses in the TA representation of the articulatory movements. The estimation of articulatory targets underlying the movements due to these factors allows us to capture the trend of the variability, which can be then summarized into a single contextually invariant target by the analysis of parameter distribution [21,22]. This approach of using an invariant target in inverse mapping is different from the Directions Into Velocities of Articulator (DIVA) framework that defines articulatory targets as regions [32,33]. The DIVA framework relies on a neural network to map the associations between acoustic and articulatory data. In this sense, DIVA is largely a mapping method. In the TA framework, there is only one invariant articulatory target corresponding to a specific functional condition. The contextual variability due to the transition from one target to another is modeled as a transient response which is a by-product of the transition. This means that for each phonetic unit a single target or a single compound target can be learned from its many context-sensitive realizations. The feasibility of this approach has been seen in our recent work on F_0 modeling [21-23].

Another critical strategy in the present implementation of the TA model is a novel segmentation method, that is, a segmental interval is defined as the time period during which its canonical pattern is unidirectionally approached [19]. As a result, the point where a segment best approximates its canonical pattern is marked as its offset rather than onset or center, as shown in Figure 3.

Table 1 shows the list of articulatory target parameters that define vocal tract configurations. Beside these positional parameters, VocalTractLab also requires the specification of articulatory strength for each TA movement. In the target approximation framework [21,22,28,31], this strength determines how fast the articulatory target is approached. The adjustment of articulatory strength would directly affect the rate of articulatory movement and indirectly affect the rate of formant changes in the spectrograms. The modeling process therefore needs to learn, for each vowel, a vocal tract shape associated with 18 articulatory parameters, as shown in Table 1, and a strength

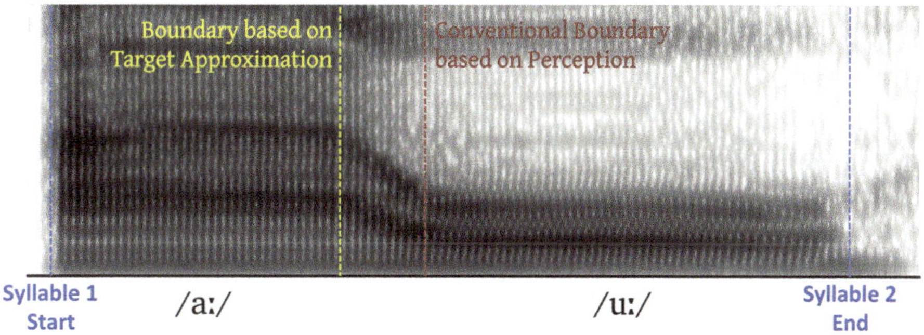

Figure 3 Syllable boundary marking based on target approximation framework comparing to that from perception.

parameter. Because each utterance consists of two vowels, 38 parameters need to be learned in total in each simulation run.

2.5 Optimization via analysis-by-synthesis

One of the critical issues in modeling the learning process is to determine the level of representation of the observable data. In this paper, we explicitly assume that the calculation of the comparison is done only at the acoustic level. This may not entirely cover the range of inputs that a child receives in the actual learning process, which may also involve orofacial features [34-36]. But it allows us to systematically and separately test the effectiveness of information that may be present in acoustic data independent of the visual features. Therefore, parameters of the visible articulators such as the lips and the jaw are acoustically optimized. Note that this strategy does not prevent future studies from including visual information as additional training input.

The representation of the acoustic kinematics should be sufficiently detailed to allow accurate analysis-by-synthesis but not so detailed as to make the computation infeasible. For segmental learning, we need to identify a spectral representation that best captures the articulatory changes and reflects human speech perception. A good candidate is Mel-frequency cepstral coefficients (MFCCs), which have been successfully used in speech recognition and HMM-based synthesis [37]. In this paper, MFCCs of the surface acoustics of both natural and synthetic speech are calculated using a standard setting in Praat [18], and the difference between the two are used as errors in the optimization of the articulatory targets.

A set of articulatory targets is searched for each segmental interval, whose boundaries are manually defined before optimization. The optimization process is assumed to follow a simple gradient descent search algorithm with random adjustments. The purpose of this simple design is so that the result of the study would only reflect the effectiveness of the overall strategy rather than that of the optimization algorithm *per se*. Further improvement of the optimization process can be done in the future study by testing various optimization modules that implement the same input and output interfaces. For each segment, articulatory targets in the form of vocal tract shapes are optimized iteratively to minimize the total sum of square errors, E, of MFCC between original and synthesized sounds, which can be described as follows:

$$E = \sum_{i=1}^{n} \sum_{j=1}^{m} \left(c_{ij} - \hat{c}_{ij}\right)^2$$

where n is the number of acoustic feature timeframes, m is the number of MFCC coefficients, c_{ij} is the jth cepstral coefficient of the ith frame in the natural utterance, and $\hat{c}_{ij}$ is the jth cepstral coefficient of the ith frame in the synthetic utterance.

Some articulatory parameters may not be entirely independent of others, however. For example, the tongue parameters have been found to be positively correlated with each other in articulatory movements for certain places of articulation, such as alveolar, palatal, and velar [38]. This correlation suggests that there is a constraint weakly tying these parameters together so that the changes in one parameter also affect other parameters, depending on the physiological locations. Such an embodiment relationship can be used to help the optimization process so

Table 1 List of articulatory target parameters of VocalTractLab

Parameter	Description	Parameter	Description
HX, HY	Hyoid positions	VO	Velic opening
JX	Horizontal jaw position	TTX, TTY	Tongue tip positions
JA	Jaw angle	TBX, TBY	Tongue blade positions
LP	Lip protrusion	TCX, TCY	Tongue body positions
LD	Lip distance	TS1-TS4	Tongue side elevation 1-4
VS	Velum shape		

that the parameter adjustment is more realistic. In this paper, we modeled this embodiment constraint by co-adjusting nearby articulators. For example, whenever the tongue blade parameters (TBX/TBY) were adjusted, those of tongue tip and tongue body (TTX/TTY, TCX/TCY) were also modified by a small amount (20%) in the direction of the main adjustment.

2.6 Numerical assessment and perceptual evaluation

After obtaining the optimal target values, the accuracy of the estimated articulatory targets was assessed by comparing the formant tracks of the synthesized utterances with the original formant tracks. Time-normalized formant tracks (F1-F3) of both synthesized and original utterances were extracted using FormantPro [39], a Praat script for large-scale systematic analysis of continuous formant movements. The comparison was done by measuring for each syllable the root mean square error (RMSE) between the synthesized and original utterances.

To assess the synthesis quality, a listening experiment was conducted with native Thai participants to identify the synthetic vowels and evaluate their naturalness. Target parameters of the same vowel optimized through the analysis-by-synthesis strategy are averaged together across multiple contexts as the underlying representation of that vowel. A monosyllabic word of each individual vowel was predictively (since no monosyllables were used in the training) synthesized using the estimated articulatory parameters. The standard vocal tract configuration of a male German speaker provided by VocalTractLab version 2.1 [26] was used to generate the stimuli. It should be noted that while the original vocal tract configuration is derived from a German speaker, the articulatory parameters and the synthesis process are language dependent. As controls, the natural stimuli of the same words were recorded by the same speaker used in the training corpus at a sampling rate of 22.05 kHz and 16-b resolution. Recording was done in a sound-treated room at the King Mongkut's University of Technology Thonburi. Both natural and synthetic stimuli had their intensities normalized to 70 dB using Praat. In total, there were 18 stimuli.

Twenty native Thai listeners participated in the experiment, which was conducted with the ExperimentMFC function of Praat. The stimulus words were randomly presented to the listeners, who were asked to first identify the Thai word they heard and then select a naturalness score on a five-level scale from terrible (1) to excellent (5). Listeners were allowed to listen to the stimuli as many times as they preferred.

3. Results

3.1 Synthesis accuracy of estimated vowel targets

Table 2 shows the mean formant RMSE of each vowel compared between the synthetic and original utterances.

Table 2 Mean formant RMSEs in percentage of each vowel in each syllable

Vowel	First syllable			Second syllable		
	F1	F2	F3	F1	F2	F3
/aː/	9.8	4.0	6.0	8.7	4.2	6.2
/iː/	8.1	3.7	8.3	6.9	5.4	7.1
/uː/	4.9	8.4	6.9	6.9	7.9	8.0
/eː/	2.8	5.1	4.1	4.3	5.1	4.6
/ɛː/	9.9	7.8	3.7	7.6	6.6	5.2
/ɯː/	7.3	6.2	4.3	7.1	5.3	5.1
/ɤː/	4.1	3.2	2.3	5.0	3.9	3.9
/oː/	6.4	4.2	7.8	6.0	4.4	8.6
/ɔː/	8.3	2.9	5.9	6.8	3.8	6.4

The percentage value is calculated relative to the original formant averages.

Low RMSEs can be observed for all vowels compared to the average RMSE levels reported in the previous studies [40]. While there are no significant difference in RMSEs of F1 and F2 between the first and second syllables (F1, $t(8) = -0.528$, $p = 0.306$; F2, $t(8) = -0.403$, $p = 0.349$), RMSE of F3 is slightly higher in the second syllable than in the first (F3, $t(8) = -2.328$, $p = 0.024$). It should also be noted that a certain level of RMSE values might be attributed to the difference in the vocal tract anatomy of the speaker and the template used in learning. By further comparing these synthesized vowels to the empirical findings reported in [41], we find that both F1 and F2 of synthesized vowels are comparable to those of the natural ones as shown in Figure 4. This indicates that the optimization can effectively estimate the underlying articulatory targets of both syllables.

Figure 5 shows examples of spectrograms of a vowel sequence in the corpus and that of a synthetic one generated with optimized articulatory targets. Further comparisons of formant frequency contours of exemplar utterances (obtained with FormantPro [39]) are shown in Figure 6. Note that each vowel is annotated to terminate at a point where its target is best achieved so that the formants in each segment move unidirectionally toward an ideal pattern. Smooth formant transitions from one vowel to another can be observed in the synthetic utterances (lower panels in Figure 5, solid lines in Figure 6), just as in the natural utterances (upper panels in Figure 5, dotted lines in Figure 6). Visual inspection of spectrograms of all other cases also confirmed the accuracy in representing the formant patterns of the estimated vocal tract shapes. The smooth synthetic formant movements are thanks to the TA dynamics of all the articulators involved. Note that while there are certain mismatches, for example, in F3 of /uː/ in /uːeː/ as shown in Figure 6, between the original and synthesized formant frequencies, these mismatches are evened out once they

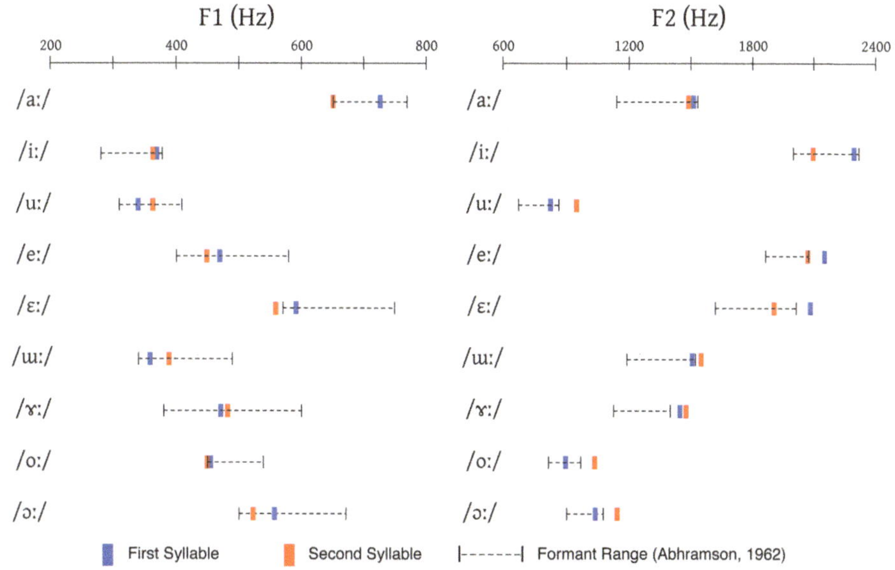

Figure 4 Comparison of F1 and F2 formant frequencies of synthetic vowels and the empirical formant ranges [41].

are averaged together with other cases (e.g., for /u:/, /a:u:/ and /u:ɤ:/ as shown in Figure 6). Further evaluation of synthesis quality will be later shown by a listening test.

3.1 Vocal tract shapes of Thai vowels

Figure 7 shows the articulatory targets as vocal tract shapes of Thai vowels reconstructed from the optimized articulatory targets by averaging the parameters across multiple instances of each vowel in both syllable locations. The shapes are largely consistent with previously reported Thai phonetic descriptions. Both vowel backness and vowel height of estimated vocal tract shapes are consistent with their theoretical locations [17]. Using these vocal tract shapes, we can synthesize vowel sequences either in isolation or in context. It should be noted that these shape targets are estimated without direct knowledge of the actual articulatory position, but through the utilization of the articulatory synthesizer and the acoustic optimization, starting from the vocal tract shape of a schwa. Distinctive vocal tract estimates that contain the gradient in terms of tongue backness and vocal tract openness as shown in Figure 7 suggest that the strategy is effective in obtaining the articulatory parameters capable of accurately generating the acoustic responses. These underlying targets of Thai vowels will be further used in the perceptual evaluation.

3.3 Synthesis quality

Figure 8 shows the results of the listening experiment on synthetic Thai vowels. Listeners could identify vowels equally well for both natural and synthetic vowels, with no significant difference between the two ($t(8) = 1.25$,

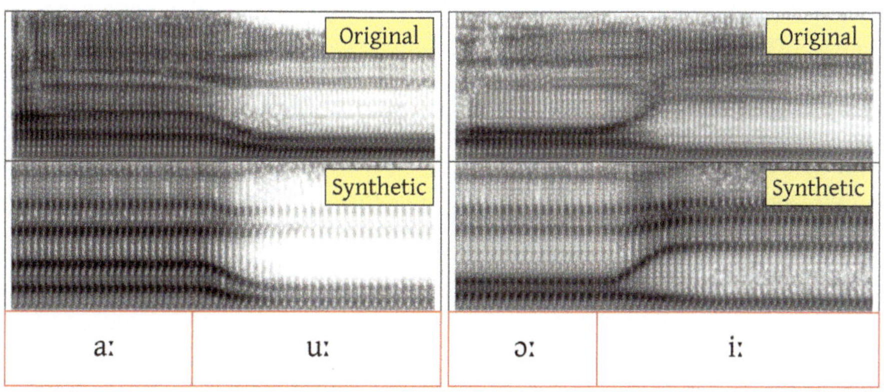

Figure 5 Spectrogram comparisons of example original and synthetic utterances.

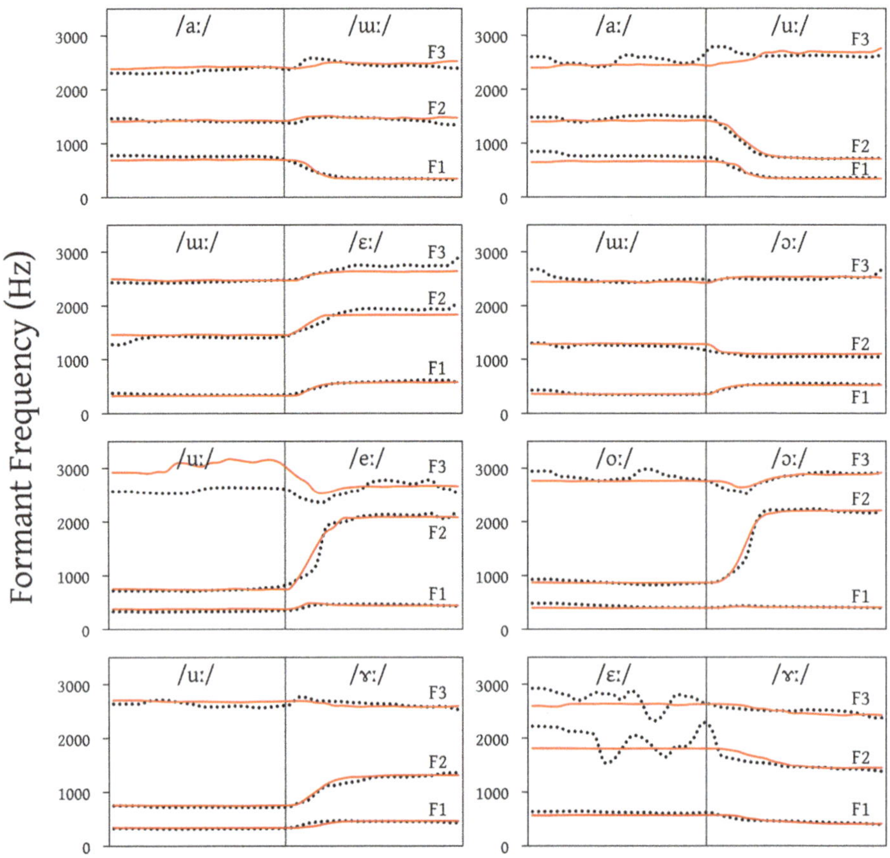

Figure 6 Time-normalized formant frequency contour comparisons of example original (dotted lines) and synthetic (solid lines) utterances. Each utterance consists of two syllables, each marked with the vowel symbol and separated by the boundary as the vertical line.

$p = 0.247$). All misidentifications were confusions with the neighboring vowels. It should be noted that in this study, one misidentified case is equivalent to 5% in the identification rate. This indicates the effectiveness of the proposed method for estimation of articulatory targets in representing vowels. The lowest identification rate of synthetic vowels is that of /u:/ which was perceived by three listeners as /o:/, while the perception of the natural /u:/ was perfect. This is possibly because the estimated velum parameters are continuous rather than discrete values, which makes it difficult for an optimization algorithm to get a full closure of the velum opening. Since the natural /u:/ does not require the usage of the nasal track, the nasality in the synthetic /u:/ vowel may cause it to be confusable with /o:/, which contains a higher degree of nasality due to the openness of the vowel. Naturalness scores of natural vowels were generally better than those of synthetic vowels ($t(8) = 3.24$, $p = 0.012$). However, once we compare the range of the naturalness score, listeners identified both of them in roughly the same score ranges, 4.2 to 5 for natural stimuli and 3.9 to 5 for synthetic stimuli. This result indicates that the present method could generate close-to-natural vowels with underlying articulatory targets learned from natural speech. Also, the slightly lower naturalness scores may be partially related to the specific voice quality used in generating the synthetic vowels, which we did not adjust to match the voice quality of the real speaker.

4. Discussion

The results of the present study have shown that it is possible to estimate the underlying articulatory targets of vowels from surface acoustics of continuous speech and predefined annotated segmental boundaries as the input, with an articulatory synthesizer controlled by target approximation models as the learner, and analysis-by-synthesis optimization as the training regimen. The numerical assessment as shown in Table 2 and Figure 4 indicates that the learned targets can be used to consistently synthesize the acoustic data that closely approximate the natural utterances. The visual impression of the synthesis examples as shown in Figures 5 and 6 suggests a good match in the dynamics of the acoustic data. The perceptual evaluation shows that the underlying articulatory targets learned this way can be used

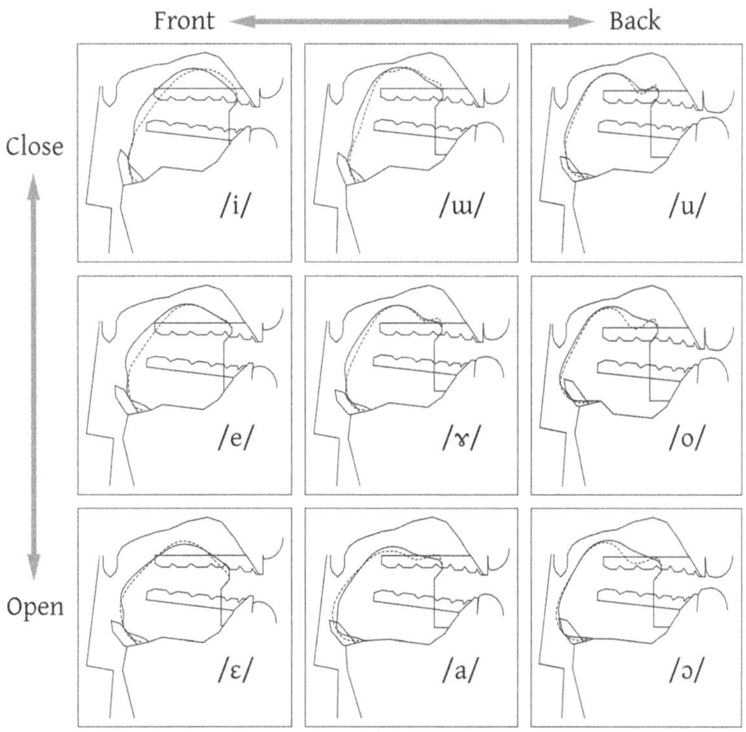

Figure 7 Vocal tract shapes reconstructed from estimated articulatory target of each Thai vowel. Dotted curves represent the tongue side elevations displayed relative to tongue body.

to generate isolated vowels that approximate to the original both perceptually as shown in Figure 8 and in terms of articulation as shown in Figure 7. All these results indicate that the estimated articulatory parameters closely represent the underlying targets of the Thai vowels.

One advantage of the present approach over the previous attempts is the decoupling of the observed data and the speech production mechanism. Compared to the mapping approaches [6-14], the present study does not use any actual articulatory data in the optimization process but

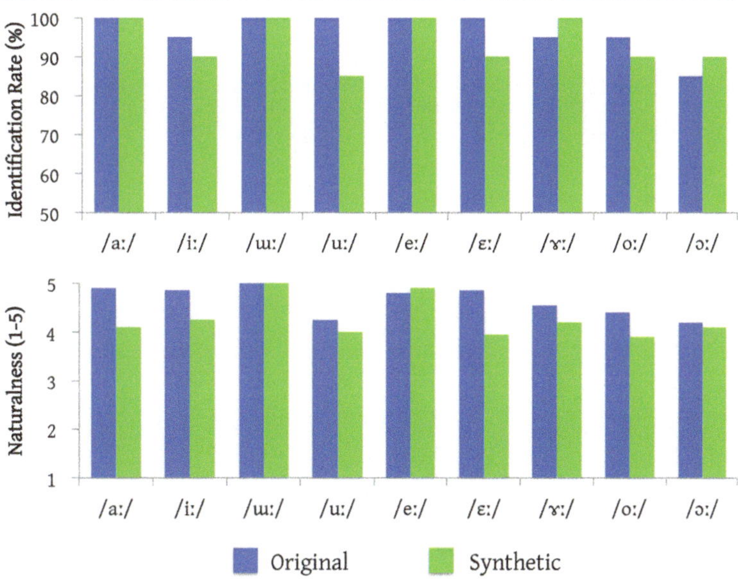

Figure 8 Mean vowel identification rate (top) and naturalness score (bottom) for both original and synthetic stimuli.

utilizes the knowledge of speech production mechanisms incorporated in the articulatory synthesis. This also provides an option that further studies may integrate the visible articulator data into the scheme to reduce the total degree of freedom. Another advantage of the present approach over the mapping approach is on the emergence of new data that are unknown to the system. The mapping approach would have difficulties in immediately applying the learned model to the new data since it has to be trained on a specific speaker. In contrast, the present approach has shown that even with the German vocal tract configuration, the system can learn Thai vowels that are numerically and perceptually accurate. Such decoupling of the articulatory mechanisms from the linguistic information enables us to apply this approach to different languages.

Another main feature of the present approach is the use of the TA model. With the TA model, the optimization only needs to estimate the targets and let the transient responses of the articulatory trajectories be calculated by the TA model. This significantly reduces the degrees of freedom of the estimation to only a set of articulatory parameters, along with a time constant, which need to be optimized, instead of having to map from every frame of acoustic data to the articulatory estimates. The usage of the TA model also allows us to simulate smooth transitions in the acoustic data as observed in natural utterances. While this feature may be present in different forms in the previous attempts such as generalized smoothness constraint [14], state transition in HMM [6,7], or smoothing algorithm [11-14], the direct incorporation of TA model into the articulatory synthesizer provides a simple yet effective strategy in the simulation of articulatory dynamics.

Some mapping studies that are based on the TD model [9,10] do take dynamic gestural control into consideration. The TD model provides a mechanism for generating movements of tract variables. It uses a critically damped second-order system to describe the movement. In TD, gestural movements are assumed to be always completed and adjacent gesture movements are assumed to be overlapped. This is in contrast to TA, which assumes that targets are not always reached, and allows remaining momentum at the end of a target approximation movement to be transferred to the next interval as its initial conditions.

Further development of the framework for organizing trained targets is still needed. First, no consonants have been simulated, so the effect of gestural overlap between consonant and vowel has not yet been modeled. Strategies have yet to be developed to simulate the learning of the fully overlapped CV gestures [19]. Second, the incorporation of a timing model in the articulatory synthesis is also needed, as timing specifications of the segments are required prior to the generation process. Third, the visible articulatory data can be directly incorporated into the learning strategy. This will further reduce the degree of freedom of the optimization process and may further improve the effectiveness of the system. Finally, the acoustic-to-articulatory inversion in the current study is not fully complete, as the segmentation of continuous utterances into discrete unidirectional movements is done manually. The underlying assumption is that the learning of perceptual segmentation is achieved prior to the learning of the articulatory targets. But the validity of this assumption is not fully established and has to be addressed in future studies.

5. Conclusions

In this study, we explored the estimation of articulatory targets of Thai vowels as a distal learning task using a model-based analysis-by-synthesis strategy. Articulatory targets as vocal tract parameters of each vowel were iteratively optimized by minimizing the acoustic error between original and synthetic utterances. The estimated vocal tract shape targets were used to synthesize the acoustical vowels, and the perceptual evaluation confirmed the synthesis quality. These results demonstrate that distal learning with an articulatory synthesizer that incorporates knowledge of speech production mechanisms is an effective strategy for the simulation of speech production acquisition.

Competing interests

The authors declare that they have no competing interests.

Acknowledgements

We would like to thank the Royal Academy of Engineering (UK) for financial support through the Newton International Fellowship Alumni follow-on funding and the Thai Research Fund (Thailand) through the Research Grant for New Researcher (TRG5680096 to SP).

Author details

[1]Department of Computer Engineering, Faculty of Engineering, King Mongkut's University of Technology Thonburi, Bangkok 10140, Thailand. [2]Department for Phoniatrics, Pedaudiology and Communication Disorders, University Hospital Aachen and RWTH Aachen University, Aachen 52074, Germany. [3]Department of Speech, Hearing and Phonetic Sciences, University College London, London WC1N 1PF, UK.

References

1. P Mermelstein, Articulatory model for the study of speech production. J. Acoust. Soc. Am. **53**(4), 1070–1082 (1973). doi:10.1121/1.1913427
2. EL Saltzman, KG Munhall, A dynamical approach to gestural patterning in speech production. Ecol. Psychol. **1**, 333–382 (1989). doi:10.1207/s15326969eco0104_2
3. Y Xu, Speech melody as articulatorily implemented communicative functions. Speech Commun. **46**, 220–251 (2005). doi:10.1016/j.specom.2005.02.014
4. J Sun, L Deng, An overlapping-feature-based phonological model incorporating linguistic constraints: applications to speech recognition. J. Acoust. Soc. Am. **111**, 1086 (2002). doi:10.1121/1.1420380
5. Z Ling, K Richmond, J Yamagishi, R Wang, Integrating articulatory features into HMM-based parametric speech synthesis. IEEE Audio Speech Lang. Process. **17**(6), 1171–1185 (2009). doi:10.1109/TASL.2009.2014796

6. G Hofer, J Yamagishi, H Shimodaira, Speech-driven lip motion generation with a trajectory HMM, in *Proceedings of the 9th Annual Conference of the International Speech Communication Association* (Interspeech, Brisbane, 2008). 22–26 September 2008, pp. 2314–2317
7. M Tamura, S Kondo, T Masuko, T Kobayashi, Text-to-visual speech synthesis based on parameter generation from HMM, in *Proceedings of the 1998 IEEE International Conference on Acoustics, Speech and Signal Processing* (ICASSP 98), Seattle, WA. 12–15 May 1998, pp. 3745–3748
8. B Uria, S Renal, K Richmond, A deep neural network for acoustic-articulatory speech inversion, in *Proceedings of the NIPS 2011 Workshop on Deep Learning and Unsupervised Feature Learning*, Sierra Nevada, Spain. 16 December 2011. http://www.cstr.ed.ac.uk/downloads/publications/2011/articulatory_inversion.pdf
9. V Mitra, H Nam, CY Espy-Wilson, E Saltzman, L Goldstein, Retrieve tract variables from acoustics: a comparison of different machine learning strategies. IEEE J. Sel. Topics Signal Process. **4**(6), 1027–1045 (2010). doi:10.1109/JSTSP.2010.2076013
10. H Nam, V Mitra, M Tiede, M Hasegawa-Johnson, C Espy-Wilson, E Saltzman, L Goldstein, A procedure for estimating gestural scores from speech acoustics. J. Acoust. Soc. Am. **132**(6), 3980–3989 (2012). doi:10.1121/1.4763545
11. J Schroeter, MM Sondhi, Dynamic programming search of articulatory codebooks, in *Proceedings of the 1989 IEEE International Conference on Acoustics, Speech and Signal Processing* (ICASSP, 1989). vol. 1, Glasgow, UK, 23–26 May 1989, pp. 588–591
12. S Ouni, Y Laprie, Modeling the articulatory space using a hypercube codebook for acoustic-to-articulatory inversion. J. Acoust. Soc. Am. **118**(1), 444–460 (2005). doi:10.1121/1.1921448
13. B Potard, Y Laprie, S Ouni, Incorporation of phonetic constraints in acoustic-to-articulatory inversion. J. Acoust. Soc. Am. **123**(4), 2310–2323 (2008). doi:10.1121/1.2885747
14. PK Ghosh, S Narayanan, A generalized smoothness criterion for acoustic-to-articulatory inversion. J. Acoust. Soc. Am. **128**(4), 2162–2172 (2010). doi:10.1121/1.3455847
15. S Panchapagesan, A Alwan, A study of acoustic-to-articulatory inversion of speech by analysis-by-synthesis using chain matrices and the Maeda articulatory model. J. Acoust. Soc. Am. **129**(4), 2144–2162 (2011). doi:10.1121/1.3514544
16. R McGowan, Recovering articulatory movement from formant frequency trajectories using task dynamics and a genetic algorithm: preliminary model test. Speech. Commun. **14**, 19–48 (1994). doi:10.1016/0167-6393(94)90055-8
17. K Tingsabadh, AS Abhramson, Thai. J Int. Phon. Assoc. **22**(1), 24–48 (1993). doi:10.1017/S0025100300004746
18. P Boersma, Praat, a system for doing phonetics by computer. Glot. Int. **5**(9/10), 314–345 (2001)
19. Y Xu, F Liu, Tonal alignment, syllable structure and coarticulation: toward an integrated model. Italian. J. Linguist. **18**, 125–159 (2006)
20. S Prom-on, P Birkholz, Y Xu, Training an articulatory synthesizer with continuous acoustic data, in *Proceedings of the 14th Annual Conference of the International Speech Communication Association* (Interspeech, Lyon, 2013). 25–29 August 2013, pp. 349–353
21. S Prom-on, B Thipakorn, Y Xu, Modeling tone and intonation in Mandarin and English as a process of target approximation. J. Acoust. Soc. Am. **125**, 405–424 (2009). doi:10.1121/1.3037222
22. Y Xu, S Prom-on, Toward invariant functional representations of variable surface fundamental frequency contours: synthesizing speech melody via model-based stochastic learning. Speech Commun. **57**, 181–208 (2014). doi:10.1016/j.specom.2013.09.013
23. S Prom-on, F Liu, Y Xu, Post-low bouncing in Mandarin Chinese: acoustic analysis and computational modeling. J. Acoust. Soc. Am. **132**, 421–432 (2012). doi:10.1121/1.4725762
24. Y Xu, F Liu, Determining the temporal interval of segments with the help of F_0 contours. J. Phon. **35**, 398–420 (2007). doi:10.1016/j.wocn.2006.06.002
25. MI Jordan, DE Rumelhart, Forward models: supervised learning with a distal teacher. Cogn. Sci. **16**, 307–354 (1992). doi:10.1207/s15516709cog1603_1
26. P Birkholz, VocalTractLab 2.1 for Windows (2013). http://www.vocaltractlab.de. Accessed 17 December 2013
27. P Birkholz, Modeling consonant-vowel coarticulation for articulatory speech synthesis. PLoS One **8**(4), e60603 (2013). doi:10.1371/journal.pone.0060603
28. P Birkholz, BJ Kröger, C Neuschaefer-Rube, Model-based reproduction of articulatory trajectories for consonantal-vowel sequences. IEEE Audio, Speech and Lang. Process **19**(5), 1422–1433 (2011). doi:10.1109/TASL.2010.2091632
29. P Birkholz, BJ Kröger, C Neuschaefer-Rube, Proceedings of the 12th Annual Conference of the International Speech Communication Association (Interspeech 2011). Florence **28-31**, 2681–2684 (August 2011)
30. P Birkholz, D Jackèl, BJ Kröger, Simulation of losses due to turbulence in the time-varying vocal system. IEEE Audio, Speech and Lang. Process **15**(4), 1218–1226 (2007). doi:10.1109/TASL.2006.889731
31. Y Xu, QE Wang, Pitch targets and their realization: evidence from Mandarin Chinese. Speech. Commun. **33**, 319–337 (2001). doi:10.1016/S0167-6393(00)00063-7
32. FH Guenther, T Vladusich, A neural theory of speech acquisition and production. J. Neurolinguist **25**, 402–422 (2012). doi:10.1016/j.jneuroling.2009.08.006
33. FH Guenther, Speech sound acquisition, coarticulation, and rate effects in a neural network model of speech production. Psychol. Rev. **102**, 594–621 (1995). doi:10.1016/j.jneuroling.2009.08.006
34. JR Green, CA Moore, M Higashikawa, RW Steeve, The physiologic development of speech motor control: lip and jaw coordination. J. Speech Lang. Hear. Res. **43**, 239–255 (2000). PMCID: PMC2890218
35. JR Green, CA Moore, KJ Reilly, The sequential development of jaw and lip control for speech. J. Speech Lang. Res **45**, 66–79 (2002). PMCID: PMC2890215
36. MP Harold, SM Barlow, Effects of environmental stimulation on infant vocalizations and orofacial dynamics at the onset of canonical babbling. Infant Behav. Dev. **36**, 84–93 (2013). doi:10.1016/j.infbeh.2012.10.001
37. P Taylor, *Text-to-Speech Synthesis* (Cambridge University Press, Cambridge, 2009)
38. JR Green, YT Wang, Tongue-surface movement patterns during speech and swallowing. J. Acoust. Soc. Am. **113**(5), 2820–2833 (2009). doi:10.1121/1.1562646
39. Y Xu, FormantPro Version 1.1, http://www.phon.ucl.ac.uk/home/yi/FormantPro. Accessed 24 December 2013
40. RS McGowan, MA Berger, Acoustic-articulatory mapping in vowels by locally weighted regression. J. Acoust. Soc. Am. **126**(4), 2011–2032 (2009). doi:10.1121/1.3184581
41. AS Abramson, *The vowels and tones of standard Thai: acoustical measurements and experiments* (Indiana University Research Center in Anthropology, Folklore, and Linguistics, Pub. 20, Bloomington, 1962). http://www.haskins.yale.edu/Reprints/HL0035.pdf. Accessed 26 February 2014

6

Lightweight multi-DOA tracking of mobile speech sources

Caleb Rascon[1*], Gibran Fuentes[1] and Ivan Meza[1]

Abstract

Estimating the directions of arrival (DOAs) of multiple simultaneous mobile sound sources is an important step for various audio signal processing applications. In this contribution, we present an approach that improves upon our previous work that is now able to estimate the DOAs of multiple mobile speech sources, while being light in resources, both hardware-wise (only using three microphones) and software-wise. This approach takes advantage of the fact that simultaneous speech sources do not completely overlap each other. To evaluate the performance of this approach, a multi-DOA estimation evaluation system was developed based on a corpus collected from different acoustic scenarios named Acoustic Interactions for Robot Audition (AIRA).

Keywords: Multi-DOA; Three-mic array; Mobile sound sources; Kalman filter

1 Introduction

Estimating the direction of arrival (DOA) of a sound source is a well written-about topic in signal processing and has found a considerable amount of areas of application where its usefulness ranges from a complementary source of information to an essential part of the application. Such applications vary from steering multiple feed parabolic dish antennas [1], to source enhancement in antenna arrays [2], to fault monitoring in aircrafts [3], to intricate robotic pets [4], to close-to-life insect emulation [5].

This usefulness has also been applied in areas where sources are mostly speech, and thus, speech enhancement is required, either for the benefit of communication clarity between the users or to benefit automatic speech recognition (ASR) between the user and a computer. Examples of these applications are the design of hearing aids [6], robot audition [7-9], automatic meeting processing [10], and generic human-computer interaction via ASR (such as with mobile phones or smart homes). To carry out speech enhancement, these applications usually have a pre-processing stage in their auditory scene analysis, which involves the automatic estimation of the DOA of the active speech sources in the environment. This is known as multi-DOA estimation, and when being carried out in real acoustic environments (with audio interferences and prevalent reverberation) with several mobile active speech sources, it has been shown to be a very challenging task, even if assuming that the sound sources are in the far-field region of the microphone array (which simplifies the signal model).

A current popular solution to the multi-DOA estimation problem was presented in [11], which is arguably the starting point of two important robot audition projects: ManyEars [12] and HARK [13]. This solution requires an eight-microphone hardware solution but is able to detect accurately four moving speakers and up to seven static speakers if given enough time. This high eight-microphone requirement has been worked around by minimizing the physical footprint of the microphone array, the result of which is the small but effective 8SoundsUSB audio interface [14] which is now being applied in Willow Garage's PR2 service robot with good preliminary results [15]. However, even with a small physical footprint, the amount of microphones makes it difficult to employ this solution in applications where space is limited, such as hearing aids or mobile phones. In addition, to carry out the multi-DOA estimation phase of the process, it required an off-site computer because of its processing and memory requirements [16], which, although it may

*Correspondence: caleb.rascon@iimas.unam.mx
Instituto de Investigaciones en Matematicas Aplicadas y en Sistemas, Universidad Nacional Autonoma de Mexico, Circuito Escolar S/N, 04510 Mexico, Mexico

be able to be countered with today's technology, can be considered high for some applications.

Other multi-DOA estimation approaches using fewer microphones (four) and less processing and memory requirements involve the use of Kalman filters to smooth the estimated trajectory of the sound source in a noisy environment [17,18], and with a bit more microphones (five), they are even able to do so in a 3D search space [19]. However, they are only able to track one mobile source. Important attempts have been successful in being able to estimate the location of multiple sources using very few microphones (two to three) [20,21], but the sources are assumed to be static. A more detailed literature review is provided in Section 2.

In this paper, we improve upon our earlier approach, presented in [22] and applied in [23]. We also evaluate its performance in a much more detailed, systematic manner, which involves the capture and use of an evaluation corpus called Acoustic Interactions for Robot Audition (AIRA) and F1-based evaluation metrics, and doing so in several acoustic scenarios.

The approach divides the problem into two phases. The first phase estimates one DOA, even in a multiple-source environment, by 1) applying a coherence threshold between the DOAs estimated by each microphone pair and by 2) taking advantage of the incomplete overlap of simultaneous speech. The second phase keeps track of the DOA of several sound sources by assigning the incoming DOA of the first phase to either a) an existing track or b) a new track, depending on its angular distance to existing tracks. A Kalman filter is applied to each track for the resulting multi-DOA estimation and for track smoothing.

This current version of the approach is now able to 1) track multiple mobile sound sources 2) while employing a light hardware setup (only three microphones in a triangular array) and with a small computational resource requirement. Both these features, as far as we know, are not present in conjunction in any of the current multi-DOA estimation approaches, which is discussed in detail in Section 2.1. Such combination, we believe, would be of interest to several areas of application, as well the audio processing community in general.

This work is organized as follows: Section 2 provides background on the technical details to carry out multi-DOA estimation, as well as discusses the issue of the amount of microphones versus the number of sources estimated. Section 3 presents some background on the nature of simultaneous speech and how the proposed approach takes advantage of it. Section 4 details the proposed system. Section 5 presents the evaluation method we employed to measure the proposed system's performance. Section 6 discusses the results, and Section 7 provides the conclusions and future work.

2 Background on multi-DOA estimation

One of the most widely used acoustic feature for DOA estimation is the inter-microphone time difference (ITD). It is the delay of a sound from one microphone to the other. Its calculation is usually based on the cross-correlation vector (CCV) between the two captured signals. One of the simplest ways to calculate the CCV is by calculating a Pearson-based correlation factor for each delay value in the CCV, described in Equation 1.

$$CCV[k] = \frac{\sum_i (x_i - m_x)(y_{i-k} - m_y)}{\sqrt{\sum_i (x_i - m_x)^2}\sqrt{\sum_i (y_{i-k} - m_y)^2}} \quad (1)$$

where x and y are the two discrete signals being compared, i is the time index, k is the point at which y is being linearly shifted (delayed) and the correlation is being calculated, and m_x and m_y are the mean values of x and y, respectively. The ITD is the k value of the highest correlation measure in the CCV.

A good example of the use of this Pearson-based cross-correlation method for DOA estimation in a robot audition application is presented in [24], where it provided limited results. Unfortunately, issues have arisen when using this method in reverberative and noisy environments [25], as the CCV calculations insert bias errors in such circumstances, which result in incorrect ITD estimations. However, as it would be seen in Section 4, this can be compensated with a combination of a form of redundancy and calculating the CCV using a variation of the generalized cross-correlation with phase transform (GCC-PHAT) [26].

The GCC-PHAT methodology has been widely recognized as one of the primary techniques for ITD calculation because of its robustness against reverberation [27,28]. Because of this, it has been extensively applied for ITD calculation in a wide variety of microphone array scenarios. For example, in [29], for the authors to be able to carry out real-time spatial rendering of different acoustic scenarios with a wide variety of sound sources, their DOA was required to be estimated via first calculating the ITD using the GCC-PHAT methodology. In addition, variations on the GCC-PHAT have been employed, such as in [30] where the authors were able to estimate jointly the DOA and pitch of two moving sources using a linear array of six microphones in reverberative simulated scenarios. Another variation was presented in [31], where the authors build an acoustic map of a room using a 13-microphone linear array, based on the GCC-PHAT technique, to directly estimate the ITD of multiple users. A detailed description of the GCC-PHAT methodology is provided in Section 4.3.1.

Having calculated an ITD, the direction of the source (DOA) can be estimated using a variety of methods. One popular method assumes that the sound source is far enough from the microphone array (aka. in the far-field region) and that there is no diffracting objects inside the microphone array (aka. the free-field assumption) so that the sound wave can be approximated to a planar-type wave. This simplifies the ITD-to-DOA relation to an arcsin function (details of which will be provided in Section 4.3.2), but this introduces other issues in the ITD-DOA relation. In Figure 1, the DOA is plotted against the ITD calculated from a two-microphone array (assuming far-field and free-field situations). As it can be seen in the −50° to 50° range, the ITD-DOA relation seems close to being uniform. However, outside that range, the relation loses its near-uniformity and the angular resolution is reduced, which causes major errors when estimating angles that are located in the sides of the microphone array [9]. To counter this issue while still maintaining the simplicity of the far-field model, in the proposed system, this loss in angular resolution is dealt with via a proper selection of the microphone pair in a triangular array from which the reported DOA is calculated, while assuming a far-field source.

Another important issue to consider is the geometry of the microphone array employed. Some geometries suffer from what is known as *ambiguity* [32], where an ITD may belong to more than one DOA. As it can also be seen in Figure 1, a two-mic array, by only estimating DOAs in the −90° to 90° range, is not able to differentiate if an ITD will be used to calculate the DOA that is coming from the front or the back of the array. This can be surmounted by implementing 'artificial ears' and a hidden Markov model (HMM) monaural mechanism [33] to then be able to detect if the sound source is coming from either side of the array, but it has been deemed impractical, as any physical change to the ear (physically or its relative position to the microphone) or to the acoustic scenario requires re-training of the HMM. This can also be tackled by a two-phase strategy: a first pair of signals could be used to estimate an initial DOA, the audio acquisition system could then rotate briefly, and then another pair of signals could be acquired to estimate a second DOA. A comparison between the DOAs would result in an angle estimation in the −179° to 180° range (countering the front-or-back ambiguity, which, in this case, could be considered as a trivial ambiguity). Unfortunately, this approach has its own set of issues: it would require considerably more time than when using one DOA estimate, the required rotation would hinder mobility requirements in some applications, and the sound source could be moving as well, rendering the DOA comparison mute. Another possibility would be to enlarge the microphone array so that it surrounds the source, as in [29,31]. To do so, however, the microphone array needs to encompass a significant amount of the space used by the users, which may be impractical in some applications.

In the proposed approach, a triangular array is employed from which various DOAs are calculated; by using a redundancy measure, the aforementioned ambiguity issue is circumvented. This is detailed in Section 4.

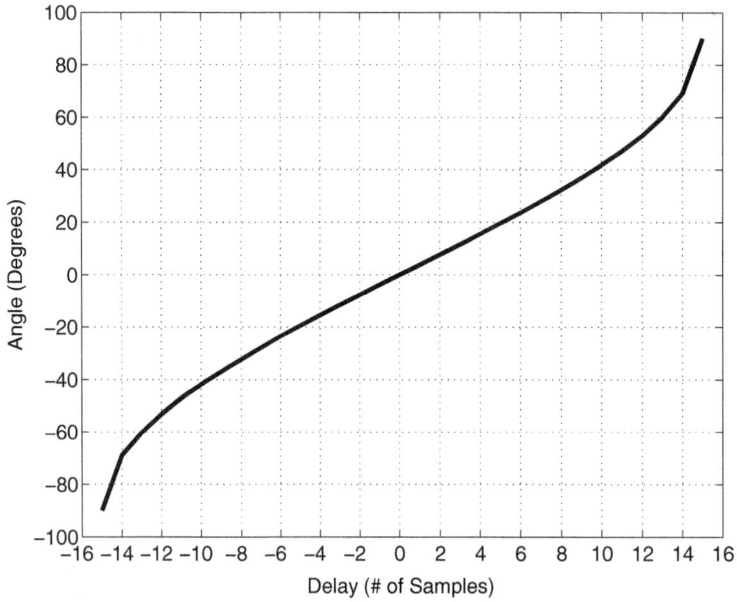

Figure 1 DOA vs. ITD. The graph shows DOA (or *Angle*) in degrees vs. ITD (or *Delay*) in number of samples and how it deviates severely from a uniform relation in the areas near the sides of the 1D array.

2.1 Amount of microphones employed vs number of sources estimated

It is important to note that the optimal amount of microphones to employ for multi-DOA estimation is a matter of discussion that is outside the scope of this work, but for ease of description, we are referring as a 'hardware-light' setup if it employs as much as four microphones, since there are portable commercial external audio interfaces that are able to handle that many microphones.

On the one hand, there is a large amount of approaches developed that carry out robust single-mobile-DOA estimation using a hardware-light setup [17-19]. On the other hand, there are important approaches in carrying out multi-mobile-DOA estimation which employ hardware-heavy solutions (which require specialized audio interfaces). It is important to mention that hardware-heavy approaches are practical for some applications; there are those that aim to conduct the audio acquisition once and process it offline, such as acoustical imaging [34] where it is not unusual to see 1,000-microphone arrays being used and no real-time localization is required.

The latter type of approaches (hardware-heavy) could be divided into two types of approaches: beamforming-based techniques and advanced statistical techniques.

An effective beamforming-based technique for multi-mobile-DOA estimation can be found in [35,36] and is somewhat representative of its type of approach. The idea is to create a noise map of the environment (or an acoustic map [31]) and then, by using metrics such as energy levels, propose possible sources of sound and their respective DOAs. It, technically, carries out a basic form of sound separation throughout all possible directions and then 'decides' which directions are valid sound sources and which are not. A good example of this decision is presented in [37] where, assuming there are only two sound sources in a meeting room, the beamforming technique 'decides' which sound source is assigned to which sound source class. These decisions are carried out at specific intervals, with which several methodologies can be employed for tracking purposes, such as Kalman filters [18] or particle filtering [38]. Another technique that is somewhat related to this type of approach, discussed and refined in [39], is known as the position-pitch (PoPi) plane, where, instead of mapping energy values onto the directional plane, a pitch spectrum is estimated per direction, providing possibilities to jointly estimate the DOA and pitch of sound sources, which, in turn, provides additional information to locate more than one source in the same direction.

As a whole, the beamforming type of approach has a pervasive issue: to increase precision and the quality of the validity metrics, it requires to obtain a high-resolution noise/acoustic map in both the amount of directions to search for and the quality of the separated sound from such directions. The sound quality requirement of the high-resolution map in turn requires a large quantity of microphones, since the quality of the separated sound is mainly defined by the signal-to-interference ratio (SIR), which is bound by the amount of microphones employed. This bound is summarized as 'the more microphones, the higher the quality'. Another important issue with this type of approach is that a balance needs to be struck between high-resolution maps and amount of computation resources required, since a high resolution results in a big search space where valid sound sources are to be found.

The techniques that approach the multi-mobile-DOA estimation problem by applying advanced statistics mostly rely on some variation of the popular technique known as Multiple Signal Classification (MUSIC) algorithm [1]. It carries out multi-static-DOA estimation by projecting the received signals in a DOA subspace, based on their eigenvectors, similar to principal component analysis (PCA). Although it has been widely reported that its performance decreases considerably in the presence of reverberation [25], it has been continuously enhanced in both resolution and computational costs [7], in handling mobile sources [40], as well as in handling an office-type amount of reverberation [10]. However, an important issue is that it is only able to estimate the DOA of as many sources as one less the amount of microphones (e.g., one source with two microphones, two sources with three microphones, etc.). This is because having more sources than microphones invokes the well-known 'more variables than observations' issue in PCA-based methods. In fact, it could be argued that if any sound interference may unexpectedly appear in the acoustic scene (which is not out of the ordinary in some scenarios) and, thus, increment the amount of sources in the acoustic scene over its upper limit, it could present instability issues in all of the provided estimations. Thus, to avoid this issue, the rule of thumb when using this type of techniques is 'just in case: the more microphones, the better'.

As it can be seen, there is an apparent tendency of employing hardware-heavy solutions to carry out multi-mobile-DOA estimation, while employing hardware-light solutions for single-mobile-DOA estimation. The proposed approach is a hardware-light solution that is able to carry out multi-mobile-DOA estimation, which we believe may be of interest to several areas of application and the audio processing community.

3 Background on simultaneous speech signals

As mentioned in Section 1, the proposed system assumes that the sound sources are from speech sound sources. This is taken advantage of, as detailed in the following section, for the purpose of multi-DOA estimation.

The scenario in which multiple speech sources are active simultaneously is bound to occur. Such is the case of crowded places (such as bars, restaurants, open-cubicle offices, etc.), where people are not participating in the same conversation but are near one another. However, it has been seen that users do not talk over each other with a 100% overlap. In fact, when analyzing speech recognition, 'spurts' of non-overlapping speech has been observed to last up to 500 ms [41]. For example, in Figure 2, two randomly chosen recordings from the DIMEx100 Corpus [42] are chosen. It can be seen that, even though these recordings are not from the same conversation (which is a very difficult scenario for multi-DOA estimation), when overlayed over each other, there are some portions with no overlap between them.

This means that a single-DOA estimation solution that is not only robust against multiple sources, but *is also sufficiently fast to 'catch' these single-source windows*, would be able to provide reliable results of single sources even in multiple-simultaneous-speech scenarios. However, because of the stochastic nature of the presence of single-user time blocks in the simultaneous audio timeline, such results would be provided in a sporadic fashion. Thus, these DOAs would be required to be 'grouped' afterwards and then be proposed as sound source directions, as it is carried out in the proposed system, detailed in the following section.

4 Proposed system

The proposed system uses a 'divide and conquer' strategy to solve the multi-DOA estimation challenge. It first provides a reverberation-robust DOA estimation of a single source, even in multi-source environments. It then takes advantage of the fact that even with simultaneous speech sources, there is not a 100% overlap between them. This means that the provided single-DOA estimations are being estimated from different sources but are 'given' in a stochastic manner. Thus, the objective of the next stage is to make sense of the incoming single-DOA estimations, by associating them to tracks or spawn new ones. The associated DOAs are then used to estimate a DOA for each track via Kalman filtering.

The proposed system is comprised by three modules:

1. *Audio acquisition and pre-processing.* Obtains audio data from the microphones and provides it to the single-DOA estimation module.
2. *Single-DOA estimation.* Estimates, from the audio data, a fast-but-reliable DOA estimation of a single sound source in the environment.
3. *Multi-DOA tracking.* Arranges the incoming single-DOA estimations from the single-DOA estimation phase into groups that are to be reported as sound sources with a Kalman-filtered DOA.

The data flow in the whole proposed system is summarized in Figure 3.

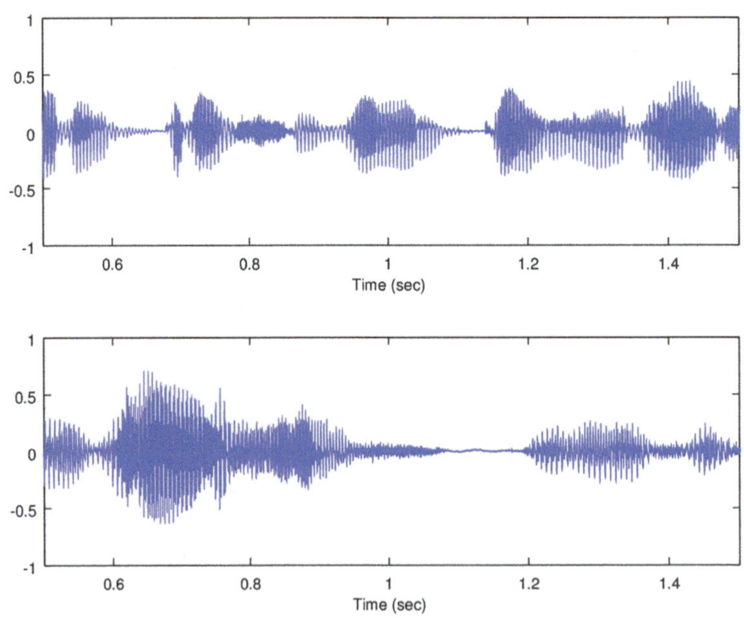

Figure 2 Non-overlapping simultaneous speech. An illustrative example of how even in simultaneous speech there is not a 100% overlap, a fact that can be taken advantage for multi-DOA.

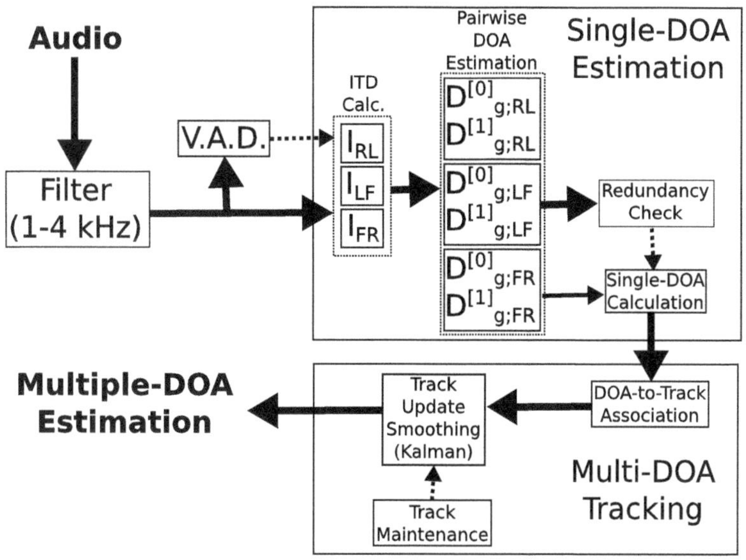

Figure 3 The flow of data throughout the proposed system. The audio data flows through a band-pass filter, voice activity detection (VAD) stage, the single direction-of-arrival (DOA) estimation (which includes multiple inter-microphone time-difference (Multi-ITD) calculation, a redundancy step, and single-DOA calculation), and the multiple direction-of-arrival (DOA) tracking (which includes DOA-to-track association, track updating, and track maintenance).

4.1 Hardware settings

To avoid the problems that arise when estimating a DOA using 1D microphone arrays (described in Section 2) and, at the same time, to maintain a hardware-light configuration, an equilateral triangular array is used, as shown in Figure 4. As it can be seen, the array maintains the free-field assumption, i.e., there are no objects inside the array that may diffract the sound waves coming into the the three microphones. Figure 5 presents one of the scenarios in which the system was tested.

4.2 Audio acquisition and pre-processing

As described in the last section, an equilaterial triangular array is used. Thus, the audio acquisition in the proposed system requires that the audio from three microphones be acquired simultaneously, in real time. For this purpose, the

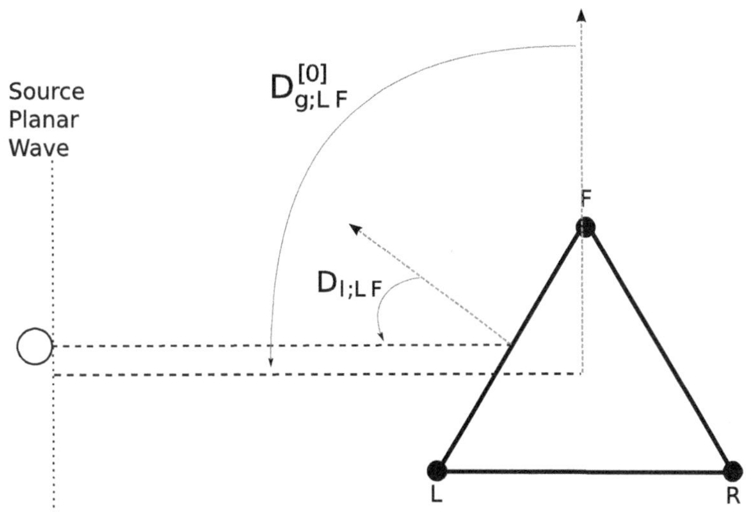

Figure 4 Microphone array setup of the proposed system and its coordinate system. The setup uses an array with three microphones, in a 2D array positioned horizontally. There are two coordinate systems being used: the local coordinate system per microphone array $D_{l;xy}$ that uses the front of the pair as its reference, and the global coordinate system $D^{[0]}_{l;xy}$ that uses the front of the whole array as reference.

Figure 5 The anechoic chamber testing scenario of the proposed system. This scenario set in the anechoic chamber was one in which the system was tested. It shows that there are no diffracting objects inside the microphone array.

JACK Audio Connection Toolkit [43] was employed for audio capture. It can sample at rates of 44.1 and 48 kHz, providing a good resolution for ITD calculations, without slowing down other software modules running in the same computer.

4.2.1 Band-pass filter

Since the focus of the multi-DOA estimation system is to estimate the direction of speech sources, a general infinite impulse response band-pass filter is used at the beginning of the process to remove general ambient noise that is outside the human speech frequency bands (between 1 and 4 kHz). Thus, sound sources that are outside of this frequency range are rejected. This filter was designed using a single Butterworth-based second-order solution.

4.2.2 Voice activity detection

To trigger the single-DOA estimation described in the next section, voice activity detection (VAD) needs to be carried out. A state-based VAD is proposed which can be in one of two possible states: SILENCE and NON-SILENCE, whose switch is triggered when the current energy values (obtained by averaging the values of the three microphones of a whole sample window) cross a pre-specified threshold (t_{change}) above the noise energy value (e_{noise}; obtained by averaging the most recent energy values before the state change, e_{hist}). When in NON-SILENCE, a sub-state ACTIVE can be achieved by surpassing an additional threshold (t_{vad}). The reasoning behind the ACTIVE sub-state is that in this manner, a 'precedent' of the noise energy value is set and such sub-state is only activated when the audio energy is over that value during the whole time the system is in the NON-SILENCE state. e_{noise} is 'reset' every time there is a state change, providing robustness in dynamic scenarios. Some considerations are now presented and discussed (all energy values are in the linear scale of 0 dBFS):

e_{hist} : The amount of energy values stored in e_{hist} will determine the VAD flexibility to changes in the noise level while also determining how much information is used to estimate the environmental noise. In [44], it was shown that moderately long time segments (750 ms) were preferable for estimating real types of noise. If using windows of 4,800 samples, and sample rates of up to 48 kHz, it is advisable to use more than eight energy values to store in e_{hist} for realistic scenarios.

t_{change} : Since t_{change} defines if a sound should be detected or not, its value is strongly linked to the signal-to-noise ratio (SNR) of the signals that are desired to be detected. A small value will detect more signals, but more potential noise will go through; a high value will be more strict but potentially will not let actual source signals through. It is advised to use an SNR value over the amplitude ratio of 1 (1 dB) but lower than 2 (6 dB), to be adequate in noisy environments (since high SNRs cannot be expected in such circumstances). Thus, having measured the average noise level in the environment and having chosen a desired SNR value, a value of t_{change} (in the linear scale of 0 dBFS) can be calculated using:

$$t_{\mathrm{change}} = 10^{(\mathrm{SNR}+N)/20} - 10^{N/20} \tag{2}$$

where N is the noise level in dBFS and SNR is the desired signal-to-noise ratio in dB. In addition, to be certain that the activated sample windows are of a signal of considerable energy, the ACTIVE sub-state should be triggered with a t_{vad} such that $t_{\mathrm{vad}} \gtrsim t_{\mathrm{change}}$.

4.3 Single-DOA estimation

Once the VAD reaches an ACTIVE state, the single-DOA estimation is triggered. This phase is divided into an ITD calculation, a pairwise DOA estimation stage, redundancy check, and single-DOA calculation described herein.

4.3.1 ITD calculation

In this stage, ITD calculation is carried out between signals captured in pairs of microphones, based on a variation of the reverberation-robust generalized cross-correlation with phase transform [26].

Equation 1 is a time-based version of the Pearson-based cross-correlation method. Its frequency-based equivalent is presented in Equation 3.

$$\mathrm{CCV}_F[f] = X[f]\, Y[f]^* \tag{3}$$

where X and Y are the Fourier transforms of the hamming-windowed x and y signals of Equation 1, respectively; f is the frequency bin; and the * operator stands for the complex conjugate operation.

The resulting cross-spectrum CCV_F relates to the cross-correlation vector CCV via the Fourier transform, such that $CCV_F = \mathcal{F}(CCV)$. Thus, the size of the resulting CCV will be dependent on the size of the sample window.

The generalized cross-correlation (GCC) applies a weighting function ($\psi[f]$) to Equation 3 to improve CCV calculation. In the proposed approach, the weighting function used is shown in Equation 4.

$$\psi[f] = \begin{cases} \frac{1}{|X[f]Y[f]^*|}, & \text{if } f_{\min} < f < f_{\max} \\ 0, & \text{otherwise} \end{cases} \quad (4)$$

where, if the first condition is reached (in our case, $f_{\min}$ is 1 kHz and $f_{\max}$ is 4 kHz), $\psi[f]$ applies the phase transform [27,28] (canceling out the magnitudes of both signals to estimate delta functions in the CCV domain). If not, the information from unused frequency bands is filtered to reduce estimation errors when carrying out $CCV = \mathcal{F}^{-1}(CCV_F)$. It is important to note that this rectangular cross-spectrum filtering introduces in the cross-correlation a leak in the CCV space, which means that some CCV bins may suffer some distortion. However, this type of distortion does not increase the value of the distorted bins to the point of being higher than the highest values of the CCV, and since the k index with the highest correlation value in the CCV ($k_{\max}$) is provided as the ITD of the two signals (I_{xy}), this is not an issue. As it can be gathered, the weighting is uniform inside the $[f_{\min}, f_{\max}]$ range, which can result in providing the same weights to a noisy frequency bin as to one that belongs to a speech source. Although automatic weighting in this step could be possible, it would require online noise estimation which is not carried out in this version of the proposed system. It is definitely being considered for future versions, though.

In addition, to avoid providing ITDs of reverb-bounces or noise sources, this part of the system only provides an ITD if its correlation value is higher than the CCV mean ($\overline{CCV}$) plus a pre-specified threshold (C_{thresh}), thus $I_{xy} = k_{\max}$. If $CCV[k_{\max}]$ is not higher than that, it is considered a 'noisy' ITD, and no ITD value is provided, thus $I_{xy} = \text{NULL}$. The selection of C_{thresh} should consider three points: 1) Although the theoretical lower bound for C_{thresh} is 0, choosing this value would result in erroneous ITD calculations in silent sample windows that are considered active after the sound source has become inactive, as only ambient noise would be present. 2) Although its theoretical upper bound is $1 - \overline{CCV}$, correlation values near 1 in real circumstances are rare; thus, a more practical generally applicable upper bound is 1. 3) A worst case scenario would be to have a large amount of values very close to $CCV[k_{\max}]$ (expected with noise/high reverberation), which in turn would result in $\overline{CCV}$ having a value very close to $CCV[k_{\max}]$. Thus, to avoid filtering out correct ITD estimations in such scenarios, C_{thresh} should be very small compared to its upper bound and very close to its lower bound, meaning $0 \lesssim C_{\text{thresh}} \ll 1$.

It is important to note that, even though the IIR filter that precedes this step is rejecting information from outside the 1- to 4-kHz range, the resolution of the CCV (aka. the spatial resolution) is not hindered, since, in this case, it is highly dependent on the sample window size. Thus, a large window size (such as 4,800) is preferable for the purpose of having a high spatial resolution, in conjunction with a high sample rate (such as 48 kHz) to not hinder the overall responsiveness of the system.

4.3.2 Pairwise DOA estimation

Using the ITD calculation procedure, three ITDs are calculated between sample windows obtained from microphones R and L (I_{RL}), L and F (I_{LF}), and F and R (I_{FR}). If any ITD is NULL, the window set is discarded. Otherwise, a local DOA ($D_{l;xy}$) is calculated from each ITD (I_{xy}) using Equation 5.

$$D_{l;xy} = \arcsin\left(\frac{I_{xy}\, V_{\text{sound}}}{F_{\text{sample}}\, d}\right) \quad (5)$$

where x and y are identifiers of the signals received from the microphones R, L, or F; V_{sound} is the speed of sound (343 m/s); F_{sample} is the sampling rate (in Hz); and d is the distance between microphones (in m). All DOAs (referred to with the base notation D) hereafter are expressed in degrees.

It is important to note that this model assumes a planar-type sound wave, which in turn assumes that the sound source is located in the far-field region of the microphone array. If r_{source} is the distance of the sound source to the center of the array and r_{mic} is the distance of any microphone to the center of an equilateral array, then the far-field requirement is $r_{\text{source}} \gg r_{\text{mic}}$ [45]. This means that small array sizes ($d = 2r_{\text{mic}}\cos(30°) < 0.25$ m) are preferable to satisfy the far-field assumption in appropriate conversation scenarios ($r_{\text{source}} \approx 0.70$ m). In the case of the proposed system, because of the use of large sample windows, the CCV maintains a high resolution even with small array sizes. However, since the following calculations only require the resulting $D_{l;xy}$ to carry on, other models (such as near-field techniques) could be used in place of Equation 5 without requiring any further modifications to the proposed system and is definitely cause for future work.

Having calculated $D_{l;xy}$, a pair of global DOAs ($D^{[0]}_{g;xy}$ and $D^{[1]}_{g;xy}$) are then calculated using Equations 6, 7, and 8.

As described in Figure 4, while the local DOA $D_{l;xy}$ provides an angle from the perspective of the microphone pair alone, the global DOAs provide an angle from the perspective of the whole array. Specifically, $D^{[0]}_{g;xy}$ is the global DOA calculated as if $D_{l;xy}$ was coming from the front of the xy microphone pair, and $D^{[1]}_{g;xy}$ as if it was coming from the back.

$$D^{[0]}_{g;\mathrm{RL}} = \begin{cases} 0^\circ & \text{, if } D_{l;\mathrm{RL}} = 0^\circ \\ -D_{l;\mathrm{RL}} & \text{, otherwise} \end{cases}$$
$$D^{[1]}_{g;\mathrm{RL}} = \begin{cases} -180^\circ + D_{l;\mathrm{RL}} & \text{, if } D_{l;\mathrm{RL}} > 0^\circ \\ 180^\circ + D_{l;\mathrm{RL}} & \text{, otherwise} \end{cases} \tag{6}$$

$$D^{[0]}_{g;\mathrm{LF}} = \begin{cases} D_{l;\mathrm{LF}} - 240^\circ & \text{, if } (120^\circ - D_{l;\mathrm{LF}}) > 180^\circ \\ 120^\circ - D_{l;\mathrm{LF}} & \text{, otherwise} \end{cases}$$
$$D^{[1]}_{g;\mathrm{LF}} = D_{l;\mathrm{LF}} - 60^\circ \tag{7}$$

$$D^{[0]}_{g;\mathrm{FR}} = \begin{cases} D_{l;\mathrm{FR}} + 240^\circ & \text{, if } (-120^\circ - D_{l;\mathrm{FR}}) > 180^\circ \\ -120^\circ - D_{l;\mathrm{FR}} & \text{, otherwise} \end{cases}$$
$$D^{[1]}_{g;\mathrm{FR}} = D_{l;\mathrm{FR}} + 60^\circ \tag{8}$$

4.3.3 Redundancy check

The three global DOA pairs $\left(\left[D^{[0]}_{g;\mathrm{RL}}, D^{[1]}_{g;\mathrm{RL}}\right], \left[D^{[0]}_{g;\mathrm{LF}}, D^{[1]}_{g;\mathrm{LF}}\right], \left[D^{[0]}_{g;\mathrm{FR}}, D^{[1]}_{g;\mathrm{FR}}\right]\right)$ are used to check if the three ITDs are from a sound source located in the same angle sector as a type of rejection step. To do this, the average of the differences between the DOA pairs (E_{pqr}) is calculated using Equation 9.

$$E_{pqr} = \frac{|D^{[p]}_{g;\mathrm{RL}} - D^{[q]}_{g;\mathrm{LF}}| + |D^{[q]}_{g;\mathrm{LF}} - D^{[r]}_{g;\mathrm{FR}}| + |D^{[r]}_{g;\mathrm{FR}} - D^{[p]}_{g;\mathrm{RL}}|}{3} \tag{9}$$

where p, q, and r, each, can be either 0 or 1. This provides eight possible E_{pqr}, the lowest of which is considered the *incoherence* of the sample window set. The global DOA set that is represented in the minimum E_{pqr}, meaning $\left[D^{[p]}_{g;\mathrm{RL}}, D^{[q]}_{g;\mathrm{LF}}, D^{[r]}_{g;\mathrm{FR}}\right]$, is proposed as the DOA set for this window sample set.

A pre-specified *incoherence threshold* (E_{thresh}; measured in degrees of separation between the DOAs) is used to reject sample window sets. A high incoherence implies that there is an absence of 'consensus' in the sample window set for a DOA estimation, usually caused by having too much noise on the CCVs because of interferences in the acoustic scene or more than one source being active. This rejection step serves as a type of redundancy check *per sampling window set*. The selection of the value of E_{thresh} depends on the minimum ITD-DOA resolution of each microphone pair, since the global DOAs may provide an angle calculated in different regions of Equation 5, which in turn depends on F_{sample} and d. This minimum resolution is calculated by obtaining the difference between the DOA calculated with the highest possible ITD and the DOA obtained with the second highest possible ITD. In addition, ITD errors can occur in the range of ± 1 bins. Thus, the worst case scenario is calculating a DOA in the minimum resolution range while having an ITD error. To this effect, a value for E_{thresh} that is not too small to reject window sets that may be sufficiently coherent, while rejecting those that are, is found around twice the minimum resolution.

As an example, Figure 6 shows two examples of two different data sets. The example shown in Figure 6a shows the scenario of two sources being active, which produces a highly incoherent set with no $\left[D^{[p]}_{g;\mathrm{RL}}, D^{[q]}_{g;\mathrm{LF}}, D^{[r]}_{g;\mathrm{FR}}\right]$ combination 'pointing' to a clear direction. On the other hand, Figure 6b shows the scenario of having only one source active (for a brief moment, the other source became inactive), which results in having a data set configuration with very low incoherence, which is $\left[D^{[1]}_{g;\mathrm{RL}}, D^{[0]}_{g;\mathrm{LF}}, D^{[1]}_{g;\mathrm{FR}}\right]$, with the combination $p = 1, q = 0, r = 1$.

4.3.4 Single-DOA calculation

If the sample window set is considered coherent, its reported DOA value (θ) is chosen from the member of

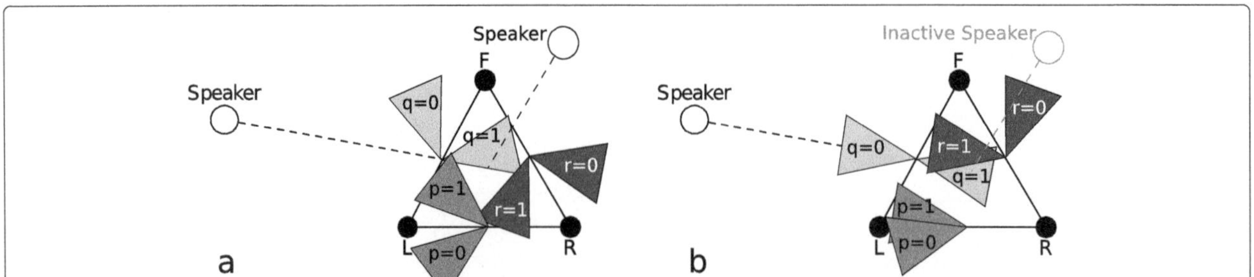

Figure 6 Incoherence example. **(a)** An example of a highly incoherent data set is presented, due to having two active sources. **(b)** A data set with very low incoherence is shown, with configuration $p = 1, q = 0, r = 1$, due to one of the sources being inactive at that brief moment.

its DOA set $\left[D^{[p]}_{g;\mathrm{RL}}, D^{[q]}_{g;\mathrm{LF}}, D^{[r]}_{g;\mathrm{FR}}\right]$ whose ITD from which it was calculated has the lowest absolute value ($I_{\mathrm{RL}}, I_{\mathrm{LF}}, I_{\mathrm{FR}}$).

This decision ensures that θ is based upon the microphone pair that is the most perpendicular to the source, and because of the equilateral nature of the triangular array, this implies that it is always estimated using a local DOA ($D_{l;xy}$) with a value inside the −30° to 30° range, well within a close-to-uniform area of Equation 5. Thus, the proposed system will always have a nearly uniform resolution between ITD and DOAs, which will result in low error rates throughout the −179° to 180° range, overcoming the issue of loss of resolution when the sound sources are located in certain directions relative to the array (detailed in Section 2).

4.4 Multi-DOA tracking

The single-DOA estimator described in the previous section only provides results when there is considerable confidence of only one sound source being detected in a small sample window (up to 100 ms).

As described in Section 3, simultaneous speech has a non-overlapping nature, which means that the single-DOA estimator would be able to provide reliable results of single sources even in multi-user scenarios. However, these results would be provided in a sporadic fashion, as the presence of single-user sample windows in the simultaneous audio timeline is stochastic.

The objective of the final phase of the proposed system is to gather these sporadic DOAs, associate them into existing tracks or new ones, and use such information to estimate their current location. This approach is based on radar tracking techniques [46] that are useful when trying to report data from different targets using consecutive radar observations.

For each sample window, 1) if deemed coherent, the single-DOA estimator will provide at most one DOA estimation, and 2) the multi-DOA tracker will update. Each update of the multi-DOA tracker goes through four steps:

Single-DOA estimation to track association. This is solved via an acceptance gate approach (similar to nearest neighbor), in which the incoming single-DOA estimation is compared to the estimated DOA of all pre-existing tracks (the manner in which these are estimated is explained in the 'Track update' step). The single-DOA estimation is associated to its nearest track, only if the angular distance between them is below a pre-specified threshold (the same threshold used for redundancy check in the ITD estimation). If no track is within that threshold, a new track is spawned with only the single-DOA estimation associated to it.

Track update/smoothing. All tracks are updated, regardless of having new associations or not. This is carried out by estimating the current DOA by solely using the DOAs that are associated with it, and, thus, reporting multiple simultaneous estimations per sample window. To model the movement of the speech sources, we consider linear dynamic and measurement models, with the state-space representation presented in Equation 10:

$$\begin{aligned} \mathbf{s}_{t+1} &= \mathbf{A}s_t + \mathbf{w} \\ \mathbf{m}_t &= \mathbf{H}s_t + \mathbf{v} \end{aligned}, \tag{10}$$

where $\mathbf{s}_t$ is the dynamic state vector, $\mathbf{m}_t$ is the measurement vector, $\mathbf{A}$ is the transition matrix, $\mathbf{H}$ is the measurement matrix, and $\mathbf{w} \sim \mathcal{N}(0, Q)$ and $\mathbf{v} \sim \mathcal{N}(0, R)$ are zero-mean process and measurement noises with covariances Q and R, respectively. The state vector $\mathbf{s}_t = [s_t, m_t, \dot{s}_t, \dot{m}_t]^T$ contains the Cartesian coordinate and velocity of the DOA, and the transition and measurement matrices are defined in Equation 11:

$$\mathbf{A} = \begin{bmatrix} 1 & 0 & S_w & 0 \\ 0 & 1 & 0 & S_w \\ 0 & 0 & 1 & 0 \\ 0 & 0 & 0 & 1 \end{bmatrix}, \quad \mathbf{H} = \begin{bmatrix} 1 & 0 \\ 0 & 1 \\ 0 & 0 \\ 0 & 0 \end{bmatrix} \tag{11}$$

where S_w is the size of the sample window in seconds. The estimation of the system state is carried out via the Kalman filter technique [47]. It was chosen since it provides an optimal and efficient solution for linear Gaussian models, such as the one being used here (see [46] for details). In addition, it reduces the influence of 'noisy' DOAs associated with the track, which is helpful in the case of close sources.

Track initiation. If the single-DOA estimation was not associated to any pre-existing track, a new track is spawned with this single-DOA estimation associated with it. However, the track is labeled as *tentative* until it has enough DOAs associated with it. What defines the final responsiveness of the system is the amount of DOAs associated with a new track such that it can be labeled *confirmed* (n_{doa}). To avoid providing 'noisy' tracks, more than 1 DOA is recommended. Since the single DOAs are provided stochastically, an upper bound for this value is difficult to set; however, values close to 1 will provide greater responsiveness. Thus, $n_{\mathrm{doa}} \gtrsim 1$ is a good rule of thumb.

Track maintenance. If a speech source goes inactive, it is important for the system to stop reporting it, to avoid false positives. However, it is also important to be robust against situations in which the speech source went inactive temporarily. Thus, a track is labeled again as *tentative* if a number of sample windows have gone by without any new association (n_{missed}). As described in Section 3, 'spurts' of non-overlapping speech has been observed to last up to 500 ms [41]. Thus, a value of sample windows that represents such an amount of time provides a good balance between responsiveness and low false positives.

5 Evaluation methodology

Four evaluation strains were carried out, each with its own objective, using an evaluation corpus based on the DIMEx100 Corpus [42], called AIRA.

The following parameter values were used during all evaluations:

- Sampling rate: 48 kHz
- S_w: 0.1 s (4,800 samples)
- e_{hist}: calculated using ten energy values
- C_{thresh}: 0.03645 (175 in an unnormalized FFT vector of 4,800 samples).
- E_{thresh}: 30°
- n_{doa}: 2
- n_{missed}: 5

The results of these evaluations are discussed in Section 6.

5.1 Anechoic chamber with static sources

To evaluate if the algorithm is able to detect the number and DOA of simultaneous speech sources, several experiments were carried in a 5.3 m × 3.7 m × 2.8 m full-anechoic chamber of the Laboratorio de Acústica y Vibraciones of the Centro de Ciencias Aplicadas y Desarrollo Tecnológico (CCADET) of the Universidad Nacional Autónoma de México (UNAM), a detailed description of which can be found in [48].

In this evaluation setting, the number of sources was varied from 1 to 4 in several positions in the whole horizontal degree range. For each number of sources, ten sessions were carried in which randomly chosen voice recordings from the DIMEx100 Corpus were reproduced for 30 s at 0 dB gain through studio-grade monitors acting as speech sources placed at 1 m away from the center of the triangular array. The microphones were set at 0.18 m apart, with 0 dB gain in the audio interface. The average sound level received at each microphone during voice activity was around −22 dBFS. Since the amount of noise inside the anechoic chamber was very low, the values for t_{change} and t_{vad} required to satisfy $t_{\text{change}} \gtrsim 0$, and $t_{\text{vad}} \gtrsim t_{\text{change}}$ (0.0005 was chosen for both). Figure 5 shows a photograph of the testing environment.

Two evaluation metrics were calculated:

Number and location of sources For each session, a sound source was considered as 'estimated' if it was detected for at least 25% of the duration of the session (30 s). If a source is estimated within a ± 15° range of the expected direction of an actual sound source, it is considered a true positive. If a sound source is estimated outside that range from an actual sound source, it is considered a false positive. If an actual sound source is not estimated during the experiment, it is considered a false negative. Using these metrics, the precision, recall, and F1 scores [49] (Chapter 8) of the proposed system's ability to detect sound sources are calculated.

Average error Once estimated, an average absolute error is calculated for every sound source that is deemed true positive, from the direction it is actually located.

In Table 1, these scores are provided per set of experiments.

5.2 Real acoustic setting with static sources

In the past evaluation, the evaluation was carried out in a controlled acoustic setting. However, it is important to measure the performance of the proposed approach in a real acoustic setting. An open-cubicle office was chosen for this test: 5.9 m × 7.9 m × 2.1 m (photo in Figure 7), with a typical indoor reverberation (RT_{60} = 0.47 s) and some presence of noise (−45 dBFS with 0 dB gain in microphones). A desired SNR for the VAD phase was set at 2.75 dB (for high sensitivity); thus, a t_{change} of 0.002 (in the linear scale of 0 dBFS) was used to achieve that, and a t_{vad} of 0.004 was used, satisfying $t_{\text{vad}} \gtrsim t_{\text{change}}$.

Table 1 Evaluation results reported in F1-type measure and average errors in the anechoic chamber

# of sources	Recall (%)	Precision (%)	F1 (%)	Average errors (in degrees)	Actual DOAs (in degrees)
1	100	100	100	1.22	45
2	100	100	100	7.36	−30
				3.49	90
3	90.00	100	94.74	7.15	−30
				3.50	90
				2.80	−150
4	72.22	100	83.87	0.04	0
				1.25	90
				0.61	180
				0.53	−90

Figure 7 Office A photograph. A photograph of the office A test scenario. The microphone array was set in the middle open space.

The same type of evaluation as in the 'Anechoic chamber with static sources' evaluation was carried out in this acoustic setting, while having the microphones set at 0.21 m apart and varying the number of simultaneous sources from 1 to 4 that were placed 1 m away from the center of the array. The average SNR captured during these experiments was of 21 dB. The results are shown in Table 2.

5.3 Real acoustic setting with mobile sources

To evaluate the performance of the proposed approach with mobile sound sources, pre-specified randomly chosen sentences from the DIMEx100 Corpus read by expert volunteers were recorded while they were moving in a pre-specified manner around the recording station in another open-cubicle office of size 10.5 m × 5.9 m × 2.1 m, which is divided into three 3.5-m-wide spaces that are acoustically connected. However, in acoustic terms, it is very similar to the acoustic scenario of the 'Real acoustic setting' evaluation. Their main difference is that there are walls much closer to the microphone array (photo in Figure 8), amplifying moderately the effects of reverberation ($RT_{60} = 0.51$ s), and it had slightly more noise (−41 dBFS with 0 dB gain in microphones). The SNR captured during these experiments was 17 dB. The same t_{change} and t_{vad} values as in the 'Real acoustic setting' evaluation were used. The microphones were also set 0.21 m apart.

An approximation of the movement of each source was calculated. Then an estimated acoustic scene description file was created that describes an approximation of a linear trajectory around the audio acquisition base, with start and stop times and DOAs. Because of this limitation, the movement of the sound source was limited to simple 'go from left to right' (or vice versa) trajectories so their approximations can be considered representative of their real trajectories.

Figures 9 and 10 present a representative plot of the tracking carried out by the proposed approach with one and two mobile speech sources respectively presented over the expected behavior (plotted as dashed lines).

An evaluation similar to the 'Anechoic chamber with static sources' evaluation was carried out, only that in this case the average errors, true positives, false positives, and false negatives were reported from the reference of the expected trajectory previously described, instead of a static value. These scores are provided in Table 3.

5.4 Resource requirements

The implementation of the proposed approach occupies a 14.6 MB memory footprint, which includes a graphical interface for data visualization. In addition, the proposed system occupies up to 7% of the CPU resources when active. To put this in perspective, the ManyEars application [12], running in the same machine, occupies

Table 2 Evaluation results reported in F1-type measure and average errors in an office-type acoustic setting with static sources

# of sources	Recall (%)	Precision (%)	F1 (%)	Average errors (in degrees)	Actual DOAs (in degrees)
1	100	100	100	2.77	0
2	65.00	100	78.79	8.25	−30
				0.07	90
3	56.67	100	72.34	2.53	−30
				0.32	90
				0.67	−150
4	45.00	94.74	61.02	1.89	0
				1.27	−90
				0.50	180
				3.97	90

Figure 8 Office B photograph. A photograph of the office B test scenario. The microphone array was set in the table in the middle open space.

a 38.6 MB memory footprint and up to 61% of the CPU resources. This implies that the proposed approach is carrying multi-DOA estimation, with high F1 scores and low error rates, using a very small resource footprint.

6 Results discussion

As it can be seen in Tables 1, 2, and 3, in most circumstances, the average error is very low ($<\pm 5°$). This is the result of the fact that the resulting θ from the single-DOA estimator is calculated from the ITD of the microphone pair most perpendicular to the sound direction. Considering the triangular geometry of the microphone array, this results in having this ITD well within the close-to-uniform range area of Figure 1, providing a uniform resolution for the resulting θ.

Another observation is that the performance of the proposed system decreases as more sources are in the environment, which is to be expected. However, the decrease was not as pronounced when in the anechoic chamber. This implies that reverberation/noise is an important influence in the proposed system. When observing the precision and recall metrics, it can be observed that the addition of more sources affects the recall metric much more then it does the precision metric. An explanation for this is that the chance of single-user sample windows occurring will tend to decrease when the number of sources there are in the environment increases, resulting in false negatives. However, varying the number of sources does not affect the precision scores as much (if at all), which implies that once a sound source direction is reported, it is highly likely that it is an actual sound source. This is evidence of the effectiveness of 1) the coherence-based redundancy measures and 2) the application of the variations to the phase transform in the GCC calculation. Both measures filter out noisy sources and reverberation bounces and, thus, provide a relatively high amount of true positives.

It can also be seen that there was a near 20% decrease in the F1 metric when moving from an anechoic environment (Table 1) to a realistic one (Table 2), per test. For example, the F1 score for two sources went from 100%

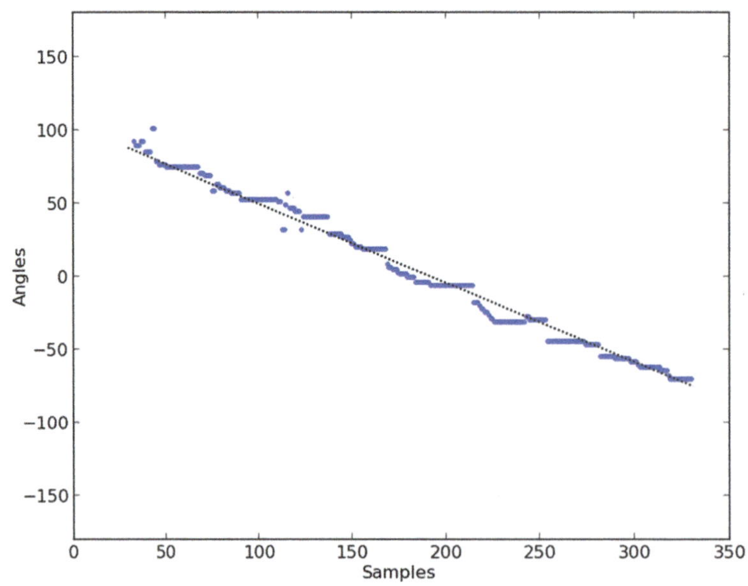

Figure 9 Office B with one source. A representative plot of the tracking carried out by the proposed approach of the acoustic scene office B with a source moving from 90°, passing through 0° and ending at −80°.

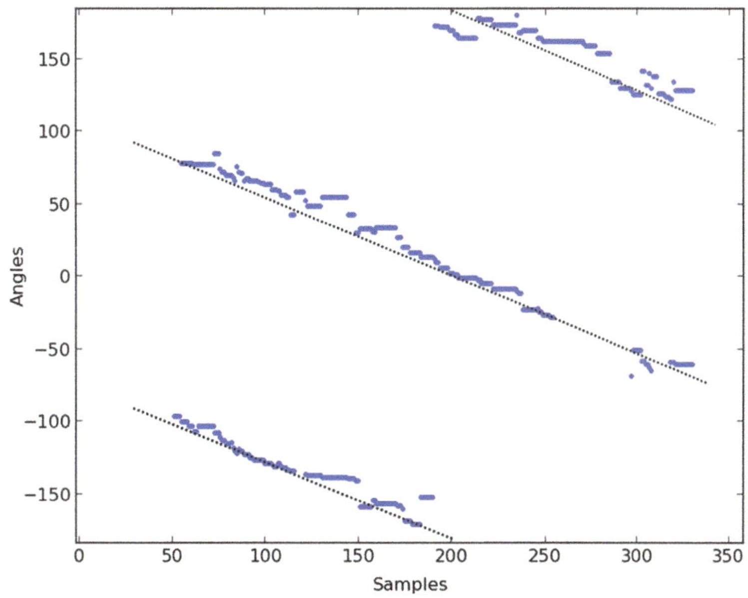

Figure 10 Office B with two sources. A representative plot of the tracking carried out by the proposed approach of the acoustic scene office B with a source moving from 90°, passing through 0° and ending at −80°, and another source moving −90°, passing through 180° and ending at 100°.

in an anechoic chamber to 78.79% in the office environment. This 20% decrease is consistent throughout most of the multi-DOA experiments, which, although significant, can be considered reasonable since the difference between both scenarios is quite notable. Moreover, it is worth mentioning that the system performed admirably in all the scenarios with one source, which is the baseline of a good multi-DOA estimation solution. Likewise, it is important to note that the system was able to carry out multi-DOA estimations with more static sources than microphones in a realistic environment.

In addition, when going from a realistic environment with static sources (Table 2) to one with mobile sources (Table 3), the decrease in the F1 metric is not as significant, which implies that the movement of the sources does not affect the performance of the proposed system as much as the noise/reverberation does (if at all). Furthermore, as evidenced in Figures 9 and 10 and Table 3, although the tracks are considerably 'noisy', it can be seen that the proposed approach is able to track the movement of the sound sources in a realistic acoustic environment and, more importantly, it is doing so with more than one mobile source, using a few-microphone solution, and with a small resource footprint requirement.

It is important to note that the recall score definitely calls for improvement. However, we believe that the proposed system, in the overall sense, has struck a balance that is ideal for real acoustic conditions and speech sources, since noise and reverberation are consistently 'tuned out' by the system while keeping track of the sound sources that are coherent and active. In fact, even if at the moment it may call for improvement, the level of performance the system is showing can currently benefit applications such as complementing the ASR of a service robot serving as a waiter in a restaurant [23], or in the analysis of the acoustic scene of a cocktail party, or in the design of hearing aids.

7 Conclusions

Multiple direction-of-arrival estimation can benefit a large array of audio applications. Current approaches tend to either be able to track multiple mobile sound sources

Table 3 Evaluation results reported in F1-type measure and average errors in an office-type acoustic setting with mobile sources

# of sources	Recall (%)	Precision (%)	F1 (%)	Average errors (in degrees)	Estimated trajectory
1	100	100	100	6.56	90° → 0° → −80°
2	60.00	92.31	72.73	7.10	90° → 0° → −80°
				8.73	−90° → 180° → 100°

using hardware-heavy solutions or use lightweight techniques to track one mobile sound source.

The proposed approach is based upon our earlier work and is now able to track, in reverberant scenarios, multiple moving sound sources while being hardware-light. We believe this is of interest in several areas of applications as well the audio processing community.

The technique carries out multi-DOA estimation by taking advantage of the fact that simultaneous speech sources do not completely overlap over each other and, thus, employs a 'divide and conquer' strategy. It first performs a quick-but-robust single-DOA estimation with single-source sample windows and then proposes sound source tracks from incoming single-DOA estimations. For each track, a Kalman filter is created to estimate the movement of the sound source the track represents.

To evaluate the performance of the proposed approach in a real-life acoustic scene, but in a consistently repeatable manner, the AIRA corpus was created in conjunction with an evaluation methodology based on the F1 score. It presented very good results in terms of the precision metric, thanks to its redundancy measures, the variations of the phase transform while carrying out GCC, and that it is always estimating a direction in the close-to-uniform range of the ITD-to-DOA function. These results confirmed that the proposed approach is able to track more than one source while using a small amount of microphones, in several acoustic scenarios.

For future work, we plan to improve upon the results presented here in terms of the recall metric, as well as in the manner of reducing its noisiness and responsiveness. One way to do this is by carrying an automatic weighting to remove noisy influences during the calculation of the GCC-PHAT. Another is to investigate other types of ITD-to-DOA models, specifically near-field techniques so that the proposed system is usable in more types of applications. In addition, it is of interest to develop a much more sophisticated DOA-to-track association technique, since it is very probable that such associations will break when having sound sources cross each other; however it is also important to maintain the low footprint requirement in such associations, which, unfortunately, is not satisfied by current techniques (such as the ones based on particle filters). We also plan to implement a stricter evaluation methodology for moving sources, providing a more representative evaluation metric for their tracking. Although keeping the number of microphones low is an important topic of this work, we plan to explore the addition of microphones to this solution for the DOA estimation in a 3D space to broaden the range of fields of where this approach can be applied. Moreover, we will explore its implementation in some of the aforementioned applications, specifically in the area of hearing aid design, where the small amount of microphones has been a limiting requirement. Another applicable area is that of Bioacoustics, by carrying census of animal species that interact with each other by song (birds, marine mammals, etc.) in acoustic scenarios that are far more dynamic and complex than the ones presented in this work and do not assume the presence of speech sources.

Competing interests
The authors declare that they have no competing interests.

Acknowledgements
The authors thank the support of CONACYT through the project CB-178673 and PAPIIT-UNAM with the project IN107513. In addition, the authors would like to thank the Golem Group and their lead researcher, Dr. Luis Pineda, who provided many insights and resources to carry these findings out. Furthermore, we would like to specifically thank the support of Varinia Estrada, a member of the Golem Group, as well as Dr. Santiago Jesus Perez Ruiz, of the Laboratorio de Acústica y Vibraciones of the Centro de Ciencias Aplicadas y Desarrollo Tecnológico of the Universidad Nacional Autónoma de México, for their invaluable support in the recollection of the AIRA Corpus.

References

1. R Schmidt, Multiple emitter location and signal parameter estimation. IEEE Trans. Antennas Propagation. **34**(3), 276–280 (1986)
2. L Griffiths, C Jim, An alternative approach to linearly constrained adaptive beamforming. IEEE Transac. Antennas Propagation. **30**(1), 27–34 (1982)
3. MG Smith, KB Kim, DJ Thompson, Noise source identification using microphone arrays. Pro. Inst. Acoust. **29**(5) (2007)
4. R Liu, Y Wang, Azimuthal source localization using interaural coherence in a robotic dog: modeling and application. Robotica. **28**(7), 1013–1020 (2010)
5. AD Horchler, RE Reeve, BH Webb, RD Quinn, in *Sound Localization, 11th International Conference on Advanced Robotics, (ICAR '03)*. Robot phonotaxis in the wild: a biologically inspired approach to outdoor sound localization (Coimbra, Portugal, 30 June 2003), pp. 1749–1756
6. ME Lockwood, DL Jones, RC Bilger, CR Lansing, WD O'Brien Jr, BC Wheeler, AS Feng, Performance of time- and frequency-domain binaural beamformers based on recorded signals from real rooms. J. Acoust. Soc. Am. **115**(1), 379–391 (2004)
7. K Nakamura, K Nakadai, HG Okuno, A real-time super-resolution robot audition system that improves the robustness of simultaneous speech recognition. Adv. Robot. **27**(12), 933–945 (2013)
8. J-M Valin, J Rouat, F Michaud, in *Proc. IEEE/RSJ Int. Conf. Intelligent Robots and Systems*. Enhanced robot audition based on microphone array source separation with post-filter (Sendai, Japan, 28 September 2004), pp. 2123–2128
9. K Nakadai, HG Okuno, H Kitano, in *Proceedings IEEE International Conference on Spoken Language Processing, 2002*. Real-time sound source localization and separation for robot audition, (2002), pp. 193–196
10. F Asano, K Yamamoto, J Ogata, M Yamada, M Nakamura, Detection and separation of speech events in meeting recordings using a microphone array. EURASIP J. Audio Speech Music Process. **2007**(027616) (2007)
11. J-M Valin, S Yamamoto, J Rouat, F Michaud, K Nakadai, HG Okuno, Robust recognition of simultaneous speech by a mobile robot. IEEE Trans. Robot. **23**(4), 742–752 (2007)
12. F Grondin, D Létourneau, F Ferland, V Rousseau, F Michaud, The ManyEars open framework. Autonomous Robots. **34**(3), 217–232 (2013)
13. K Nakadai, T Takahashi, HG Okuno, H Nakajima, Y Hasegawa, H Tsujino, Design and implementation of robot audition system HARK - open source software for listening to three simultaneous speakers. Adv. Robot. **24**(5–6), 739–761 (2010)
14. D Abran-Cote, M Bandou, A Beland, G Cayer, S Choquette, F Gosselin, F Robitaille, DT Kizito, F Grondin, D Letourneau, USB synchronous multichannel audio acquisition system. http://sourceforge.net/projects/eightsoundsusb/files/Technical%20Paper/USB%20Synchronous%20Multichannel%20Audio%20Acquisition%20System.pdf. Accessed 26 June 2014

15. A Canoso, Giving ears to PR2 with 8Sounds and ManyEars. http://www.willowgarage.com/blog/2013/04/12/giving-ears-pr2-8sounds-and-manyears. Accessed 12 April 2013
16. J-M Valin, F Michaud, B Hadjou, J Rouat, in *Proceedings of IEEE International Conference on Robotics and Automation*. Localization of simultaneous moving sound sources for mobile robot using a frequency- domain steered beamformer approach, vol. 1 (New Orleans, LA, USA, 26 April 2004), pp. 1033–1038
17. Z Liang, X Ma, X Dai, Robust tracking of moving sound source using multiple model Kalman filter. Appl. Acoust. **69**(12), 1350–1355 (2008)
18. H Sun, P Yang, H Sun, L Zu, in *Control, Automation and Systems Engineering (CASE), 2011 International Conference On*. Outliers based Double-Kalman filter for sound source localization (Singapore, 30 July 2011), pp. 1–4
19. D Bechler, MS Schlosser, K Kroschel, in *Intelligent Robots and Systems, 2004. (IROS 2004). Proceedings. 2004 IEEE/RSJ International Conference On*. System for robust 3D speaker tracking using microphone array measurements, vol. 3 (Sendai, Japan, 28 September 2004), pp. 2117–21223
20. S Mohan, ME Lockwood, ML Kramer, DL Jones, Localization of multiple acoustic sources with small arrays using a coherence test. J. Acoust. Soc. Am. **123**(4), 2136–2147 (2008)
21. NTN Tho, S Zhao, DL Jones, in *Acoustics, Speech and Signal Processing (ICASSP), 2014 IEEE International Conference On*. Robust DOA estimation of multiple speech sources (Florence, Italy, 4 May 2014), pp. 2287–2291
22. C Rascon, L Pineda, in *IAENG Transactions on Engineering Technologies, Special Issue of the World Congress on Engineering and Computer Science 2012. Lecture Notes in Electrical Engineering*. Multiple direction-of-arrival estimation for a mobile robotic platform with small hardware setup, vol. 247 (Springer Netherlands, 2014)
23. C Rascon, I Meza, G Fuentes, L Salinas, LA Pineda, Integration of the multi-DOA estimation functionality to human-robot interaction. Int. J. Adv. Robot. Syst. **12**(8) (2015)
24. JC Murray, H Erwin, S Wermter, in *AI Workshop on NeuroBotics*. Robotics sound-source localization and tracking using interaural time difference and cross-correlation (Ulm, Germany, 20 September 2004)
25. (D Wang, GJ Brown, eds.), *Computational Auditory Scene Analysis: Principles, Algorithms, and Applications*. (IEEE Press/Wiley, Interscience, 2006). http://www.casabook.org
26. MS Brandstein, HF Silverman, in *Acoustics, Speech, and Signal Processing, 1997. ICASSP-97., 1997 IEEE International Conference On*. A robust method for speech signal time-delay estimation in reverberant rooms, vol. 1 (Munich, Bavaria, Germany, 21 April 1997), pp. 375–3781
27. H Teutsch, *Modal Array Signal Processing: Principles and Applications of Acoustic Wavefield Decomposition. Lecture Notes in Control and Information Sciences*, vol. 348. (Springer, Berlin Heidelberg, 2007)
28. J Benesty, J Chen, Y Huang, *Microphone Array Signal Processing. Springer Topics in Signal Processing*, vol. 1. (Springer, Berlin Heidelberg, 2008)
29. E Gallo, N Tsingos, G Lemaitre, 3D-audio matting, postediting, and rerendering from field recordings. EURASIP J. Adv. Signal Process. **2007**(47970) (2007)
30. S Gerlach, J Bitzer, S Goetze, S Doclo, Joint estimation of pitch and direction of arrival: improving robustness and accuracy for multi-speaker scenarios. EURASIP J. Audio Speech Music Process. **2014**(1) (2014)
31. A Brutti, M Omologo, P Svaizer, Multiple source localization based on acoustic map de-emphasis. EURASIP J. Audio Speech Music Process. **2010**(147495) (2010)
32. A Manikas, C Proukakis, Modeling and estimation of ambiguities in linear arrays. IEEE Trans. Signal Process. **46**(8), 2166–2179 (1998)
33. A Saxena, AY Ng, in *ICRA'09: Proceedings of the 2009 IEEE International Conference on Robotics and Automation*. Learning sound location from a single microphone (IEEE Press Piscataway, 2009), pp. 4310–4315
34. WS Gan, *Acoustical Imaging: Techniques and Applications for Engineers*. (John Wiley & Sons, Ltd, 2012)
35. J-M Valin, F Michaud, J Rouat, D Letourneau, in *Intelligent Robots and Systems, 2003. (IROS 2003). Proceedings. 2003 IEEE/RSJ International Conference On*. Robust sound source localization using a microphone array on a mobile robot, vol. 2 (Las Vegas, USA, 27 October 2003), pp. 1228–12332
36. J-M Valin, F Michaud, J Rouat, Robust localization and tracking of simultaneous moving sound sources using beamforming and particle filtering. Robot. Autonomous Syst. **55**(3), 216–228 (2007)
37. R Chakraborty, C Nadeu, T Butko, Source ambiguity resolution of overlapped sounds in a multi-microphone room environment. EURASIP J. Audio Speech Music Process. **2014**(18) (2014)
38. P Pertila, T Korhonen, A Visa, Measurement combination for acoustic source localization in a room environment. EURASIP J. Audio Speech Music Process. **2008**(278185) (2008)
39. M Wohlmayr, M Kepesi, in *Proceedings of the 8th Annual Conference of the International Speech Communication Association*, ed. by ISCA. Joint position-pitch extraction from multichannel audio (Antwerp, Belgium, 27 August 2007), pp. 1629–1632
40. T Otsuka, K Nakadai, T Ogata, HG Okuno, in *INTERSPEECH'11*. Bayesian extension of MUSIC for sound source localization and tracking (Florence, Italy, 15 August 2011)
41. E Shriberg, A Stolcke, D Baron, in *Proceedings of Eurospeech 2001*. Observations on overlap: findings and implications for automatic processing of multi-party conversation (Scandinavia, Aalborg, Denmark, 3 September 2001), pp. 1359–1362
42. LA Pineda, H Castellanos, J Cuétara, L Galescu, J Juárez, J Llisterri, P Pérez, L Villaseñor, The Corpus DIMEx100: transcription and evaluation. Lang. Res. Eval. **44**, 347–370 (2010)
43. P Davis, JACK connecting a world of audio. http://jackaudio.org. Accessed 27 April 2015
44. C Ris, S Dupont, Assessing local noise level estimation methods: application to noise robust {ASR}. Speech Commun. **34**(1–2), 141–158 (2001)
45. J Huang, T Supaongprapa, I Terakura, F Wang, N Ohnishi, N Sugie, A model-based sound localization system and its application to robot navigation. Robot. Autonomous Syst. **27**(4), 199–209 (1999)
46. KV Ramachandra, *Kalman Filtering Techniques for Radar Tracking*. (Marcel Dekker, Florida, USA, 2000)
47. RE Kalman, A new approach to linear filtering and prediction problems. Trans. ASME–J. Basic Eng. **82**(Series D), 35–45 (1960)
48. R Ruiz-Boullosa, A Perez-Lopez, Some acoustical properties of the anechoic chamber at the Centro de Instrumentos, Universidad Nacional Autonoma de Mexico. Appl. Acoust. **56**(3), 199–207 (1999)
49. CD Manning, P Raghavan, H Schutze, *Introduction to Information Retrieval*. (Cambridge University Press, US, 2008)

7

Practical design of delta-sigma multiple description audio coding

Jack Leegaard[1], Jan Østergaard[1*], Søren Holdt Jensen[1] and Ram Zamir[2]

Abstract

It was recently shown that delta-sigma quantization (DSQ) can be used for optimal multiple description (MD) coding of Gaussian sources. The DSQ scheme combined oversampling, prediction, and noise-shaping in order to trade off side distortion for central distortion in MD coding. It was shown that asymptotically in the dimensions of the resampling, prediction, and noise-shaping filters as well as asymptotically in the quantizer dimensions, all rate-distortion points on the symmetric quadratic Gaussian MD rate-distortion function could be achieved. In this work, we show that this somewhat theoretical framework is suitable for practical low-delay MD audio coding. In particular, we design a practical MD audio coder with two descriptions and provide simulations on real audio data. The simulations demonstrate that even when using low-dimensional noise-shaping, prediction, and resampling filters, it is possible to obtain good quality audio in the presence of packet losses. Simulations on real audio reveal that, contrary to existing designs, it is straightforward to obtain a large number of trade-off points between side distortion and central distortion, which makes the proposed coder suitable for a wide range of applications.

Keywords: Audio coding; Multiple descriptions; Noise shaping; Predictive coding

1 Introduction

There is a growing interest in achieving reliable streaming of high-quality audio over networks for digital audio broadcast services, internet radio, youtube, and similar multimedia streaming services. High-quality streaming can be achieved by using, e.g., error-correcting codes and by allowing large delays, large bandwidths, or dedicated/prioritized networks. However, for certain applications, long delays cannot be tolerated. For example, for interactive services such as voice over IP or musicians playing together via the internet, it is crucial that the delay is kept at a minimum. Indeed, for the latter case, it has been noted that delays less than 5 ms are often required [1].

Conventional broadband connections in homes are generally asymmetric in the sense that their downlink capacity is much greater than their corresponding uplink capacity. While this is good for common internet usage such as browsing, it is not ideal for interactive high-quality streaming services, where instead a more symmetric strategy would be advantageous. To reduce the required bandwidth for audio streaming services, it is common to exploit efficient audio compression methods. The *de facto* standard for lossy compression of music is the family of advanced audio coding (AAC) algorithms, which have been standardized by ISO and IEC as part of the MPEG-2 and MPEG-4 specifications [2-4]. AAC is used for audio compression for digital TV as well as digital audio broadcast (DAB) in several countries. AAC achieves better quality than MP3 and allows for high sampling rates and multiple channels [5]. It is based on the modified discrete cosine transform (MDCT), which is able to provide a high-frequency resolution by using long delays [5]. For low-delay coding, transform coders are generally not as efficient as parametric (model based) coders [6]. Recently, low-delay parametric audio coders based on linear prediction [6,7] and generalized noise-shaped quantization [8] have been proposed. With such techniques, it is possible to achieve delays less than 5 ms while maintaining high-quality audio. In fact, even a few sample delays can be achieved by compromising the efficiency of the perceptual model [6-8].

*Correspondence: jo@es.aau.dk
[1]Department of Electronic Systems, Aalborg University, Niels Jernes Vej 12, Aalborg 9220, Denmark
Full list of author information is available at the end of the article

In order to achieve a certain degree of robustness towards packet dropouts or channel failures, it is possible to use error-correcting codes. If the tolerable delay or bandwidth is large enough, then these codes can be extremely efficient. An alternative is to use joint source-channel coding techniques, where a certain amount of source redundancy is introduced to help the channel/source decoder at the receiving end. A particular case is multiple description (MD) coding [9], where the source is encoded into multiple (partially) redundant packets, which can be decoded independently of each other. Moreover, if several packets are received, then they are able to refine each other and, thus, improve the reconstruction quality. MDs generalize the concepts of repetition coding and layered source coding (i.e., successive refinements), where in the former case, a packet is simply duplicated, and in the latter case, packets are (nearly) independent of each other[a]. In [10], it was shown that a tolerable music quality could be achieved even on unreliable networks having more than 30% packet dropouts by using MDs audio coding between two and four descriptions (packets). The MD audio coding schemes presented in [10] and [11] are both based on transform coders and therefore not able to achieve ultra low delay. Recently, a predictive strategy for high-quality audio MD coding was presented in [12] and a noise-shaped strategy in [8], which are both able to achieve very low delays.

Common for existing MD audio coders are that the audio coding part and the MD part are somewhat separated and only a few trade-off points on the operational MD rate-distortion function can be obtained. On the other hand, in this paper, we use a unified strategy based on predictive coding and noise shaping that allows an almost continuous trade-off between rate and distortions of the descriptions. In particular, in [13], an inherent connection between oversampled delta-sigma quantization and MD coding for Gaussian sources was discovered. By oversampling the signal, a certain amount of redundancy is introduced, and by proper noise shaping of the quantization noise a trade-off between the reconstruction qualities due to receiving different subsets of descriptions is possible, cf. Figure 1. The work in [13] treated white Gaussian signals and noise shaping and was extended to the case of predictive coding and noise shaping of colored Gaussian sources in [14,15].

In this paper, we construct an efficient high-quality low-delay MD coder based on the principles of oversampling, predictive coding, and noise shaping. Specifically, we utilize the theoretic construction proposed for Gaussian sources in [13-15] and show that it naturally extends to real audio signals. We restrict attention to two descriptions and a symmetric setup, where the rate-distortion performance of the individual descriptions are identical. It is worth emphasizing that contrary to Gaussian source coding, it is crucial in audio coding that the temporal envelope is kept smooth when filter parameters are updated. This complicates the transition from theory to practice. The contributions of this paper are therefore dedicated to the practical design of the coder, and we refer the reader to the aforementioned literature for the theoretical foundations. The present paper discusses the design of filters for resampling, prediction, and noise shaping, in addition to coding the parameters[b]. The proposed design is evaluated on several audio signals sampled at 48 kHz, and the performance due to receiving different subsets of descriptions is assessed. Moreover, the proposed coder is simulated in an environment with packet losses. It is shown that good quality music is achievable with delays less than 5 ms.

In comparison, the low-delay noise-shaped coder in [8] reveals a significant reduction of the coding rate, which is mainly due to the inclusion of individual description prediction loops.

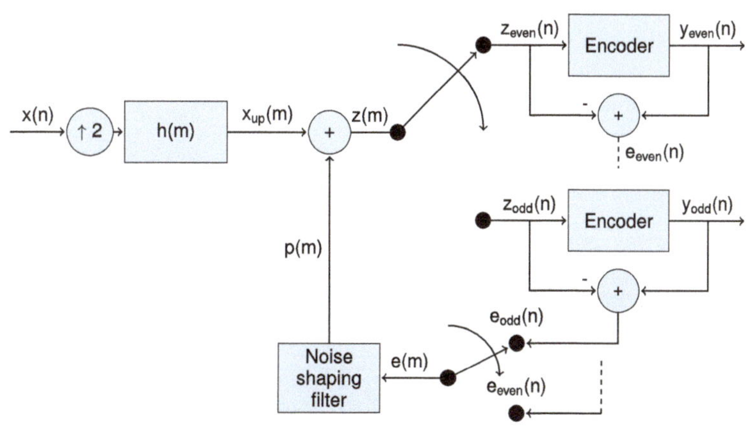

Figure 1 Schematics of the MD noise-shaped predictive encoder [13-15].

2 Background

The MD predictive noise-shaped coder proposed in [13-15] consists of *sampling rate conversion, encoders, noise-shaping, and decoding.* We briefly describe these components below and refer to [13-15] for further details.

2.1 Sampling rate conversion

In the MD noise-shaped predictive coder proposed in [13-15], the original signal $x(n)$ is first oversampled by a factor of two to obtain $x_{up}(m)$ as shown in Figure 1. We use the indices n and m to refer to samples of signals in the original and upsampled domains, respectively. Sampling rate conversion of discrete-time audio signals has been widely studied in the audio engineering literature, cf. [16-18] to name a few. Theoretically, if the signal is bandlimited, then changing its sampling rate is a reversible process, as long as the resulting sample rate is greater than the Nyquist frequency of the signal, i.e., greater than twice the signals' bandwidth.

2.2 Encoder

The upsampled signal is split into even and odd samples (see Figure 1), and the even (odd) stream is fed to the even (odd) encoder. Each encoder can be cast in the framework of 'noisy' single-description prediction, which in case of Gaussian signals was treated in [19]. In [19], it was shown that optimal encoding is achieved with minimum mean square error (MMSE) prediction. This result was used in [14] to show that the optimal encoders should be MMSE prediction filters for the colored 'even' and 'odd' Gaussian signals, respectively. Interestingly, in [15], it was furthermore suggested that the encoders could actually be any existing standard compression schemes. Indeed, in [20], it was proposed to combine delta-sigma quantization with standard JPEG coding schemes to form compression algorithms for MD coding of still images. Similarly, in this work, we could choose to use e.g., standard AAC compression schemes. A key motivation for choosing a standard coder is of course the fact that one avoids the trouble of having to design an efficient audio coder, but perhaps more interestingly, the individual descriptions would then also be completely standard-compliant with existing technology. Only if both descriptions are to be jointly considered at the receiving end, the decoder needs to be slightly altered.

2.3 Noise shaping

We now turn our attention to the feedback filter whose purpose is to perform noise-shaping of the coding noise [21]. As illustrated in Figure 1, the coding noises $e_{\text{even}}(n)$ and $e_{\text{odd}}(n)$ from the two encoders are interlaced to form the error signal $e(m)$ having the same sampling frequency as $x_{\text{up}}(m)$. The error signal $e(m)$ is filtered by the noise-shaping filter and then added to $x_{\text{up}}(m)$ and thereby closing the feedback loop. The purpose of the noise-shaping filter in conventional oversampled quantization (e.g., delta-sigma quantization) is to shape the noise away from the in-band spectrum and thereby reduce the energy of the noise, which is imposed upon the signal [21]. In MD coding, on the other hand, the purpose is to shape the noise so that a proper trade-off is achieved regarding the distortion due to receiving a single packet versus receiving both packets [13]. Indeed, it is illustrative to consider what happens when only a single packet is received. In this case, since we split the signal into even and odd samples, we have in fact downsampled the oversampled signal by a factor of two (without first applying an anti-aliasing filter). But since the source spectrum only covers half the frequency spectrum of $x_{\text{up}}(m)$ (due to oversampling), there will not be any source aliasing. On the other hand, due to interlacing the noise samples, the noise spectrum covers the full frequency range, see Figures Eight and Twelve in [13] for an illustration of the noise spectra. Thus, the noise spectrum will be aliased. In particular, the out-of-band noise spectrum will be aliased and superimposed upon the in-band source + noise spectrum. The effect is that no matter how the noise is shaped, the full noise spectrum will be imposed upon the source spectrum. On the other hand, if both packets are received, the oversampled signal can be recovered without noise aliasing and, thus, it is possible to apply a low-pass filter and get rid of the out-of-band noise. To summarize, on one hand, we would like to minimize the total noise energy in order to reduce the distortion when receiving only a single packet. On the other hand, we would like to put as much noise in the high-frequency spectrum and thereby reduce the amount of noise in the in-band spectrum, in order to minimize the distortion when receiving both packets. It is also interesting to note that the entropy rate of the quantizer, under high-resolution conditions, is independent of the noise-shaping filter and given by the ratio of the power σ_X^2 of the input signal and the power of the excitation noise (the input $e(m)$ to the noise-shaping filter), i.e.,

$$R \approx \frac{1}{2}\log_2\left(\frac{\sigma_X^2}{\sigma_e^2}\right).$$

2.4 Decoder

At the receiver, the received signals are decoded as shown in Figure 2. If only a single packet is received, the corresponding decoder is applied. If both packets are received, the even and odd signals are first individually reconstructed and then interlaced to form an approximation of $x_{\text{up}}(m)$. Then, a low-pass filter is applied to get rid of the out-of-band noise, and the signal is downsampled by two. We note that there is no noise-shaping loop at the decoder.

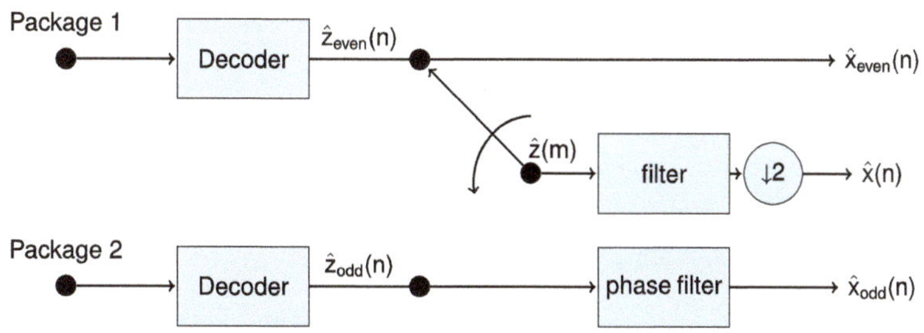

Figure 2 Schematics of the MD noise-shaped predictive decoder [13-15].

3 Practical construction of the MD audio coder

In the following subsections, the design of the individual parts of the coder is described.

3.1 Sampling rate conversion

Ideal oversampling can theoretically be obtained by inserting zeros between every sample of the original signal and then apply an ideal low-pass filter (i.e., convolution by an infinite-length sinc function). The resulting signal $x_{\text{up}}(m)$ has twice as many samples as $x(n)$, and for $m = 2n$ and m even, we have that $x_{\text{up}}(m) = x(n)$ (if we ignore possible integer time-delays due to filtering). Thus, we recover the original signal simply by taking all the even samples of $x_{\text{up}}(m)$. The odd samples of $x_{\text{up}}(m)$ are phase-shifted versions of $x(n)$. Of course, in practice, we cannot use ideal low-pass filtering since this will result in prohibitively large delays. Due to using finite filters, a certain degree of aliasing is unavoidable.

In this work, we use an FIR filter $h(z)$ with linear phase obtained via the window method (Chebyshev) as the interpolation filter [22]. Specifically, we insert zeros between every sample of $x(n)$ and apply the filter to obtain the upsampled signal $x_{\text{up}}(m)$ [16]. Figure 3 shows the performance in MSE as a function of the filter length. The solid lines are for a unit-variance audio signal 'Abba'[c] and the dashed lines (with circles) are for a zero-mean unit-variance white Gaussian signal. As foreshadowed above, the error on the even samples due to resampling is negligible for filter orders greater than $N = 18$. On the other hand, the odd samples are highly affected by the resampling operation, which is due to frequency aliasing[d]. An estimate of the power spectral density (PSD) of the Abba signal is shown in Figure 4. It is clear that it is not flat, and, thus, the impact of aliasing is much less than for the Gaussian case.

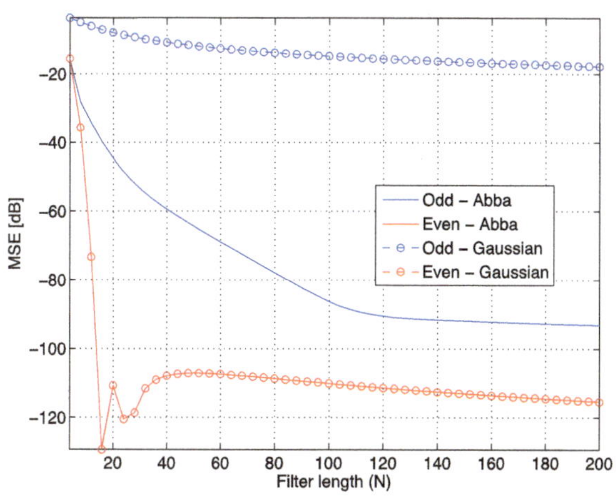

Figure 3 Error due to resampling using finite length filters.

3.2 Closed-loop prediction

Let us momentarily forget about the feedback loop shown in Figure 1 (since this will be treated in the following subsection) and simply consider that $x_{\text{up}}(m)$ is to be split into its even and odd samples, which are denoted by $z_{\text{even}}(n)$ and $z_{\text{odd}}(n)$ in Figure 1. For clarity, we have redrawn the encoder with more details in Figure 5. The even signal $z_{\text{even}}(n)$ is now fed into a compression algorithm (encoder), and the compressed signal $y_{\text{even}}(n)$ is transmitted to the decoder. The odd signal $z_{\text{odd}}(n)$ is processed in

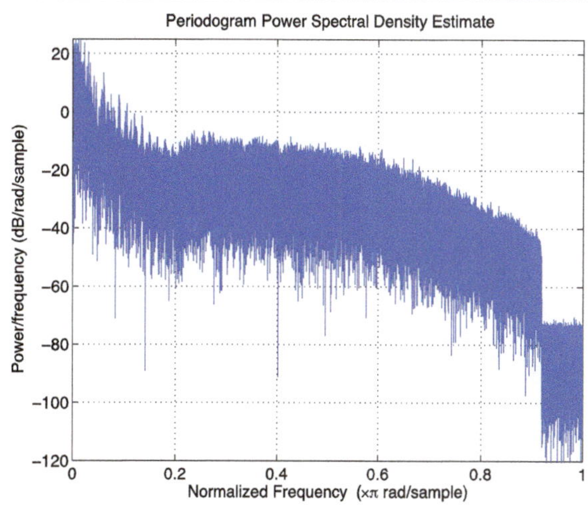

Figure 4 Estimate of the one-sided power spectral density of the music signal Abba.

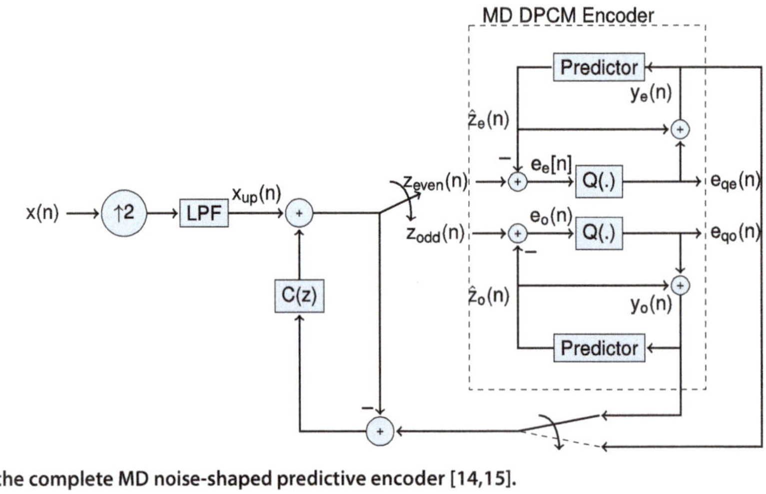

Figure 5 **Schematics of the complete MD noise-shaped predictive encoder [14,15].**

a similar way. Thus, the even samples then constitute one of the packets in the MD coder, and the odd samples constitute the other packet. The number of samples to include in each packet depends upon several factors and will be treated in the sequel.

3.2.1 Linear predictive coding

The encoders will in this work be given by forward linear prediction coding. In particular, in order to encode the even signal $z_{\text{even}}(n)$, we design a linear predictor based on the even unquantized samples $z_{\text{even}}(n)$. We use a forward linear predictor, which as usual is obtained by minimizing the prediction error in the least squares sense, cf. [23] for details. The predictor performs closed-loop prediction, i.e., the quantizer is contained within the prediction loop [19]. To do so, we consider a block of samples and use these for estimating the prediction filter. The filter needs to be encoded and transmitted to the decoder. Thus, there is a trade-off between the rate required for coding the filter coefficients, the update rate of the filter, and the rate required for coding the prediction error. A general approach to choosing a proper rate distribution between model parameters and signal was considered in [24].

3.2.2 Coding the prediction error

Even though the prediction filters are updated only once per block of samples, quantization of the prediction error is performed on a sample-by-sample basis. Thus, we need to use scalar quantization, and for simplicity, we will use scalar uniform quantization [25]. We therefore only need to design the proper step-size Δ of the quantizer. To obtain the bitrates of the coder, we first run the predictor using a fixed step size Δ on a large data set of mixed audio having a sampling frequency of 48 kHz. Then, a scalar (Huffman) entropy coder is designed on the quantized output of the predictor [26]. Thus, we are using a static and memoryless entropy coder. Finally, the predictor is tested on an audio segment (in this case, it consists of jazz music), which is not part of the training material. Figure 6 shows the resulting coding rate due to using a scalar uniform quantizer with a step-size Δ followed by a scalar (Huffman) entropy coder. The corresponding MSE due to changing the step size of the quantizer is shown in Figure 7. In these simulations, we update the two linear predictive coding (LPC) filters once in each block of 128 samples. Since the audio signals have a sampling frequency of 48 khz, then if the bitrate is say 5 bits/sample, the resulting rate for coding the prediction error is 240 kbps per packet.

3.2.3 Predictor order

In predictive audio coding, it is common to use predictors of orders greater than 10 [6]. However, in our case, the

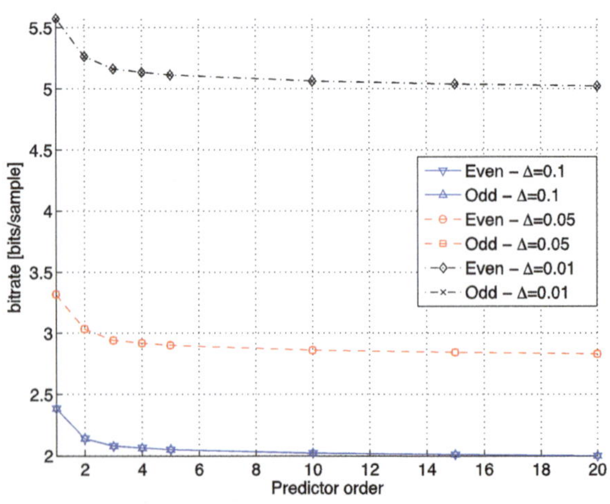

Figure 6 **Bitrates due to forward linear prediction followed by encoding of the prediction error.**

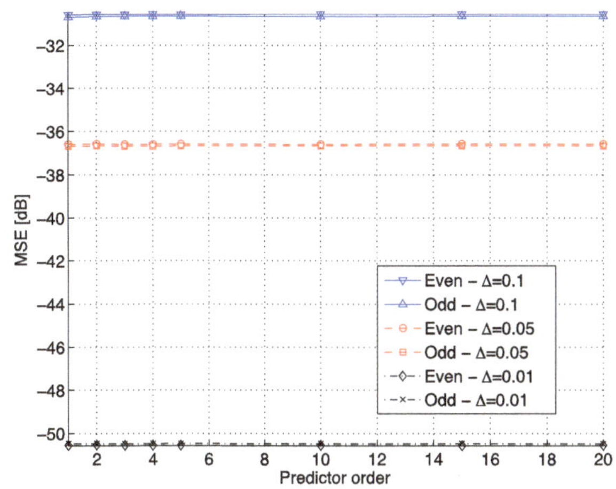

Figure 7 MSE due to forward linear prediction and encoding the prediction error.

outer loop introduces noisy feedback, which to a certain degree reduces the predictor capabilities. For example, let $\Delta = 0.01$, and construct a 10th-order noise-shaping filter using the design in Equation 1 provided in the next subsection. Then, the performance in terms of rate and distortion of the predictor as a function of its order is shown in Figures 6 and 7. The bitrates illustrated in the figures correspond to the rates required for encoding the prediction residuals. The actual predictor coefficients have not been coded in these simulations. The simulations are repeated for a wide range of predictor orders. It may be noticed that increasing the order from 1 to 5 significantly decreases the required bitrate for coding the residual, whereas using an order above 10 does not lead to significant improvements. On the other hand, the resulting MSE is approximately unaffected by the predictor order.

3.3 Noise shaping

The purpose of the noise-shaping filter is to shape the quantization noise appropriately in the frequency domain [21]. Ideally, the frequency response of the noise-shaping filter should be a two-step function, which in the in-band frequency range has power δ^{-1} and in the out-of-band frequency range has power δ [13]. Thus, if both descriptions are received, one is able to filter out the out-of-band noise and thereby obtain a resulting noise power that is proportional to δ^{-1}. On the other hand, if only a single description is received, then due to aliasing, the resulting noise power is proportional to $\delta + \delta^{-1}$. Furthermore, fixing the levels as δ and δ^{-1}, respectively, guarantees that their geometric mean is one, which basically fixes the coding rate while allowing one to trade-off side distortion for central distortion [13].

In practice, we need to approximate the two-step response using a finite length feedback filter. The optimal design of the noise-shaping filter $c(z)$ for any filter order p was given in [13] as:

$$\mathbf{c} = -\left(\mathbf{G} + 2\lambda \mathbf{I}\right)^{-1} \mathbf{g} \tag{1}$$

where $\mathbf{c} = (c_1, \ldots, c_p)^T$ are the filter coefficients, $\mathbf{g} = (\mathrm{sinc}(1/2), \mathrm{sinc}(2/2), \ldots, \mathrm{sinc}(p/2))^T$, and $\mathbf{G}$ is the matrix with entries $\mathbf{G}_{i,j} = \mathrm{sinc}((i-j)/2), i,j = 1, \ldots, p$. In (1), λ denotes the trade-off between central and side distortion. Choosing $\lambda = 1$ indicates that the central and side distortion are given the same weight. In this case, the central distortion will on average be around 3 dB smaller than the side distortion. On the other hand, choosing $\lambda \ll 1$ reduces the central distortion at the price of increasing the side distortion. This is illustrated in Figures 8 and 9 for the case of $p = 10$ and $p = 30$, respectively. In these simulations, $\Delta \in \{0.01, 0.05, 0.1\}$. It may be noticed that larger Δ yields larger distortions as expected. It can also be seen that for large λ, the central distortion is approximately $-10\log_{10}(\lambda)$ dB smaller than the side distortion.

3.3.1 *Coding the predictor coefficients*

There exists a vast amount of literature on efficient encoding of LPC coefficients, cf. [27] and the references therein. Here, we will use a common approach and transform the LPC coefficients to line spectral frequencies (LSF) coefficients [28]. This is done partly due to the fact that it is easy to guarantee stability of the inverse filter in the LSF domain and partly due to fact that they are easier to encode efficiently [29]. From Figure 6, it is clear that using a filter order above 10 does not significantly decrease the coding rate for the prediction error over what is possible with a 10th order filter. We therefore proceed using a 10th order predictor, which is first converted to 10 LSF coefficients. The 10 LSF coefficients are then quantized using a scalar quantizer with a step size of $\pi/64$. The quantized coefficients are then split into three subvectors of length

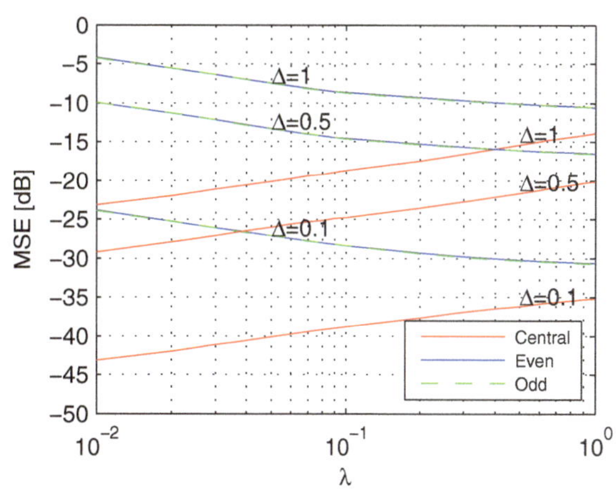

Figure 8 Performance for the Abba signal when using a noise shaping filter of order $p = 10$.

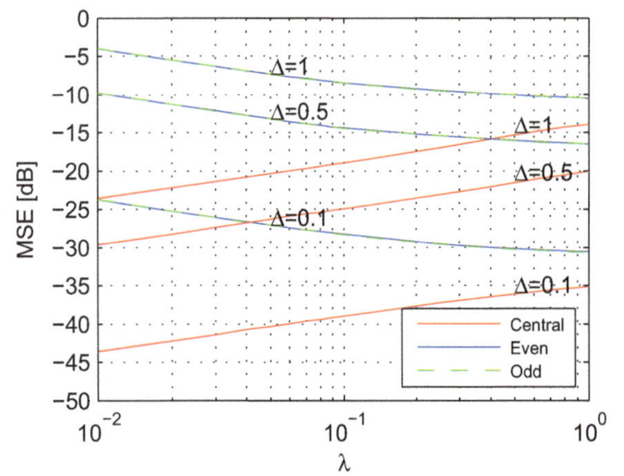

Figure 9 Performance for the Abba signal when using a noise shaping filter of order $p = 30$.

3, 3, and 4, respectively. Finally, each subvector is independently vector Huffman coded. The resulting bitrates are shown in Tables 1 and 2, where LSF_i denotes the ith subvector. In these simulations, the window size of the predictor is 128 samples. It may be noticed from Table 2 that the average coding rate is approximately 20.8 bits per LSF vector. Since the sampling frequency is 48 kHz and the block size is 128 samples, the resulting average bitrate for coding the LSF vectors is 7.8 kbps per packet.

3.4 Decoding

Decoding the received audio packets is more challenging than in conventional audio coding since the encoder does not know which packets (if any) the decoder receives. In fact, since the encoder forms two packets, the decoder will at each time instance enter one of 16 different states, which depends upon its previous state, see Table 3. If the decoder remains in one of the states on the 'diagonal' in Table 3, i.e., states 1, 6, 11, or 16, it is straightforward to guarantee a smooth transition between blocks. The problem occurs when the decoder switches to the other states. We solve these issues in the sequel.

Table 1 Bitrates for coding the (even/odd) subvectors of the LSF vector

Audio	LSF_1	LSF_2	LSF_3
Jazz	6.63 / 6.62	6.50 / 6.47	5.63 / 5.58
Harpsichord	6.77 / 6.77	7.29 / 7.25	8.11 / 8.14
Speech	7.99 / 8.00	7.35 / 7.32	6.97 / 7.01
Pop	6.98 / 6.98	5.80 / 5.78	5.26 / 5.23
Rock	6.92 / 6.93	7.86 / 7.85	7.85 / 7.86
Average	7.06 / 7.06	6.96 / 6.93	6.77 / 6.76

LSF_i denotes the ith subvector.

Table 2 Bitrates (in bits/vector) for coding the (even/odd) LSF vectors

Audio	LSF vector
Jazz	18.76 / 18.67
Harpsichord	22.17 / 22.17
Speech	22.32 / 22.33
Pop	18.04 / 17.99
Rock	22.64 / 22.63
Average	20.79 / 20.76

3.4.1 State 1

This is the case with no packet dropouts. The decoder simply processes the two descriptions as described in Subsection 2.4. Both descriptions are first individually reconstructed, then interlaced, and finally downsampled to produce a single high-quality reconstruction. Thus, the states of the LPC filters at the side decoders as well as the state of the low-pass filter at the central decoder are all properly updated, which results in smooth transitions between consecutive blocks.

3.4.2 States 2 and 3

Assume that the decoder is in state 1 (i.e., it has received both packets) but then in the next time slot it only receives the odd packet and thereby enters state 3^e. Then, no new even LPC filter coefficients are received, and the even LPC filter state (memory) is therefore not properly updated. The odd samples are phase-shifted by a 1/2 sample compared to the original signal, and the odd predictor is therefore not identical to the even predictor. Moreover, since both packets are not received, the low-pass filter at the central decoder is not applied and its state (memory) is therefore not updated.

Figure 10 illustrates the effect on the reconstructed signal due to the decoder switching from state 1 to 3 at sample 128 and from state 3 to 1 at sample 640. In this example, both packets have been received prior to the frame beginning at the first vertical line after which the even packet is lost and only the odd packet is therefore received. The second vertical line denotes the point where both packets are again received. In order to construct the central reconstruction and thereby make sure that the state of low-pass filter at the central decoder is updated,

Table 3 The 16 (next) states the decoder can enter depending upon the decoders (current) state information

Current/next	Central	Even	Odd	None
Central	1	2	3	4
Even	5	6	7	8
Odd	9	10	11	12
None	13	14	15	16

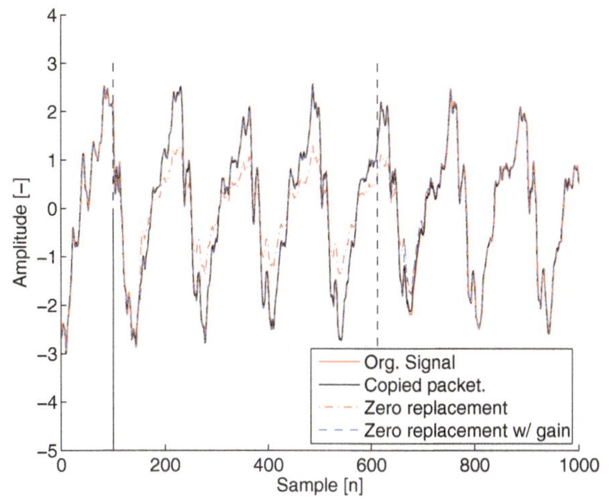

Figure 10 Illustration of the boundary effects due to the decoder switching.

a naive approach is simply to replace the lost packet by zeros. However, since this effectively means that only a single packet is used, the central reconstruction suffers from a decrease in energy as can be seen in Figure 10 (the dash-dotted line).

An obvious solution is to scale the received Odd packet by two and thereby counteract the loss of energy. Unfortunately, while less severe, an audible notch around sample 152 is still present in the reconstructed signal (illustrated by the dashed line in Figure 10), see also Figure 11. To solve the issue, we let the even packet be equal to the odd packet, which yields a smooth boundary transition (illustrated by the black line in Figure 10). In this case, the even LPC filter states are updated with sample values closer to the desired. Interestingly, while the latter method (packet copying) yields more visually and acoustically pleasing boundary transitions, the former method (zeroing even packet and scaling odd packet) actually results in a smaller overall MSE, i.e., -38.8 versus -38.3 dB, respectively, for the case of 1% packet losses. In the example described above, we used LPC filter orders of 5, prediction block sizes of 512 samples, a resampling filter order of 200, noise-shaping filters of order 10, $\lambda = 1/100$, and $\Delta = 1/100$.

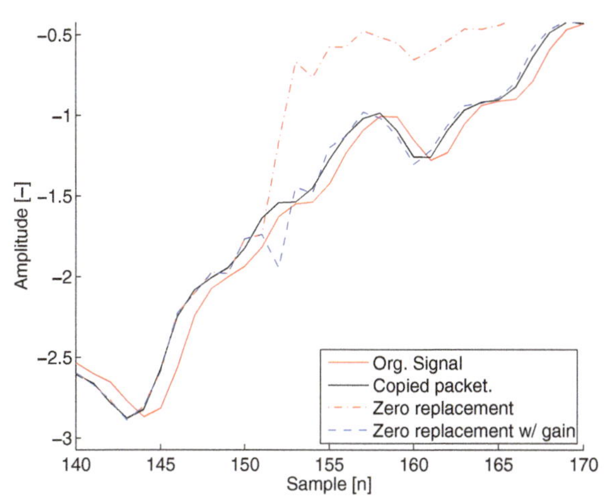

Figure 11 Same setup as in Figure 10 but here zoomed-in on small interval.

3.4.3 State 13

In this state, all buffers are zero, which corresponds to the initial state of the system. The decoder is then operated as in state 1.

3.4.4 States 14 and 15

As was the case for state 13, all buffers are also zero here. If the current state is 14 (15), the decoder is then in the next state operated as in state 2 (3).

3.4.5 States 4, 8, 12, and 16

In these states, no packets are received by the decoder. We then simply replace both packets by zeros and update the states of the LPC filters and low-pass filter accordingly.

4 Simulation study

In this section, we provide simulation studies of the proposed coder. We simulate an environment with packet losses of 0.1%, 1%, and 10%. We restrict the quantization step sizes to $\Delta \in \{0.01, 0.05\}$, the block size upon which the predictor is used to $\{64, 128, 256, 512, 1024, 2048\}$, and the LPC filter order to $p_{\text{lpc}} \in \{5, 10\}$. Finally, in all simulations, the low-pass filters used for resampling are of order 200, the noise-shaping filter is of order 10, and the noise-shaping ratio $\lambda = 0.01$.

4.1 Study 1

In this study, we quantize the residual but we do not quantize the predictor (LPC) coefficients. The test data consists of five audio segments containing *rock, jazz, pop, speech*, and *harpsichord* music, respectively. Each segment is sampled at 48 kHz and with a duration of 10 s.

Table 4 Relationship between the ITU-R 5-grade scale and ODG [30]

Impairment	ITU-R 5-grade scale	ODG
Imperceptible	5.0	0.0
Perceptible but not annoying	4.0	-1.0
Slightly annoying	3.0	-2.0
Annoying	2.0	-3.0
Very annoying	1.0	-4.0

Table 5 Average ODG at 0.1% packet losses

Block size	64	128	256	512	1,024	2,048
$\Delta = 0.01$						
$p_{\mathrm{lpc}} = 5$	-0.77	-0.82	-0.81	-0.22	-0.14	-0.20
$p_{\mathrm{lpc}} = 10$	-0.18	-0.21	-0.21	-0.20	-0.16	-0.18
$\Delta = 0.05$						
$p_{\mathrm{lpc}} = 5$	-1.54	-1.44	-1.43	-1.02	-0.99	-1.04
$p_{\mathrm{lpc}} = 10$	-1.06	-0.95	-0.95	-1.04	-1.06	-1.12

Table 7 Average ODG at 10% packet losses

Block size	64	128	256	512	1,024	2,048
$\Delta = 0.01$						
$p_{\mathrm{lpc}} = 5$	-2.77	-2.16	-1.20	-0.87	-0.60	-0.60
$p_{\mathrm{lpc}} = 10$	-2.61	-1.88	-1.45	-1.08	-0.72	-0.67
$\Delta = 0.05$						
$p_{\mathrm{lpc}} = 5$	-3.31	-3.00	-2.49	-2.44	-2.19	-2.16
$p_{\mathrm{lpc}} = 10$	-3.25	-2.97	-2.93	-2.85	-2.61	-2.35

We use objective difference grades (ODG) instead of MSE in order to better reflect the perceived quality of the reconstructed audio signals. For an explanation of the relationship between the ITU-R 5-grade scale and ODG, see Table 4 and [30]. To obtain the ODG scores, we use a Matlab implementation of the PEAQ standard [31]. The resulting ODG are shown in Tables 5, 6, and 7. In the tables, we have averaged the ODG scores over all audio segments.

From the tables, it is clear that decreasing the packet loss rate or the step size of the quantizers increases the quality as expected. It is also interesting to note that using a longer block size appears to improve the performance.

4.2 Study 2

We now quantize as well as entropy code the residual and the LPC coefficients and will therefore be able to obtain bitrates as well as ODG scores. For training the entropy coders, we use a collection of different music genres constituting about 5 min of audio sampled at 48 kHz. The test data is the same as above and is not part of the training data. The resulting bitrates (expressed in bits/sample) for the even and odd descriptions are shown in Figure 12. The corresponding ODGs are shown in Figures 13 and 14 for $\Delta = 0.05$ and $\Delta = 0.01$, respectively. Interestingly, the bitrate (per sample) as well as the ODG are improving as a function of the block size upon which the predictor is applied. Intuitively, one would think that a fixed-order predictor would be better on shorter segments of the signal. We ascribe this phenomenon to the fact that the performance of the current predictor depends upon the predictor applied in the previous block due to the filter's memory (i.e., we reuse the state of the past predictor). Thus, for short blocks, a substantial part of the prediction of the block is influenced by the history of the previous predictor. This phenomenon is particularly pronounced in the case of large packet-loss rates, where the ODG is significantly improved by going from block sizes of, e.g., 64 to 512 samples.

4.3 Comparison to existing works

It is interesting to compare the performance of the proposed coder to the noise-shaped MD coder presented in [8]. The coder in [8] is based upon the principle of moving horizon (MH) estimation, which is also known as model predictive control when applied in closed-loop control [32]. The scheme in [8] does not use source prediction, and it is therefore expected to perform worse than the proposed coder, which combines noise shaping with source prediction. It is also important to note that the scheme in [8] is of ultra low delay, which means that it can operate on small block sizes. Indeed, in the present simulation, we use a block size of one sample. Instead of oversampling, the MDs in [8] are based on the index assignment construction derived in [33]. The simulation results are presented

Table 6 Average ODG at 1% packet losses

Block size	64	128	256	512	1,024	2,048
$\Delta = 0.01$						
$p_{\mathrm{lpc}} = 5$	-0.99	-0.94	-0.40	-0.46	-0.43	-0.37
$p_{\mathrm{lpc}} = 10$	-0.42	-0.43	-0.32	-0.35	-0.36	-0.34
$\Delta = 0.05$						
$p_{\mathrm{lpc}} = 5$	-1.87	-1.81	-1.73	-1.32	-1.17	-1.09
$p_{\mathrm{lpc}} = 10$	-1.50	-1.39	-1.23	-1.28	-1.25	-1.16

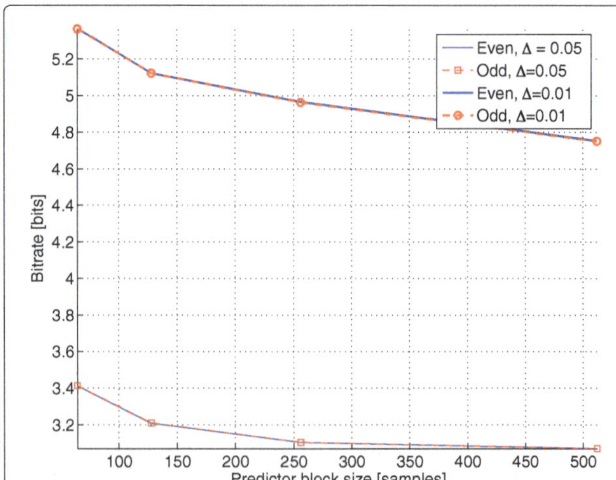

Figure 12 Bitrates as a function of predictor block size and for $\lambda = 1/100$.

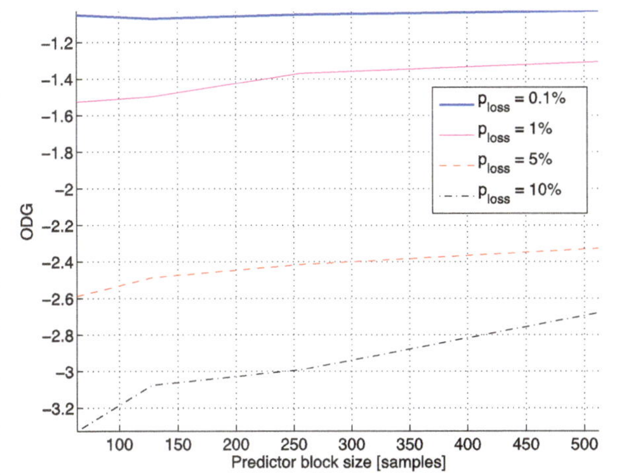

Figure 13 ODG as a function of predictor block size for $\Delta = 0.05$ and for $\lambda = 1/100$.

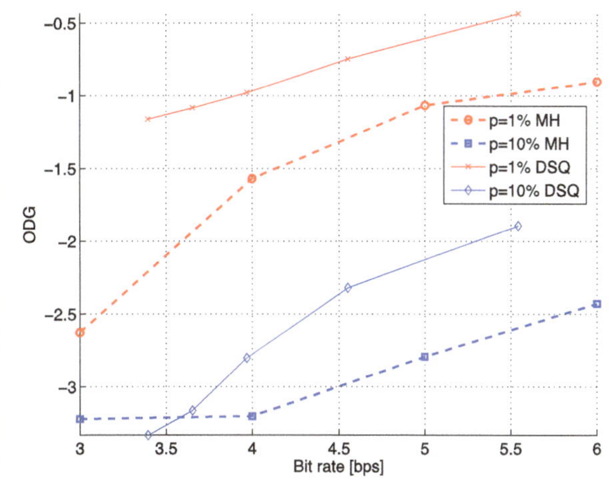

Figure 15 ODG as a function of bitrates for the proposed DSQ scheme and the MD scheme of [8].

in Figure 15. The *jazz* music signal has been used, and two packet loss scenarios have been simulated, a high loss (10% packet losses) and a low loss (1% packet losses) scenario. For the proposed coder, we vary Δ in the interval 0.01 to 0.05 in steps of 0.01. The total bitrate consists of the rates required for coding the LSF coefficients as well as the prediction residual. It can be see in the figure that the proposed coder is able to efficiently exploit its prediction loops and thereby reduce the bitrate over what is possible with the MH design. In these simulations, the proposed coder uses a block size of 64 samples for the prediction. Further improvement is possible by increasing the block size.

5 Conclusions

We presented a practical design of a low-delay MD audio coder, which is able to provide a certain degree of robustness towards packet losses. The proposed coder combined oversampling and noise shaping with source prediction. The oversampling process creates two source descriptions in order to counteract possible packet losses on the network. The prediction loop removes source redundancy and thereby reduces the coding rate, whereas the noise-shaping process controls the distortion due to receiving subsets of the descriptions. The quantized prediction residual was entropy coded using a static and memoryless Huffman coder. In practical simulations on real audio, it was shown that it is enough to use LPC filters of order 10 (estimated from blocks of 64 samples), noise-shaping filters of order 10, resampling filters of order 200, and bitrates of approximately 4 bits per sample (per description) in order to achieve good quality (ODG better than -1) music in the presence of 1% packet losses.

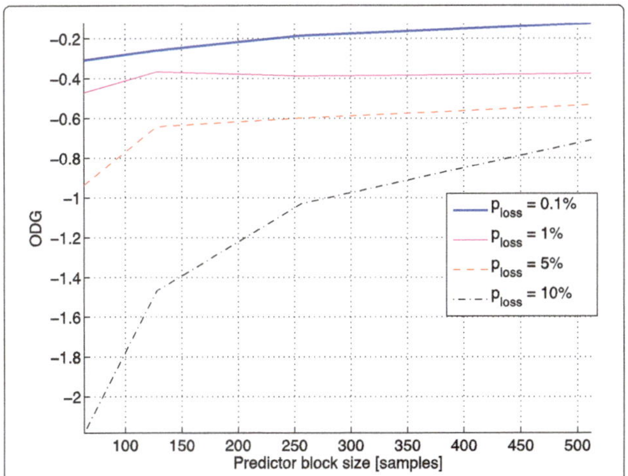

Figure 14 ODG as a function of predictor block size for $\Delta = 0.01$ and for $\lambda = 1/100$.

Endnotes

[a]In layered source coding, the source is usually split into a base layer and at least one refinement layer. While the base layer can be used by itself, the refinement layers are usually no good without the base layer.

[b]For reproducibility, the complete source code for the proposed coding scheme is electronically available at http://kom.aau.dk/~jo.

[c]The audio signal 'Abba' is a 10-s clip of the song 'Head Over Heals' by Abba - sampled at 44.1 kHz.

[d]In order to correctly estimate the error, we need to correct the phase shift of the odd samples. This is done by once more filtering the odd samples with the same filter. Of course, for subjective listening tests, we do not need to correct the phase.

[e]The effect of receiving different numbers of descriptions from frame to frame, corresponds in some sense to (noisy) non-uniform sampling in MDs, cf. [34].

Competing interests
The authors declare that they have no competing interests.

Authors' information
The work of J. Leegaard was performed while he was affiliated with the Department of Electronic System, Aalborg University. He is now with the Department of Architecture, Design and Media Technology, Aalborg University.

Author details
[1]Department of Electronic Systems, Aalborg University, Niels Jernes Vej 12, Aalborg 9220, Denmark. [2]Department of Electrical Engineering-Systems, Tel Aviv University, Ramat Aviv, Tel Aviv 69978, Israel.

References

1. C Chafe, M Gurevich, G Leslie, S Tyan, Effect of time delay on ensemble accuracy, in *Proc Intl. Soc. Musical Acoustics; Nara*, (2004)
2. International Standard ISO/IEC 11172-3 (MPEG), Information technology — coding of moving pictures and associated audio for digital storage media up to about 1.5 mbit/s. Part 3: Audio (1993)
3. International Standard ISO/IEC 13818-7, Information technology – generic coding of moving pictures and associated audio information – Part 7: Advanced Audio Coding (AAC) (2006)
4. International Standard ISO/IEC 14496-3:2005/Amd 2, MPEG 4 Audio profile - high efficiency advanced audio coding (2006)
5. M Bosi, RE Goldberg, *Introduction to Digital Audio Coding and Standards* (Kluwer Academic Publishers, 2003)
6. GDT Schuller, B Yu, D Huang, B Edler, Perceptual audio coding using adaptive pre-and post-filters and lossless compression. IEEE Trans. Speech Audio Process. **10**(6), 379–390 (2002)
7. G Simkus, M Holters, U Zoler, Ultra-low delay lossy audio coding using DPCM and block companded quantization, in *Australian Communications Theory Workshop (AusCTW)* (2013), pp. 43–46
8. J Østergaard, DE Quevedo, J Jensen, Real-time perceptual moving-horizon multiple-description audio coding. IEEE Trans. Signal Process. **59**(9), 4286–4299 (2011)
9. VK Goyal, Multiple description coding: compression meets the network. IEEE Signal Process. Mag. **18**(5), 74–93 (2001)
10. J Østergaard, OA Niamut, J Jensen, R Heusdens, Perceptual audio coding using n-channel lattice vector quantization, in *IEEE Int. Conf. Acoustics, Speech, and Signal Processing, vol. 5; Toulouse* (2006), pp. 197–200
11. R Arean, J Kovacevic, VK Goyal, Multiple description perceptual audio coding with correlating transform. IEEE Trans. Speech Audio Process. **8**, 140–145 (2000)
12. G Schuller, J Kovacevic, F Masson, VK Goyal, Robust low-delay audio coding using multiple descriptions. IEEE Trans. Speech Audio Process. **13**, 1014–1024 (2005)
13. J Østergaard, R Zamir, Multiple description coding by dithered delta-sigma quantization. IEEE Trans. Inform. Theor. **55**(10), 4661–4675 (2009)
14. Y Kochman, J Østergaard, R Zamir, Noise-shaped predictive coding for multiple descriptions of a colored gaussian source, in *IEEE Data Compression Conference (DCC)* (Snowbird Utah, 2008), pp. 362–371
15. J Østergaard, Y Kochman, R Zamir, Colored gaussian multiple descriptions: spectral-domain characterization and time-domain design. Submitted to IEEE Transactions on Information Theory (2010). Electronically available on arXiv.org: http://arxiv.org/abs/1006.2002
16. RE Crochiere, LR Rabiner, Interpolation and decimation of digital signals — a tutorial review. Proc. IEEE **69**(3), 300–331 (1981)
17. JO Smith, P Gossett, A flexible sampling-rate conversion method, in *Proceedings of the International Conference on Acoustics, Speech, and Signal Processing; San Diego* (1984)
18. AJ Russel, PE Beckmann, Efficient arbitrary sampling rate conversion with recursive calculation of coefficients. IEEE Trans. Signal Process. **50**, 854–865 (2002)
19. R Zamir, Y Kochman, U Erez, Achieving the gaussian rate-distortion function by prediction. IEEE Trans. Inform. Theor. **54**(7), 3354–3364 (2008)
20. M Palgy, J Østergaard, R Zamir, Multiple description image/video compression using oversampling and noise shaping in the DCT domain, in *IEEE 26th Convention of Electrical and Electronics Engineers in Israel* (Eilat Israel, 2010)
21. SK Tewksbury, RW Hallock, Oversampled, linear predictive and noise-shaping coders of order $n > 1$. IEEE Trans. Circ. Syst. **CAS-25**(7), 436–447 (1978)
22. TW Parks, JH McClellan, Chebyshev approximation for nonrecursive digital filters with linear phase. IEEE Trans. Circ. Theor. **ct-19**, 189–194 (1972)
23. D O'Shaughnessy, Linear predictive coding. IEEE Potentials **7**(1), 29–32 (1988)
24. J Klejsa, WB Kleijn, Rate distribution between model and signal for multiple descriptions, in *IEEE International Conference on Acoustics, Speech and Signal Processing; Taipei*, (2009), pp. 2489–2492
25. RM Gray, DL Neuhoff, Quantization. IEEE Trans. Inform. Theor. **44**(6), 2325–2383 (1998)
26. DA Huffman, A method for the construction of minimum-redundancy codes. Proc. IRE **40**(9), 1098–1101 (1952)
27. WB Kleijn, KK Paliwal (eds.), *Speech Coding and Synthesis, 1st edn.* (Elsevier, 1995)
28. F Itakura, Line spectrum representation of linear predictor coefficients of speech signals. J. Acoust. Soc. Amer. **57** (1975)
29. FK Soong, B Juang, Optimal quantization of LSP parameters. IEEE Trans. Speech Audio Process. **1**, 15–24 (1993)
30. ITU-R Recommendation BS.1387, Perceptual Evaluation of Audio Quality (PEAQ) (1998)
31. P Kabal, An examination and interpretation of itu-r bs.1387: perceptual evaluation of audio quality. Technical report, McGill University. Version 2 (2003)
32. GC Goodwin, MM Seron, JAD Dona, *Constrained Control and Estimation: An Optimisation Approach.* (Springer, 2005)
33. J Østergaard, J Jensen, R Heusdens, n-channel entropy-constrained multiple-description lattice vector quantization. IEEE Trans. Inform. Theor. **52**(5), 1956–1973 (2006)
34. A Mashiach, J Østergaard, R Zamir, Sampling versus random binning for multiple descriptions of a bandlimited source, in *IEEE Information Theory Workshop; Seville*, (2013)

8

Basis approach to estimate the instantaneous frequencies in multicomponent AM-FM signals

Mohammadali Sebghati[1], Hamidreza Amindavar[1*] and James A Ritcey[2]

Abstract

In this paper, an analytical approach to estimate the instantaneous frequencies of a multicomponent signal is presented. A non-stationary signal composed of oscillation modes or resonances is described by a multicomponent AM-FM model. The proposed method has two main stages. At first, the signal is decomposed into its oscillation components. Afterwards, the instantaneous frequency of each component is estimated. The decomposition stage is performed through the basis expansion exploiting orthogonal rational functions in the complex plane. Orthogonal rational bases are generalized to expand linear time-varying systems. To decompose the non-stationary signal, its equivalent time-varying system is sought. The time-varying poles of this system are required to construct appropriate basis functions. An adaptive data segmentation algorithm is provided for this purpose. The effect of noise is scrutinized analytically and evaluated experimentally to verify the robustness of the new method. The performance of this method in extraction of embedded instantaneous frequencies is asserted by simulations on both synthetic data and real-world audio signal.

Keywords: Instantaneous frequency estimation; AM-FM modeling; Multicomponent signals; Oscillation modes; Orthogonal rational functions

1 Introduction

Non-stationary signals which are a compound of constituents with time-varying amplitudes and frequencies can be characterized by amplitude-modulated frequency-modulated (AM-FM) models. This modeling is attended to express genuine signals in communications [1,2], acoustic and speech processing [3-6], biomedical signal processing [7], and image processing [8]. To estimate instantaneous amplitudes (IAs) and instantaneous frequencies (IFs) embedded in a multicomponent signal [9], the first step is decomposing it into oscillation components. This procedure is termed 'demodulation', 'separation' or 'decomposition' in the literature. Each component should represent a valid oscillation for which the definition of instantaneous frequency is physically meaningful [10].

All methods of multicomponent AM-FM signal decomposition and parameter estimation can be categorized into non-parametric and parametric methods. One class of non-parametric approaches is based on the joint time-frequency processing. Widespread time-frequency distributions (TFDs) such as short-time Fourier transform (STFT), Wigner-Ville distribution, and Choi-Williams distribution are employed [11]. These methods are limited by the well-known compromise between time and frequency resolution. Moreover, the cross-terms appear troublesome. The energy separation algorithm (ESA) which uses a non-linear differential operator, called Teager-Kaiser energy operator (TKEO), is another method [12]. The energy separation algorithm which tracks the energy of the source producing the signal is originally applicable for monocomponent AM-FM signals. Nevertheless, they are modified for application in multicomponent cases by designing a bank of filters. Multiband-ESA (MESA) that consists of bandpass filtering followed by monocomponent energy separation is introduced based on this concept [13]. The separation performed by bandpass filtering is proper when components are spectrally far enough.

Huang et al. [14] proposed an iterative technique known as the empirical mode decomposition (EMD). This technique is an algorithmic way to extract oscillation modes embedded in the signal, named intrinsic mode functions

*Correspondence: hamidami@aut.ac.ir
[1] Electrical Engineering Department, Amirkabir University of Technology, Tehran 15914, Iran
Full list of author information is available at the end of the article

(IMFs). Each IMF gives a valid IF which is estimated by applying the Hilbert transform (HT) [14]. The proposed algorithm for the implementation of EMD called the sifting process showed several drawbacks such as sensitivity to perturbation and mode mixing problem [15]. To overcome these difficulties, the original EMD is modified, and a new algorithm is developed which is named ensemble empirical mode decomposition (EEMD) [15]. This method is indeed the iteration of EMD for noise-added signals. In each iteration, controlled white noise is added to the data, and EMD is applied. Each individual trial may generate noisy results, but the noise is canceled out by taking the average of the results. Thus, true solution is the ensemble mean of enough trials. Although each trial produces a set of IMFs, the sum of IMFs is not necessarily an IMF. An empirical solution for this issue is suggested in [15]. EMD and EEMD methods suffer from the lack of analytic foundation. Some research has attempted to establish and improve analytical aspects of these empirical approaches [16]. In [17], an alternative algorithm for EMD is introduced based on iterating certain filters, such as Toeplitz filters. The results of iterative filtering are similar to those of the conventional sifting process. Although the authors of [17] do not claim superiority for their method, it lays down a mathematical framework for an alternative approach to EMD. The convergence of iterative filtering EMD is studied in [18]. A variant of EMD to decompose multiscale data is proposed in [19]. This work provides some theoretical understanding of EMD for a class of multiscale data and introduces two algorithms, Newton-Raphson-based EMD and ODE-based EMD, as the variations of the sifting process. The decomposition of multiscale data based on EMD is pursued in [20] inspired by the compressed sensing theory. The sparsest representation of multiscale data is sought within the largest possible dictionary constructed of IMFs. The problem is formulated as a non-linear L^1 optimization, and an iterative algorithm is proposed to solve it. Noise and perturbation in data may cause numerical instability in this method. Daubechies et al. developed a method which captures the philosophy of EMD and decomposes special functions in a defined class [21]. This method employs a combination of wavelet analysis and reallocation technique called synchrosqueezing transform which aim to sharpen a time-frequency representation. Synchrosqueezed wavelet transform is also investigated for signal sampling and denoising applications in multicomponent signal analysis [22]. In [23], an algorithm for AM-FM parameter estimation is proposed based on the iterated application of the Hilbert transform to amplitude envelopes obtained by adaptively low-pass filters. Furthermore, the IF of AM-FM components can be calculated by *a posteriori* adaptive segmentation of the acquired phase signal. Another iterative AM-FM decomposition is suggested in [24] using the quasi-harmonic model (QHM) for quasi-harmonic signals such as voiced speech.

There are various parametric approaches to extract IFs of multicomponent signals. One common approach is based on signal segmentation, while some simplifying assumptions such as constant frequency seem logical in each segment. Then, an estimator is designed to estimate the model parameters segment by segment. In [25], the maximum windowed likelihood (MWL) criterion is used to estimate the AM-FM components. The high nonlinearity of this method makes the necessary optimization difficult. Another parametric approach is based on the statistical modeling of the signal according to its statistical attributes and assumptions. Speech signals are statistically modeled as AM-FM signals, and the extended Kalman filter (EKF) is applied for demodulation [3]. The idea of EKF is also exploited in [26]. Polynomial phase signal (PPS) modeling is another parametric approach which is employed for AM-FM signals [27].

The interpretations and estimation of instantaneous frequencies embedded in a multicomponent signal have been controversial [28]. Three different approaches are proposed to estimate the IFs after the decomposition stage. In the first approach, the Hilbert transform is exploited to get the analytic signal whose phase is differentiated to find the IF [10]. The energy operator (TKEO) is utilized in the second approach [12], and the third one is based on TFD [11,29]. Different definitions of IF are considered in these approaches, and consequently, their results are not necessarily equivalent. In [28] and [30], the different definitions and estimation methods are compared and discussed. The main contribution of this paper is to develop a novel approach based on the expansion of time-varying systems by orthogonal rational functions. The method introduced in [31] is extended and improved to be applied as the essence of the new method for IF estimation. An adaptive segmentation procedure in the proposed algorithm allows us to estimate the IFs locally. The decomposition is performed using orthogonal rational functions.

2 Problem statement

Multicomponent signals are first introduced in [9]. A multicomponent AM-FM model describes a non-stationary signal as the combination of oscillation terms with time-varying amplitudes and frequencies:

$$x(t) = \sum_{k=1}^{N_c} A_k(t) e^{j\theta_k(t)}, \tag{1}$$

where N_c is the number of components. $A_k(t)$ and $\theta_k(t)$ are time-varying envelope and time-varying phase of the

kth component respectively, and the instantaneous frequency denoted by $f_k(t)$ is defined from $\theta_k(t)$:

$$f_k(t) = \frac{d\theta_k(t)}{dt}. \tag{2}$$

The general model in (1) can be interpreted as the signal expansion by a generalized complex exponential basis, which are exponential functions with time-varying amplitudes and frequencies. The decomposition of the multicomponent AM-FM signal is investigated through this point of view. Therefore, we are going to find an appropriate basis to expand the non-stationary signal $x(t)$:

$$x(t) = \sum_{k=1}^{K} c_k g_k(t). \tag{3}$$

The functions $\{g_k; k = 1, 2, \cdots, K\}$ should represent the oscillation modes in signal, for which the instantaneous frequency is meaningfully definable. Accordingly, each term represents a valid IF of the multicomponent signal. The main idea to attain such decomposition is expanding the corresponding system of the AM-FM signal in the complex plane. Since the transfer function of a realistic linear system has a rational representation, it can be expanded by orthogonal rational functions in the complex z-plane. Returning back to the time-domain, each rational function is equivalent to a generalized exponential basis and represents one valid oscillation term or resonance. To perform this procedure, we should specify the generating system of the AM-FM signal. The corresponding system of a non-stationary signal is modeled by a linear time-varying (LTV) system [32]. LTV models have been applied to describe non-stationary signals [33,34]. Our proposed method is developed based on this approach of modeling. Let us consider the discrete-time AM-FM signal $x[n]$, obtained by time sampling of $x(t)$ at the rate of f_s. Its generating system is modeled as a LTV system, which can be described by a bivariate function to characterize the input-output linear relationship [32]. Hence, a bivariate discrete-time impulse response, $h[m,n]$, is considered, where n and m are two independent time instants, representing the time variable of the signal and the time variable of the system, respectively. Taking the Z-transform of $h[m,n]$ with respect to the time variable of the signal, $H(m,z)$ is obtained which denotes the generating system of $x[n]$. The orthogonal rational basis has been investigated for the decomposition of linear time-invariant (LTI) systems [35] and should be generalized to expand the time-varying generating system of the AM-FM signal as follows:

$$H(m,z) = \sum_{k=1}^{K} C_k(m) G_k(m,z), \tag{4}$$

where $\{G_k(m,z); k = 1, 2, \cdots, K\}$ is a rational basis. Fortunately, to find the orthogonal rational functions for system expansion, it is not necessary to have the system's transfer function with all the details. The knowledge of poles or logical assumptions about them are sufficient to extract a proper basis [35].

3 Orthogonal rational functions

The decomposition step in the proposed method is indeed an expansion of the AM-FM signal, which is accomplished through the incorporating expansion of the equivalent time-varying system by orthogonal rational functions. Generally, the knowledge about the poles is sufficient as *a priori* information to describe the desired space being spanned by a rational basis. Let us consider a set of M time-varying poles, $\{\xi_k[m], k = 1, \ldots, M\}$. We can make a first-order IIR transfer function by each pole. So, a basis set is constructed including all specified poles, but not orthogonal. The Blaschke products [35] formed by these poles are two-dimensional functions,

$$B_0(m,z) = 1,$$
$$B_k(m,z) = \prod_{i=1}^{k} \frac{1 - \bar{\xi}_i[m]\, z}{z - \xi_i[m]}, \quad k = 1, 2, \ldots, M. \tag{5}$$

Applying the Gram-Schmidt procedure on these rational functions with respect to z, two-dimensional functions are obtained:

$$G_k(m,z) = \frac{\sqrt{1 - |\xi_k[m]|^2}}{z - \xi_k[m]} \prod_{i=1}^{k-1} \left(\frac{1 - \xi_i[m]^*\, z}{z - \xi_i[m]} \right), \quad k = 1, \ldots, M. \tag{6}$$

This is the same routine for finding Takenaka-Malmquist functions [36]. Now, these functions are generalized to two-dimensional functions for the LTV system expansion. The resultant functions in (6) are orthogonal with respect to z in the complex domain. The inner product of each pair of these functions is a function of time, m:

$$d_{kl}[m] = \langle G_k(m,z), G_l(m,z) \rangle_z = \frac{1}{2\pi j} \oint_C G_k(m,z) G_l^*\left(m, \frac{1}{z^*}\right) \frac{dz}{z}. \tag{7}$$

Utilizing the Cauchy integral implies that $d_{kl}[m]$ is zero at each snapshot, m, for $k \neq l$. Taking the inverse Z-transform of $G_k(m,z)$ produces $g_k[m,n]$. Since the Z-transform and its inverse are homomorphic transforms, these functions would preserve orthogonality with respect to n. Two problems remain to be studied. At first, the underlying time-varying poles of the corresponding LTV system should be determined. Secondly, univariate terms should be extracted from bivariate functions, $g_k[m,n]$, to express the oscillation modes of $x[n]$. These issues are resolved simultaneously by adaptive segmentation which is addressed in the following section.

4 Time-varying modeling

The concept of poles and zeros are also generalized for linear time-varying systems. Concerning the stability and behavior of the LTV systems, several definitions of poles or eigenvalues of such systems have been proposed [37], depending on the characterization method of the LTV system. The notion of time-varying poles in this paper is founded on the time-varying autoregressive model. Parametric models for LTI systems can be generalized for LTV ones by imposing time-varying parameters on the model. The AM-FM signal, $x[n]$, is modeled by a time-varying autoregressive (TVAR) of order M:

$$x[n] = -\sum_{m=1}^{M} a_m[n]\, x[n-m] + \nu[n]. \tag{8}$$

$\{a_m[n], m = 1,\ldots,M\}$ are the time-varying parameters, and $\nu[n]$ is the zero-mean innovation process, also addressed as a modeling error. The most general case of this model is where the parameters are completely uncorrelated at each time sample. Therefore, each time sample of $x[n]$ would be represented by M unknown coefficients; hence, it is not a practical approach. Based on a common practical assumption, the non-stationary signal is approximately regarded locally stationary or quasi-stationary. This assumption implies that the parameters of the TVAR model are correlated, and the coefficients are supposed to be constant in subintervals of the total time span, referred to as segments. This model is called a block stationary AR model [34]. For multicomponent AM-FM signals whose IAs and IFs are slowly time-varying or piecewise-constant, the segmentation strategy is applicable. By virtue of this assumption, a real multicomponent AM-FM signal over its support is considered as a superposition of temporarily more limited signals with constant frequencies. These intervals can generally have various lengths, and different methods from fixed-length windowing to adaptive segmentation algorithms are introduced to determine the borders of the segments [23]. In the proposed method, the segmentation is performed adaptively from the aspect of TVAR parameter estimation.

4.1 Segmentation procedure

The entire signal of N samples is segmented into L blocks with various lengths:

$$x_\ell[n] = x[n], \quad n_{\ell-1} \le n < n_\ell, \tag{9}$$

where $\ell = 1,\ldots,L$ and $n_0 = 0$. The TVAR coefficients are supposed to be constant in each segment. The mean square error in the ℓth segment is given by

$$J_\ell = \frac{1}{n_\ell - n_{\ell-1}} \sum_{n=n_{\ell-1}}^{n_\ell - 1} \left| x[n] + \sum_{m=1}^{M} a_{\ell,m} x[n-m] \right|^2, \tag{10}$$

where $\{a_{\ell,m}, m = 1,\ldots,M\}$ are the TVAR coefficients of the ℓth segment. The boundaries of each segment are determined such that the error J_ℓ remains under a specified threshold. The segmentation algorithm operates as follows. At the start of each stage, the length of the current segment (say ℓ) is considered the minimum possible length, equal to the order of the TVAR model, i.e., $n_\ell = n_{\ell-1} + M$. The TVAR coefficients, $a_{\ell,m}$, are estimated by the recursive least squares (RLS) technique, and the error in (10) is computed. If it is still greater than the prespecified threshold, the length of the segment increases by one sample, and the calculations are repeated. This procedure continues by one-sample increment in each stage until the error falls below the threshold. Now, the boundaries and the length of the current segment are determined, and the procedure starts over the next time sample for another segment establishment. This algorithm runs through the entire signal repeatedly and stops at the end of the data batch. The question arises here about the threshold setting and how it can affect the accuracy of the IF estimation. This issue is scrutinized in the succeeding subsection separately. Once the TVAR parameters are estimated, the corresponding time-varying poles, denoted by $\{\xi_k[m], k = 1,\ldots,M\}$, are obtained by applying the Z-transform of (8) with respect to n.

4.2 Error analysis

It is noteworthy to mention the relation between the error caused by segmentation and the error of IF estimation. This analysis leads us to select a reliable error threshold in the adaptive segmentation procedure. Let us consider a discrete-time AM-FM component:

$$x[n] = A[n]\, e^{j\theta[n]}. \tag{11}$$

The error in the instantaneous phase imposed by the TVAR modeling in each segment, denoted by ϵ_θ, sets off an error in signal:

$$\widehat{x}[n] = A[n]\, e^{j(\theta[n] + \epsilon_\theta[n])}. \tag{12}$$

For very small phase errors, the following approximation is considered using the Maclaurin series:

$$e^{j(\theta[n]+\epsilon_\theta[n])} \approx \left(1 - \frac{\epsilon_\theta^2[n]}{2} + j\epsilon_\theta[n]\right) e^{j\theta[n]}. \tag{13}$$

Substituting this approximation in (12), we have

$$e[n] = x[n] - \widehat{x}[n] = \varepsilon[n]\, A[n]\, e^{j\theta[n]}, \tag{14}$$

where

$$\varepsilon[n] = \frac{\epsilon_\theta^2[n]}{2} - j\epsilon_\theta[n]. \tag{15}$$

The error $e[n]$ whose instantaneous amplitude is absolute error of modeling is also an AM-FM signal:

$$|e[n]| = |\varepsilon[n]\, A[n]|. \tag{16}$$

The error in phase is now transduced to the error of amplitude. Let us define a time-varying threshold denoted by $\eta[n]$ such that $|e[n]|$ is restrained lower than it, i.e., $|e[n]| < \eta[n]$. If we substitute $|e[n]|$ by (16), the following inequality holds:

$$|\varepsilon[n]| < \frac{\eta[n]}{|A[n]|}. \tag{17}$$

So, the absolute error of phase depends on the signal envelope. This means that for a fixed threshold, where $\eta[n]$ is constant over the entire signal, larger phase errors can occur when IA becomes smaller. Therefore, the threshold should vary adaptively, adjusted to the envelope of the observed signal. In other words, the locally normalized error for each segment is a proper threshold. Since the IA evolves slowly, its mean or minimum amount during the segment can be utilized for normalization. The normalized threshold is denoted by $\bar{\eta}$ for brevity:

$$\bar{\eta} = \frac{\eta[n]}{\mathrm{mean}\{|A[n]|\}}. \tag{18}$$

Thus, the inequality (17) is practically used as the following one:

$$|\varepsilon[n]| < \bar{\eta}. \tag{19}$$

The square of $|\varepsilon[n]|$ is obtained from (15):

$$|\varepsilon[n]|^2 = \frac{\epsilon_\theta^4[n]}{4} + \epsilon_\theta^2[n]. \tag{20}$$

Exploiting this relation in the inequality (19) and performing some mathematical reformulations result in a bound for the phase error:

$$|\epsilon_\theta[n]|^2 < 2\left(\sqrt{1+\bar{\eta}^2} - 1\right). \tag{21}$$

When $\bar{\eta} \to 0$, the right-hand side of the above inequality is approximately equal to $\bar{\eta}^2$. Keeping the phase error (ϵ_θ) under control, the error of IF is consequently controlled. By definition (2), IF is the derivative of instantaneous phase, which is a difference equation in the discrete-time situation:

$$\omega[n] = \frac{\theta[n] - \theta[n-1]}{T_s}, \tag{22}$$

where $\omega[n] = 2\pi f[n]$ is the instantaneous frequency in radian per second, and T_s denotes the sampling time. In the worst case, the maximum phase errors of two consecutive instants accumulate. Thus, the maximum error of IF is $2\epsilon_\theta f_s$. For example, if $\bar{\eta} = 10^{-3}$, then from (21), the maximum phase error is almost 10^{-3}, and the absolute error of IF is at most 0.2% of the sampling frequency. This error can be controlled by arbitrary selection of $\bar{\eta}$. A smaller threshold leads to wider segments, in which the assumption of constant frequency is no longer respected. Our experiments verified that the condition of piecewise-constant frequency for slowly varying IFs is satisfied for $\bar{\eta}$ in the order of $10^{-3} \sim 10^{-2}$.

5 Estimation framework

The main algorithm of IF estimation takes the extracted time-varying poles to construct $G_k(m,z)$ in (6). Then, bivariate functions, $\{g_k[m,n], k = 1,\ldots,M\}$, are produced by taking the inverse Z-transform. Now, the one-dimensional basis is extracted from the existing bivariate functions to achieve a one-dimensional expansion for $x[n]$ as in (3). The basis $g_k[n]$ is constructed by the concatenation of truncated pieces of bivariate functions, $g_k[m,n]$, based on the result of the segmentation procedure:

$$g_k[n] = \sum_{\ell=1}^{L} W_\ell[n]\, g_k[m_\ell, n], \tag{23}$$

where L is the number of total segments, and $W_\ell[n]$ is an arbitrary weighting window over the ℓth segment. It is supposed that during this block, the corresponding pole remains equal to $\xi_k[m_\ell]$. Each resultant function, $g_k[n]$, is a valid oscillation mode for which IF is definable. Thus, the estimation of the embedded IFs is achieved through the IF estimation of the extracted functions. To estimate the IF, linear regression of the phase for each segment is computed by applying the weighted linear least squares technique on a first-order polynomial model. The abrupt changes in the phase which can affect the IF estimation severely are an important issue. While the consecutive segments derived from different rows of $g_k[m,n]$ are concatenated, there may be some phase discontinuities over the resultant bases in the junctions of segments. Such discontinuities in the phase trajectory cause serious deficiency in the IF estimation which appears as spikes over the resultant IFs. To remedy this problem, a proper data window such as the Hamming window is chosen as $W_\ell[n]$ in (23), which controls the effect of borderline samples. The Hamming window is commonly utilized as an analysis window in audio and speech processing [24,27].

When the signal is contaminated by noise, the time-varying poles estimated from noisy observations are misplaced. Thus, the estimated IF incurs more error due to the error in the estimation of poles imposed by the noise. This issue is alleviated by increasing the order of the TVAR model. Each resonance of a clean signal is represented by a pair of time-varying poles; hence, the order of the TVAR model (M) is twice the number of components (N_c). Nonetheless, to improve the estimation of time-varying poles in the presence of noise, we should have $M > 2N_c$. Therefore, extraneous poles appear besides the valid poles. A minimum distance classifier [38] is applied to assign the poles of the resonances in each segment and distinguish them from the invalid poles. The perturbation of the poles due to the estimation error of TVAR coefficients is investigated mathematically in the succeeding section. The steps of the proposed algorithm are summarized as follows:

1. Adaptive segmentation of the AM-FM signal based on TVAR modeling and computation of underlying time-varying poles.
2. Assignment of the poles to the components by minimum distance classifier.
3. Employing the time-varying poles to construct oscillation terms, $g_k[n]$.
4. Fitting a linear model to the phase of each segment of $g_k[n]$ to estimate the IF.

In this novel method of IF extraction, the Hilbert transform that is a global operator is not employed. Additionally, a linear model is applied to the phase of components segment by segment in spite of differentiating throughout. It makes the proposed method less sensitive to phase changes. Therefore, the adaptive segmentation is advantageous for both decomposition and frequency estimation.

6 Pole perturbation

The coefficients of the TVAR model (8), estimated through the RLS technique, are affected by noise. The perturbation in these coefficients leads to the perturbation in the resultant time-varying poles. Let $p(\ell, z)$ be the polynomial of AR model in the ℓth segment whose roots are time-varying poles:

$$p(\ell, z) = \sum_{i=0}^{M} a_{\ell,i} z^{-i} = \prod_{k=1}^{M} \left(1 - \xi_k[m_\ell] z^{-1}\right), \tag{24}$$

where the coefficients are normalized, i.e., $a_{\ell,0} = 1$. If the perturbation $\Delta a_{\ell,i}$ occurs in the ith coefficient, $a_{\ell,i}$, the polynomial (24) changes to

$$\tilde{p}(\ell, z) = p(\ell, z) + \Delta a_{\ell,i} z^{-i}. \tag{25}$$

The roots of this new polynomial differ from $\xi_k[m_\ell]$ by $\Delta\xi_k[m_\ell]$, which may be real or complex. Let us denote the perturbed roots by $\tilde{\xi}_k[m_\ell] = \xi_k[m_\ell] + \Delta\xi_k[m_\ell]$:

$$\tilde{p}(\ell, \tilde{\xi}_k[m_\ell]) = p\left(\ell, \tilde{\xi}_k[m_\ell]\right) + \Delta a_{\ell,i} \left(\tilde{\xi}_k[m_\ell]\right)^{-i} = 0. \tag{26}$$

From now on, the kth pole and its perturbation are denoted by ξ_k and $\Delta\xi_k$, respectively, and their arguments are neglected for brevity. When $\Delta\xi_k \to 0$, the above equation is simplified by employing Taylor's expansion:

$$p'(\ell, \xi_k)\, \Delta\xi_k + \Delta a_{\ell,i} \xi_k^{-i} = 0, \tag{27}$$

where $p'(\ell, z)$ is the first derivative of $p(\ell, z)$ with respect to z. This equation expresses the linear relation between perturbation in poles and perturbation in coefficients:

$$\Delta\xi_k = -\frac{\xi_k^{-i}}{p'(\ell, \xi_k)} \Delta a_{\ell,i}. \tag{28}$$

Considering $\Delta a_{\ell,i}$ and $\Delta\xi_k$ as random variables, their variances, respectively $\sigma^2_{a_{\ell,i}}$ and $\sigma^2_{\xi_k}$, are related linearly,

$$\sigma^2_{\xi_k} = \left| \frac{\xi_k^{-i}}{p'(\ell, \xi_k)} \right|^2 \sigma^2_{a_{\ell,i}}. \tag{29}$$

Obtaining $p'(\ell, \xi_k)$ from (24), the ratio of the variances is given by

$$\frac{\sigma^2_{\xi_k}}{\sigma^2_{a_{\ell,i}}} = \frac{1}{\left| \xi_k^{i-1} \prod_{\substack{j=1 \\ j \neq k}}^{M} \left(1 - \xi_k \xi_j^{-1}\right) \right|^2}. \tag{30}$$

This ratio indicates the sensitivity of the poles to the perturbation of the coefficients. Smaller ratio represents less sensitivity or more robustness of pole. When all poles are inside the unit circle, i.e., $|\xi_k| < 1, k = 1, \ldots, M$, the ratio in (30) takes its maximum value for $i = M$. This means that the perturbation in the coefficient of the highest order in the polynomial has the most effect on misplacing the poles. This maximum value for each pole is defined as its variance ratio:

$$\gamma_k = \frac{\sigma^2_{\xi_k}}{\sigma^2_{a_{\ell,M}}} = \frac{1}{\left| \xi_k^{M-1} \prod_{\substack{j=1 \\ j \neq k}}^{M} \left(1 - \xi_k \xi_j^{-1}\right) \right|^2}. \tag{31}$$

γ_k is the variance ratio of ξ_k which indicates the robustness of this pole. This parameter only depends on the positions of the underlying poles, which are determined by AM-FM signal characteristics.

7 Experimental evaluation

The proposed method is implemented on both synthetic and real-world signals, and its performance is compared to the results of two previously introduced methods.

7.1 Synthetic data

A two-component AM-FM signal is considered as below:

$$x(t) = \sum_{k=1}^{2} \exp(\alpha_k t) \exp\left[j2\pi\left(\beta_{k,3}t^3 + \beta_{k,2}t^2 + \beta_{k,1}t\right)\right]. \quad (32)$$

The parameters of the first and the second components are $\alpha_1 = -2, \beta_{1,1} = 10^3, \beta_{1,2} = 5 \times 10^3, \beta_{1,3} = -20 \times 10^3$ and $\alpha_2 = -5, \beta_{2,1} = 5 \times 10^3, \beta_{2,2} = -3 \times 10^3, \beta_{2,3} = 0$. The two following IFs are embedded in this AM-FM signal:

$$f_1(t) = -60t^2 + 10t + 1 \text{ kHz}, \qquad f_2(t) = -6t + 5 \text{ kHz}. \quad (33)$$

The signal in (32) is sampled at $T_s = 50$ μs intervals, so the sampling frequency is $f_s = 20$ kHz. Since the absolute values of the instantaneous frequencies increase over time, a limited span of signal is observed to avoid aliasing. The algorithm is run over $N = 2,000$ samples of the signal. The corresponding system has four time-varying complex poles $\{\xi_1[m], \ldots, \xi_4[m]\}$ in conjugate pairs. There are two conjugate pairs, and each pair represents one oscillation component. Consequently, four complex functions, $g_1[n], \ldots, g_4[n]$, are extracted whose phases yield desired IFs. $g_1[n]$ and $g_2[n]$ are common resonances, which means that their IFs have equal absolute values, but opposite signs. $g_3[n]$ and $g_4[n]$ determine the second IF likewise. The adaptive segmentation procedure divides the data batch into $L = 45$ segments with different lengths. The threshold of error in the segmentation procedure is set to 0.001. The coefficients of the TVAR model and, accordingly, the time-varying poles are estimated through the RLS algorithm with the forgetting factor of 0.98. Figure 1 demonstrates the estimated IFs denoted by $\hat{f}_i; i = 1, 2$ besides the original IFs. $\hat{f}_1$ and $\hat{f}_2$ track the original IFs very closely. The step-like variation, produced by segmentation, is obvious. For quantitative evaluation, the relative mean absolute error (MAE) of IF estimation is computed for $\hat{f}_1$ and $\hat{f}_2$, and illustrated in Tables 1 and 2, respectively.

The proposed method is compared with two previous methods, EEMD-HT and QHM. The EEMD-HT is a nonparametric method which decomposes the components by the EEMD algorithm and estimates the IFs of resultant IMFs utilizing the Hilbert transform [14]. The QHM is a parametric method which has been appraised on speech signals [24]. The result of decomposition of the original signal by the EEMD procedure is displayed in Figure 2. The relative standard deviation of added noise is 0.2, and the ensemble number for each run is 100. Nine IMFs are derived while there are just two embedded oscillation modes. The first and the second IMFs are expected oscillation modes, and the others are false IMFs generated due to deficiencies in the iterative algorithm. EEMD initially extracts faster oscillations. Thus, the first IMF has higher frequency and represents our second component, and reversely, the second IMF corresponds to the first component. The valid IMFs are selected and assigned subjectively or based on the result of estimation [15]. In this simulation, the estimated IFs of the first and the second IMFs are closer to f_2 and f_1, respectively. MAEs of the estimation of IFs are recorded in Tables 1 and 2. The estimation errors of QHM are also provided in these tables for comparison. The analysis window of 64 samples and hop step of one sample is considered for this algorithm.

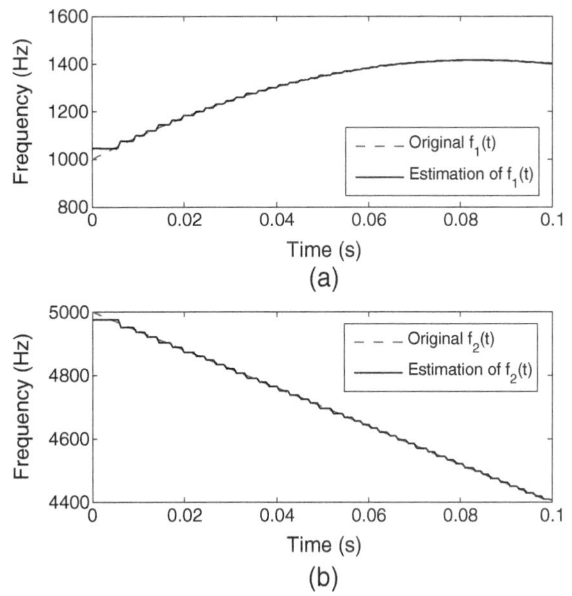

Figure 1 The original instantaneous frequency and the estimation result. (a) First component. **(b)** Second component.

The simulation is repeated in the presence of additive white-Gaussian noise, and the errors for different SNRs are depicted in the same tables. Each value is the average of 100 iterations. The proposed method outperforms the other algorithms, especially in stronger noise. These results verify the robustness of the proposed algorithm. Although the QHM has smaller error for clean signal, it is more sensitive to noise. The EEMD, which is more robust than EMD, is still affected by the perturbations on

Table 1 Relative MAE (%) for the estimation of the first instantaneous frequency, $f_1(t)$

SNR	EEMD-HT	QHM	Proposed method
∞	0.57	0.06	0.53
30 dB	1.49	0.83	1.62
10 dB	79.68	44.03	40.74

Table 2 Relative MAE (%) for the estimation of the second instantaneous frequency, $f_2(t)$

SNR	EEMD-HT	QHM	Proposed method
∞	0.17	0.03	0.09
30 dB	0.46	0.24	0.19
10 dB	21.54	8.30	3.23

the amplitude of the data. Therefore, EEMD-HT has the worst performance in the presence of noise. For the clean signal (SNR $= \infty$), the order of the TVAR model, M, is equal to 4. Since higher error is imposed on pole estimation for stronger noise, M should increase to alleviate this problem. We have $M = 24$ for SNR = 30 dB and SNR = 10 dB. Figure 3 depicts the MAE of IF estimation with respect to M for both components. It demonstrates that the error decreases remarkably as the order increases. However, there are more poles for higher orders which raise the error of classification routine and consequently the total error of estimation. Thus, there is a compromise to select the optimum order. In Figure 3, MAEs of $\hat{f}_1$ and $\hat{f}_2$ are minimum at $M = 24$ and $M = 28$, respectively.

The effect of noise is illustrated by the perturbation variance ratios, $\gamma_1, \ldots, \gamma_4$ (31) for each segment in Figure 4. The variance ratios of the poles in a conjugate pair are very close to each other or even equal in some segments. γ_3 and γ_4, corresponding to ξ_3 and ξ_4 or the second resonance, are smaller than γ_1 and γ_2. This means that the poles corresponding to the first oscillation component are more sensitive to the noise. This observation justifies why the MAE for $\hat{f}_1$ (Table 1) deteriorates more than the MAE of $\hat{f}_2$ (Table 2) due to the noise.

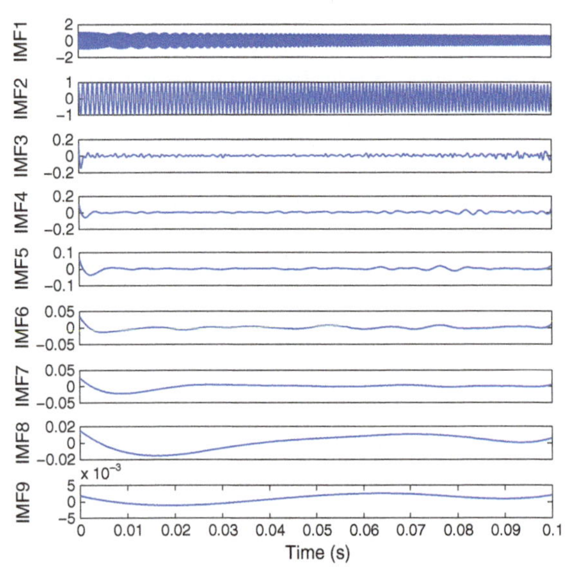

Figure 2 Extracted IMFs by EEMD from AM-FM signal.

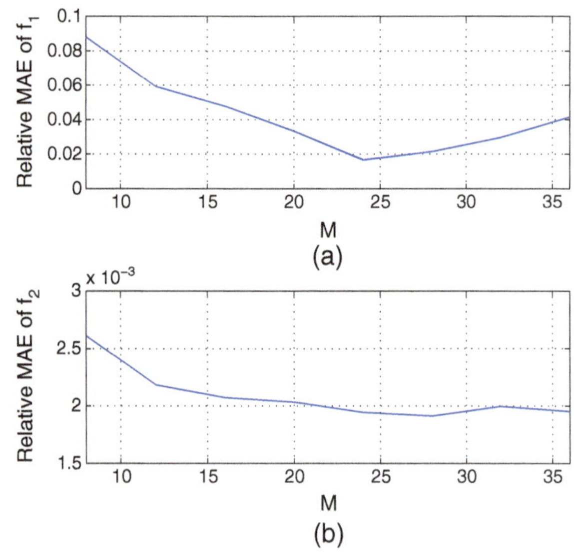

Figure 3 Relative mean absolute error of IF estimation vs. the order of the TVAR model, *M*. (a) First component. **(b)** Second component.

The proposed method can distinguish components and estimate embedded close IFs, even when they have crossovers. To illustrate this capability, the simulation of a similar two-component signal is repeated, except that the frequency of the second component, f_2, is reduced by 4.8 kHz, i.e., $\beta_{2,1} = 1,200$ in (32). In this case, two IFs are closer and intersect each other. Both IFs are depicted in Figure 5. The proposed method is implemented in the same condition, and the results of estimation are exhibited

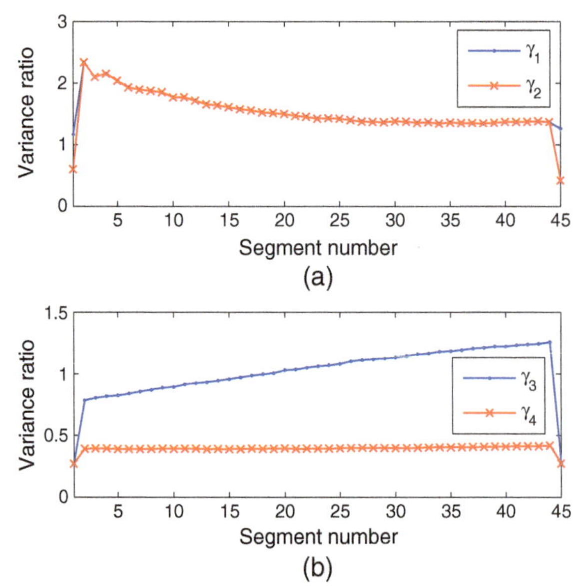

Figure 4 Perturbation variance ratios in each segment. (a) ξ_1 and ξ_2. **(b)** ξ_3 and ξ_4.

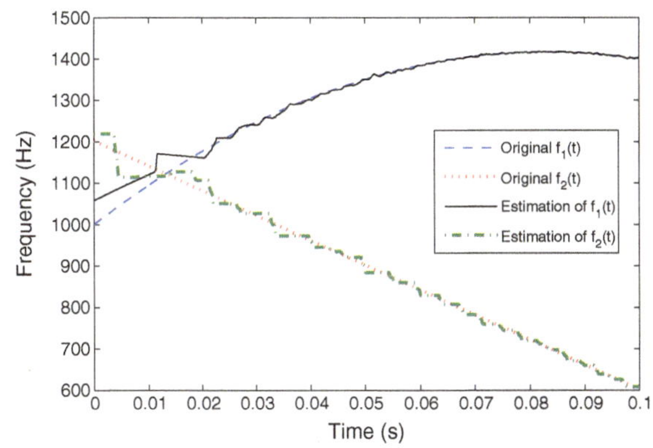

Figure 5 **The original instantaneous frequencies and the estimation results, when two IFs intersect each other.**

in Figure 5, besides the original IFs. This method can track both IFs, and the intersection in their trajectories is handled properly. The degradation of estimation, especially around the crossover junction, is obvious by comparing Figures 5 and 1, but the trajectories are not missed and the estimated IFs are improved with time. This advantage arises from the step of pole classification in the proposed algorithm.

7.2 Real-world signal

Acoustic studies have revealed that many natural acoustic signals such as the sounds of songbirds and oceanic mammals fit into the AM-FM model [27,39]. A duration of 68 ms of a song of two songbirds, a canary, and a kinglet, recorded at 22.05 kHz sampling frequency, is considered for the experiment. The spectrum of this signal estimated by STFT with sliding window of 128 samples, an overlap of 100 samples, and frequency resolution bin of 2 Hz is depicted in Figure 6. This signal is composed of two AM-FM components and real-world disturbing signals such as ambient and instrumental noises. Four complex poles and consequently four bases are extracted which present two embedded dominant oscillations. Through the adaptive segmentation procedure, the signal is divided into $L = 76$ unequal blocks. The threshold of the segmentation procedure, $\bar{\eta}$, is set to 0.01. The extracted IFs are illustrated in the same figure of the STFT spectrum by black marks. The estimated IF tracks the frequency of dominant oscillations. In Figure 7, the IFs are demonstrated in addition to the results of EEMD-HT and QHM. Nine IMFs are extracted through EEMD, but the first and the second ones are valid resonances whose IFs are displayed. The relative standard deviation of added noise is set to 0.2, and the number of ensembles is 100. It takes several samples for EEMD-HT to resolve two IFs. Furthermore, the local variations and spikes in the results of EEMD-HT and QHM are noticeable. By contrast, the IFs of the proposed method are piecewise-constant, and no spike appears due to the segmentation. The lengths of the segments and consequently the variations of the resultant IFs (flatness or volatility) are controllable by the threshold selected in the segmentation procedure.

8 Conclusion

The analytical approach developed for extraction of oscillation modes of multicomponent non-stationary signals is founded on several facts which are summarized herein. Firstly, the decomposition of a multicomponent signal is considered as a signal expansion and investigated from this point of view. Secondly, the equivalent time-varying system of the non-stationary signal is modeled by TVAR and expanded by the orthogonal rational functions. Finally, oscillation modes are constructed, employing these bivariate rational functions, and their IFs are estimated. For this purpose, the evolution of the time-varying

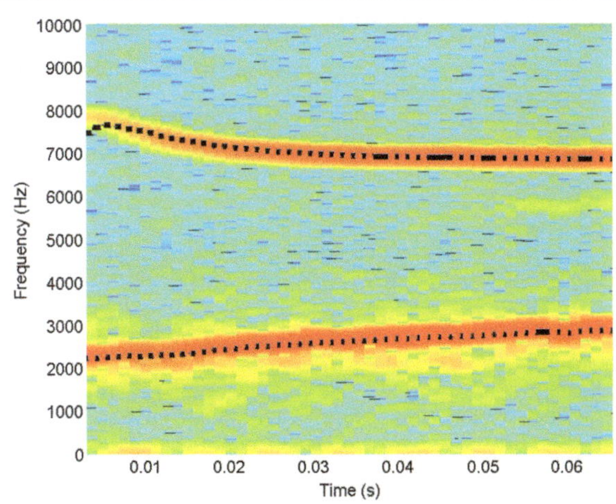

Figure 6 **STFT of birds' song and the trajectories of the estimated IFs.**

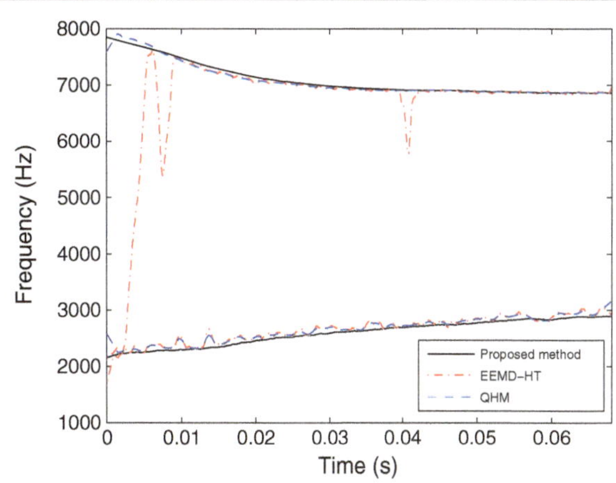

Figure 7 **The results of IF estimation on the selected piece of the birds' song.**

poles is considered piecewise-constant which imposes the same approximation in frequency estimation. The threshold parameter in the adaptive segmentation algorithm controls this approximation and governs the flatness of estimated IF against volatility.

The order of the TVAR model or equivalently the number of time-varying poles is another parameter which is set properly to confront the noise. Because this method utilizes the poles of the generating system of the AM-FM signal, the distortions on the amplitude, like noise, affect the estimation results mildly in comparison to those of the empirical methods. Although the noise degrades the estimation of underlying poles, it can be controlled by the order of the TVAR model. Simulations reveal the superiority of this method in the presence of noise. The controlling parameters in the proposed method yield some degree of freedom to adjust it for different realistic applications. Its capability to extract the embedded IFs in audio signals is illustrated.

Competing interests
The authors declare that they have no competing interests.

Author details
[1]Electrical Engineering Department, Amirkabir University of Technology, Tehran 15914, Iran. [2]Electrical Engineering Department, University of Washington, Seattle, WA 98195, USA.

References
1. B Santhanam, P Maragos, Multicomponent AM-FM demodulation via periodicity-based algebraic separation and energy-based demodulation. IEEE Trans Commun. **48**(3), 473–490 (2000)
2. S Jang, P Loughlin, AM-FM interference excision in spread spectrum communications via projection filtering. EURASIP J Appl. Signal Process. **4**, 239–248 (2001)
3. S Lu, P Doerschu, Nonlinear modeling and processing of speech based on sums of AM-FM formant models. IEEE Trans. Signal Process. **44**(4), 773–782 (1996)
4. M Bester, P Collen, G Richard, B David, Estimation of frequency for AM-FM models using the phase vocoder framework. IEEE Trans. Signal Process. **56**(2), 505–517 (2008)
5. M Grimaldi, F Cummins, Speaker identification using instantaneous frequencies. IEEE Trans. Audio Speech Lang. Process. **16**(6), 1097–1111 (2008)
6. Y Kubo, S Okawa, A Kurematsu, K Shirai, Temporal AM-FM combination for robust speech recognition. Speech Commun. **53**, 716–725 (2011)
7. C Loizou, V Murray, M Pattichis, I Seimenis, M Pantziaris, C Pattichis, Multiscale amplitude-modulation frequency-modulation (AM-FM) texture analysis of multiple sclerosis in brain MRI images. IEEE Trans. Inform. Technol. Biomed. **15**(1), 119–129 (2011)
8. V Murray, P Rodriguez, M Pattichis, Multiscale AM-FM demodulation and image reconstruction methods with improved accuracy. IEEE Trans. Image Process. **19**(5), 1138–1152 (2010)
9. L Cohen, What is a multicomponent signal? Paper presented at the of IEEE international conference on acoustic, speech and signal processing (ICASSP) (USA, 23–26 Mar 1992), pp. 113–116
10. B Boashash, Estimating and interpreting the instantaneous frequency of a signal-part1: fundamentals. Proc. IEEE. **80**(4), 520–538 (1992)
11. B Lovell, R Williamson, B Boashash, The relationship between instantaneous frequency and time-frequency representations. IEEE Trans. Signal Process. **41**(3), 1458–1461 (1993)
12. P Maragos, F Kaiser, T Quatieri, Energy separation in signal modulations with application to speech analysis. IEEE Trans. Signal Process. **41**(10), 3024–3051 (1993)
13. A Bovik, P Maragos, T Quatieri, AM-FM energy detection and separation in noise using multiband energy operator. IEEE Trans. Signal Process. **41**(12), 3245–3265 (1993)
14. N Huang, Z Shen, S Long, M Wu, H Shih, Q Zheng, N Yen, C Tung, H Liu, The empirical mode decomposition and the Hilbert spectrum for nonlinear and non-stationary time series analysis. Proc. R. Soc. London. **454**, 903–995 (1998)
15. Z Wu, N Huang, Ensemble empirical mode decomposition: a noise-assisted data analysis method. Adv. Adaptive Data Anal. **1**(1), 1–41 (2009)
16. E Deléchelle, J Lemoine, O Niang, Empirical mode decomposition: an analytical approach for sifting process. IEEE Signal Process. Lett. **12**(11), 764–767 (2005)
17. L Lin, Y Wang, H Zhou, Iterative filtering as an alternative algorithm for empirical mode decomposition. Adv. Adaptive Data Anal. **1**(4), 543–560 (2009)
18. Y Wang, Z Zhou, On the convergence of iterative filtering empirical mode decomposition. Excursions in Harmonic Anal. **2**, 157–172 (2013)
19. T Hou, M Yan, A variant of the EMD method for multi-scale data. Adv. Adaptive Data Anal. **1**(4), 483–516 (2009)
20. T Hou, Z Shi, Adaptive data analysis via sparse time-frequency representation. Adv. Adaptive Data Anal. **3**(1), 1–28 (2011)
21. I Daubechies, J Lu, H Wu, Synchrosqueezed wavelet transforms: an empirical mode decomposition-like tool. Appl. Comput. Harmonic Anal. **30**(2), 243–261 (2011)
22. S Meignen, T Oberlin, S McLaughlin, A new algorithm for multicomponent signals analysis based on synchrosqueezing: with an application to signal sampling and denoising. IEEE Trans. Signal Process. **60**(11), 5787–5798 (2012)
23. F Gianfelici, G Biagetti, P Crippa, C Turchetti, Multicomponent AM-FM representations: an asymptotically exact approach. IEEE Trans. Audio Speech Lang. Process. **15**(3), 823–837 (2007)
24. Y Pantazis, O Rosec, Y Stylianou, Adaptive AM-FM signal decomposition with application to speech analysis. IEEE Trans. Audio Speech Lang. Process. **19**(2), 290–300 (2011)
25. S Gazor, R Rashidi Far, Adaptive maximum windowed likelihood multicomponent AM-FM signal decomposition. IEEE Trans. Audio Speech Lang. Process. **14**(2), 479–491 (2006)
26. W Pai, C Doerschuk, Statistical AM-FM models, extended Kalman filter demodulation, Cramer-Rao bounds and speech analysis. IEEE Trans. Signal Process. **48**(8), 2300–2313 (2000)
27. M Jabloun, F Leonard, M Vieira, N Martin, A New flexible approach to estimate the IA and IF of nonstationary signals of long-time duration. IEEE Trans. Signal Process. **55**(7), 3633–3644 (2007)
28. N Huang, Z Wu, S Long, K Arnold, Z Chen, K Blank, On instantaneous frequency. Adv. Adaptive Data Anal. **1**(2), 177–229 (2009)
29. N Khan, B Boashash, Instantaneous frequency estimation of multicomponent nonstationary signals using multiview time-frequency distributions based on the adaptive fractional spectrogram. IEEE Signal Process. Lett. **20**(2), 157–160 (2013)
30. D Vakman, On the analytic signal, the Teager-Kaiser energy algorithm, and other methods for defining amplitude and frequency. IEEE Trans. Signal Process. **44**(4), 791–797 (1996)
31. M Sebgahti, H Amindavar, A novel analytical approach to orthogonal bases extraction from AM-FM signals. Paper presented at the IEEE international conference on acoustic, speech and signal processing (ICASSP) (Czech Republic, 22–27 May 2011), pp. 3820–3823
32. H D'Angelo, *Linear Time-Varying Systems, Analysis and Synthesis*. (Allyn and Bacon, Boston, 1970)
33. N Martin, An AR spectral analysis of non-stationary signals. Signal Process. **10**, 61–74 (1986)
34. J Hopgood, P Rayner, Blind single channel deconvolution using nonstationary signal processing. IEEE Trans. Speech Audio Process. **11**(5), 476–488 (2003)
35. P Heuberger, P Van den Hof, B Wahlberg, *Modeling and Identification with Rational Orthogonal Basis Functions*. (Springer, London, 2005)
36. B Ninness, F Gustafsson, A unifying construction of orthonormal bases for system identification. IEEE Trans. Automatic Control. **42**(4), 515–521 (1997)

37. R O'Brien Jr., P Iglesias, On the poles and zeros of linear time-varying systems. IEEE T Circuits-I. **48**(5), 565–577 (2001)
38. R Duda, P Hart, D Stork, *Pattern Classification*, 2nd edn. (Wiley, New York, 2000)
39. Q Meng, M Yuan, J Zhao, Empirical AM-FM decomposition of auditory signals. J. Acoust. Soc. Am. **131**(4) (2012). doi:10.1121/1.4708214

Voice conversion using speaker-dependent conditional restricted Boltzmann machine

Toru Nakashika[1*†], Tetsuya Takiguchi[2†] and Yasuo Ariki[2†]

Abstract

This paper presents a voice conversion (VC) method that utilizes conditional restricted Boltzmann machines (CRBMs) for each speaker to obtain high-order speaker-independent spaces where voice features are converted more easily than those in an original acoustic feature space. The CRBM is expected to automatically discover common features lurking in time-series data. When we train two CRBMs for a source and target speaker independently using only speaker-dependent training data, it can be considered that each CRBM tries to construct subspaces where there are fewer phonemes and relatively more speaker individuality than the original acoustic space because the training data include various phonemes while keeping the speaker individuality unchanged. Each obtained high-order feature is then concatenated using a neural network (NN) from the source to the target. The entire network (the two CRBMs and the NN) can be also fine-tuned as a recurrent neural network (RNN) using the acoustic parallel data since both the CRBMs and the concatenating NN have network-based representation with time dependencies. Through voice-conversion experiments, we confirmed the high performance of our method especially in terms of objective evaluation, comparing it with conventional GMM, NN, RNN, and our previous work, speaker-dependent DBN approaches.

Keywords: Voice conversion; Conditional restricted Boltzmann machine; Deep learning; Recurrent neural network; Speaker-specific features

1 Introduction

In recent years, voice conversion (VC), a technique used to change specific information in the speech of a source speaker to that of a target speaker while retaining linguistic information, has been garnering much attention in speech signal processing. VC techniques have been applied to various tasks, such as speech enhancement [1], emotion conversion [2], speaking assistance [3], and other applications [4,5]. Most of the related work in VC focuses not on f0 conversion but on the conversion of spectrum features, and we conform to that in this report as well.

Various statistical approaches to VC have been studied so far, including those discussed in [6,7]. Among these approaches, the Gaussian mixture model (GMM)-based mapping method [8] is widely used, and a number of improvements have been proposed. Toda et al. [9] introduced dynamic features and the global variance (GV) of the converted spectra over a time sequence. Helander et al. [10] proposed transforms based on partial least squares (PLS) to prevent the over-fitting problem encountered in standard multivariate regression. There have also been approaches that do not require parallel data since they use a GMM adaptation technique [11], eigenvoice GMM [12,13] or probabilistic integration model [14]. Other approaches based on statistical approaches have been proposed; Jian et al. [15] used canonical correlation analysis for the VC, and Takashima et al. [16] proposed a VC technique using exemplar-based non-negative matrix factorization (NMF).

However, most of the conventional VC methods, including the GMM-based approaches, rely on 'shallow' voice conversion based on linear (or piecewise linear) transformation. That means a source speech was converted in the original feature space directly or in the shallow architecture with a few hidden layers. To capture the characteristics of speech more precisely, it is necessary to have a deeper non-linear architecture with more hidden

*Correspondence: nakashika@me.cs.scitec.kobe-u.ac.jp
†Equal contributors
1 Graduate School of System Informatics, Kobe University, 1-1 Rokkodai, Nada-ku, 657-8501 Kobe Japan
Full list of author information is available at the end of the article

layers. The shape of the vocal tract is generally non-linear, so non-linear voice conversion is more compatible with human speech. One example of a non-linear VC method is proposed by Narendranath et al. [17] and Desai et al. [18] based on neural networks (NN). In the GMM-based approaches, the conversion is achieved so as to maximize the conditional probability calculated from a joint probability of source speech and target speech, which is trained beforehand. On the other hand, NN-based approaches directly train the conditional probability, which converts the feature vector of a source speaker to that of a target speaker. It is often reported that such a discriminative approach performs better than a generative approach, such as GMM, in speech recognition and synthesis as well as in VC [19,20]. For these reasons, NN-based approaches achieve relatively high performance if the training samples are carefully prepared [18].

These approaches often suffer from over-smoothing or over-fitting problems. GMM-based approaches represent acoustic features using multiple Gaussian distributions, which are estimated by averaging observations with similar context descriptions in the training. Therefore, the outputs of the GMM distribute near the modes (means) of the Gaussians, which causes problems with over-smoothing. Furthermore, over-fitting problems arise when we give more Gaussian mixtures due to precise estimation of the observed distribution. In the NN-based approaches, the model is often over-fitted due to its complexity because it exaggerates small fluctuations in the unknown data if the number of training data is not enough relative to the number of parameters.

In order to alleviate the over-smoothing effect in a GMM-based method, some methods have been proposed so far, such as the global variance model [21], a minimizing-divergence model [22], and post-filtering [23]. An exemplar-based VC system using non-negative matrix factorization (NMF) has also been proposed to tackle the over-smoothing problems [16,24]. In our earlier work [25], we proposed a new VC technique that copes with the over-fitting problems in NN-based approaches, using a combination of speaker-dependent restricted Boltzmann machines (RBMs) [26] (or deep belief nets (DBN) [27]) that captures high-order features in an unsupervised manner and a concatenating NN. It is reported that these graphical models are better at representing the distribution of high-dimensional observations with cross-dimension correlations than GMM in speech synthesis [28] and in speech recognition [29]. Since Hinton et al. introduced an effective training algorithm for the DBN in 2006 [27], the use of deep learning rapidly spread in the field of signal processing, as well as in speech signal processing. An RBM (or DBN) has been used, for example, for hand-written character recognition [27], 3-D object recognition [30], machine transliteration [31], and so on.

In this paper, we extend our earlier work in [25] to systematically capture time information as well as latent (deep) relationships between a source speaker's and a target speaker's features in a single network. This is accomplished by combining speaker-dependent conditional restricted Boltzmann machines (CRBMs) and a concatenating NN.

A CRBM is a non-linear probabilistic model used to represent time series data consisting of three factors: (i) an undirected model between binary latent variables and the current visible variables, (ii) a directed model from the previous visible variables to the current visible variables, and (iii) a directed model from the previous visible variables to the latent variables. In our approach, we first train two exclusive CRBMs for the source and the target speakers independently using segmented training data prepared for each speaker, then train a NN using the projected features, and finally, fine-tune the networks as a single network for VC. Because the training data for the source speaker CRBM include various phonemes particular to the speaker, the speaker-dependent network tries to capture the abstractions to maximally express the training data that have abundant speaker individuality information and less phoneme-related information. Furthermore, the network captures time-series features with the directed models (ii) and (iii), enabling it to discover temporal correlations at the same time. Therefore, we expect that if feature conversion is conducted in such time-related individuality-emphasized high-order spaces, it is much easier to convert voice features than if the original spectrum-based space is used.

Similar research can be found in [32] and [33]. Wu et al. employed a CRBM to capture the linear and non-linear relationship between the source and the target features [32]. Chen et al. also used a RBM to model the joint spectral distribution instead of using the conventional joint density GMM [33]. Unlike these approaches, which is based on a joint model, our method trains two exclusive RBMs for each speaker, aiming to capture speaker-specific conversion-friendly features. We will discuss the differences between these approaches and the proposed method in Section 4.

The rest of the article is organized as follows. In Section 2, we briefly review the fundamental techniques, (RBMs and CRBMs) before explaining our method. The proposed VC system is presented in Section 3, and we compare the proposed method with existing related work in Section 4. We describe the various experiments and VC results in Section 5, and we conclude the article in Section 6.

2 Preliminaries

Our voice conversion system uses CRBMs to capture high-order conversion-friendly features. A RBM, the fundamental model of the CRBM, was first introduced as a method of representing binary-valued data [34,35], and it later came to be used to deal with real-valued data (such as acoustic features) known as a Gaussian-Bernoulli RBM (GBRBM) [36]. However, it has been reported that the original GBRBM had some difficulties because the training of the parameters was unstable [27,37,38]. Later, an improved learning method for GBRBM was proposed by Cho et al. [39] to overcome the difficulties. We briefly review RBMs and CRBMs in this section, introducing their improved versions.

2.1 RBM

A RBM is an undirected graphical model that defines the distribution of visible variables with binary hidden (latent) variables. In literature dealing with the improved GBRBM [39], the joint probability $p(\boldsymbol{v},\boldsymbol{h})$ of real-valued visible units $\boldsymbol{v} = [v_1, \cdots, v_I]^T, v_i \in \mathbb{R}$ and binary-valued hidden units $\boldsymbol{h} = [h_1, \cdots, h_J]^T, h_j \in \{0,1\}$ is defined as follows:

$$p(\boldsymbol{v},\boldsymbol{h}) = \frac{1}{Z} e^{-E(\boldsymbol{v},\boldsymbol{h})} \tag{1}$$

$$E(\boldsymbol{v},\boldsymbol{h}) = \left\| \frac{\boldsymbol{v}-\boldsymbol{b}}{2\boldsymbol{\sigma}} \right\|^2 - \boldsymbol{c}^T\boldsymbol{h} - \left(\frac{\boldsymbol{v}}{\boldsymbol{\sigma}^2}\right)^T \boldsymbol{W}\boldsymbol{h} \tag{2}$$

$$Z = \sum_{\boldsymbol{v},\boldsymbol{h}} e^{-E(\boldsymbol{v},\boldsymbol{h})}, \tag{3}$$

where $\| \cdot \|^2$ denotes L2 norm. $\boldsymbol{W} \in \mathbb{R}^{I\times J}$, $\boldsymbol{\sigma} \in \mathbb{R}^{I\times 1}$, $\boldsymbol{b} \in \mathbb{R}^{I\times 1}$, and $\boldsymbol{c} \in \mathbb{R}^{J\times 1}$ are model parameters of the GBRBM, indicating the weight matrix between visible units and hidden units, the standard deviations associated with Gaussian visible units, a bias vector of the visible units, and a bias vector of hidden units, respectively. The fraction bar in Equation 2 denotes the element-wise division.

Because there are no connections between visible units or between hidden units, the conditional probabilities $p(\boldsymbol{h}|\boldsymbol{v})$ and $p(\boldsymbol{v}|\boldsymbol{h})$ form simple equations as follows:

$$p(h_j = 1|\boldsymbol{v}) = \mathcal{S}\left(c_j + \left(\frac{\boldsymbol{v}}{\boldsymbol{\sigma}^2}\right)^T \boldsymbol{W}_{:j}\right) \tag{4}$$

$$p(v_i = v|\boldsymbol{h}) = \mathcal{N}\left(v \mid b_i + \boldsymbol{W}_{i:}\boldsymbol{h}, \sigma_i^2\right), \tag{5}$$

where $\boldsymbol{W}_{:j}$ and $\boldsymbol{W}_{i:}$ denote the jth column vector and the ith row vector, respectively. $\mathcal{S}(\cdot)$ and $\mathcal{N}(\cdot|\mu,\sigma^2)$ indicate an element-wise sigmoid function and Gaussian probability density function with the mean μ and variance σ^2, respectively.

For parameter estimation, the log-likelihood of a collection of visible units $\mathcal{L} = \log \prod_n p(\boldsymbol{v}_n)$ is used as an evaluation function. Differentiating partially with respect to each parameter, we obtain:

$$\frac{\partial \mathcal{L}}{\partial W_{ij}} = \left\langle \frac{v_i h_j}{\sigma_i^2} \right\rangle_{\text{data}} - \left\langle \frac{v_i h_j}{\sigma_i^2} \right\rangle_{\text{model}} \tag{6}$$

$$\frac{\partial \mathcal{L}}{\partial b_i} = \left\langle \frac{v_i}{\sigma_i^2} \right\rangle_{\text{data}} - \left\langle \frac{v_i}{\sigma_i^2} \right\rangle_{\text{model}} \tag{7}$$

$$\frac{\partial \mathcal{L}}{\partial c_j} = \langle h_j \rangle_{\text{data}} - \langle h_j \rangle_{\text{model}}, \tag{8}$$

where $\langle\cdot\rangle_{\text{data}}$ and $\langle\cdot\rangle_{\text{model}}$ indicate expectations of input data and the inner model, respectively. However, it is generally difficult to compute the second term, so, typically, the expected reconstructed data $\langle\cdot\rangle_{\text{recon}}$ computed by Equations 4 and 5 is used instead [27].

In the improved GBRBM, the variance parameter σ_i^2 is replaced as $\sigma_i^2 = e^{z_i}$ so as to constrain the variance to a non-zero value and provide stability in training the parameters. Under this modification, the gradient with respect to z_i becomes:

$$\begin{aligned}\frac{\partial \mathcal{L}}{\partial z_i} = \; & e^{-z_i} \left\langle \frac{(v_i - b_i)^2}{2} - v_i \boldsymbol{W}_{i:}\boldsymbol{h} \right\rangle_{\text{data}} \\ & - e^{-z_i} \left\langle \frac{(v_i - b_i)^2}{2} - v_i \boldsymbol{W}_{i:}\boldsymbol{h} \right\rangle_{\text{model}}.\end{aligned} \tag{9}$$

Using Equations 6, 7, 8, and 9, each parameter can be updated by stochastic gradient descent with a fixed learning rate and a momentum term.

2.2 CRBM

A CRBM is an extended version of RBM proposed by Taylor et al. [40] and is suitable for the representation of time series data. In addition to the use of an undirected model as in RBM, CRBM also employs directed models from a collection of previous visible units $\mathcal{V}'^{(t)} = \{\boldsymbol{v}^{(p)}\}_{p=t-P}^{t-1}$, $\boldsymbol{v}^{(p)} = \left[v_1^{(p)}, \cdots, v_I^{(p)}\right]^T, v_i^{(p)} \in \mathbb{R}$ to binary hidden units $\boldsymbol{h}^{(t)} = \left[h_1^{(t)}, \cdots, h_J^{(t)}\right]^T, h_j^{(t)} \in \{0,1\}$ and to the current visible units $\boldsymbol{v}^{(t)} = \left[v_1^{(t)}, \cdots, v_I^{(t)}\right]^T, v_i^{(t)} \in \mathbb{R}$ at the current frame t, where P is the number of previous frames from the current frame taken into account. In this model, there are three types of parameters to be estimated: $\boldsymbol{W}_{v'_p v} \in \mathbb{R}^{I\times I}$ (a directed weight matrix from $\boldsymbol{v}^{(t-p)}$ to $\boldsymbol{v}^{(t)}$), $\boldsymbol{W}_{v'_p h} \in \mathbb{R}^{I\times J}$ (a directed weight matrix from $\boldsymbol{v}^{(t-p)}$ to $\boldsymbol{h}^{(t)}$), and $\boldsymbol{W}_{vh} \in \mathbb{R}^{I\times J}$ (an undirected weight matrix between $\boldsymbol{v}^{(t)}$ and $\boldsymbol{h}^{(t)}$). These weights are estimated using contrastive divergence in a similar manner to an RBM by maximizing the likelihood $\mathcal{L} = \log \prod_t p\left(\boldsymbol{v}^{(t)}|\mathcal{V}'^{(t)}\right)$, where:

$$p\left(\boldsymbol{v}^{(t)}|\mathcal{V}'^{(t)}\right) = \frac{1}{Z}\sum_{\boldsymbol{h}^{(t)}} e^{-E\left(\boldsymbol{v}^{(t)},\boldsymbol{h}^{(t)}|\mathcal{V}'^{(t)}\right)} \tag{10}$$

$$Z = \sum_{\boldsymbol{v}^{(t)},\boldsymbol{h}^{(t)}} e^{-E\left(\boldsymbol{v}^{(t)},\boldsymbol{h}^{(t)}|\mathcal{V}'^{(t)}\right)}. \tag{11}$$

Inspired by the improvement learning method of a GBRBM, we define the energy function E in this paper as follows:

$$E\left(\boldsymbol{v}^{(t)},\boldsymbol{h}^{(t)}|\mathcal{V}'^{(t)}\right) \tag{12}$$
$$= \left\|\frac{\boldsymbol{v}^{(t)}-\boldsymbol{b}'^{(t)}}{2\sigma}\right\|^2 - \boldsymbol{c}'^{(t)T}\boldsymbol{h}^{(t)} - \left(\frac{\boldsymbol{v}^{(t)}}{\sigma^2}\right)^T \boldsymbol{W}_{vh}\boldsymbol{h}^{(t)}$$

$$\boldsymbol{b}'^{(t)} = \boldsymbol{b} + \sum_p \boldsymbol{W}^T_{v'_p v}\boldsymbol{v}^{(t-p)} \tag{13}$$

$$\boldsymbol{c}'^{(t)} = \boldsymbol{c} + \sum_p \boldsymbol{W}^T_{v'_p h}\boldsymbol{v}^{(t-p)}. \tag{14}$$

We obtain the following partial differential equations to the log-likelihood $\mathcal{L}$:

$$\frac{\partial\mathcal{L}}{\partial\left(\boldsymbol{W}_{v'_p v}\right)_{i'i}} = \left\langle\frac{v_i^{(t)}v_{i'}^{(t-p)}}{\sigma_i^2}\right\rangle_{\text{data}} - \left\langle\frac{v_i^{(t)}v_{i'}^{(t-p)}}{\sigma_i^2}\right\rangle_{\text{model}} \tag{15}$$

$$\frac{\partial\mathcal{L}}{\partial\left(\boldsymbol{W}_{v'_p h}\right)_{i'j}} = \left\langle v_{i'}^{(t-p)}h_j^{(t)}\right\rangle_{\text{data}} - \left\langle v_{i'}^{(t-p)}h_j^{(t)}\right\rangle_{\text{model}} \tag{16}$$

The other parameters related to the undirected model ($\boldsymbol{W}_{vh}$, $\boldsymbol{b}$, $\boldsymbol{c}$, and $\boldsymbol{\sigma}$ (or $\boldsymbol{z}$)) are also calculated from Equations 6, 7, 8, and 9 by proper substitution of variables. Once the parameters are estimated, forward inference (the conditional probability of $\boldsymbol{h}^{(t)}$ given $\boldsymbol{v}^{(t)}$ and $\mathcal{V}'^{(t)}$) and backward inference (the conditional probability of $\boldsymbol{v}^{(t)}$ given $\boldsymbol{h}^{(t)}$ and $\mathcal{V}'^{(t)}$) are respectively written as:

$$p\left(h_j^{(t)} = 1|\boldsymbol{v}^{(t)},\mathcal{V}'^{(t)}\right)$$
$$= \mathcal{S}\left(c_j + \boldsymbol{v}^{(t)T}\boldsymbol{W}_{vh_{:j}} + \sum_p \boldsymbol{v}^{(t-p)T}\boldsymbol{W}_{v'_p h_{:j}}\right) \tag{17}$$

$$p\left(v_i^{(t)} = v|\boldsymbol{h}^{(t)},\mathcal{V}'^{(t)}\right)$$
$$= \mathcal{N}\left(v|b_i + \boldsymbol{h}^{(t)T}\boldsymbol{W}^T_{vh_{i:}} + \sum_p \boldsymbol{v}^{(t-p)T}\boldsymbol{W}_{v'_p v_{:j}}, \sigma_i^2\right) \tag{18}$$

3 Proposed voice conversion

In general, the fewer phonological and the more individuality-emphasized features a source input includes for a speaker, the easier it is to convert the source features to target features. This paper proposes voice conversion using such features.

Figure 1 shows an overview of our proposed voice conversion system where we set $P = 1$. In our approach, we independently train CRBMs for each speaker beforehand as shown in Figure 1a. Variables $\boldsymbol{x}^{(t)}$ and $\boldsymbol{y}^{(t)}$ ($\boldsymbol{x}^{(t-1)}$ and $\boldsymbol{y}^{(t-1)}$) represent acoustic feature vectors (e.g., visible units in a CRBM), such as Mel-frequency cepstral coefficient (MFCC), at frame t (at frame $t-1$) for a source speaker and a target speaker, respectively.

For the source speaker, for instance, the parameter matrix $\boldsymbol{W}_{xh}$ is, along with $\boldsymbol{W}_{x'h}$ and $\boldsymbol{W}_{x'x}$, estimated so as to maximize the probability of T chained training samples $p(\boldsymbol{x}) = \prod_{t=1}^T p\left(\boldsymbol{x}^{(t)}|\boldsymbol{x}^{(t-1)}\right)$. Using these matrices, an input vector $\boldsymbol{x}^{(t)}$ at frame t given the previous vector $\boldsymbol{x}^{(t-1)}$ is projected into the speaker-dependent latent space that captures speaker-individualities. The latent features $\boldsymbol{h}_x^{(t)}$ can be calculated using mean-field approximation as follows:

$$\boldsymbol{h}_x^{(t)} = \mathcal{S}\left(\boldsymbol{W}_{xh}\boldsymbol{x}^{(t)} + \boldsymbol{W}_{x'h}\boldsymbol{x}^{(t-1)} + \boldsymbol{c}_x\right) \tag{19}$$

from Equation 17, where $\boldsymbol{c}_x$ is a bias vector of forward inference for the source speaker. Because each unit in the hidden vector $\boldsymbol{h}_x^{(t)}$ is independent from the others (due to the nature of RBM), it captures the *common* characteristics in the visible units. The training data usually include various phonemes and unvarying speaker-specific features; thus, we expect that the extracted features in $\boldsymbol{h}_x^{(t)}$ represent speaker-individual information. Since we estimate the time-related matrices $\boldsymbol{W}_{x'h}$ and $\boldsymbol{W}_{x'x}$ jointly with the static term $\boldsymbol{W}_{xh}$ as shown in Equation 12 using the training data, they capture time-related information and $\boldsymbol{W}_{xh}$ can focus on capturing other static information. This means that the obtained features in the hidden units $\boldsymbol{h}_x^{(t)}$ also help to capture time-related speaker-individualities. The above discussion applies to the target speaker, and the hidden vector for the target $\boldsymbol{y}^{(t)}$ is obtained in the same manner as in Equation 19:

$$\boldsymbol{h}_y^{(t)} = \mathcal{S}\left(\boldsymbol{W}_{yh}\boldsymbol{y}^{(t)} + \boldsymbol{W}_{y'h}\boldsymbol{y}^{(t-1)} + \boldsymbol{c}_y\right) \tag{20}$$

where $\boldsymbol{c}_y$ is a bias vector for the target speaker.

In our approach, we convert such individuality-emphasized features (from $\boldsymbol{h}_x^{(t)}$ to $\boldsymbol{h}_y^{(t)}$) using a NN that has $L+2$ layers (L is the number of hidden layers; typically, L is 0 or 1) as shown in Figure 1b. To train the NN, we use the parallel training set $\{\boldsymbol{x}_t, \boldsymbol{y}_t\}_{t=0}^{T'}$ where T' is the number of frames of the parallel data[a]. During the training stage of the NN, the projected vectors of the source speaker's acoustic features $\boldsymbol{h}_x^{(t)}$ are used as inputs, and the projected vectors of the corresponding target speaker's features $\boldsymbol{h}_y^{(t)}$ are used as outputs. Weight parameters of the NN $\{\boldsymbol{W}_l, \boldsymbol{d}_l\}_{l=0}^{L}$ are estimated to minimize the error between the output $\eta\left(\boldsymbol{h}_x^{(t)}\right)$ and the target vector $\boldsymbol{h}_y^{(t)}$

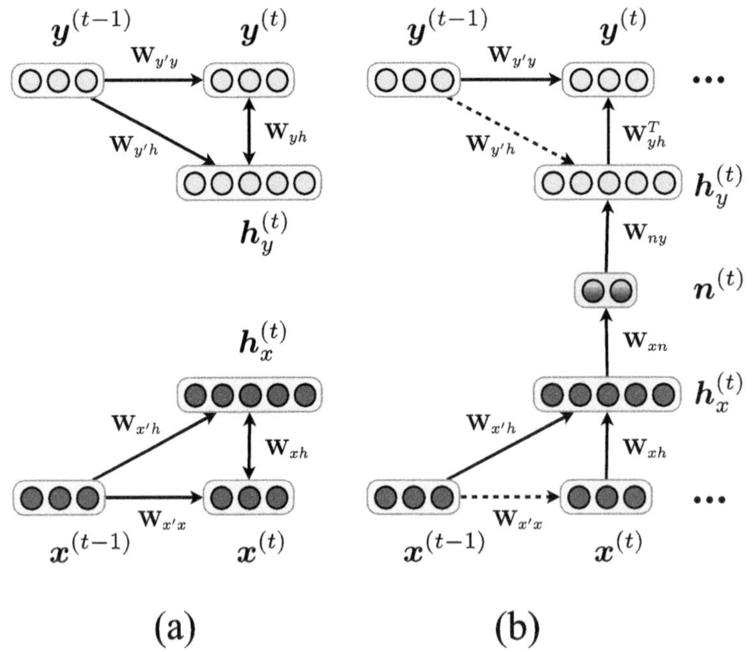

Figure 1 A flow chart of the proposed voice conversion system. (a) CRBMs for a source speaker (below) and a target speaker . **(b)** Our proposed voice conversion architecture combining the two pre-trained speaker-dependent CRBMs with a concatenating NN.

as is typical for a NN. Once the weight parameters are estimated, an input vector $\boldsymbol{h}_x^{(t)}$ is converted to:

$$\eta\left(\boldsymbol{h}_x^{(t)}\right) = \bigodot_{l=0}^{L} \eta_l\left(\boldsymbol{h}_x^{(t)}\right) \tag{21}$$

$$\eta_l\left(\boldsymbol{h}_x^{(t)}\right) = \mathcal{S}\left(\boldsymbol{W}_l \boldsymbol{h}_x^{(t)} + \boldsymbol{d}_l\right) \tag{22}$$

where $\bigodot_{l=0}^{L}$ denotes the composition of $L+1$ functions. For instance, $\bigodot_{l=0}^{1} \eta_l(\boldsymbol{z}) = \mathcal{S}(\boldsymbol{W}_1\mathcal{S}(\boldsymbol{W}_0\boldsymbol{z} + \boldsymbol{d}_0) + \boldsymbol{d}_1)$ for a NN with one hidden layer.

To convert the output of the NN to the acoustic features of the target speaker, we simply use backward inference of a CRBM using Equation 18, resulting in:

$$p\left(\boldsymbol{y}^{(t)}|\boldsymbol{h}_y^{(t)}, \boldsymbol{y}^{(t-1)}\right) = \mathcal{N}\left(\boldsymbol{y}|\boldsymbol{W}_{yh}^T\boldsymbol{h}_y^{(t)} + \boldsymbol{W}_{y'y}\boldsymbol{y}^{(t-1)} + \boldsymbol{b}_y, \boldsymbol{\sigma}_y^2\right) \tag{23}$$

where $\boldsymbol{b}_y$ and $\boldsymbol{\sigma}_y$ are a bias vector of backward inference for the target speaker, respectively. Generalizing and summarizing the above discussion, a voice conversion function of our method from a source acoustic vector $\boldsymbol{x}^{(t)}$ to a target vector $\boldsymbol{y}^{(t)}$ at frame t, given the previous vectors $\mathcal{X}'^{(t)} = \{\boldsymbol{x}^{(t-p)}\})_{p=1}^{P}$ and $\mathcal{Y}'^{(t)} = \left\{\boldsymbol{y}^{(t-p)}\right\}_{p=1}^{P}$, is written as:

$$\boldsymbol{y}^{(t)} = \bigodot_{k=0}^{L+2} f_{(k)}\left(\boldsymbol{W}_{(k)}\boldsymbol{x}^{(t)} + \boldsymbol{a}_{(k)}(\mathcal{X}'^{(t)}, \mathcal{Y}'^{(t)})\right) \tag{24}$$

where $\boldsymbol{W}_{(k)}$ and $\boldsymbol{a}_{(k)}\left(\mathcal{X}'^{(t)}, \mathcal{Y}'^{(t)}\right)$ denote elements of a set of our model parameters $\Theta = \{\mathcal{W} \cup \mathcal{A}\}$:

$$\mathcal{W} = \left\{\boldsymbol{W}_{(k)}\right\}_{k=0}^{L+2} \tag{25}$$

$$= \left\{\boldsymbol{W}_{xh}, \boldsymbol{W}_0, \cdots, \boldsymbol{W}_L, {\boldsymbol{W}_{yh}}^T\right\} \tag{26}$$

$$\mathcal{A} = \left\{\boldsymbol{a}_{(k)}(\mathcal{X}'^{(t)}, \mathcal{Y}'^{(t)})\right\}_{k=0}^{L+2} \tag{27}$$

$$= \left\{\boldsymbol{c}_x + \sum_p \boldsymbol{W}_{x'h}\boldsymbol{x}^{(t-p)}, \boldsymbol{d}_0, \right. \tag{28}$$

$$\left. \cdots, \boldsymbol{d}_L, \boldsymbol{b}_y + \sum_p \boldsymbol{W}_{y'y}\boldsymbol{y}^{(t-p)}\right\}, \tag{29}$$

and $\left\{f_{(k)}\right\}_{k=0}^{L+2} = \{\mathcal{S}, \mathcal{S}, \cdots, \mathcal{S}, \mathcal{I}\}$, where $\mathcal{I}$ indicates an identity function.

The conversion function shown in Equation 24 implies a dynamic model of a $(L+4)$-layer network with sigmoid activated functions. Therefore, regarding it as a recurrent neural network (RNN),[b] we can fine-tune[c] each parameter of the entire network by back-propagation through time (BPTT) [41] using the acoustic parallel data. Specifically, each parameter is re-updated so as to minimize the total error ϵ in a gradient-descent-based approach, which is defined as:

$$\epsilon = \sum_{1 \geq t \geq T} \epsilon^{(t)} = \frac{1}{2} \sum_{1 \geq t \geq T} \left(\boldsymbol{y}^{(t)} - \boldsymbol{\nu}^{(t)}\right)^2, \tag{30}$$

where $\boldsymbol{v}^{(t)}$ denotes the output of RNN at frame t. The gradient with respect to θ, which is a parameter in the highest recursive hidden layer, for instance, can be written as follows:

$$\frac{\partial \epsilon}{\partial \theta} = \sum_{1 \le t \le T} \frac{\partial \epsilon^{(t)}}{\partial \theta} \tag{31}$$

$$\frac{\partial \epsilon^{(t)}}{\partial \theta} = \sum_{1 \le k \le t} \left(\frac{\partial \epsilon^{(t)}}{\partial \boldsymbol{h}_y^{(t)}} \frac{\partial \boldsymbol{h}_y^{(t)}}{\partial \boldsymbol{h}_y^{(k)}} \frac{\partial^+ \boldsymbol{h}_y^{(k)}}{\partial \theta} \right) \tag{32}$$

$$\frac{\partial \boldsymbol{h}_y^{(t)}}{\partial \boldsymbol{h}_y^{(k)}} = \prod_{t \ge i > k} \frac{\partial \boldsymbol{h}_y^{(i)}}{\partial \boldsymbol{h}_y^{(i-1)}} \tag{33}$$

$$= \prod_{t \ge i > k} W_{y'y} \left(1 - \mathcal{S}\left(\boldsymbol{h}_y^{(i-1)}\right)\right), \tag{34}$$

where $\frac{\partial^+ \boldsymbol{h}^{(k)}}{\partial \theta}$ refers to the immediate partial derivative of the hidden units $\boldsymbol{h}^{(k)}$ with respect to θ (i.e., $\boldsymbol{h}^{(k-1)}$ is regarded as a constant with respect to θ).

As Equation 24 indicates, we need a current acoustic vector from a source speaker and previous vectors from both a source speaker and a target speaker to estimate the target speaker's current acoustic vector. However, we never know the correct previous vector of the target speaker, so in practice, we use the last converted (estimated) vectors as the previous target vector iteratively, starting from a zero vector. We confirmed that this approach worked well through our preliminary experiments.

Meanwhile, a conventional GMM-based approach [9] with M Gaussian mixtures converts the source features $\boldsymbol{x}$ as:

$$\boldsymbol{y} = \sum_{m=1}^{M} P(m|\boldsymbol{x}) \left(\boldsymbol{\Sigma}_{yx}^{(m)} \boldsymbol{\Sigma}_{xx}^{(m)-1} \left(\boldsymbol{x} - \boldsymbol{\mu}_x^{(m)} \right) + \boldsymbol{\mu}_y^{(m)} \right) \tag{35}$$

$$P(m|\boldsymbol{x}) = \frac{w^{(m)} \mathcal{N}\left(\boldsymbol{x}; \boldsymbol{\mu}_x^{(m)}, \boldsymbol{\Sigma}_{xx}^{(m)}\right)}{\sum_{m=1}^{M} w^{(m)} \mathcal{N}\left(\boldsymbol{x}; \boldsymbol{\mu}_x^{(m)}, \boldsymbol{\Sigma}_{xx}^{(m)}\right)} \tag{36}$$

where $w^{(m)}$, $\boldsymbol{\mu}_{\cdot}^{(m)}$ and $\boldsymbol{\Sigma}_{\cdot}^{(m)}$ are the weight, the corresponding mean vectors, and the corresponding covariance matrices to the speaker of the mth mixture, respectively, showing it to be an additive model of piecewise linear functions. Our approach using Equation 24 is based on the composite function of multiple different non-linear functions feeding time-series data. Therefore, it is expected that our composite model can represent more complex relationships than the conventional GMM-based method and other static network approaches [18,25].

4 Related work

It is worth noting that we compare our method with the conventional method proposed by Wu et al. in [32], that also employed a CRBM for VC. Figure 2 shows the comparison of graphical models among three methods. Wu's method directly uses CRBM to estimate the target features $\boldsymbol{y}^{(t)}$ from the input $\boldsymbol{x}^{(t)}$ along with the latent features $\boldsymbol{h}^{(t)}$ to capture the linear and non-linear relationship between the source and the target features (Figure 2b). On the other hand, our method (Figure 2c) uses two CRBMs for each of the source and the target speakers to obtain their

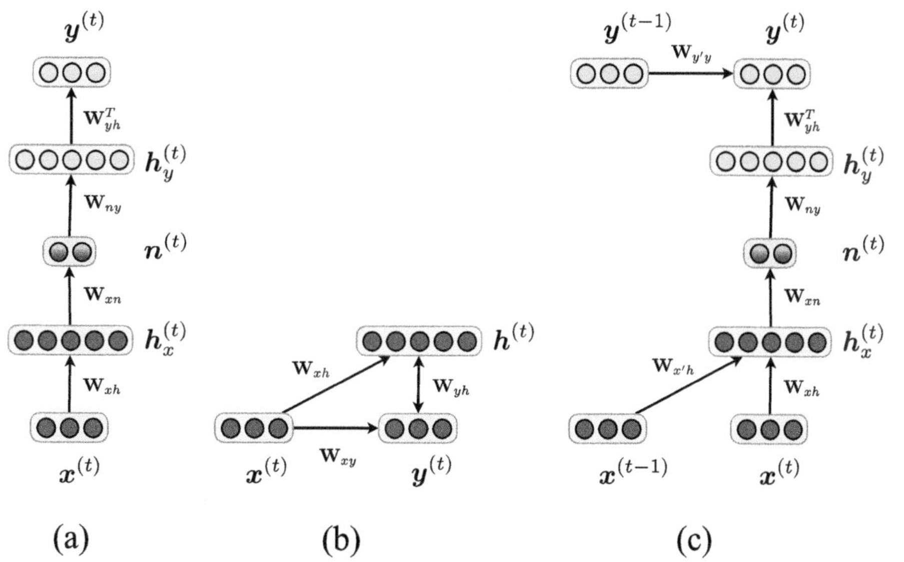

Figure 2 Model structures of the related systems. (a) Our earlier work, speaker-dependent RBM, **(b)** CRBM proposed in [32], and **(c)** our proposed method, speaker-dependent CRBM.

latent features $h_x^{(t)}$ and $h_y^{(t)}$, capturing time-related information (from $t-1$ to t frames). Connecting the latent features using a NN, the entire conversion network of our method consequently forms a deep architecture. Our previous approach [25] has a deep network similar to that of the proposed method (Figure 2a); however, the difference is that it involves time-related relationships in the network.

Since the acoustic signals we are targeting are time-series data, the model that captures time-related information will provide us with the better performance.

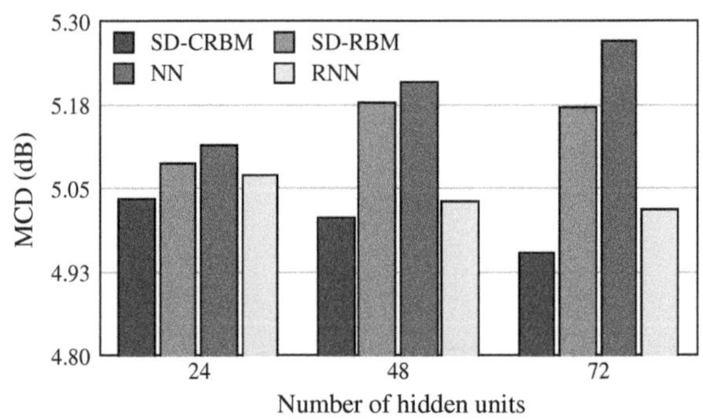

Figure 3 Changing hidden units. The values show averaged Mel-cepstral distortion with varying numbers of hidden units J for all network-based methods ($N = 20,000$).

5 Experiments

5.1 Conditions

In our experiments, we conducted voice conversion using the ATR Japanese speech database [42], comparing our method (speaker-dependent restricted Boltzmann machines; say 'SD-CRBM') with four methods: the well-known GMM-based approach ('GMM'), conventional NN-based voice conversion [18] ('NN'), our previous work [25] ('SD-RBM') and, for a reference, a recurrent neural network with randomly-initialized weights ('RNN'). In order to evaluate our method under various circumstances, we tested male-to-female (the source and the target speakers are identified with MMY and FTK in the database, respectively), female-to-female (FKN and FTK), and male-to-male (MMY and MHT) patterns.

For an input vector, we calculated 24-dimensional MFCC features from 513-dimensional STRAIGHT spectra [43] using the filter-theory [44] to decode the MFCC back to STRAIGHT spectra in the synthesis stage. Each speech signal was sampled at 12 kHz and windowed with a 25-ms Hamming window every 10 ms. Unlike our previous work [25], we processed the obtained MFCC with zero component analysis (ZCA) whitening [38], where we confirmed it worked better than without whitening, especially for 'NN.' The parallel data of the source/target speakers processed by dynamic programming were created from 216 word utterances in the dataset and were used for the training of each method (note that two CRBMs for 'SD-CRBM' and two RBMs for 'SD-RBM' can be trained without the necessity of using parallel data, although we used the same parallel training data for the CRBMs and the RBMs in this research.)

The network-based approaches ('SD-CRBM', 'SD-RBM', 'NN', and 'RNN') were trained using gradient descent with a learning rate of 0.01 and momentum of 0.9, with the number of epochs being 400. The parameters of 'NN' and 'RNN' were initialized randomly. All the network-based methods had four layers including an input layer, two hidden layers, and an output layer. Other configurations, such as the number of hidden units, will be discussed in the following section.

For the GMM-based approach, we used diagonal covariance matrices without global variance and dynamic features.

For the objective test, 15 sentences (about 60 s long) that were not included in the training data were arbitrarily selected from the database (identified with SDA01∼SDA15). For the objective evaluation, we used Mel-cepstral distortion (MCD) to measure how close the converted vector is to the target vector in Mel-cepstral space. The MCD is defined as below:

$$\text{MCD}\,[dB] = \frac{10}{\ln 10}\sqrt{2\sum_{d=1}^{24}(c_d - c'_d)^2} \qquad (37)$$

where c_d and c'_d denote the dth original target MFCC and the converted MFCC, respectively. The smaller the value of MCD is, the closer the converted spectra are to the target spectra. We calculated the MCD for each frame in the training data and averaged the MCD values for the final evaluation.

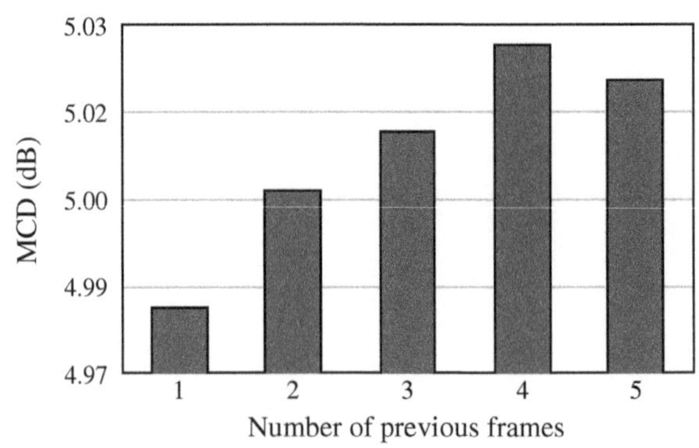

Figure 4 Changing previous frames. The values show averaged Mel-cepstral distortion with varying numbers of previous frames P to be taken into account for our method ($N = 20,000$).

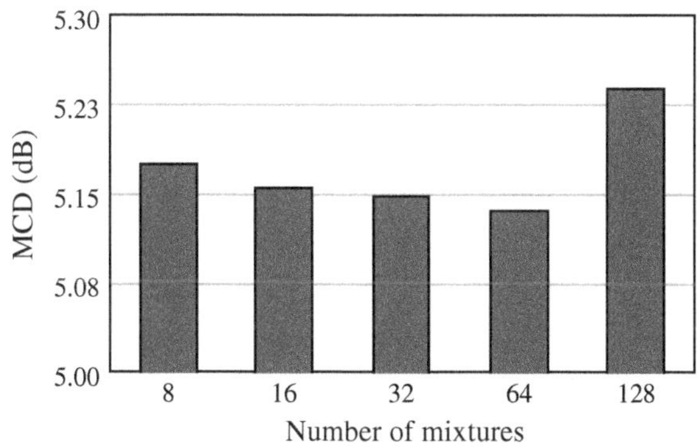

Figure 5 Changing mixtures. The values show averaged Mel-cepstral distortion with varying numbers of mixtures M for GMM method ($N = 20,000$).

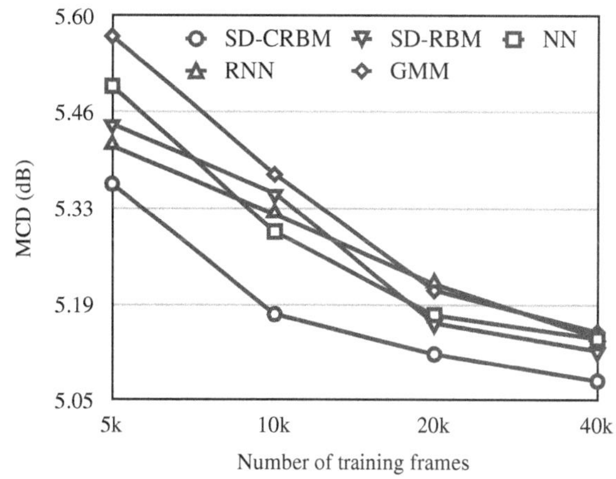

Figure 7 Female-to-female voice conversion results. The values show averaged Mel-cepstral distortion for each method with varying amounts of training data.

For the subjective evaluation, ABX listening tests were conducted, where nine participants listened to five pairs of converted speech signals (from a development set, which was used for the determination of model parameters) produced using our approach ('SD-CRBM') and the converted speech signals produced by the other methods ('SD-RBM', 'NN', 'RNN', and 'GMM') along with an original target speech signal (generated from analysis-by-synthesis). We evaluated the models, which were trained using $N = 5,000$ or $N = 20,000$ training frames. They then selected the *better* one in terms of speaker identity (how well they can recognize the speaker from the converted speech) and speech quality (how clear and natural the converted speech is).

5.2 Determining appropriate parameters

In this section, we report preliminary experiments in which we tested various models with different hyper parameters to determine the appropriate ones. All models were trained using $N = 20,000$ frames from the male-to-female training data and evaluated using a development set of five sentences (identified with SDA16~SDA20 in the database) that were not included in either the training set or the test set.

5.2.1 Network-based methods

Here, we will see how our approach works as the number of hidden units J in each hidden layer changes, comparing it to four network-based methods ('SD-CRBM', 'SD-RBM', 'NN', and 'RNN'). In this preliminary experiment, three architectural patterns were tested, where $J = 24$, 48, and 72. We used $L = 0$, which forms a four-layer network for all methods (for example, when $J = 48$ is

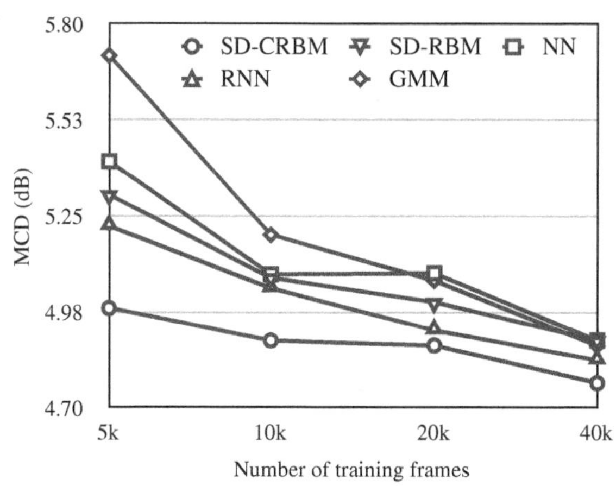

Figure 6 Male-to-female voice conversion results. The values show averaged Mel-cepstral distortion for each method with varying amounts of training data.

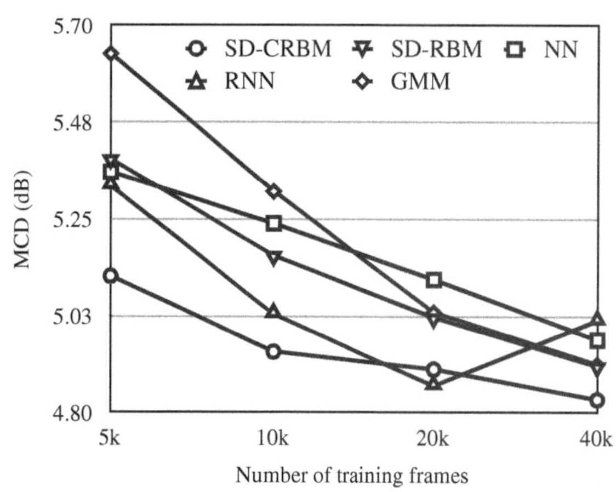

Figure 8 Male-to-male voice conversion results. The values show averaged Mel-cepstral distortion for each method with varying amounts of training data.

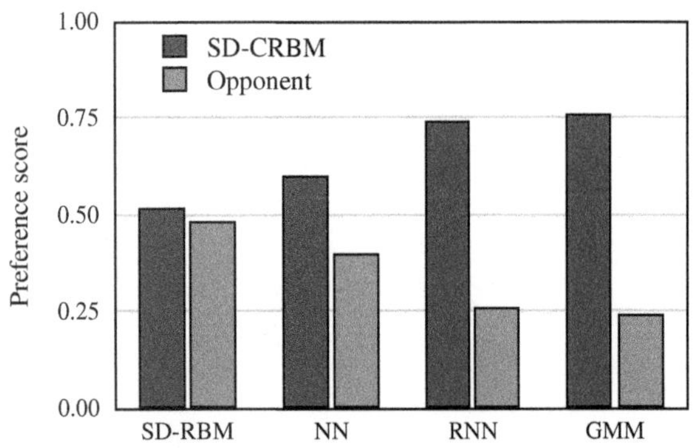

Figure 9 Subjective preference scores w.r.t. speaker identity (in case $N = 5,000$). Our method 'SD-CRBM' was compared to four other methods: 'SD-RBM', 'NN', 'RNN', and 'GMM'.

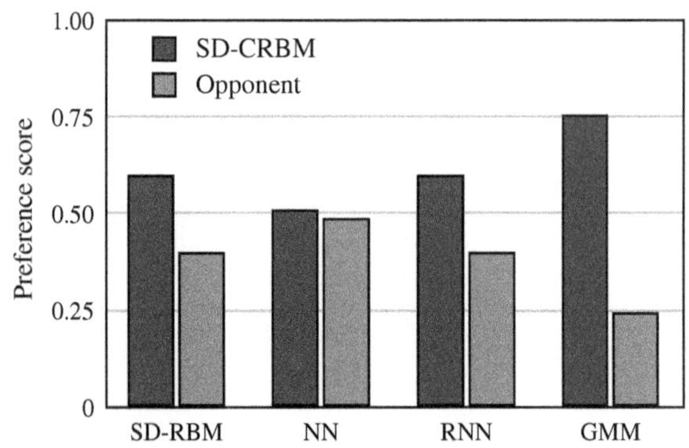

Figure 11 Subjective preference scores w.r.t. speaker identity (in case $N = 20,000$). Our method 'SD-CRBM' was compared to four other methods: 'SD-RBM', 'NN', 'RNN', and 'GMM'.

used, the numbers of units in 'NN' from the input/source layer to the output/target layer become 24, 48, 48, and 24 in order). For 'SD-CRBM', we set $P = 1$ (1 delay for 'RNN' as well), which means we take into account only one previous frame.

Figure 3 compares the averaged MCD obtained for each architecture. As shown in Figure 3, our method 'SD-CRBM' performed the best of all the methods for each case. The interesting point is that the more hidden units the network has, the better performance it provides for 'SD-CRBM' and 'RNN', while it is the other way around for 'SD-RBM' and 'NN'. This is considered to be due to over-fitting to the training data for 'SD-RBM' and 'NN' when the number of parameters is large (e.g., $J = 72$), while 'SD-CRBM' and 'RNN' still required parameters to fit the models that capture time-series data.

For the remaining experiments in this paper, the best architectures for each method were used, i.e., $J = 24$ for 'SD-RBM' and 'NN', and $J = 72$ for 'SD-CRBM' and 'RNN'.

5.2.2 The number of previous frames

We further investigated the performance of our method 'SD-CRBM' with the hidden units of $J = 72$, changing the number of previous frames in the CRBM as $P = 1, 2, 3, 4, 5$. The evaluation results are described in Figure 4, showing the averaged MCDs obtained from each case. As shown in Figure 4, we could not necessarily obtained a better performance as the number of previous frames increased. One reason is that the neighbor source vectors previous to the current one contained similar information, and only a few source vectors were required to estimate the current target vector. Therefore, the poor performance with the larger number of previous frames (e.g., $P = 4$) was caused because the parameter estimation became more difficult as the redundant parameters increased.

In the remaining experiments, we used $P = 1$, which provided the best performance in the preliminary experiment.

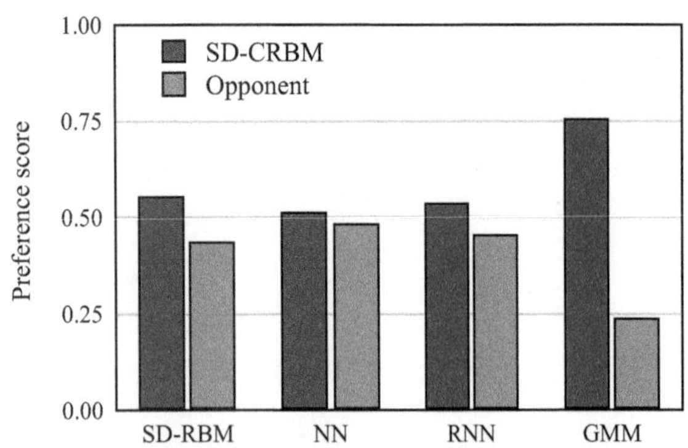

Figure 10 Subjective preference scores w.r.t. speech quality (in case $N = 5,000$). Our method 'SD-CRBM' was compared to four other methods: 'SD-RBM', 'NN', 'RNN', and 'GMM'.

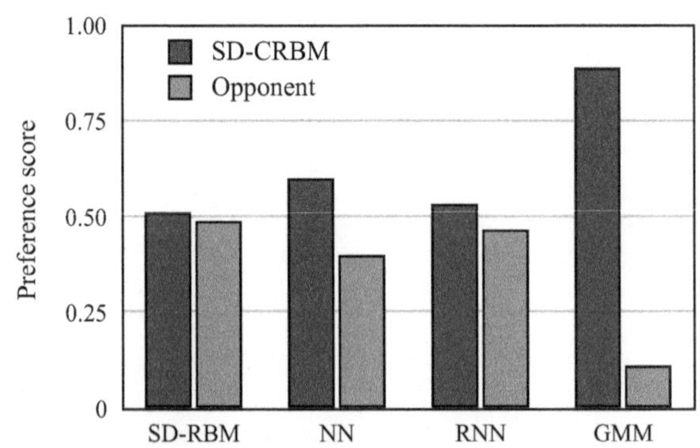

Figure 12 Subjective preference scores w.r.t. speech quality (in case $N = 20,000$). Our method 'SD-CRBM' was compared to four other methods: 'SD-RBM', 'NN', 'RNN', and 'GMM'.

Table 1 *p* values between our method and each method w.r.t. speaker identity in case $N = 5,000$

	SD-RBM	NN	RNN	GMM
p	0.2796	0.1636	*0.0013*	*0.0032*

The values that satisfy $p < 0.1$ are in italics.

Table 3 *p* values between our method and each method w.r.t. speaker quality in case $N = 5,000$

	SD-RBM	NN	RNN	GMM
p	0.2796	0.4343	0.3096	*0.0032*

The values that satisfy $p < 0.1$ are in italics.

5.2.3 *GMM-based method*

For the GMM-based voice conversion ('GMM'), we tried and evaluated five mixtures ($8, 16, 32, 64,$ and 128 mixtures) to determine an appropriate number of mixtures. Figure 5 shows the averaged MCDs over the development set when using the GMM with various mixtures. As shown in the figure, the GMM with 64 mixtures performed the best of all. Therefore, we used mixtures of 64 for 'GMM' in the evaluation experiments described in Section 5.3.

5.3 Evaluation

In this section, we evaluate our method ('SD-CRBM') comparing it with four methods ('SD-RBM', 'NN', 'RNN', and 'GMM') using objective and subjective criteria for each pair of speakers, by changing the number of training frames as $N = 5,000$, 10,000, 20,000, and 40,000.

5.3.1 *Results*

Figures 6, 7, and 8 summarize the experimental results for the test data, comparing each method with respect to objective criteria for male-to-female, female-to-female, and male-to-male voice conversion, respectively. As shown in these Figures, the MCDs decreased as the number of training data increased in most cases (regardless of the gender or the method). Furthermore, our approach outperformed the other methods in every case, except for the case where $N = 20,000$ in the male-to-male experiment.

Figures 9 and 10 show the results of subjective evaluation comparing each method in terms of speaker identity and speaker quality, respectively, when we use training samples of $N = 5,000$. Figures 11 and 12 also show the subjective evaluation results when we use training samples of $N = 20,000$. We also list the *p* values produced by pairwise *t*-testing for each experiment in Tables 1 and 2, and in Tables 3 and 4, in terms of speaker identity and speech quality, respectively. As shown in Figures 11 and 12, our method performed better than each opponent method in regard to mean preference score in terms of both speaker identity and speech quality. However, as shown in Tables 1, 2, 3, and 4, we could not, unfortunately, obtain a significant difference between our method and the other methods in some cases (e.g., 'NN' with respect to (w.r.t.) speaker identity, and 'SD-RBM' and 'RNN' w.r.t. speech quality when $N = 20,000$ training frames were used). We obtained significant differences at a significance level of 0.1 in the other cases.

5.3.2 *Discussion*

In objective criteria, our approach ('SD-CRBM') outperformed the other methods, including the popular GMM-based voice conversion method, in most cases. In subjective criteria as well, we obtained significantly better performance compared with each opponent, in terms of speaker identity and/or speech quality (to be specific, in terms of both speaker identity and speech quality for 'GMM', in terms of only speech quality for 'NN', in terms of only speaker identity for 'SD-RBM' and 'RNN'). The reason for the improvement is attributed to the fact that our time-involving, high-order conversion system using CRBMs is able to capture and convert the abstractions of speaker individualities better than the other methods. In particular, as shown in Figures 6, 7, and 8, our approach achieved high performance in MCD criteria. This is because the CRBMs captured time-series data more appropriately and alleviated estimation errors.

One interesting point is that 'NN' and 'RNN', which were based on random initialization in weight parameters, produced unstable performance (e.g., the MCD by 'NN' increased even as the number of training frames increased from 10,000 to 20,000 in male-to-female conversion, and the MCD by 'RNN' also increased as the number of training data changed from 20,000 to 40,000 in male-to-male conversion). This is caused by a fall into local minima starting from the randomly-initialized weights. Figure 13 shows some of the converged weights in the network, comparing 'RNN' and 'SD-CRBM', where

Table 2 *p* values between our method and each method w.r.t. speaker identity in case $N = 20,000$

	SD-RBM	NN	RNN	GMM
p	*0.0913*	0.4417	*0.0913*	*0.0001*

The values that satisfy $p < 0.1$ are in italics.

Table 4 *p* values between our method and each method w.r.t. speech quality in case $N = 20,000$

	SD-RBM	NN	RNN	GMM
p	0.4417	*0.0913*	0.3299	*0.0000*

The values that satisfy $p < 0.1$ are in italics.

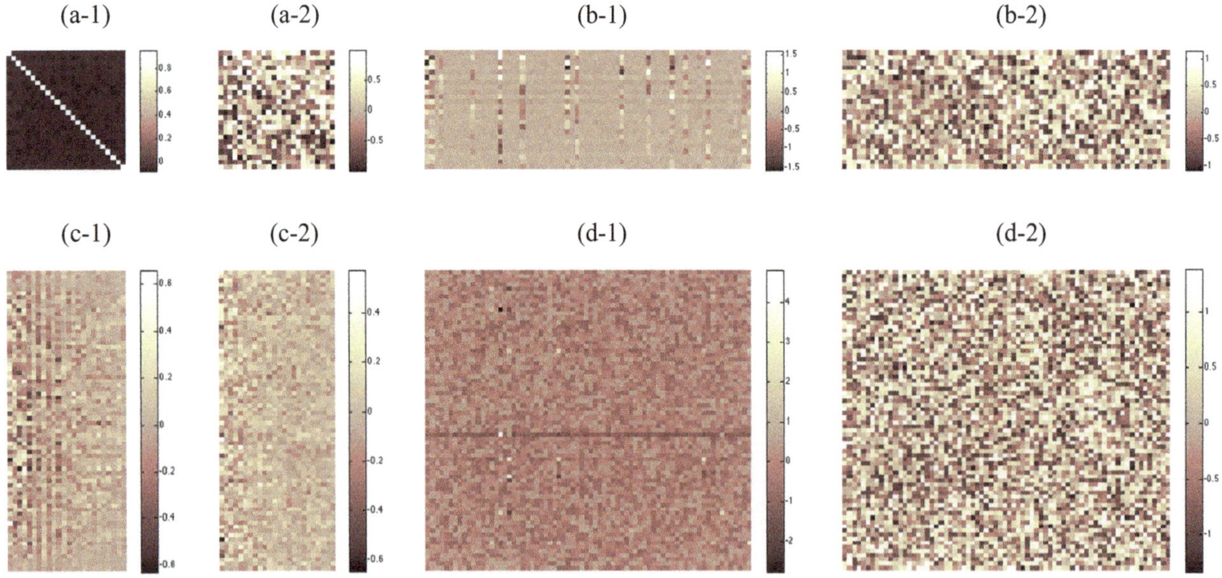

Figure 13 Estimated weights of the pre-trained RNN (··1) and the randomly-initialized RNN (··2). After 400 epochs ($N = 40,000, J = 72$, male-to-female). (**a**) The weights from the previous target vector to the current target vector $\boldsymbol{W}_{y'y}$, (**b**) the weights from the second hidden layer to the current target vector $\boldsymbol{W}_{yh}$, (**c**) the weights from the current source vector to the first hidden layer $\boldsymbol{W}_{xh}$, and (**d**) the weights from the first hidden layer to the second hidden layer $\boldsymbol{W}_0$.

the weights were pre-trained using speaker-dependent CRBMs and a concatenating NN followed by fine-tuning using RNN. As shown in Figure 13, the weights in 'RNN' were almost meaningless and messy; meanwhile, we see that the weights in 'SD-CRBM' had a sparse structure and operative bases. In general, an acoustic feature vector at the last previous frame ($\boldsymbol{v}^{(t-1)}$) is very similar to the feature vector at the current frame ($\boldsymbol{v}^{(t)}$), and, therefore, we expect that the conversion matrix from $\boldsymbol{v}^{(t-1)}$ to $\boldsymbol{v}^{(t)}$ may be close to an identity matrix. The recurrent weight obtained by our approach shown in Figure 13a-1 indicates this fact.

6 Conclusion

We presented a voice conversion method that combines speaker-dependent CRBMs and a NN to extract speaker-individual information for speech conversion. Through experiments, we confirmed that our approach is effective, especially in terms of MCD, compared with the well-known conventional GMM-based approach, a NN-based approach, and our own previous work, SD-RBM, (and recurrent neural network for a reference), regardless of the gender in conversion.

We also conducted ABX experiments for subjective evaluation. The results showed that the performance of our method was not always significantly different in comparison to NN, RNN, and SD-RBM; however, it did perform significantly better than these methods in terms of either speaker identity or speech quality. In the future, we will work to improve our method so that it obtains better results in regard to the sense of hearing.

Competing interests

The authors declare that they have no competing interests.

Author details

[1]Graduate School of System Informatics, Kobe University, 1-1 Rokkodai, Nada-ku, 657-8501 Kobe Japan. [2]Organization of Advanced Science and Technology, Kobe University, 1-1 Rokkodai, Nada-ku, 657-8501 Kobe, Japan.

References

1. A Kain, MW Macon, in *Proceedings of IEEE International Conference on Acoustics, Speech and Signal Processing (ICASSP)*. Spectral voice conversion for text-to-speech synthesis, (1998), pp. 285–288
2. C Veaux, X Robet, in *Proceedings of Interspeech*. Intonation conversion from neutral to expressive speech, (2011), pp. 2765–2768
3. K Nakamura, T Toda, H Saruwatari, K Shikano, Speaking-aid systems using GMM-based voice conversion for electrolaryngeal speech. Speech Commun. **54**(1), 134–146 (2012)
4. L Deng, A Acero, L Jiang, J Droppo, X Huang, in *Proceedings of IEEE International Conference on Acoustics, Speech, and Signal Processing (ICASSP)*. High-performance robust speech recognition using stereo training data, (2001), pp. 301–304
5. A Kunikoshi, Y Qiao, N Minematsu, K Hirose, in *Proceedings of Interspeech*. Speech generation from hand gestures based on space mapping, (2009), pp. 308–311
6. R Gray, Vector quantization. ASSP Mag. IEEE. **1**(2), 4–29 (1984)
7. H Valbret, E Moulines, J-P Tubach, Voice transformation using PSOLA technique. Speech Commun. **11**(2), 175–187 (1992)
8. Y Stylianou, Cappé O, E Moulines, Continuous probabilistic transform for voice conversion. IEEE Trans. Speech Audio Process. **6**(2), 131–142 (1998)

9. T Toda, AW Black, K Tokuda, Voice conversion based on maximum-likelihood estimation of spectral parameter trajectory. IEEE Trans. Audio Speech Lang. Process. **15**(8), 2222–2235 (2007)
10. E Helander, T Virtanen, J Nurminen, Gabbouj, Voice conversion using partial least squares regression. IEEE Trans. Audio Speech Lang. Process. **18**(5), 912–921 (2010)
11. C-H Lee, C-H Wu, in *Proceedings of Interspeech*. Map-based adaptation for speech conversion using adaptation data selection and non-parallel training, (2006), pp. 2254–2257
12. T Toda, Y Ohtani, K Shikano, in *Proceedings of Interspeech*. Eigenvoice conversion based on gaussian mixture model, (2006), pp. 2446–2449
13. D Saito, Yamamoto K, N Minematsu, K Hirose, in *Proceedings of Interspeech*. One-to-many voice conversion based on tensor representation of speaker space, (2011), pp. 653–656
14. D Saito, S Watanabe, A Nakamura, N Minematsu, in *Proceedings of Interspeech*. Probabilistic integration of joint density model and speaker model for voice conversion, (2010), pp. 1728–1731
15. Z Jian, Z Yang, in *Proceedings of International Conference on Software Engineering, Artificial Intelligence, Networking, and Parallel/Distributed Computing*. Voice conversion using canonical correlation analysis based on Gaussian mixture model, (2007), pp. 210–215
16. R Takashima, T Takiguchi, Y Ariki, in *IEEE Spoken Language Technology Workshop (SLT)*. Exemplar-based voice conversion in noisy environment, (2012), pp. 313–317
17. M Narendranath, HA Murthy, S Rajendran, B Yegnanarayana, Transformation of formants for voice conversion using artificial neural networks. Speech Commun. **16**(2), 207–216 (1995)
18. S Desai, EV Raghavendra, B Yegnanarayana, AW Black, K Prahallad, in *Proceedings of IEEE International Conference on Acoustics, Speech and Signal Processing (ICASSP)*. Voice conversion using artificial neural networks, (2009), pp. 3893–3896
19. Y-J Wu, H Kawai, J Ni, R-H Wang, in *Proceedings of IEEE International Conference on Acoustics, Speech, and Signal Processing (ICASSP)*. Minimum segmentation error based discriminative training for speech synthesis application, (2004), p. 629
20. E McDermott, TJ Hazen, J Le Roux, A Nakamura, S Katagiri, Discriminative training for large-vocabulary speech recognition using minimum classification error. IEEE Trans. Audio Speech Lang. Process. **15**(1), 203–223 (2007)
21. T Tomoki, K Tokuda, A speech parameter generation algorithm considering global variance for HMM-based speech synthesis. IEICE Trans. Inform. Syst. **90**(5), 816–824 (2007)
22. Z-H Ling, L-R Dai, Minimum Kullback-Leibler divergence parameter generation for HMM-based speech synthesis. IEEE Trans. Audio Speech Lang. Process. **20**(5), 1492–1502 (2012)
23. Z-H Ling, Y-J Wu, Y-P Wang, L Qin, R-H Wang, in *Blizzard Challenge Workshop*. USTC system for blizzard challenge 2006 an improved HMM-based speech synthesis method, (2006)
24. Z Wu, T Virtanen, T Kinnunen, ES Chng, H Li, in *Proceedings of the 8th ISCA Speech Synthesis Workshop*. Exemplar-based voice conversion using non-negative spectrogram deconvolution, (2013), pp. 221–226
25. T Nakashika, R Takashima, T Takiguchi, Y Ariki, in *Proceedings of Interspeech*. Voice conversion in high-order eigen space using deep belief nets, (2013), pp. 369–372
26. P Smolensky, Information processing in dynamical systems: foundations of harmony theory. Parallel Distributed Process. **1**, 194–281 (1986)
27. GE Hinton, S Osindero, Y-W Teh, A fast learning algorithm for deep belief nets. Neural Comput. **18**(7), 1527–1554 (2006)
28. Z-H Ling, L Deng, D Yu, Modeling spectral envelopes using restricted Boltzmann machines and deep belief networks for statistical parametric speech synthesis. IEEE Trans. Audio Speech Lang. Process. **21**(10), 2129–2139 (2013)
29. A-R Mohamed, GE Dahl, G Hinton, Acoustic modeling using deep belief networks. Audio Speech Lang. Process. IEEE Trans. **20**(1), 14–22 (2012)
30. V Nair, G Hinton, 3-D object recognition with deep belief nets. Adv. Neural Inform. Process. Syst. **22**, 1339–1347 (2009)
31. T Deselaers, S Hasan, O Bender, H Ney, in *Proceedings of the Fourth Workshop on Statistical Machine Translation*. A deep learning approach to machine transliteration, (2009), pp. 233–241
32. Z Wu, ES Chng, H Li, in *Proceedings of the IEEE China Summit and International Conference on Signal and Information Processing (ChinaSIP)*. Conditional restricted Boltzmann machine for voice conversion, (2013)
33. C Ling-Hui, L Zhen-Hua, S Yan, D Li-Rong, in *Proceedings of Interspeech*. Joint spectral distribution modeling using restricted Boltzmann machines for voice conversion, (2013), pp. 3052–3056
34. DH Ackley, GE Hinton, TJ Sejnowski, A learning algorithm for Boltzmann machines. Cogn. Sci. **9**(1), 147–169 (1985)
35. Y Freund, D Haussler, Unsupervised learning of distributions of binary vectors using two layer networks. Adv, Neural Inform. Process. Syst. **4**, 912–919 (1991)
36. GE Hinton, RR Salakhutdinov, Reducing the dimensionality of data with neural networks. Science. **313**(5786), 504–507 (2006)
37. G Hinton, in *Tech. Rep. Department of Computer Science*. A practical guide to training restricted Boltzmann machines (University of Toronto, 2010)
38. A Krizhevsky, G Hinton, *Learning multiple layers of features from tiny images*. (Computer Science Department, University of Toronto, Tech. Rep, 2009)
39. K Cho, A Ilin, T Raiko, in *Artificial Neural Networks and Machine Learning–ICANN 2011*. Improved learning of gaussian-bernoulli restricted Boltzmann machines, (2011), pp. 10–17
40. GW Taylor, GE Hinton, ST Roweis, in *Advances in Neural Information Processing Systems*. Modeling human motion using binary latent variables, (2006), pp. 1345–1352
41. R Pascanu, T Mikolov, Y Bengio, On the difficulty of training recurrent neural networks. (2012)
42. A Kurematsu, K Takeda, Y Sagisaka, S Katagiri, H Kuwabara, K Shikano, ATR japanese speech database as a tool of speech recognition and synthesis. Speech Communication. **9**(4), 357–363 (1990)
43. H Kawahara, M Morise, T Takahashi, R Nisimura, T Irino, H Banno, in *Proceedings of IEEE International Conference on Acoustics, Speech and Signal Processing (ICASSP)*. Tandem-straight: a temporally stable power spectral representation for periodic signals and applications to interference-free spectrum, f0, and aperiodicity estimation, (2008), pp. 3933–3936
44. B Milner, X Shao, in *Proceedings of Interspeech*. Speech reconstruction from mel-frequency cepstral coefficients using a source-filter model, (2002), pp. 2421–2424

Speaker adaptation based on regularized speaker-dependent eigenphone matrix estimation

Wen-Lin Zhang[1*], Wei-Qiang Zhang[2], Dan Qu[1] and Bi-Cheng Li[1]

Abstract

Eigenphone-based speaker adaptation outperforms conventional maximum likelihood linear regression (MLLR) and eigenvoice methods when there is sufficient adaptation data. However, it suffers from severe over-fitting when only a few seconds of adaptation data are provided. In this paper, various regularization methods are investigated to obtain a more robust speaker-dependent eigenphone matrix estimation. Element-wise l_1 norm regularization (known as lasso) encourages the eigenphone matrix to be sparse, which reduces the number of effective free parameters and improves generalization. Squared l_2 norm regularization promotes an element-wise shrinkage of the estimated matrix towards zero, thus alleviating over-fitting. Column-wise unsquared l_2 norm regularization (known as group lasso) acts like the lasso at the column level, encouraging column sparsity in the eigenphone matrix, i.e., preferring an eigenphone matrix with many zero columns as solution. Each column corresponds to an eigenphone, which is a basis vector of the phone variation subspace. Thus, group lasso tries to prevent the dimensionality of the subspace from growing beyond what is necessary. For nonzero columns, group lasso acts like a squared l_2 norm regularization with an adaptive weighting factor at the column level. Two combinations of these methods are also investigated, namely elastic net (applying l_1 and squared l_2 norms simultaneously) and sparse group lasso (applying l_1 and column-wise unsquared l_2 norms simultaneously). Furthermore, a simplified method for estimating the eigenphone matrix in case of diagonal covariance matrices is derived, and a unified framework for solving various regularized matrix estimation problems is presented. Experimental results show that these methods improve the adaptation performance substantially, especially when the amount of adaptation data is limited. The best results are obtained when using the sparse group lasso method, which combines the advantages of both the lasso and group lasso methods. Using speaker-adaptive training, performance can be further improved.

Keywords: Eigenphones; Speaker adaptation; Regularization methods; Sparse group lasso

1 Introduction

Model space speaker adaptation is an important technique in modern speech recognition system. The basic idea is that given some adaptation data, the parameters of a speaker-independent (SI) system are transformed to match the speaking pattern of an unknown speaker, resulting in a speaker-adapted (SA) system. In this paper, we focus on the speaker adaptation of a conventional hidden Markov model Gaussian mixture model (HMM-GMM)-based speech recognition system. To deal with the scarcity of the adaptation data, parameter sharing schemes are usually adopted. For example, in the eigenvoice method [1], the SA models are assumed to lie in a low-dimensional speaker subspace. The subspace bases are shared among all speakers, and a speaker dependent coordinate vector is estimated for each unknown speaker. The maximum likelihood linear regression (MLLR) method [2] estimates a set of linear transformations to transform an SI model into an SA model. The transformation matrices are shared among different HMM state components.

Recently, a novel phone subspace-based method, the eigenphone-based method, was proposed [3]. In contrast to the eigenvoice method, the phone variations of a speaker are assumed to be in a low-dimensional subspace,

*Correspondence: zwlin_2004@163.com
[1] Zhengzhou Information Science and Technology Institute, Zhengzhou 450000, China
Full list of author information is available at the end of the article

called the phone variation subspace. The coordinates of the whole phone set are shared among different speakers. During speaker adaptation, a speaker-dependent eigenphone matrix representing the main phone variation patterns for a specific speaker is estimated. In [4], the 'eigenphone' is first introduced as a set of linear basis vectors of the phone space used in conjunction with eigenvoices. The set of linear basis vectors are obtained by a Kullback-Leibler divergence minimization algorithm for a closed set of training speakers. Estimation of the eigenphones for unknown speakers is not studied. Kenny's eigenphone method is a multi-speaker modeling technique rather than a speaker adaptation technique in the usual sense. In our method, the speaker-independent phone coordinate matrix is obtained by principal component analysis (PCA), and speaker adaptation is performed by estimating a set of eigenphones for each unknown speaker using the maximum likelihood criterion.

Due to its more elaborate modeling, the eigenphone method outperforms both the eigenvoice and the MLLR method, when sufficient amounts of adaptation data are available. However, with limited amounts of adaptation data, the estimation shows severe over-fitting, resulting in very bad adaptation performance [3]. Even with a fine tuned Gaussian prior, the eigenphone matrix estimated by the maximum *a posteriori* (MAP) criterion still does not match the performance of the eigenvoice method.

In machine learning, regularization techniques are widely employed to address the problem of data scarcity and model complexity. Recently, regularization has been widely adopted in speech processing and speech recognition applications. For instance, l_1 and l_2 regularization have been proposed for spectral denoising in speech recognition [5]. In [6], similar regularization methods are adopted to improve the estimation of state-specific parameters in the subspace Gaussian mixture model (SGMM). In [7], l_1 regularization is used to reduce the nonzero connections of deep neural networks (DNNs) without sacrificing speech recognition performance. In [8], it was found that group sparse regularization can offer significant gains over efficient techniques like the elastic net (combining of l_1 and l_2 regularization) in noise robust speech recognition.

In this paper, we investigate the regularized estimation of the speaker-dependent eigenphone matrix for speaker adaptation. Three regularization methods and their combinations are applied to improve the robustness of the eigenphone-based method. The l_1 norm regularization can be used to constrain the sparsity of the matrix, which can reduce the number of free parameters of each eigenphone, thus improving the robustness of the adaptation. The squared l_2 norm can prevent each eigenphone from being too large, yielding better generalization of the adapted model. Each column in the eigenphone matrix corresponds to one eigenphone and hence is a basis vector of the phone variation subspace. Thus, the number of nonzero columns determines the dimension of the phone variation subspace. The column-wise unsquared l_2 norm regularization forces some columns of the matrix to be zero, thus effectively preventing the dimensionality of the phone variation subspace to grow beyond what is necessary. In this paper, all these regularization methods, as well as two combinations of them, namely, the elastic net and sparse group lasso, are presented in a unified framework. Accelerated proximal gradient descent is adopted to solve the mathematical optimization problems in a flexible way.

In [9], a speaker-space compressive sensing method is used to perform speaker adaptation using an overcomplete speaker dictionary in case of limited amount of adaptation data. In this paper, we discuss the phone-space speaker adaptation method, which obtains good performance when the adaptation data is sufficient. Various regularization methods are applied to improve performance in case of insufficient adaptation data. Although the speaker-space and phone-space methods can be combined using a hierarchical Bayesian framework [10], we will not pursue that in this paper.

In the next section, a brief overview of the eigenphone-based speaker adaptation method is given, a simplified method for row-wise estimation of the eigenphone matrix in case of diagonal covariance matrices is derived, and the comparisons between the eigenphone method and various existing methods are presented. A unified framework of regularized eigenphone estimation is proposed in Section 3. Various regularization methods and their combinations are discussed in detail. The optimization of the eigenphone matrix using an accelerated incremental proximal gradient decent algorithm is given in Section 4. In Section 5, different regularization methods are compared through experiments on supervised speaker adaptation of a Mandarin syllable recognition system and unsupervised speaker adaptation of an English large vocabulary speech recognition system using the *Wall Street Journal* (WSJ; New York, NY, USA) corpus. Finally, conclusions are given in Section 6.

2 Review of the eigenphone-based speaker adaptation method

2.1 Eigenphone-based speaker adaptation

Given a set of speaker-independent HMMs containing a total of M mixture components across all states and models and a D-dimensional speech feature vector, let $\boldsymbol{\mu}_m$, $\boldsymbol{\mu}_m(s)$, and $\boldsymbol{u}_m(s) = \boldsymbol{\mu}_m(s) - \boldsymbol{\mu}_m$ denote the SI mean vector, the SA mean vector, and the phone variation vector for speaker s and mixture component m, respectively. In eigenphone-based speaker adaptation method, the phone variation vectors $\{\boldsymbol{u}_m(s)\}_{m=1}^{M}$ are assumed to be located in a speaker-dependent $N(N << M)$ dimensional phone variation subspace. The eigenphone decomposition of the phone variation matrix can be expressed by the following equation [3]:

$$\begin{aligned} \boldsymbol{U}(s) &= [\boldsymbol{u}_1(s)\ \boldsymbol{u}_2(s) \cdots \boldsymbol{u}_M(s)] \\ &\approx [\boldsymbol{v}_0(s)\ \boldsymbol{v}_1(s)\ \boldsymbol{v}_2(s) \cdots \boldsymbol{v}_N(s)] \begin{bmatrix} 1 & 1 & 1 & \dots & 1 \\ l_{11} & l_{21} & l_{31} & \dots & l_{M1} \\ l_{12} & l_{22} & l_{32} & \dots & l_{M2} \\ \vdots & \vdots & \dots & \vdots & \vdots \\ l_{1N} & l_{2N} & l_{3N} & \dots & l_{MN} \end{bmatrix} \\ &= \boldsymbol{V}(s) \cdot \boldsymbol{L}, \end{aligned} \quad (1)$$

where $\boldsymbol{v}_0(s)$ and $\{\boldsymbol{v}_n(s)\}_{n=1}^N$ denote the origin and the bases of the phone variation subspace of speaker s, respectively, $\left[l_{m1}\ l_{m2}\ \cdots\ l_{mN}\right]^{\mathrm{T}}$ is the corresponding coordinate of mixture component m. We call $\{\boldsymbol{v}_n(s)\}_{n=0}^N$ the eigenphones of speaker s.

Equation 1 can be viewed as the decomposition of the phone variation matrix $\boldsymbol{U}(s)$ to the multiplication of two low-rank matrices $\boldsymbol{L}$ and $\boldsymbol{V}(s)$. Note that the phone coordinate matrix $\boldsymbol{L}$ is shared among all speakers, and the eigenphone matrix $\boldsymbol{V}(s)$ is speaker dependent. Given $\boldsymbol{L}$, speaker adaptation can be performed by estimating $\boldsymbol{V}(s\prime)$ for each unknown speaker $s\prime$ during adaptation.

Suppose there are S speakers in the training set. Concatenating each column of all training speaker phone variation matrices $\{\boldsymbol{U}(s)\}_{s=1}^S$, we can obtain

$$\begin{aligned} \boldsymbol{U} &= \begin{bmatrix} \boldsymbol{U}(1) \\ \boldsymbol{U}(2) \\ \vdots \\ \boldsymbol{U}(S) \end{bmatrix} = \begin{bmatrix} \boldsymbol{u}_1(1) & \boldsymbol{u}_2(1) & \cdots & \boldsymbol{u}_M(1) \\ \boldsymbol{u}_1(2) & \boldsymbol{u}_2(2) & \cdots & \boldsymbol{u}_M(2) \\ \vdots & \vdots & \ddots & \vdots \\ \boldsymbol{u}_1(S) & \boldsymbol{u}_2(S) & \cdots & \boldsymbol{u}_M(S) \end{bmatrix} \\ &\approx \begin{bmatrix} \boldsymbol{v}_0(1) & \boldsymbol{v}_1(1) & \boldsymbol{v}_2(1) & \cdots & \boldsymbol{v}_N(1) \\ \boldsymbol{v}_0(2) & \boldsymbol{v}_1(2) & \boldsymbol{v}_2(2) & \cdots & \boldsymbol{v}_N(2) \\ \vdots & \vdots & \vdots & \ddots & \vdots \\ \boldsymbol{v}_0(S) & \boldsymbol{v}_1(S) & \boldsymbol{v}_2(S) & \cdots & \boldsymbol{v}_N(S) \end{bmatrix} \times \begin{bmatrix} 1 & 1 & 1 & \dots & 1 \\ l_{11} & l_{21} & l_{31} & \dots & l_{M1} \\ l_{12} & l_{22} & l_{32} & \dots & l_{M2} \\ \vdots & \vdots & \vdots & \ddots & \vdots \\ l_{1N} & l_{2N} & l_{3N} & \dots & l_{MN} \end{bmatrix} = \boldsymbol{V} \cdot \boldsymbol{L}. \end{aligned} \quad (2)$$

Note that the nth column of $\boldsymbol{V}$, which is the concatenation of the nth eigenphones for all speakers $\{\boldsymbol{v}_n(s)\}_{s=1}^S$, can be viewed as a basis vector of the column vectors of matrix $\boldsymbol{U}$. The mth column of the phone coordinate matrix $\boldsymbol{L}$ corresponds to the coordinate vector for mixture component m. Hence, $\boldsymbol{L}$ implicitly contains the correlation information for different Gaussian components, which is speaker independent. From Equation 2, it can be observed that $\boldsymbol{L}$ can be calculated by performing PCA on the columns of matrix $\boldsymbol{U}$.

During speaker adaptation, given some adaptation data, the eigenphone matrix $\boldsymbol{V}(s)$ is estimated using the maximum likelihood criterion. Let $\boldsymbol{O}(s) = \{\boldsymbol{o}(s,1), \boldsymbol{o}(s,2), \cdots, \boldsymbol{o}(s,T)\}$ denote the sequence of feature vectors of the adaptation data for speaker s. Using the expectation maximization (EM) algorithm, the auxiliary function to be minimized is given as follows:

$$Q(\boldsymbol{V}(s)) = \frac{1}{2} \sum_t \sum_m \gamma_m(t) \left[\boldsymbol{o}(s,t) - \boldsymbol{\mu}_m(s)\right]^{\mathrm{T}} \times \boldsymbol{\Sigma}_m^{-1} \left[\boldsymbol{o}(s,t) - \boldsymbol{\mu}_m(s)\right], \quad (3)$$

where $\boldsymbol{\mu}_m(s) = \boldsymbol{\mu}_m + \boldsymbol{v}_0(s) + \sum_{n=1}^N l_{mn}\boldsymbol{v}_n(s)$, and $\gamma_m(t)$ is the posterior probability of being in mixture m at time t given the observation sequence $\boldsymbol{O}(s)$ and the current estimation of the SA model.

Suppose the covariance matrix $\boldsymbol{\Sigma}_m$ is diagonal. Let $\sigma_{m,d}$ denote its dth diagonal element and $o_d(s,t)$, $\mu_{m,d}$, and $v_{n,d}(s)$ represent the dth component of $\boldsymbol{o}(s,t)$, $\boldsymbol{\mu}_m$, and $\boldsymbol{v}_n(s)$, respectively. After some mathematical manipulation, Equation 3 can be simplified to

$$Q(\boldsymbol{V}(s)) = \frac{1}{2} \sum_d \sum_t \sum_m \gamma_m(t) \sigma_{m,d}^{-1} \left[o'_{m,d}(s,t) - \hat{\boldsymbol{l}}_m^{\mathrm{T}} \boldsymbol{v}_d(s)\right]^2, \quad (4)$$

where $o'_{m,d}(s,t) = o_d(s,t) - \mu_{m,d}$, $\hat{\boldsymbol{l}}_m = [1, l_{m1}, l_{m2}, \ldots, l_{mN}]^{\mathrm{T}}$, and $\boldsymbol{v}_d(s) = [v_{0,d}(s), v_{1,d}(s), v_{2,d}(s), \ldots, v_{N,d}(s)]^{\mathrm{T}}$, which is the dth row of the eigenphone matrix $\boldsymbol{V}(s)$.

Define

$$\boldsymbol{A}_d = \sum_t \sum_m \gamma_m(t) \sigma_{m,d}^{-1} \hat{\boldsymbol{l}}_m \hat{\boldsymbol{l}}_m^{\mathrm{T}}$$

and

$$\boldsymbol{b}_d = \sum_t \sum_m \gamma_m(t) \sigma_{m,d}^{-1} o'_{m,d}(s,t) \hat{\boldsymbol{l}}_m.$$

Equation 4 can be further simplified to

$$Q(\boldsymbol{V}(s)) = \frac{1}{2} \sum_d \left[\boldsymbol{v}_d(s)^{\mathrm{T}} \boldsymbol{A}_d \boldsymbol{v}_d(s) - \boldsymbol{b}_d^{\mathrm{T}} \boldsymbol{v}_d(s)\right] + \text{Const.} \quad (5)$$

Setting the derivative of (5) with respect to $\boldsymbol{v}_d(s)$ to zero yields $\hat{\boldsymbol{v}}_d(s) = \boldsymbol{A}_d^{-1}\boldsymbol{b}_d$. Because of the independence of different feature dimensions, $\{\hat{\boldsymbol{v}}_d(s)\}_{d=1}^D$ can be calculated in parallel very efficiently.

It is well known that many conventional speaker adaptation methods, such as MLLR and the eigenvoice method, work substantially better in combination with speaker-adaptive training (SAT) [11]. The above eigenphone-based speaker adaptation method can also be combined with

SAT. Initially, an SI model $\boldsymbol{\Lambda}^{\mathrm{SI}}$ is trained using all the training data. Then, the speaker-adapted model $\boldsymbol{\Lambda}^{s}$ for each training speaker s is obtained using conventional speaker adaptation methods such as MLLR + MAP. The phone coordinate matrix $\boldsymbol{L}$ is calculated using PCA on the columns of the training speaker phone variation matrix $\boldsymbol{U}$ (Equation 2). Let $\boldsymbol{\Lambda}^{\mathrm{C}}$ denote the canonical model; then, the eigenphone-based SAT procedure can be summarized as follows:

1. Initialize $\boldsymbol{\Lambda}^{\mathrm{C}}$ with $\boldsymbol{\Lambda}^{\mathrm{SI}}$.
2. Given $\boldsymbol{\Lambda}^{\mathrm{C}}$ and $\boldsymbol{L}$, estimate the eigenphone matrices $\boldsymbol{V}(s)$ for each training speaker s using the corresponding speaker-dependent training data.
3. Given $\{\boldsymbol{V}(s)\}_{s=1}^{S}$ and $\boldsymbol{L}$, re-estimate the canonical model $\boldsymbol{\Lambda}^{\mathrm{C}}$ using all training data.
4. Repeat steps 2 and 3 for a predefined count K.

Note that in step 3, the first-order statistic $\boldsymbol{s}_m$ and second-order statistic $\boldsymbol{S}_m$ of Gaussian component m are calculated as

$$\boldsymbol{s}_m = \sum_s \sum_t \gamma_m(t) \boldsymbol{o}_m(s,t) \quad (6)$$

$$\boldsymbol{S}_m = \sum_s \sum_t \gamma_m(t) \boldsymbol{o}_m(s,t) \boldsymbol{o}_m^{\mathrm{T}}(s,t), \quad (7)$$

where $\boldsymbol{o}_m(s,t) = \boldsymbol{o}(s,t) - \boldsymbol{V}(s)\hat{\boldsymbol{l}}_m$.

2.2 Comparison with existing methods

Various adaptation methods have been proposed in the past 2 decades, which can be classified into three broad categories: MAP [12], MLLR [13], and speaker subspace-based methods [1]. In conventional MAP adaptation, with a conjugate prior distribution, the SA model parameters are estimated using the maximum *a posteriori* criterion. The main advantage of MAP adaptation is its good asymptotic property, which means that the MAP estimate approaches the maximum likelihood (ML) estimate when the adaptation data is sufficient. But, it is a local update of the model parameters, in which only model parameters observed in the adaptation data can be modified from their prior means. The number of free parameters in MAP adaptation is fixed to $M \cdot D$. Large amounts of adaptation data are required to obtain good performance. The chance of over-fitting is controlled by the prior weight. The larger the prior weight, the lower the chance of over-fitting.

Instead of estimating the SD model directly, the MLLR method estimates a set of linear transformations to transform an SI model into a new SA model. Using a regression class tree, the Gaussian components are grouped into classes with each class having its own transformation matrix. The number of regression classes (denoted by R) can be adjusted automatically according to the amount of adaptation data. There are RD^2 free parameters in MLLR, which are much fewer than those of the MAP method. Hence, the MLLR method has lower data requirements. However, its asymptotic behavior is poor, as performance improvement saturates rapidly as the adaptation data increases. The chance of over-fitting is closely related to the number of regression classes N used. The number of free parameters can only be an integer multiple of D^2, which restricts its flexibility.

Unlike MAP and MLLR, speaker subspace-based approaches deal with the speaker adaptation problem in a different way. These assume that all SD models lie in a low-dimensional manifold, so that speaker adaptation is no more than the estimation of the local or global coordinate of the new SD model. A representative of these methods is the eigenvoice method [1], where the low-dimensional manifold is a linear subspace and a set of linear bases (called eigenvoices), which capture most of the variance of the SD model parameters, can be obtained by principal component analysis. During speaker adaptation, the coordinate of a new SD model is estimated using the maximum likelihood criterion. The number of free parameters in the eigenvoice method is equal to the dimension (K) of the speaker subspace, which is much fewer than that of the MAP and MLLR methods. So, the eigenvoice method can yield good performance even when only a few seconds of adaptation data is provided. The chance of over-fitting is related to K, which can be adjusted according to amount of adaptation data using a heuristic formula or regularization method [9]. However, due to the strong subspace constraint, its performance is poor compared with that of the MLLR or MAP method when there is a sufficient amount of adaptation data.

In the eigenphone method, a phone variation subspace is assumed. Each speaker-dependent eigenphone matrix $\boldsymbol{V}(s)$ is of size $(N+1) \times D$, containing more free parameters than the eigenvoice method. By adjusting the dimensionality (N) of the phone variation subspace, the number of free parameters can be varied by an integer multiplier of D. So, the eigenphone method is more flexible than the MLLR method. When a sufficient amount of adaptation data is available, better performance can be obtained with a large N (typically $N = 100$). However, when the amount of adaptation data is limited, performance degrades quickly. The recognition rate can even fall bellow that of the unadapted SI model. In order to alleviate the over-fitting problem, a Gaussian prior is assumed and an MAP adaptation method is derived in [3]. In this paper, we address this problem using an explicit matrix regularization function.

The advantages of the MAP method, the eigenvoice method, and the eigenphone method can be combined using a probabilistic formulation and the Bayesian principle, resulting in a hierarchical Bayesian adaptation method [10]. This paper focuses on using various matrix regularization methods to improve the performance of the eigenphone method in case of insufficient amount of adaptation data. In the following sections, we omit the speaker identifier s for brevity, i.e., we write $\boldsymbol{V}$ for $\boldsymbol{V}(s)$ and $\boldsymbol{v}_n$ for $\boldsymbol{v}_n(s)$.

3 Regularized eigenphone matrix estimation

The center of the eigenphone adaptation method is the robust estimation of the eigenphone matrix $\boldsymbol{V}$. This type of problem, i.e., the estimation of an unknown matrix from some observation data, has appeared frequently across many diverse fields. Regularization proved to be a valid method to overcome the data scarcity. For robust eigenphone matrix estimation, the regularized objective function to be minimized is as following:

$$Q\prime(\boldsymbol{V}) = Q(\boldsymbol{V}) + J(\boldsymbol{V}), \tag{8}$$

where $J(\boldsymbol{V})$ denotes a regularization function (known as regularizer) for $\boldsymbol{V}$.

In this paper, we consider the following general regularization function:

$$J(\boldsymbol{V}) = \lambda_1||\boldsymbol{V}||_1 + \lambda_2||\boldsymbol{V}||_2^2 + \lambda_3\sum_{n=0}^{N}||\boldsymbol{v}_n||_2, \tag{9}$$

where $||\boldsymbol{V}||_1 = \sum_{n=0}^{N}||\boldsymbol{v}_n||_1$ and $||\boldsymbol{V}||_2^2 = \sum_{n=0}^{N}||\boldsymbol{v}_n||_2^2$ denote the l_1 norm and squared l_2 norm of matrix $\boldsymbol{V}$. $||\boldsymbol{v}_n||_1$ and $||\boldsymbol{v}_n||_2$ denote the l_1 norm and l_2 norm of column vector $\boldsymbol{v}_n$. λ_1, λ_2, and λ_3 are nonnegative weighting factors for the matrix l_1 norm, squared l_2 norm, and column-wise unsquared l_2 norm, respectively.

Different norms have different effects of regularization. Equation 9 is a mixed norm regularizer, with many well-known regularizers as special cases of it. The general form has the advantage that we can solve the various regularization problems in a unified framework using a single algorithm.

The l_1 norm is the standard convex relaxation of the l_0 norm. The l_1 norm regularizer ($J(\boldsymbol{V})$ with $\lambda_1 > 0$ and $\lambda_2 = \lambda_3 = 0$) is sometimes referred to as lasso [14], which can drive an element-wise shrinkage of $\boldsymbol{V}$ towards zero, thus leading to a sparse matrix solution. l_1 norm regularization has been widely used as an effective parameter selection method in compressive sensing, signal recovery etc.

The squared l_2 norm regularizer ($J(\boldsymbol{V})$ with $\lambda_2 > 0$ and $\lambda_1 = \lambda_3 = 0$) is referred to as ridge regression [15] or weight decay in the literature. This penalizes large value components of the parameters, enabling more robust estimation and prevents model over-fitting.

The column-wise l_2 norm regularizer ($J(\boldsymbol{V})$ with $\lambda_3 > 0$ and $\lambda_1 = \lambda_2 = 0$) is a variant of the group lasso [16], which acts like lasso at the group level: due to the non-differentiability of $\boldsymbol{v}_n$ at 0, the entire group of parameters may be set to zero at the same time [16]. Here a 'group' corresponds to one column of the matrix $\boldsymbol{V}$, and the group lasso is a good surrogate for column sparsity. Previous experiments on eigenphone-based speaker adaptation have shown that when the amount of adaptation data is sufficient, the number of eigenphones should be large. When less adaptation data is available, fewer eigenphones should be used, i.e., many eigenphones should be zero. In this situation, the optimal eigenphone matrix should show 'group sparsity', where a group corresponds to an eigenphone vector. Hence, group lasso is a good choice for eigenphone matrix regularization. Each eigenphone is of dimension D, and there are N eigenphones in the N-dimensional phone variation subspace. If we combine all eigenphones to form a dictionary, the dictionary is over-complete when $N > D$. Learning such an over-complete dictionary requires a large amount of adaptation data. The group lasso regularizer removes unnecessary eigenphones from the dictionary according to the amount of adaptation data available. When insufficient data is provided, the resulting dictionary may not be complete due to the effect of nondifferentiable column-wise l_2 norm penalties.

Each type of norm regularizer has its own strong points, and the combination of them through the generic form $J(\boldsymbol{V})$ (9) is expected to obtain better performance. Two typical variants of this are the elastic net [17] and sparse group lasso (SGL) [18].

The elastic net regularizer combines l_1 and l_2 norm regularization through linear combination and can be written as $J(\boldsymbol{V})$ with $\lambda_1 > 0$, $\lambda_2 > 0$, and $\lambda_3 = 0$. It has been successfully applied to many fields of speech processing and recognition, such as spectral denoising [5], sparse exemplar-based representation for speech [19], and robust estimation of parameters for the SGMM [6] and DNN [7].

The combination of group lasso and the original lasso is referred as SGL [18], which corresponds to $J(\boldsymbol{V})$ with $\lambda_1 > 0$, $\lambda_3 > 0$, and $\lambda_2 = 0$. The basic considerations are as follows: the group lasso selects the best set of parameters through nonzero columns of matrix $\boldsymbol{V}$ and the lasso regularization can further reduce free parameters of each column, resulting in a column-wise sparse and within-column sparse matrix. Sparse group lasso looks very similar to the elastic net regularizer but differs in that the column-wise l_2 norm term is not squared, which makes it not differentiable at 0. However, we will show in Section 4 that within each nonzero group (i.e., eigenphone) it gives an 'adaptive' elastic net fit.

4 Optimization

There is no closed form solution to the regularized objective function (8). Many numerical methods have been proposed in the literature to solve the regularization problem. For example, a gradient projection method has been proposed in [20] to solve the lasso and elastic net problem for sparse reconstruction. The software tool SLEP [21] implements the sparse group lasso formulation using a version of the fast iterative shrinkage-thresholding algorithm (FISTA) [22]. Recently, a more efficient algorithm using accelerated generalized gradient descent method has been proposed [18]. In this paper, for robust eigenphone matrix estimation using the regularization function $J(V)$, we propose an accelerated version of the incremental proximal descent algorithm [23,24], which is fast and flexible and can be viewed as a natural extension of the incremental gradient algorithm [25] and the FISTA algorithm [22].

For a convex regularizer $R(V)$, $V \in \mathbb{R}^{N \times D}$, the proximal operator [26] is defined as

$$\mathbf{prox}_R(V) = \arg\min_X \frac{1}{2}||X - V||_2^2 + R(X). \quad (10)$$

The proximal operator for the l_1 norm regularizer is the soft thresholding operator

$$\mathbf{prox}_{\gamma||\cdot||_1}(V) = sgn(V) \circ (|V| - \gamma)_+, \quad (11)$$

where $\circ$ denotes the Hadamard product of two matrices, $(\boldsymbol{x})_+ = max\{\boldsymbol{x}, 0\}$. The sign function (sgn), product, and maximum are all taken component-wise.

The proximal operator for the squared l_2 norm regularizer is the multiplicative shrinkage operator

$$\mathbf{prox}_{\gamma||\cdot||_2^2}(V) = \frac{1}{1+2\gamma}V. \quad (12)$$

For the column-wise group sparse regularizer, the proximal operator $\mathbf{prox}_{\gamma||\cdot||_2}$ is given by the shrinkage operation on each column of the parameter matrix $\boldsymbol{v}_n$ as follows [26]:

$$\mathbf{prox}_{\gamma||\cdot||_2}(\boldsymbol{v}_n) = (1 - \frac{\gamma}{||\boldsymbol{v}_n||_2})_+\boldsymbol{v}_n. \quad (13)$$

The proximal operator (13) is sometimes called the block soft thresholding operator. In fact, when $||\boldsymbol{v}_n||_2 > \gamma$, the resulting nth column will be nonzero, and it can be written as

$$\mathbf{prox}_{\gamma||\cdot||_2}(\boldsymbol{v}_n) = \frac{1}{1 + \frac{\gamma}{||\boldsymbol{v}_n||_2 - \gamma}}\boldsymbol{v}_n. \quad (14)$$

Comparing (14) with (12), it can be seen that for nonzero columns, the group sparse lasso is equivalent to the squared l_2 norm regularization with a weighting factor of $\frac{\gamma}{2(||\boldsymbol{v}_n||_2 - \gamma)}$. The larger $||\boldsymbol{v}_n||_2$, the smaller the weighting factor of the squared l_2 norm. So within each nonzero column, the weighting factor of the equivalent l_2 norm is kind of adaptive.

In fact, the proximity operator of a convex function is a natural extension of the notion of a projection operator onto a convex set. The incremental proximal descent algorithm [24] could be viewed as a natural extension of the iterated projection algorithm, which activates each convex set modeling a constraint individually by means of its projection operator. In this paper, an accelerated version of the incremental proximal descent algorithm is introduced for the estimation of the eigenphone matrix $\boldsymbol{V}$, which is summarized in Algorithm 1.

Algorithm 1 Accelerated Incremental Proximal Descent Algorithm for Regularized Eigenphone Matrix Estimation

1: $k = 0$, $\eta^{(0)} = 1.0$, $t^{(0)} = t^{(-1)} = 1.0$
2: $\boldsymbol{V}^{(0)} = \boldsymbol{V}^{(-1)} = \mathbf{0}$
3: **repeat**
4: $\quad \boldsymbol{Y}^{(k)} = \boldsymbol{V}^{(k)} + \frac{t^{(k-1)}-1}{t^{(k)}}\left(\boldsymbol{V}^{(k)} - \boldsymbol{V}^{(k-1)}\right)$
5: $\quad$ **repeat** $\quad \triangleright$ Search for a suitable step size $\eta^{(k)}$
6: $\quad\quad \boldsymbol{V}^{(k+1)} = \boldsymbol{Y}^{(k)} - \eta^{(k)}\nabla Q(\boldsymbol{Y}^{(k)})$
7: $\quad\quad \boldsymbol{V}^{(k+1)} \leftarrow \mathbf{prox}_{\eta^{(k)}\lambda_1||\cdot||_1}\left(\boldsymbol{V}^{(k+1)}\right)$
8: $\quad\quad \boldsymbol{V}^{(k+1)} \leftarrow \mathbf{prox}_{\eta^{(k)}\lambda_2||\cdot||_2^2}\left(\boldsymbol{V}^{(k+1)}\right)$
9: $\quad\quad \boldsymbol{V}^{(k+1)} \leftarrow \mathbf{prox}_{\eta^{(k)}\lambda_3||\cdot||_2}\left(\boldsymbol{V}^{(k+1)}\right)$
10: $\quad\quad \Delta Q'^{(k+1)} = Q'\left(\boldsymbol{V}^{(k+1)}\right) - Q'\left(\boldsymbol{V}^{(k)}\right)$
11: $\quad\quad$ **if** $\Delta Q'^{(k+1)} > 0$ **then**
12: $\quad\quad\quad \eta^{(k)} \leftarrow \theta\eta^{(k)}$
13: $\quad\quad$ **end if**
14: $\quad$ **until** $\Delta Q'^{(k+1)} \leq 0$
15: $\quad \eta^{(k+1)} = \eta^{(k)}$, $t^{(k+1)} = \frac{1+\sqrt{1+4(t^{(k)})^2}}{2}$
16: $\quad k \leftarrow k+1$
17: **until** $\left|\Delta Q'^{(k)}\right| / \left|Q'\left(\boldsymbol{V}^{(k-1)}\right)\right| < \epsilon.$
18: **return** $\boldsymbol{V} = \boldsymbol{V}^{(k-1)}$.

In Algorithm 1, $\nabla Q(\boldsymbol{V})$ is the gradient of (5), which can be easily calculated from $\nabla Q(\boldsymbol{v}_d) = -\boldsymbol{A}_d\boldsymbol{v}_d + \boldsymbol{b}_d$. Step 6 is the normal gradient descent step of the original objective function $Q(\boldsymbol{V})$. In steps 7, 8 and 9, the proximal operators of the element-wise l_1 norm, squared l_2 norm, and column-wise group sparse regularizer are applied in sequence. The initial descent step size $\eta^{(0)}$ is simply set to 1.0. From step 10 to 14, we calculate the change of the regularized objective function (8) as $\Delta Q'^{(k+1)}$ and reduce the current step size $\eta^{(k)}$ by a factor of θ ($0 < \theta < 1$, i.e., $\theta = 0.8$) until $\Delta Q'^{(k+1)}$ is below zero.

To accelerate the convergence speed, a momentum term [27] is included in step 4. For fastest convergence, $t^{(k)}$ should increase as fast as possible. In step 15, $t^{(k)}$ is updated using the formula proposed by [22]. Note that when $k = 0$, $\frac{t^{(k-1)}-1}{t^{(k)}} = 0$; when $k \to \infty$, $\frac{t^{(k-1)}-1}{t^{(k)}} \to 1$. This gives the nice property that when $\boldsymbol{V}$ approaches its optimal value, the momentum term increases towards 1, which prevents unnecessary oscillations during the iteration process, thus improves convergence speed. The whole procedure is iterated until the relative change of

(8) is smaller than some predefined threshold $\epsilon = 10^{-5}$ (step 17). In our experiments, the typical number of outer iterations is around 200. After finding suitable step sizes $\eta^{(k)}$ in the first k iterations (typically $k < 10$), there is almost no change in step size for the following outer iterations. Once $k > 10$, the average number of inner iterations is nearly a constant 1. For each iteration, there are only a few element-wise matrix addition, multiplication and thresholding operations, together with an evaluation of the objection function $Q'^{(k)}$. Using any modern linear algebra software package, an efficient implementation of Algorithm 1 can be obtained.

Algorithm 1 is also very flexible. If $\lambda_2 = 0$ and $\lambda_3 = 0$, step 8 and 9 can be omitted, resulting in an accelerated version of the iterative shrinkage-thresholding (IST) [28] algorithm for solving the lasso problem. If only one of steps 7, 8, and 9 is retained, it reduces to FISTA [22] for solving lasso, ridge regression, and group lasso problems, respectively. If $\lambda_3 = 0$ or $\lambda_2 = 0$, the algorithm becomes the accelerated generalized gradient descent method for solving the elastic net and sparse group lasso problems [18], respectively.

5 Experiments

This section presents an experimental study to evaluate the performance of various regularized eigenphone speaker adaptation methods on a Mandarin Chinese continuous speech recognition task provided by Microsoft [29] (Redmond, WA, USA) and the WSJ English large vocabulary continuous speech recognition task. Supervised and unsupervised speaker adaptation using a varying amount of adaptation data were evaluated. For both tasks, we compare the proposed methods with various conventional methods. In all methods, only the Gaussian means are updated. When comparing the results of the two methods, statistical significance tests were performed using the suite of significance tests implemented by NIST [30]. Three significance tests were applied, including the matched pair (MP) sentence segment (word error) test, the signed paired (SI) comparison test (speaker word accuracy rate), and the Wilcoxon (WI) signed rank test (speaker word accuracy rate). We use this to define significant improvement at a 5% level of significance. All experiments were based on the standard HTK (v 3.4.1) tool set [31]. Detailed experimental setups and results are presented below for each task.

5.1 Experiments on the Mandarin Chinese task

5.1.1 Experimental setup

Supervised speaker adaptation experiments were performed on the Mandarin Chinese continuous speech recognition task provided by Microsoft [29]. The training set contains 19,688 sentences from 100 speakers with a total of 454,315 syllables (about 33 h total). The testing set consists of 25 speakers, and each speaker contributes 20 sentences (the average length of a sentence is 5 s). The frame length and frame step size were set as 25 and 10 ms, respectively. Acoustic features were constructed from 13 dimensional Mel-frequency cepstral coefficients (MFCC) and their first and second derivatives. The basic units for acoustic modeling are 27 initial and 157 tonal final units of Mandarin Chinese as described in [29]. Monophone models were first created using all 19,688 sentences. Then, all possible cross-syllable triphone expansions based on the full syllable dictionary were generated, resulting in 295,180 triphones. Out of these triphones, 95,534 triphones actually occur in the training corpus. Each triphone was modeled by a 3-state left-to-right HMM without skips. After decision tree-based state clustering, the number of unique tied states was reduced to 2,392. We then use the HTK's Gaussian splitting capability to incrementally increase the number of Gaussians per state to 8, resulting in 19,136 different Gaussian components in the SI model.

Standard regression class tree-based MLLR was used to obtain the 100 training speakers' SA models. HVite was used as the decoder with a full connected syllable recognition network. All 1,679 tonal syllables are listed in the network, with any syllable allowed to follow any other syllable, or a short pause or silence. This recognition framework puts the highest demand on the quality of the acoustic models. We drew 1, 2, 4, 6, 8, and 10 sentences randomly from each testing speaker for adaptation in supervised mode; the tonal syllable recognition rate was measured among the remaining 10 sentences. To ensure statistical robustness of the results, each experiment was repeated eight times using cross-validation, and the recognition rates were averaged. The recognition accuracy of the SI model is 53.04% (the baseline reference result reported in [29] is 51.21%).

For the purpose of comparison, we carried out experiments using conventional MLLR + MAP [32], eigenvoices [1], and the ML and MAP eigenphone methods [3] with varying parameter settings. The MAP eigenphone method is equivalent to the squared l_2 norm regularized eigenphone method. Other regularization methods, namely the lasso, the elastic net, the group lasso and the sparse group lasso, were tested with a wide range of weighting factors. Experimental results are presented and compared in the following sections.

5.1.2 Speaker adaptation based on conventional methods

For MLLR + MAP adaptation, we experimented with different parameter settings. The best result was obtained at a prior weighting factor of 10 (for MAP), a regression class tree with 32 base classes and three-block-diagonal transformation matrices (for MLLR). The number of transformation matrices is adjusted automatically based on the amount of adaptation data using the default setting of HTK. For eigenvoice adaptation, the dimension K of the speaker subspace was varied from 10 to 100. For the eigenphone-based method, both the ML and MAP estimation schemes were tested. For the MAP eigenphone method, $\sigma^{(-2)}$ denotes the inverse prior variance for the eigenphone. In fact, the MAP estimation using a zero mean Gaussian prior is equivalent to the squared l_2 norm regularized estimation with $\lambda_2 = \sigma^{(-2)}$.

The experiment results of the above methods are summarized in Table 1. Significance tests show that when the amount of adaptation data is sufficient ($\geq$ 4 sentences) and the number (N) of the eigenphones is 50, the ML eigenphone method outperforms the MAP + MLLR method significantly. But, when the adaptation data is limited to 1 or 2 sentences (about 5~10 s), the performance degrades quickly due to over-fitting. The situation is worse when a high-dimensional phone variation subspace (i.e., $N = 100$) is used. Reducing the number of the eigenphones improves the recognition rate. However, even with $N = 10$, the performance is still worse than that of the SI model when the adaptation data is one sentence. MAP estimation using a Gaussian prior can alleviate over-fitting to some extent. To prevent performance degradation, a very small Gaussian prior (i.e., a large weighting factor of the squared l_2 norm regularizer) is required, which heavily limits the performance when there is a sufficient amount of adaptation data available. This suggests that the l_2 regularization can only improve the performance in the case of limited amount of adaptation data (less than two sentences, about 10 s). In order to demonstrate the performance of the various regularization methods, the subsequent experiments all employ a large number of eigenphones, 100.

5.1.3 Eigenphone speaker adaptation using lasso

The lasso regularizer ($J(V)$ with $\lambda_1 > 0$, $\lambda_2 = \lambda_3 = 0$) leads to a sparse eigenphone matrix. To measure the sparseness of a matrix, we calculate its 'overall sparsity', which is defined as the percentage of nonzero elements

Table 1 Average tonal syllable recognition rate (%) after speaker adaptation using conventional methods

Methods	Number of adaptation sentences					
	1	2	4	6	8	10
MAP + MLLR	53.32	54.93	57.83	58.50	59.65	60.16
Eigenvoice						
$K = 20$	55.32	56.38	56.61	56.90	57.11	57.05
$K = 40$	55.67	56.59	57.03	57.26	57.62	57.45
$K = 60$	55.72	57.01	57.15	57.36	57.87	57.95
$K = 80$	55.37	56.97	57.39	57.45	58.14	58.18
$K = 100$	55.20	57.11	57.24	57.53	57.91	58.39
ML eigenphone						
$N = 10$	51.45	56.71	56.95	57.41	57.87	58.12
$N = 25$	47.25	55.73	57.99	59.36	59.34	59.57
$N = 50$	33.74	51.38	58.16	59.00	59.84	60.62
$N = 100$	19.14	41.46	54.30	57.91	59.44	60.13
MAP eigenphone, $N = 50$						
$\sigma^{(-2)} = 10$	43.26	53.67	58.43	59.11	59.78	60.45
$\sigma^{(-2)} = 100$	50.08	53.69	56.71	58.35	59.21	59.80
$\sigma^{(-2)} = 1,000$	53.69	54.28	55.35	56.13	56.95	57.41
$\sigma^{(-2)} = 2,000$	53.63	54.13	54.80	55.43	56.27	56.69
MAP eigenphone, $N = 100$						
$\sigma^{(-2)} = 10$	27.91	44.63	53.78	57.39	59.61	60.70
$\sigma^{(-2)} = 100$	45.24	50.31	55.77	57.55	59.34	60.30
$\sigma^{(-2)} = 1,000$	53.29	54.22	55.75	56.78	57.41	58.29
$\sigma^{(-2)} = 2,000$	53.92	54.28	55.52	56.34	56.55	57.74

For MLLR + MAP adaptation, we only show the best results which were obtained at a prior weighting factor of 10 (for MAP) and 32 regression classes with a three-block-diagonal transformation matrix (for MLLR). For eigenvoice adaptation, K denotes the number of eigenvoices. For the eigenphone-based method, N denotes the number of eigenphones. For the MAP eigenphone method, $\sigma^{(-2)}$ denotes the inverse prior variance for the eigenphone, i.e., the weighting factor λ_2 of the squared l_2 norm term.

in that matrix. The weighting factor (λ_1) of the l_1 norm is varied between 10 and 40. The experimental results are summarized in Table 2. For each experiment setting, the average overall sparsity of the eigenphone matrix among all testing speakers is shown in parentheses.

Significance tests show that compared with the ML eigenphone method, the l_1 norm regularization method can improve the performance significantly. It shows performance gain over the MAP eigenphone method under almost every testing condition. The larger the weighting factor λ_1, the more sparse the resulting eigenphone matrix becomes. When the amount of adaptation data is limited to one, two, four, or six sentences, the best results are obtained with $\lambda_1 = 20$. The relative improvements over the ML eigenphone method are 181.5%, 36.4%, 7.8%, and 2.5%, respectively. When the amount of adaptation data is increased to eight or ten sentences, a small weighting factor of 10 performs best. The resulting recognition rates are still better than that of the ML eigenphone method, with relative improvements of 1.3% and 1.1%, respectively.

5.1.4 Eigenphone speaker adaptation using elastic net

For the elastic net method, λ_1 was fixed to 10. All experiments were repeated with λ_2 changing from 10 to 2,000. The results are summarized in Table 3. Again, the average overall sparsity of the eigenphone matrix is shown in parentheses.

Unfortunately, the results in Table 3 show little improvement over the lasso method. The overall sparsity remains the same in all testing conditions. When the adaptation data is one sentence, even with a large weighting factor of $l_2 = 2,000$, the relative improvement over the lasso method is only 0.2%. We also set λ_1 to different values of 20, 30, and 40, and experimented with λ_2 varying from 10 to 2,000. Again, almost no improvement was observed over the results in Table 2. The squared l_2 regularization term seems to not work in combination with l_1 regularization.

5.1.5 Eigenphone speaker adaptation using group lasso

As pointed out in Section 3, the group lasso regularizer leads to a column-wise group sparse eigenphone matrix that is a matrix with many zero columns. To measure the column-wise group sparseness of the eigenphone matrix, we calculate its 'column sparsity', which is defined as the percentage of nonzero columns in that matrix. In the group lasso experiments, the weighting factor (λ_3) of the column-wise l_2 norm is varied between 10 and 150. The results are summarized in Table 4. For each experiment setting, the average column sparsity of the eigenphone matrix among all testing speakers is shown in parentheses.

From Table 4, it can be observed that the group lasso method improves the recognition results compared with the ML eigenphone method, especially with limited adaptation data. Under all testing conditions, its best results are better than that of the MAP eigenphone method, i.e., the squared l_2 regularization method. When the adaptation data is limited to one sentence and λ_3 is larger than 120, the recognition rate is higher than the best results of the lasso method. However, when more adaptation data is provided, the group lasso method no longer achieves better results than the lasso method. The larger the weighting factor λ_3, the larger the column sparsity of the eigenphone matrix. With two adaptation sentences or less, λ_3 should be larger than 120 to obtain a good row sparsity. With lots of adaptation data (more than four sentences), even with a large value of λ_3 of 150, the column sparsity remains very small, that is, almost no column is set to zero. For these nonzero columns, the group lasso is equivalent to the 'adaptive' l_2 regularization, and the recognition results are better than those obtained with the ML eigenphone method.

5.1.6 Eigenphone-based speaker adaptation using sparse group lasso

In the sparse group lasso experiments, we fixed λ_1 to 10 and varied λ_3 from 10 to 150, in hope that the advantages of the lasso and group lasso methods can be combined.

Table 2 Average tonal syllable recognition rate (%) after eigenphone-based speaker adaptation using lasso

λ_1	Number of adaptation sentences					
	1	2	4	6	8	10
10	52.25	56.04	58.06	59.06	60.24	61.27
	(0.61)	(0.43)	(0.23)	(0.16)	(0.12)	(0.04)
20	53.88	56.55	58.54	59.36	60.24	60.83
	(0.83)	(0.63)	(0.42)	(0.33)	(0.26)	(0.23)
30	53.63	55.96	57.70	59.19	60.05	60.81
	(0.91)	(0.74)	(0.54)	(0.44)	(0.37)	(0.34)
40	53.82	55.18	57.30	58.90	59.75	60.49
	(0.95)	(0.820)	(0.65)	(0.61)	(0.49)	(0.42)

The number of eigenphones (N) was fixed to 100. $\lambda_2 = \lambda_3 = 0$, and λ_1 was varied between 10 and 40. The average overall sparsity is shown in parentheses.

Table 3 Average tonal syllable recognition rate (%) after eigenphone-based speaker adaptation using elastic net

λ_2	Number of adaptation sentences					
	1	2	4	6	8	10
10	52.27	55.98	58.10	59.19	60.22	61.08
	(0.67)	(0.48)	(0.33)	(0.24)	(0.20)	(0.16)
40	52.27	55.98	58.14	59.17	60.18	61.08
	(0.67)	(0.48)	(0.33)	(0.24)	(0.20)	(0.16)
80	52.22	55.96	58.12	59.17	60.20	61.04
	(0.67)	(0.48)	(0.33)	(0.24)	(0.20)	(0.16)
120	52.22	55.98	58.16	59.17	60.16	61.08
	(0.67)	(0.48)	(0.33)	(0.24)	(0.20)	(0.16)
1,000	52.31	55.98	58.02	59.13	60.13	60.97
	(0.67)	(0.48)	(0.33)	(0.24)	(0.20)	(0.16)
2,000	52.35	55.98	58.02	59.13	60.16	60.97
	(0.67)	(0.48)	(0.33)	(0.24)	(0.20)	(0.16)

The number of eigenphones (*N*) was fixed to 100. $\lambda_1 = 10$, $\lambda_3 = 0$, and λ_2 was varied between 10 and 2,000. The average overall sparsity is shown in parentheses.

The results are summarized in Table 5. The average overall sparsity and column sparsity of the eigenphone matrix are shown in parentheses.

From Table 5, it can be seen that when the weighting factor λ_3 is set to 20 $\sim$ 30, the recognition results obtained by applying l_1 regularization and the column-wise l_2 regularization simultaneously are better than that of using any one of the regularizers. When the amount of adaptation data is limited to one and two sentences, the relative improvements over the lasso method are 1.63% and 0.54%, respectively. These results are comparable to that of the best results of the eigenvoice method. When more adaptation data is available, the relative improvement over the lasso method becomes smaller. However, compared with the group lasso method, the relative improvement is more significant when sufficient adaptation data is provided. The advantages of both the l_1 regularization and the column-wise l_2 regularization combine well. Significance tests show that with $\lambda_1 = 10$ and $\lambda_3 = 30$, the sparse group lasso is significantly better than all other regularization methods under all testing conditions.

An interesting phenomenon is observed in that for all experimental settings, the overall sparsity is larger than that of the lasso method, while the column sparsity remains small when $\lambda_3 \leq 40$, that is, most of the columns remain nonzero. This observation implies that comparing with the lasso method, the performance improvement using the sparse group lasso should be attributed to the column-wise adaptive shrinkage property of the column-wise unsquared l_2 norm regularizer.

5.2 Experiments on the WSJ task

This section gives the unsupervised speaker adaptation results on the WSJ 20K open vocabulary speech recognition task. A two-pass decoding strategy was adopted. For a batch of recognition data from one speaker, hypothesized

Table 4 Average tonal syllable recognition rate (%) after eigenphone-based speaker adaptation using group lasso

λ_2	Number of adaptation sentences					
	1	2	4	6	8	10
60	52.56	53.36	56.84	58.06	59.78	60.85
	(0.07)	(0.0)	(0.0)	(0.0)	(0.0)	(0.0)
90	53.84	54.51	56.90	58.37	59.86	60.45
	(0.34)	(0.02)	(0.0)	(0.0)	(0.0)	(0.0)
120	54.22	55.77	57.03	58.06	59.63	60.34
	(0.65)	(0.12)	(0.01)	(0.0)	(0.0)	(0.0)
150	54.26	55.33	56.99	57.97	59.30	60.30
	(0.84)	(0.32)	(0.03)	(0.0)	(0.0)	(0.0)

The number of eigenphones (*N*) was fixed to 100. $\lambda_1 = \lambda_2 = 0$, and λ_3 was varied between 10 and 150. The average column sparsity of the eigenphone matrix is shown in parentheses.

Table 5 Average tonal syllable recognition rate (%) after eigenphone-based speaker adaptation using sparse group lasso

λ_3	Number of adaptation sentences					
	1	2	4	6	8	10
10	53.78	56.57	58.14	59.06	60.05	60.91
	(0.61, 0.01)	(0.47, 0.0)	(0.31, 0.0)	(0.22, 0.0)	(0.18, 0.0)	(0.15, 0.0)
20	54.76	56.74	58.29	59.21	60.18	60.93
	(0.62, 0.01)	(0.45, 0.0)	(0.31, 0.0)	(0.22, 0.0)	(0.18, 0.0)	(0.15, 0.0)
30	54.55	56.86	58.55	59.53	60.20	61.25
	(0.63, 0.02)	(0.44, 0.0)	(0.32, 0.0)	(0.23, 0.0)	(0.18, 0.0)	(0.15, 0.0)
40	54.49	56.65	58.35	59.32	60.11	60.93
	(0.63, 0.05)	(0.43, 0.0)	(0.31, 0.0)	(0.23, 0.0)	(0.18, 0.0)	(0.16, 0.0)
80	54.13	56.04	57.72	58.92	59.90	60.43
	(0.78, 0.37)	(0.45, 0.02)	(0.33, 0.0)	(0.23, 0.0)	(0.19, 0.0)	(0.16, 0.0)
120	54.05	54.95	57.01	58.35	59.38	60.24
	(0.91, 0.76)	(0.58, 0.21)	(0.35, 0.01)	(0.23, 0.0)	(0.19, 0.0)	(0.16, 0.0)

The number of eigenphones (*N*) was fixed to 100. $\lambda_1 = 10.0$, $\lambda_2 = 0$, and λ_3 was varied between 10 and 150. the average overall sparsity and column sparsity of the eigenphone matrix are shown in parentheses as pairs.

transcriptions were obtained using the SI model in the first pass. Then, speaker adaptation was performed using the hypothesized transcriptions based on the SI model (without SAT) or the canonical model (with SAT). The final results were obtained in a second decoding pass using the adapted model.

The SI model was trained using the following configurations. The standard SI-284 WSJ training set was used for training, which consists of 7,138 WSJ0 utterances from 83 WSJ0 speakers and 30,275 WSJ1 utterances from 200 WSJ1 speakers. The whole training set contains about 70 h of read speech in 37,413 training utterances from 283 speakers. The acoustic features are the same as that of the Mandarin Chinese task. There were 22,699 crossword triphones based on 39 base phonemes, and these were tree-clustered to 3,339 tied states. At most 16 Gaussian components were estimated for each tied state, resulting in a total of 53,424 Gaussian components.

The WSJ1 Hub 1 development test data (denoted by 'si_dt_20' in the WSJ1 corpus) were used for evaluation. For each of the 10 speakers, 40 sentences were selected randomly for testing, resulting in 52 min of read speech in 400 utterances. HDecoder was used as the decoder and the standard WSJ 20K-vocabulary trigram language model was used for compiling the decoding graph. We use word error rate (WER) to evaluate the recognition results. The WER of the SI model is 14.71%.

Unsupervised speaker adaptation was performed with varying amounts of adaptation data. The testing data of each speaker was grouped into batches, with the batch size varying from 2 to 20 sentences. Different batches of data were used for adaptation and evaluation independently.

The following five adaptation methods were tested for comparison:

1. EV: The standard eigenvoice method.
2. MLLR: The standard MLLR method.
3. SAT + MLLR: The standard MLLR method with speaker-adaptive training.
4. EP: The eigenphone method with or without regularization.
5. SAT + EP: The eigenphone method with speaker-adaptive training.

For the eigenvoice method, the number (*K*) of the eigenvoices was varied between 20 and 150. The best results were obtained with $K = 100$ and $K = 120$ for two and four adaptation sentences, respectively. When the amount of adaptation data is more than six sentences, $K = 150$ yields the best performance. For the MLLR method, the best results were obtained with a regression class tree using 32 base classes and three-block-diagonal transformation matrices. For the eigenphone method, the dimension (*N*) of the phone variation subspace was set to 100. Different regularization methods were tested with a wide range of weighting factors[a]. Again, the best results were obtained with the SGL method with $\lambda_1 = 10$ and $\lambda_3 = 30$. The results are summarized in Table 6, where 'ML-EP' denotes the eigenphone method with maximum likelihood estimation and 'SGL-EP' denotes the SGL regularized eigenphone method. For the sake of brevity, only the best results of each method are shown in the table.

It can be seen that the eigenvoice method performs best when the amount of adaptation data is limited to two sentences. But, it cannot achieve the performance of other

Table 6 Word error rate (%) after unsupervised speaker adaptation on the WSJ task

Methods	Number of adaptation sentences					
	2	4	6	8	10	20
EV	13.88	13.82	13.76	13.68	13.64	13.58
	$K = 100$	$K = 120$	$K = 150$	$K = 150$	$K = 150$	$K = 150$
MLLR	14.44	13.86	13.70	13.56	13.43	13.22
SAT + MLLR	13.96	13.41	13.37	13.35	13.26	13.06
ML-EP	16.28	14.24	13.75	13.47	13.41	13.06
SAT + ML-EP	16.80	14.24	13.51	13.17	13.12	12.70
SGL-EP	14.05	13.72	13.52	13.41	13.37	13.00
SAT + SGL-EP	13.92	13.36	13.29	13.11	13.03	12.70

The WER of the SI model is 14.71%. For the sake of brevity, only the best results of each adaptation method are shown in the table. For MLLR, the best results were obtained at a prior weighting factor of 10 (for MAP) and 32 regression classes with a three-block-diagonal transformation matrix (for MLLR). For the eigenphone method, the number of eigenphones (N) was fixed to 100. The weighting factors of the SGL regularization method were set to $\lambda_1 = 10$ and $\lambda_3 = 30$.

methods when more adaptation data become available. The 'ML-EP' method outperforms the MLLR method when more than six adaptation sentences are used. Severe over-fitting occurs when the amount of adaptation data is less than four sentences. With sparse group lasso regularization, the robustness of the eigenphone method is improved significantly[b]. Compared with the 'ML-EP' method, the relative improvements are 13.7%, 3.7%, and 1.7% with two, four, and six adaptation sentences, respectively. With more adaptation data, the relative improvements are negligible.

Combined with SAT, significant performance improvement is observed for all testing methods, except for the 'SAT + ML-EP' method with insufficient adaptation data (less than four sentences). Again, the performance degeneration is due to severe over-fitting. The 'SAT + SGL-EP' method performs best under all testing conditions. The relative improvements over the 'SAT + MLLR' method and the 'SAT+ML-EP' method are 0.3%, 0.4%, 0.6%, 1.8%, 1.7%, and 2.8% and 17.1%, 6.2%, 1.6%, 0.5%, 0.7%, and 0.0% with 2, 4, 6, 8, 10, and 20 adaptation sentences, respectively.

6 Conclusion

In this paper, we investigate various regularization methods to improve the robustness of the estimation of the eigenphone matrix in eigenphone-based speaker adaptation. The l_1 norm regularization (lasso) introduces sparseness, which reduces the number of free parameters and improves generalization. The squared l_2 norm penalizes large values of the matrix, thus alleviating over-fitting. The column-wise unsquared l_2 norm regularization (group lasso) forces many columns of the eigenphone matrix to be zero, thus preventing the dimension of the phone variation subspace from being higher than necessary. For nonzero columns, the group lasso is equivalent to the adaptively weighted column-wise squared l_2 norm regularizer. A unified framework for solving the various regularized matrix estimation problems is presented, and the performances of these regularization methods, including two combinations of them, i.e., elastic net and sparse group lasso, are compared for a supervised speaker adaptation task as well as an unsupervised speaker adaptation task using varying adaptation data. Compared with the maximum likelihood estimation method, significant performance improvements are observed using any of the regularization methods. Among them, the sparse group lasso method yields best results, which combines the advantages of the lasso and the group lasso methods in a consistent way. The group lasso plays an important role in case of limited amounts of adaptation data, with the performance improvement attributed to its column-wise adaptive shrinkage property. With large amounts of adaptation data, lasso seems to be more important than group lasso. Combined with speaker-adaptive training, performance is further improved.

When the dimension (N) of the phone variation subspace is larger than the feature dimension D and the adaptation data is sufficient, the columns of the eigenphone matrix $\boldsymbol{V}$ form an over-complete dictionary. The corresponding coordinate vector for each Gaussian component should be sparse. However, the matrix $\boldsymbol{L}$ obtained by PCA will not necessarily show any sparsity. Future work will focus on estimation of a sparse coordinate matrix at training time to obtain more performance gain.

Endnotes

[a] λ_1, λ_2 and λ_3 were varied between 0 and 1,000 at a step size of 10, respectively.

[b] Again, all significance tests show that the differences between the results of the 'ML-EP' and 'SGL-EP' methods are significant under all testing conditions.

Abbreviations

DNN, deep neural network; EP, eigenphone; EV, eigenvoice; FISTA, fast iterative shrinkage-thresholding algorithm; HMM, hidden Markov model; HTK, hidden Markov toolkit; IST, iterative shrinkage-thresholding MAP, maximum *a posteriori*; ML, maximum likelihood MLLR, maximum likelihood linear regression; SA, speaker adapted; SAT, speaker-adaptive training; SD, speaker dependent; SGL, sparse group lasso; SGMM, subspace Gaussian mixture model; SI, speaker independent; SLEP, sparse learning with efficient projections.

Competing interests

The authors declare that they have no competing interests.

Acknowledgements

This work was supported in part by the National Natural Science Foundation of China (No. 61175017 and No. 61005019) and the National High-Tech Research and Development Plan of China (No. 2012AA011603). The authors would like to thank the anonymous reviewers and Prof. Michael T. Johnson for their valuable suggestions that improved the presentation of the paper.

Author details

[1]Zhengzhou Information Science and Technology Institute, Zhengzhou 450000, China. [2]Tsinghua National Laboratory for Information Science and Technology, Department of Electronic Engineering, Tsinghua University, Beijing 100084, China.

References

1. R Kuhn, JC Junqua, P Nguyen, N Niedzielski, Rapid speaker adaptation in eigenvoice space. IEEE Trans. Speech Audio Process. **8**(6), 695–707 (2000)
2. MJF Gales, Maximum likelihood linear transformations for HMM-based speech recognition.Comput. Speech Lang. **12**(2), 75–98 (1998)
3. WL Zhang, WQ Zhang, BC Li, *Speaker adaptation based on speaker-dependent eigenphone estimation*. (Paper presented at the IEEE Workshop on automatic speech recognition and understanding, Waikoloa, HI, 11–15 Dec 2011), pp. 48–52
4. P Kenny, G Boulianne, P Ouellet, P Dumouchel, Speaker adaptation using an eigenphone basis. IEEE Trans. Speech Acoust. Process. **12**(6), 579–589 (2004)
5. QF Tan, PG Georgiou, SS Narayanan, Enhanced sparse imputation techniques for a robust speech recognition front-end. IEEE Trans. Acoust. Speech Signal Process. **19**(8), 2418–2429 (2011)
6. L Lu, A Ghoshal, S Renals, Regularized subspace Gaussian mixture models for speech recognition. IEEE Signal Process. Lett. **18**(7), 419–422 (2011)
7. F D Yu, G Seide, L Li, *Deng, Exploiting sparseness in deep neural networks for large vocabulary speech recognition*. (Paper presented at ICASSP, Kyoto, Japan, 25–30 Mar 2012), pp. 4409–4412
8. QF Tan, SS Narayanan, Novel variations of group sparse regularization techniques with applications to noise robust automatic speech recognition. IEEE Trans. Acoust. Speech Signal Process. **20**(4), 1337–1346 (2012)
9. DQu WL Zhang, WQ Zhang, BC Li. Rapid speaker adaptation using compressive sensing.Speech Commun. **55**(10), 950–963 (2013)
10. WL Zhang, WQ Zhang, DQu BC Li, MT Johnson, Bayesian speaker adaptation based on a new hierarchical probabilistic model. IEEE Trans. Audio Speech Lang. Process. **20**(7), 2002–2015 (2012)
11. J T Anastasakos, R McDonough, J Schwartz, A Makhoul, compact model for speaker-adaptive training. Paper presented at the ICSLP, Philadelphia, PA. USA, 3–6, 1137–1140 (Oct 1996)
12. CH Lee, CH Lin, BH Juang, A study on speaker adaptation of the parameters of continuous density hidden Markov models. IEEE Trans. Acoust. Speech Signal Process. **39**(4), 806–814 (1991)
13. CJ Leggetter, PC Woodland, *Flexible speaker adaptation using maximum likelihood linear regression*. (Paper presented at the ARPA spoken language technology workshop, 22–25 Jan 1995), pp. 110–115
14. R Tibshirani, Regression shrinkage and selection via the LASSO. J. R. Stat. Soc. (Ser. B). **58**, 267–288 (1996)
15. R T Hastie, J Tibshirani, *Friedman, The Elements of Statistical Learning: Data Mining, Inference and Prediction*. (Springer, Berlin, 2005)
16. M Yuan, Y Lin, Model selection and estimation in regression with grouped variables. J. R. Stat. Soc. (Ser. B). **68**, 49–67 (2007)
17. H Zou, T Hastie, Regularization and variable selection via the elastic net. J. R. Stat. Soc. B (Stat. Methodol.) **67**(2), 301–320 (2005)
18. N Simon, J Friedman, T Hastie, R Tibshirani, A sparse-group lasso. J. Comput. Graph. Stat. **22**(2), 231–245 (2013)
19. JF Gemmeke, T Virtanen, A Hurmalainen, Exemplar-based sparse representations for noise robust automatic speech recognition. IEEE Trans. Acoust. Speech Signal Process. **19**(7), 2067–2080 (2011)
20. M Figueiredo, R Nowak, S Wright, Gradient projection for sparse reconstruction: application to compressed sensing and other inverse problems. IEEE J. Sel. Top. Signal Process. **1**(4), 586–597 (2007)
21. J Liu, S Ji, J Ye, *SLEP: Sparse Learning with Eefficient Projections*. (Arizona State University, Tempe, 2009)
22. A Beck, M Teboulle, A fast iterative shrinkage-thresholding algorithm for linear inverse problems.SIAM. J. Imaging Sci. **2**, 183–202 (2009)
23. E Richard, PA Savalle, *Estimation of simultaneously sparse and low rank matrices*. (Paper presented at the ICML, 26 June – 1 July 2012), pp. 1351–1358
24. DP Bertsekas, Incremental proximal methods for large scale convex optimization. Math. Program. **129**(2), 163–195 (2011)
25. D Blatt, AO Hero, H Gauchman, A convergent incremental gradient method with a constant step size.SIAM. J. Optim. **18**, 29–51 (2008)
26. N Parikh, S Boyd. Proximal algorithms.Foundations Trends Optimization. **1**(3), 1–108 (2013)
27. Y Nesterov, A method of solving a convex programming problem with convergence rate $O\left(\frac{1}{k^2}\right)$. Sov. Math. Doklady. **27**, 372–376 (1983)
28. I Daubechies, MD Friese, CD Mol, An iterative thresholding algorithm for linear inverse problems with a sparsity constraint.Comm. Pure Appl. Math. **57**, 1413–1457 (2004)
29. E Chang, Y Shi, J Zhou, C Huang, Speech lab in a box : a Mandarin speech toolbox to jumpstart speech related research. Aalborg, Denmark, 2799–2802 (3–7 Sept 2001)
30. The National Institute of Standards and Technology the NIST Scoring Toolkit (SCTK-2.4.0). ftp://jaguar.ncsl.nist.gov/pub/sctk-2.4.0-20091110-0958.tar.bz2. Accessed 25 Sept 2013
31. G S Young, M Evermann, T Gales, D Hain, X Kershaw, G Liu, J Moore, D Odell, V Ollason, P Valtchev, *Woodland, The HTK Book (for HTK Version 3.4)*. (Cambridge University Engineering Department, Cambridge, 2009)
32. W Digalakis, LG Neumeyer, Speaker adaptation using combined transformation and Bayesian methods. IEEE Trans. Speech Audio Process. **4**(4), 294–300 (1996)

An optical music recognition system for traditional Chinese *Kunqu* Opera scores written in *Gong-Che* Notation

Gen-Fang Chen[1*] and Jia-Shing Sheu[2]

Abstract

This paper presents an optical music recognition (OMR) system to process the handwritten musical scores of *Kunqu* Opera written in *Gong-Che* Notation (GCN). First, it introduces the background of *Kunqu* Opera and GCN. *Kunqu* Opera is one of the oldest forms of musical activity, spanning the sixteenth to eighteenth centuries, and GCN has been the most popular notation for recording musical works in China since the seventh century. Many *Kunqu* Operas that use GCN are available as original manuscripts or photocopies, and transforming these versions into a machine-readable format is a pressing need. The OMR system comprises six stages: image pre-processing, segmentation, feature extraction, symbol recognition, musical semantics, and musical instrument digital interface (MIDI) representation. This paper focuses on the symbol recognition stage and obtains the musical information with Bayesian, genetic algorithm, and *K*-nearest neighbor classifiers. The experimental results indicate that symbol recognition for *Kunqu* Opera's handwritten musical scores is effective. This work will help to preserve and popularize Chinese cultural heritage and to store *Kunqu* Opera scores in a machine-readable format, thereby ensuring the possibility of spreading and performing original *Kunqu* Opera musical scores.

Keywords: Optical music recognition; *Kunqu* Opera; *Gong-Che* Notation; MIDI; Classifiers

1. Introduction

Kunqu Opera is one of the oldest forms of opera performed in China, dating back to the end of the *Yuan* Dynasty (1271 to 1368 AD) approximately 700 years ago. It dominated Chinese theatre from the sixteenth to the eighteenth centuries (see Figure 1) and is distinct in the virtuosity of its rhythmic patterns. It has been a dominant influence on more recent forms of local opera in China and is considered the mother of all Chinese operas, such as the *Beijing* opera. In 2001, United Nations Educational, Cultural and Scientific Organization (UNESCO) proclaimed *Kunqu* Opera a Masterpiece of the Oral and Intangible Heritage of Humanity [1].

Its composers wrote the musical scores of *Kunqu* Opera in Chinese using seven or ten characters, that is, in *Gong-Che* Notation (GCN) [2]. Several *Kunqu* Opera scores exist in the ancient literature [3]. In particular, there are three famous script sets that were popular during the *Qing* Dynasty (1636 to 1911 AD): *Jiugong Dacheng Nanbei Ci Gongpu* (A comprehensive anthology of texts and notation of the Southern and Northern opera tunes in nine modes, 1746 AD, with 4,466 musical works) [4], *Nashuying Qupu* (*Nashu* Studio Theatrical Music, 1792 AD, with more than 360 drama scripts) [5], and *Nanci Dinglv* (The Law of the Southern Word, 1720 AD, with 1,342 musical works) [6].

The risk of losing this rich cultural heritage must be urgently addressed, as *Kunqu* Opera is facing competition from mass culture, as well as a lack of audience interest since the eighteenth century because of the high level of technical knowledge required. Of the 400 arias regularly sung in opera performances in the mid-twentieth century, only a few dozen are performed now, such as *Mudan Ting* (The Peony Pavilion). Many *Kunqu* Opera works are available only as original manuscripts or photocopies, and their digitization and transformation into a machine-readable format is a pressing need.

* Correspondence: chengenfang@gmail.com
[1]Department of Computer Science and Technology, Hangzhou Normal University, Hangzhou, Zhejiang 310036, China
Full list of author information is available at the end of the article

Figure 1 Photo of the *Kunqu* 'Hall of Longevity'. Playwright: Sheng Hong (1645 to 1704); actor and actress from Shanghai *Kunqu* Opera Troupe: Zheng-Ren Cai and Jing-Xian Zhang; performance venue: *Hangzhou* Theatre, China; photographer: Gen-Fang Chen; date taken: October 11, 2011.

Optical music recognition (OMR) refers to the automatic processing and analysis of images of musical notation. This process typically employs a scanner or other digital device to transform paper-based musical scores into a digital image, which is then processed, recognized, and automatically translated into a standard format for music files, such as the musical instrument digital interface (MIDI) format. OMR represents a comprehensive amalgamation of fields and methods, including musicology, artificial intelligence, image engineering, pattern recognition, and MIDI technology. OMR technology provides a way to convert paper-based scores and has numerous applications, including computer-assisted music teaching, digital music libraries, musical statistics, digital music image automatic classification, and synchronous music and audio communication.

The first published OMR work was conducted by Pruslin [7] at the Massachusetts Institute of Technology. Pruslin's system recognized a subset of Western Music Notation (WMN), primarily musical notes. Before the 1990s, many studies were performed [8,9], and research had begun to focus on handwritten formats [9-11] and ancient (or folk) music, including medieval music [12], white mensural notation [13,14], early music prints [15], Orthodox Hellenic Byzantine Music notation [16], and Greek traditional music [17]. During the same period, several commercial OMR software packages, such as Capella-scan, Optical Music easy Reader, Photo Score, Sharp Eye, Smart Score, and Vivaldi Scan, were developed. These reached the market for WMN, and their recognition rate exceeded 90% for commonly printed musical scores [18].

Some techniques and methods used in OMR technology research include projection [19], mathematical morphology [20], neural networks [21], fuzzy theory [22], genetic algorithms [23], high-level domain knowledge [24], graph grammar [25], and probability theory [26]. Most of these methods are intended for WMN scores and can convert paper-based scores to digital form (such as a MIDI sequence). Replacing the scores' manual input mode in this way will improve the rate of musical score digitization.

In this study, an OMR system for *Kunqu* Opera musical scores is presented. The remainder of this paper is organized as follows: Section 2 describes *Kunqu* Opera musical scores, Section 3 discusses the structure of the OMR system and some preliminary operations, Section 4 provides the experimental results for selected scores using the OMR system, and Section 5 offers ideas for future research.

2. *Gong-Che* Notation in *Kunqu* Opera musical scores

GCN is a type of musical notation that was developed and cultivated in East Asia, flourishing mainly in ancient China. GCN was invented in the *Tang* Dynasty (618 to 907 AD) [27] and became a popular form of music notation in China during the *Song* (960 to 1279 AD), *Ming* (1368 to 1644 AD), and *Qing* (1636 to 1911 AD) Dynasties. Spanning 1,000 years of use, it is a widely accepted musical notation in East Asia and resembles WMN in Europe. Many traditional Chinese musical manuscripts have been written in GCN [28], including the *Kunqu* Opera scripts. GCN represents musical notes using Chinese characters, with the basic GCN note pitch symbols represented by ten Chinese characters, as shown in Figure 2. The meaning of these characters is presented in Table 1.

A *Kunqu* Opera GCN score contains contents such as the title, key signature, *qupai* (*qupai* is the general term for the tune names used in traditional Chinese lyrics and music writing, and every *qupai* singing tone has its own musical form, tone, and tonality), lyrics, and notes. An example of a GCN score is provided in Figure 3. Words constitute a primary component of a GCN score. GCN is scripted in the format of traditionally written Chinese: from top to bottom and from right to left. Rhythm marks are indicated to the right of the note characters.

The title of a *Kunqu* Opera script is located in the first column, and the *qupai* is located in the second column, with the key signature in the top row. The lyrics with notes appear below the *qupai*, and the notes with the rhythm marks are to the upper right of the lyrics. A frame with four borders usually encircles all the columns. Figure 4 depicts this layout for a GCN score.

Commonly, a Chinese lyric word includes several GCN notes, as shown in Figure 5. The number of pitches in a Chinese lyric word ranges from 1 to 8, and

Figure 2 Ten GCN symbols of a pitch.

the rhythm mark number of a pitch is from 0 to 2. Thus, words, pitches, and rhythms form a complex three-dimensional (3D) spatial relationship in a GCN score, with the first dimension containing lyrics, the second dimension containing pitches, and the third dimension containing rhythm marks. The font size of a Chinese lyric word is larger than the font size of a pitch, whose font size is in turn larger than that of a rhythm mark.

GCN does not mark a note's duration but instead gives the rhythm marks at regular intervals, with each rhythm mark being indicated in the corresponding note's upper right corner. The basic GCN note rhythm mark comprises eight symbols, as shown in Table 2, and two main types of rhythms are distinguishable: '*ban*', the stronger beat, and '*yan*', the weaker beat.

The difficulties encountered in the processing of GCN musical scores are due to the following: (1) the complex 3D relationship between symbols in a score's image, (2) symbols being written in different sizes, shapes, and intensities, (3) the variation in relative size between different components of a musical symbol, (4) different symbols appearing connected to each other and the same musical symbol appearing in separated components, and (5) difficulties in converting the musical information in GCN scores to other styles of musical notation, such as WMN.

3. Description of the OMR system for *Kunqu* Opera musical scores

The OMR system for *Kunqu* Opera (KOMR), shown in Figure 6, is an off-line optical recognition system comprising six independent stages: (1) image pre-processing, (2) document image segmentation, (3) feature extraction, (4) musical symbol recognition, (5) musical semantics, and (6) MIDI representation.

Table 1 Description of a GCN score's general symbols (pitch)

Symbols	Description		
	Phoneticize in simplified Chinese	Movable doh	Solfege
合	*hé*	5	sol
四	*sì*	6	la
一	*yī*	7	ti
上	*shàng*	1	do
尺	*Chě*	2	re
工	*gōng*	3	mi
凡	*fán*	4	fa
六	*liù*	5	sol
五	*wǔ*	6	la
乙	*yǐ*	7	ti

Because of the maturity of image digitizing technology [29], paper-based *Kunqu* Opera musical scores can easily be converted into digital images. Therefore, this paper does not discuss the image acquisition stage. The system processes a gray-level bitmap from a scanner or reads directly from image files, with the input bitmap generally being 300 dpi with 256 gray levels.

3.1 Image pre-processing

Pre-processing might involve any of the standard image-processing operations, including noise removal, blurring, de-skewing, contrast adjustment, sharpening, binarization, or morphology. Many operations may be necessary to prepare a raw input image for recognition, such as the selection of an interesting area, the elimination of nonmusical elements, image binarization, and the correction of image rotation.

Many documents, particularly historical documents such as those in *Kunqu* Opera GCN scores, rely on careful pre-processing to ensure good overall system performance and, in some cases, to significantly improve the recognition performance. In this work, we briefly touch upon the basic pre-processing operations of *Kunqu* Opera GCN scores, such as binarization, noise removal, skew correction, and the selection of an area of interest. Image binarization uses the algorithm of Otsu [30], noise removal is conducted using basic morphological operations [31], and image rotation has been corrected using the least squares criterion [32]. The area of interest in a *Kunqu* score is surrounded by a frame (wide lines) (see Figure 3). The frame is a connected component containing the longest line in the score, so we use the Hough transform [33] to locate the longest line then delete the connected component with the longest line from the score, leaving the area of interest. Because these are common image-processing operations, we do not give a detailed description here.

3.2 Document segmentation and analysis of *GCN* scores

Document segmentation is a key phase in the KOMR system. With the symbols in a GCN score document having been classified, this stage first segments the document into two sections, one including symbols for the

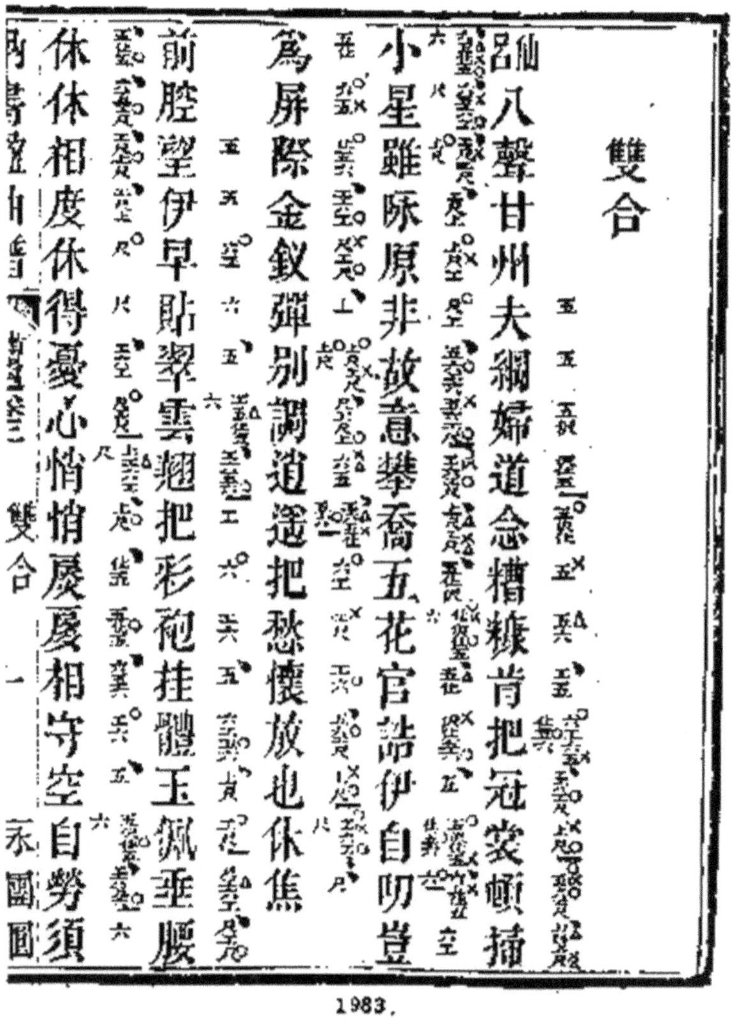

Figure 3 Musical score sheet of *Nashu Studio Theatrical Music* (p. 1983).

notes and the other for the non-notes. Elements of non-note symbols, such as the title, key signature, *qupai*, lyrics, noise, and border lines of a textural framework, are then identified and removed.

Because music is a time-based art, the arrangement of the notes is one of its most important factors. Therefore, obtaining the arrangement of the notes in a GCN score is requisite to document the segmentation. Concordant with the writing style of the GCN score, the arrangement of the notes can be organized based on high-level field knowledge of *Kunqu* Opera scores.

Several document image segmentation methods have been proposed, with the best known being *X-Y* projection [34], run length smoothing algorithm (RLSA) [35], component grouping [36], scale-space analysis [37], Voronoi tessellation [38], and the Hough transform [39]. These methods are suitable for handwritten document segmentation, such as for GCN scores, which are compatible with *X-Y* projection, RLSA, scale-space analysis, and the Hough transform [40].

A preliminary result of GCN score segmentation has been presented in [41]. A self-adaptive RLSA was used to segment the image according to an *X*-axis function (denoted by $PF(x)$) indicating the number of foreground pixels in each column of the image. This *X*-axis function uses *X*-projection to compute the number of flex points that satisfy the following conditions: $PF(x-1) < PF(x)$ and $PF(x) > PF(x+1)$ or $PF(x-1) > PF(x)$ and $PF(x) < PF(x+1)$. Next, the algorithm iteratively smoothes the function and analyzes the next smoothed function to ensure that the number of flex points in both functions is equal. Finally, the image is segmented into several sub-images based on the *X*-axis values for flex points in the function. To extract notes from the image, all connected components are identified using a conventional connected component labeling algorithm, and the minimum bounding box of all connected components is computed.

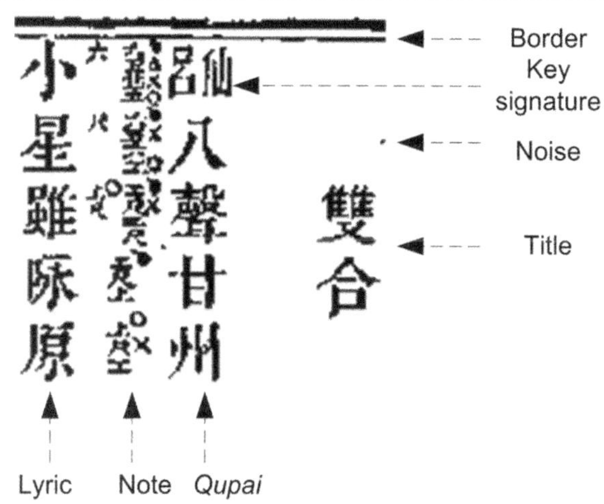

Figure 4 Sample annotation for a sub-section of Figure 3. 呂仙, *Xianlv*, a Chinese key signature in a musical work, and nine key signatures are commonly used in *Kunqu* Opera; 八聲甘州, *Bashing ganzhou*, is a *qupai*, a fixed type of rhythm or melody in Chinese poetry and opera.

Next, the algorithm matches each connected component to its corresponding sub-image. According to the experimental results in [41], the rate of correct segmentation in lines is 98.9%, and the loss rate of notes is 2.7%; however, the total error rate of all notes and lyrics is almost 22%.

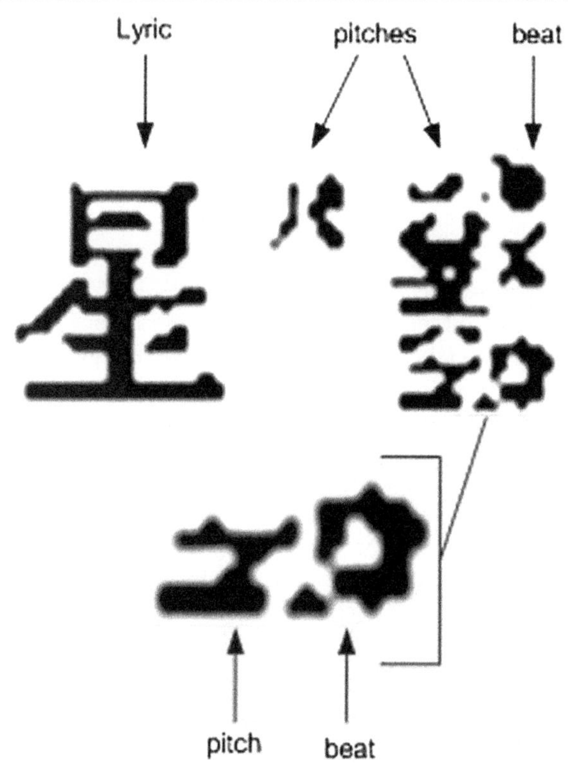

Figure 5 Sample annotation for a sub-section of Figure 4.

Table 2 Description of GCN score's general symbols (rhythm)

Symbols	Phoneticize in simplified Chinese	Description
、	*ban*	A note with *'ban'* starts a stronger beat
○	*yan*	A note with *'yan'* starts a weaker beat
△	*yaoyan*	A note with *'yaoyan'* starts a weaker beat
—	*yaoban*	Denotes a note's duration from a weaker beat to the next stronger beat
×	*ban*	A note with *'ban'* starts a stronger beat
|×	*zhengban*	Denotes the stronger beat in the middle of a bar for an 8/8 beat
L	*yaoban*	Denotes a note's duration from a weaker beat to the next stronger beat in some scores
□	*yaoyan*	A note with *'yaoyan'* starts a weaker beat in some scores

3.3 Symbol feature extraction

Selecting suitable features for pattern classes is a critical process in the OMR system. Feature representation is also a crucial step, because perfect feature data can effectively enhance the symbol recognition rate. The goal of feature extraction is to characterize a symbol to be recognized by measurements whose values are highly similar for symbols in the same category, but different for symbols in different categories. The feature data must also be invariant during relevant transformations, such as translation, rotation, and scaling.

Popular feature extraction methods for OCR and Chinese character recognition include peripheral feature [42], cellular feature [43], and so on [44]. Because symbols are written in different sizes in a GCN score, to construct a simple and intuitive approach, four types of structural features have been used in this exploratory work based on the reference [42,43]; these are suited to feature extraction of symbols in a GCN score and to compare with the recognition rate of KOMR. An $n \times m$ matrix featuring symbol data is used to obtain symbols of the same size as in the feature matrix. These are obtained using a grid to segment the symbols [45]. In Figure 7, a sample symbol with size $H \times W$ is shown in sub-graph (a), and an $n \times m$ grid is shown in sub-graph (b). In this example, $h_0 = 0, h_n = H, h_i = \left[\frac{H}{n}\right] \times i, w_0 = 0, w_m = W, w_j = \left[\frac{W}{m}\right] \times j$, and the four features of each symbol are given by the following equations:

$$T_1 = \begin{bmatrix} \ldots, \ldots, \ldots \\ \ldots, t_{i,j}, \ldots \\ \ldots, \ldots, \ldots \end{bmatrix}, t_{i,j} = \sum_{x=w_{i-1}}^{w_i} \sum_{y=h_{j-1}}^{h_j} f(x, y), 1 \le i \le m, 1 \le j \le n \quad (1)$$

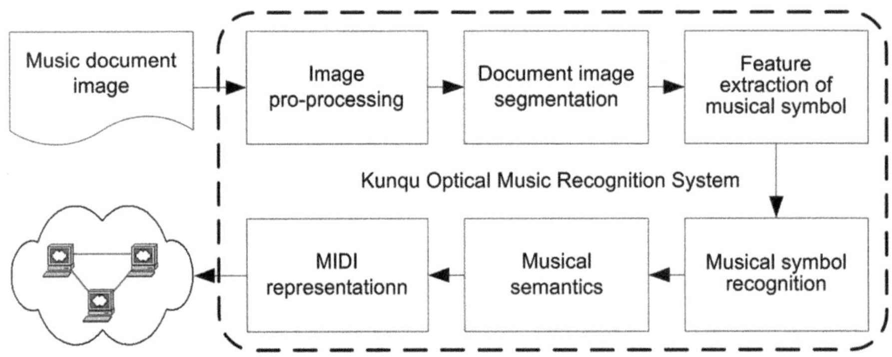

Figure 6 **The KOMR system architecture.**

$$T_2 = \begin{bmatrix} \ldots, \ldots, \ldots \\ \ldots, t_{i,j}, \ldots \\ \ldots, \ldots, \ldots \end{bmatrix}, t_{i,j} = \sum_{x=w_{i-1}}^{w_i} f(x, h_j), 1 \le i \le m, 1 \le j \le n \quad (2)$$

$$T_3 = \begin{bmatrix} \ldots, \ldots, \ldots \\ \ldots, t_{i,j}, \ldots \\ \ldots, \ldots, \ldots \end{bmatrix}, t_{i,j} = \sum_{y=h_{j-1}}^{h_j} f(w_i, y), 1 \le i \le m, 1 \le j \le n \quad (3)$$

$$T_4 = \begin{bmatrix} \ldots, \ldots, \ldots \\ \ldots, t_{i,j}, \ldots \\ \ldots, \ldots, \ldots \end{bmatrix}, t_{i,j} = \begin{cases} 0, & \text{for } 1 < i < m \text{ and } 1 < j < n \\ x, & \text{if } \sum_{k=0}^{x} f(k, h_j) = 0 \text{ and } f(x+1, h_j) = 1, \text{ for } j = 1 \\ W-x, & \text{if } \sum_{k=x}^{w_m} f(k, h_j) = 0 \text{ and } f(x-1, h_j) = 1, \text{ for } j = m \\ x, & \text{if } \sum_{k=0}^{x} f(w_i, k) = 0 \text{ and } f(w_i, x+1) = 1, \text{ for } i = 1 \\ H-x, & \text{if } \sum_{k=x}^{h_n} f(w_i, k) = 0 \text{ and } f(w_i, x-1) = 1, \text{ for } i = n \end{cases} \quad (4)$$

where $f(x, y)$ is the value of a pixel at coordinates (x, y) in an image. If $f(x, y) = 1$, then the pixel is a foreground pixel; otherwise, 0 indicates a background pixel. The T_1 feature is the number of pixels in each cell of the grid as a cellular feature, T_2 feature appears to be the number of pixels falling on the upper edge of each grid cell, T_3 is the number of pixels falling on the right edge of each grid cell, and T_4 is the number to be some measure of how many background pixels there are from the edge to the first foreground pixel, if $f(1,1) = 1$, then $t_{1,1} = 0$; if $f(n,1) = 1$, then $t_{n,1} = 0$; if $f(1,m) = 1$, then $t_{1,m} = 0$; if $f(n,m) = 1$, then $t_{n,m} = 0$; and if there is no foreground pixel in the image, then $t_{i,j} = 0$, for $1 \le i \le m$ and $1 \le j \le n$.

3.4 Musical symbol recognition

Computer recognition of handwritten characters has been intensely researched for many years. Optical character recognition (OCR) is an active field, particularly for handwritten documents in such languages as Roman [45], Arabic [46], Chinese [47], and Indian [48].

Several Chinese character recognition methods have been proposed, with the best known being the transformation invariant matching algorithm [49], adaptive confidence transform based classifier combination [50], probabilistic neural networks [51], radical decomposition [52], statistical character structure modeling [53], Markov random fields [54], and affine sparse matrix factorization [55].

The basic symbols for musical pitch (see Figure 2) in a *Kunqu* Opera GCN score are Chinese characters, but other musical pitch symbols and all rhythm symbols are specialized symbols for GCN. Thus, the methods of *Kunqu* Opera GCN score recognition refer to the techniques of both OMR and OCR. In this paper, the following three approaches to musical symbol recognition are compared.

3.4.1 K-nearest neighbor

The *K*-nearest neighbor (KNN) classifier is one of the simplest machine learning algorithms. It classifies foregrounds based on the closest training examples in the feature space and is a type of instance-based, or lazy, learning in which the function is only approximated locally, and all computation is deferred until classification [56]. The neighbors are selected from a set of foregrounds for which the correct classification is known, a set that can be considered the training set for the algorithm, though no explicit training step is required.

In the experiment described in the manuscript, we choose half of the test musical symbols as training samples; then, for the convenience of computing, we set the value of *K* to 1. The feature data of each class are obtained from the samples by calculating the average feature data of all samples. Although the distance function

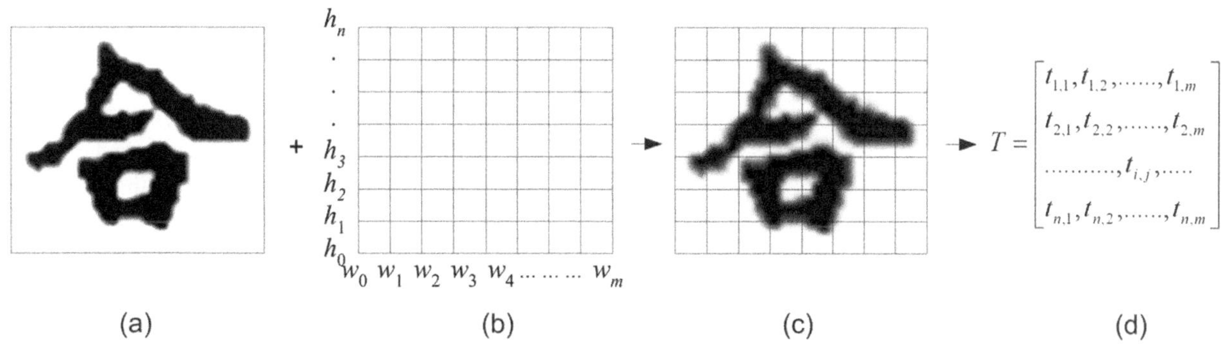

Figure 7 Matrix featuring data of a musical symbol using an $n \times m$ grid partition (a to d).

can also be learned during the training, the corresponding distance functions for the four features above are determined by the following equations:

$$f(S,T)_1 = \frac{m \times n - \sum_{i=1}^{m}\sum_{j=1}^{n}\rho, \rho = 1 \text{ if } s_{i,j} = t_{i,j}, \text{other } \rho = 0}{m \times n} \quad (5)$$

$$f(S,T)_2 = \frac{\sum_{i=1}^{m}\sum_{j=1}^{n}\left|s_{i,j} - t_{i,j}\right|}{m \times n} \quad (6)$$

$$f(S,T)_3 = \frac{\sum_{i=1}^{m}\sum_{j=1}^{n}\left|s_{i,j} - t_{i,j}\right|}{m \times n} \quad (7)$$

$$f(S,T)_4 = \frac{m \times n - \sum_{i=1}^{m}\sum_{j=1}^{n}\rho, \text{if} \alpha < \frac{t_{i,j}}{s_{i,j}} < \beta, \rho = 1, \text{other } \rho = 0}{m \times n} \quad (8)$$

where $f(S,T)_{1-4}$ is the distance function of the corresponding feature 1-4, $s_{i,j}$ is the feature matrix of the prototype class (S), and $t_{i,j}$ is the feature matrix of the test sample (T). In Equation 5, if $s_{i,j} = t_{i,j}$, $1 \le i \le m$, $1 \le j \le n$, then $\rho = 1$; otherwise, $\rho = 0$. $f(S,T)$ counts the number of unequal elements in the feature matrixes (S and T). In Equations 6 and 7, $f(S,T)$ calculates the sum of the difference between all corresponding elements in the feature matrixes (S and T). Equation 8 counts the number of non-approximate elements in S and T, using the parameters α and β as experience values. In this work, $\alpha = 0.9$ and $\beta = 1.1$.

3.4.2 Bayesian decision theory

Bayesian decision theory is used in many statistics-based methods. Classifiers based on Bayesian decision theory are simple, probabilistic classifiers that apply Bayesian decision theory with conditional independence assumptions, providing a simple approach to discriminative classification learning [57].

In this work, the conditional probabilities $P(T|S_k)_r$, $1 \le r \le 4$ with each of the four features for the Bayesian decision theory classifier are calculated as follows:

$$P(T|S_k)1 = \frac{\sum_{i=1}^{m}\sum_{j=1}^{n}\rho, \rho = 1 \text{ if } s^k_{i,j} = t_{i,j}, \text{other } \rho = 0}{m \times n} \quad (9)$$

$$P(T|S_k)2 = \frac{\sum_{i=1}^{m}\sum_{j=1}^{n}\left|s^k_{i,j} - t_{i,j}\right|}{m \times n} \quad (10)$$

$$P(T|S_k)3 \frac{\sum_{i=1}^{m}\sum_{j=1}^{n}\left|s^k_{i,j} - t_{i,j}\right|}{m \times n} \quad (11)$$

$$P(T|S_k)4 = \frac{m \times n - \sum_{i=1}^{m}\sum_{j=1}^{n}\rho, if \alpha < \frac{t_{i,j}}{s^k_{i,j}} < \beta, \rho = 1, \text{other } \rho = 0}{m \times n} \quad (12)$$

where $P(T|S_k)_{1-4}$ is the conditional probability of the corresponding feature 1-4, $S = \{S_1, \ldots, S_k, \ldots, S_c\}$ is the set of prototype classes, $s^k_{i,j}$ is the feature matrix of the kth class S_k in the set of prototype classes, and $t_{i,j}$ is the feature matrix of the test sample (T). In Equation 9, if $s^k_{i,j} = t_{i,j}$, then $\rho = 1$; otherwise, $\rho = 0$. In Equation 12, α and β are again experience values, and in this work, we set $\alpha = 0.9$ and $\beta = 1.1$.

In the experiment, the prior probability $P(S_k)$, $1 \le k \le c$ of different symbols S_k is not equal, and all prior probabilities come from the prior statistics of the training dataset. For example, the beat symbol '↘' has the prior probability 0.354511. If $P(S_k|T) = \max P(S_j|T)$, T is classified to S_k. The Bayesian rule was used. $P(T|S_k)rP(S_k)/\sum_{i=1}^{c} P(T|S_i)P(S_i), 1 \le r \le 4$ and has four expressions to estimate $P(S_k|T)$ which correspond to each of the four features.

3.4.3 Genetic algorithm

In the field of artificial intelligence, a genetic algorithm (GA) is a search heuristic that mimics the process of natural evolution. This heuristic is routinely used to generate useful solutions to optimization and search problems. GAs generate solutions to optimization problems using techniques inspired by natural evolution, such as inheritance, mutation, selection, and crossover. In GAs, the search space parameters are encoded as strings, and a collection of strings constitutes a population. The processes of selection, crossover, and mutation continue for a fixed number of generations or until some condition is satisfied. GAs have been applied in such fields as image processing, neural networks, and machine learning [58].

In this work, the key GA parameter values are as follows:

- Selection-reproduction rate: $p_s = 0.5$
- Crossover rate: $p_c = 0.6$
- Mutation rate: $p_m = 0.05$
- Population class: $C = 12$
- Number of individuals in the population: $N_p = 200$
- Maximum iteration number: $G = 300$

An individual's fitness value is determined from the following fitness function:

$$\begin{aligned} F(I_u) &= \sum_{v=1}^{R} f\left(e\left(h_{u,v}, s_{u,v}\right)\right) = \sum_{v=1}^{R} f(\rho) \\ &= \sum_{v=1}^{R} \frac{\rho}{m \times n}, \textit{if } h_{i,j} = s_{i,j}, \text{then } \rho \\ &= 1, \text{else } \rho = 0 \end{aligned} \tag{13}$$

where I_u is the kth individual, $F(I_u)$ is the fitness value of I_u, R is the number of a gene bits in I_u, $h_{i,j}$ is the jth gene of I_u, $s_{i,j}$ is the feature matrix of the set of prototype classes corresponding to $h_{i,j}$, the function $e()$ computes the value of the comparative result between $h_{i,j}$ and $s_{i,j}$, and m and n are the width and length, respectively, of a symbol's grid.

3.5 Semantic analysis and MIDI representation

After all the stages of the OMR system are complete (see Figure 6), the recognized symbols can be employed to write the score in different data formats, such as MIDI, Nyquist, MusicXML, WEDELMUSIC, MPEG-SMR, notation interchange file format (NIFF), and standard music description language (SMDL). Although different representation formats are now available, no standard exists for computer symbolic music representation. However, MIDI is the most popular digital music format in modern China, analogous to the status of the GCN format in ancient China.

This work selected MIDI for music representation, because the musical melody in a *Kunqu* Opera GCN score provides monophonic information. Thus, the symbols recognized from the GCN score can be represented by an array melody[L][2], with L representing the number of notes in the GCN score, the first dimension in the array representing the pitch of all notes, and the second dimension representing the duration of all notes.

Finally, the note information in melody[L][2] can be transformed into a MIDI file using an associated coding language, such as Visual C++, and the MIDI file of the *Kunqu* Opera GCN score can then be disseminated globally via the Internet.

4. Experimental results

A Chinese *Kunqu* Opera GCN score may contain information as Chinese characters and musical notes, such as lyrics, title, tune, pitch, rhythm, voices, preface of music, and so on. In this experiment, a dataset was randomly selected from two *Kunqu* score books [4,5]. Because OCR for Chinese characters is actively researched [47] and OMR for musical information is a developing field, we consider only the musical note information in this experiment. This includes pitches and rhythm marks (beats) but neglects other information such as lyrics (Chinese characters).

First, all scores in the set were segmented according to [41], and then the musical note information, including

Table 3 Information on the 39 selected images

	Name of works	
	Nashu Studio Theatrical Music	***Jiugong Dacheng Nanbei Ci Gongpu***
Range of pages	pp. 1975-1996	pp. 4326-4344
Range of training pages	pp. 1975-1984	pp. 4326-4335
Range of testing pages	pp. 1985-1996	pp. 4336-4344
Serial number of testing pages	No. 1-10	No. 11 ~ 19
Number of all pages	20	19
Number of pitches in all pages	3,100	2,157
Number of beats in all pages	1,413	661

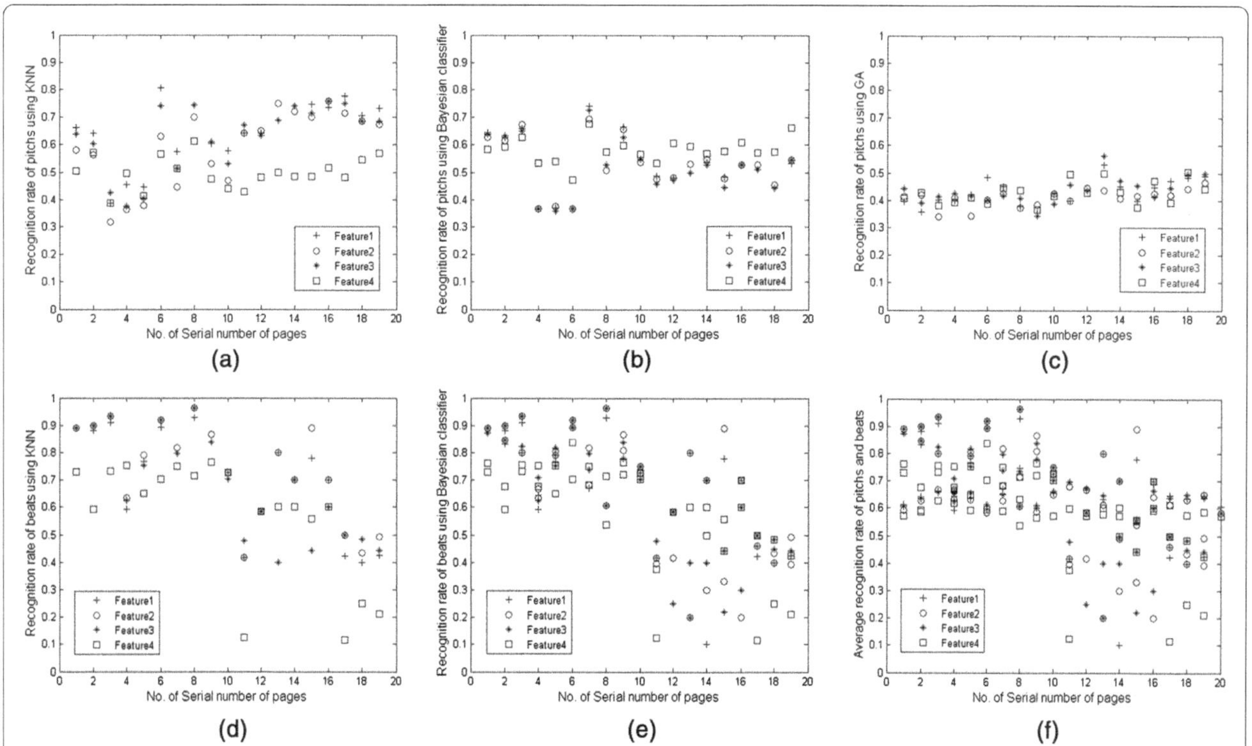

Figure 8 Recognition rate of 39 images using four features. *X*-axis: serial number of pages; *Y*-axis: **(a to c)** recognition rate of pitch using three classifiers and **(d, e)** recognition rate of beat using two classifiers. **(f)** Average recognition rate of pitch and beat.

pitches and beats (rhythm marks), was decoded. Finally, the musical information was translated to a MIDI file.

4.1 Training and test datasets

To estimate the efficiency of KOMR, experiments were conducted using images randomly selected from *Nashu Studio Theatrical Music* [5] and *Jiugong Dacheng Nanbei Ci Gongpu* [4]. Thirty-nine images were used. The primary information from these images is shown in Table 3 and includes the number of pages, pitches, and beats. The resolution was calculated as (680-720) × (880-900). Figure 3 displays a common musical score sheet from page 1983 of *Nashu* Studio Theatrical Music.

To improve the recognition rate of all classifiers, a symbol's grid was used to render a partition of resolution 16 × 16 ($n \times m$ above). Each symbol was then resized to 16 × 16 pixels and converted to a 256-dimensional vector, a feature datum of the symbol.

To create training datasets, we randomly selected images from the available datasets. Specifically, the images on pp. 1975-1984 and pp. 4326-4335 (see Table 3) constituted the training sets, and other images were included in the test sets.

4.2 Results

To evaluate the accuracy of our results, the various classifiers were compared based on their test GCN scores. The three classifiers' pitch recognition rates for the test images are shown in Figure 8, with (a) the KNN classifier (in this work, $K = 1$), (b) the Bayesian classifier, and

Table 4 Experimental statistics of the recognition rate for all pitches and all beats using features 1 to 4

Feature	Recognition rate of pitch (%)			Recognition rate of beat (%)	
	BC	GA	KNN	BC	KNN
Feature 1	52.75	43.09	64.80	55.59	70.78
Feature 2	52.59	41.05	59.35	57.10	73.46
Feature 3	51.97	43.60	62.72	58.32	68.17
Feature 4	58.24	42.72	49.85	61.68	56.63
Average rate	53.89	42.62	59.18	58.17	67.26

(c) the GA classifier. The beat recognition rates for (d) the KNN classifier and (e) the Bayesian classifier are also shown. The recognition rate obtained by KNN is higher than that of the other two classifiers.

From the results obtained for the musical notes, shown in Table 4, it can be seen that the KNN classifier outperformed the other two classifiers. Notably, the performance of the KNN classifier was clearly superior to that of the GA classifier. The recognition rate of beats is greater than that of pitches, and the results in Figure 8 indicate that the recognition rate of the beats in *Nashu Studio Theatrical Music* is greater than that in *Jiugong Dacheng Nanbei Ci Gongpu*. This is because the musical symbols for the beats in the first book included seven different symbols: ﹅ ○ △ _ X |X L, whereas there were only four different symbols in the second book: ﹅ □ _ L. Thus, each beat feature in the first book is very different from that in the second book, and the second has less discrimination. Meanwhile, Figure 8 and Table 4 suggest that the four features have little effect on the recognition rate.

Following the recognition stage, the musical information of a GCN score was saved to an array melody[L][2], which was transformed into MIDI format and stored as a MIDI file (Additional file 1). Figure 9 shows the sheet score of the MIDI file in WMN based on the recognition of musical information from Figure 3.

Figure 9 Sheet score of the MIDI file in WMN for Figure 3.

5. Conclusions

As one of UNESCO's Oral and Intangible Heritages of Humanity, *Kunqu* Opera had a deep effect on ancient Chinese entertainment and has been studied in many ways. GCN is the most widely used recording method for ancient Chinese musical information, employed by a significant number of musical works, including all *Kunqu* Opera scores.

This paper presented the six-stage KOMR system, with the key phases being image segmentation and symbol recognition. This paper focused on the recognition phase, studying the effectiveness of three classifiers to obtain musical information from the GCN score. The results indicate that a KNN classifier is most suitable for this task.

However, because the average recognition rate of KOMR is less than 90%, it is necessary to find more powerful methods that can be applied to all phases of KOMR. These problems merit further research.

This work is preliminary and tentative and will help to preserve and popularize Chinese cultural heritage. In particular, this study will assist the conversion of handwritten musical *Kunqu* Opera scores to a machine-readable format, thus ensuring the possibility of disseminating and performing the original *Kunqu* Opera musical scores.

Additional file

Additional file 1: Shuanghe.zip. The file is a MIDI format file which includes musical information from Figure 3 in the main body text.

Competing interests
The authors declare that they have no competing interests.

Acknowledgements
This work is supported by the key project of National Social Science Foundation of China Grant No. 12AZD120, the promotion project of National Culture Science and Technology in China Grant No. [2013]718-6, HangZhou Special Funds to Support Cultural Creativity project Grant No. 93, and the planning project of educational technology research in Zhejiang Province under Grant JB001.

Author details
[1]Department of Computer Science and Technology, Hangzhou Normal University, Hangzhou, Zhejiang 310036, China. [2]Department of Computer Science, National Taipei University of Education, Taipei 10671, Taiwan.

References
1. UNESCO, Masterpieces 2001 and 2003: *Kunqu* Opera. UNESCO Physics Web (2008). http://www.unesco.org/culture/ich/en/RL/00004. Accessed 12 Oct 2012
2. Z Chen, *Introduction to Gong-Che Notation*, 1st edn. (Huayue's Music Publishing House, Beijing, 2006), pp. 1–2
3. Y-H Wang, *Musical Notation of Chinese Traditional Music*, 1st edn. (Fujian Education Publishing House, Fuzhou, 2006), pp. 245–246
4. X Zhou, J Zhou, X Xu, W Wang, T Zhu, Y Xu, *A comprehensive anthology of texts and notation of the Southern and Northern opera tunes in nine modes.* new, edn (Shanghai Guji Publishing House, Shanghai, 1995), pp. 4326–4344
5. T Ye, *Nashu Studio Theatrical Music.* new, edn (Shanghai Guji Publishing House, Shanghai, 1995), pp. 1975–1996
6. S Lv, X Yang, H Liu, S Tang, *The Law of the Southern Word.* new edn. (Shanghai Guji Publishing House, Shanghai, 1995), pp. 1–10
7. D Pruslin, *Automatic recognition of sheet music, Dissertation* (Massachusetts Institute of Technology, Cambridge, 1966)
8. I Fujiinaga, *Adaptive optical music recognition, Dissertation* (Faculty of Music, McGill University, 1996)
9. K Ng, Optical music analysis for printed music score and handwritten music manuscript, in *Visual Perception of Music Notation: On-Line and Off-Line Recognition*, ed. by SE George (Idea Group Publishing, Pennsylvania, 2003), pp. 108–127
10. J Wolman, J Choi, S Asghzrzadeh, J Kahana, Recognition of handwritten music notation, in *Proceedings of the International Computer Music Conference* (San Jose, 9–13 Nov 1992), pp. 125–127
11. J Caldas Pinto, P Vieira, M Ramalho, MO Mengucci, P Pina, F Muge, Ancient music recovery for digital libraries, in *Proceedings of the 4th European Conference on Research and Advanced Technology for Digital Libraries. Lecture Notes in Computer Science* (Springer, Heidelberg, 2000), pp. 245–249
12. W McGee, P Merkley, The optical scanning of medieval music. Comput. Hum. **25**(1), 47–53 (1991)
13. LJ Tardón, S Sammartino, I Barbancho, Optical music recognition for scores written in white mensural notation. EURASIP J. Image Video Process. (2009). doi: 10.1155/2009/843401
14. NP Carter, Segmentation and preliminary recognition of madrigals notated in white mensural notation. Mach. Vis. Appl. **5**(3), 223–230 (1992)
15. L Pugin, JA Burgoyne, I Fujinaga, Goal-directed evaluation for the improvement of optical music recognition on early music prints, in *Proceedings of the 7th ACM/IEEE Joint Conference on Digital Libraries* (JCDL'07, Vancouver, 1 April 2007), pp. 303–304
16. VG Gezerlis, S Theodoridis, Optical character recognition of the Orthodox Hellenic Byzantine Music notation. Pattern Recogn. **35**(4), 895–914 (2002)
17. A Pikrakis, Recognition of isolated musical patterns in the context of Greek traditional music, in *Proceedings of the Third IEEE International Conference on Electronics, Circuits, and Systems*, vol 2nd edn. (IEEE Xplore Digital Library, Rodos, 1996), pp. 1223–1226
18. K Ng, J Barthelemy, B Ong, I Bruno, P Nesi, CIMS: coding images of music sheets. Leeds University PhysicsWeb (2005). http://music.leeds.ac.uk/people/kia-ng. Accessed 20 June 2005
19. I Fujinaga, *Optical music recognition using projections, Dissertation* (Montreal University, Canada, 1988)
20. G Chen, *The study of optical music recognition and prototype system, Dissertation* (Zhejiang University, China, 2003)
21. IG Ratiu, Self organizing neural network and direction of arrival estimation technique MUSIC 2D applied in target radiolocation recognition, in *Proceedings of the 4th WSEAS International Conference on Computer Engineering and Applications* (Stevens Point, 2011), pp. 231–236
22. HS Park, JO Yoo, SB Cho, *A context-aware music recommendation system using fuzzy Bayesian networks with utility theory. Lecture Notes in Computer Science.* vol. 4223 (Springer, Heidelberg, 2006), pp. 970–979
23. I Yoda, K Yamamoto, Automatic construction of recognition procedures for musical notes by genetic algorithm, in *International Association for Pattern Recognition Workshop on Document Analysis Systems* (World Scientific, Hackensack, 1995), pp. 153–159
24. MV Stuckelberg, C Pellegrini, M Hilario, An architecture for musical score recognition using high-level domain knowledge, in *Proceedings of the Fourth International Conference on Document Analysis and Recognition*, 2nd edn. (Ulm, 1997), pp. 813–818
25. S Baumann, G Res, A simplified attributed graph grammar for high-level music recognition, in *Proceedings of the Third International Conference on Document Analysis and Recognition*, 2nd edn. (Montreal, 1995), pp. 1080–1083
26. G Taubman, *Music hand: a handwritten music recognition system, Dissertation* (Brown University, 2005)
27. YL Yang, *Chinese Ancient Music Historical Manuscript* (People's Music Publishing House, Beijing, 2004), pp. 258–259
28. Z Tong, J Gu, Y Zhou, X Sun, *Chinese Traditional Musicology* (Fujian Education Publishing House, Fuzhou, 2003), pp. 73–75
29. A Rebelo, G Capela, JS Cardoso, Optical recognition of music symbols. IJDAR **13**, 19–31 (2010)
30. N Otsu, A threshold selection method from gray-level histogram. IEEE. Trans. Syst. Man Cybern. **9**, 62–66 (1979)

31. L Najman, H Talbot, *Mathematical Morphology* (Wiley, Hoboken, 2010), pp. 12–13
32. YH Liang, ZY Wang, A skew detection method for 2D bar code images based on the least square method, in *2006 International Conference on Machine Learning and Cybernetics, Dalian*, 2006, pp. 3974–3977. doi:10.1109/ICMLC.2006.258793
33. RC Gonzalez, RE Woods, SL Eddins, *Digital image processing using MATLAB* (Prentice Hall, 2004), pp. 393–403
34. G Nagy, S Seth, Hierarchical representation of optically scanned documents, in *Seventh International Conference on Pattern Recognition* (Montreal, 1984), pp. 347–349
35. FM Wahl, RG Wong, KY Casey, Block segmentation and test extraction in mixed text/image documents. Comput. Graph. Image Process. **20**, 375–390 (1982)
36. M Feldbach, KD Tonnies, Line detection and segmentation in historical document analysis and recognition. IJDAR **14**, 743–747 (2011)
37. R Manmatha, JL Rothfeder, A scale space approach for automatically segmenting words from historical handwritten documents. IEEE Trans. Pattern Anal. Mach. Intell. **27**(8), 1212–1225 (2005)
38. K Kise, A Sato, M Iwata, Segmentation of page images using the area Voronoi diagram. Comput. Vis. Image Underst. **70**(3), 370–382 (1998)
39. RD Duda, PE Hart, Use of the Hough transform to detect lines and curves in pictures. Commun. ACM **15**(1), 11–15 (1972)
40. N Nikolaou, M Makridis, B Gatos, N Stamatopoulos, N Papamarkos, Segmentation of historical machine-printed documents using adaptive run length smoothing and skeleton segmentation paths. Image Vis. Comput. **28**, 590–604 (2010)
41. G-F Chen, W Zhang, H Cui, Extracting notes from Chinese *Gong-Che* notation musical score image using a self-adaptive smoothing and connected component labeling algorithm. Int. J. Adv. Comput. Technol. **4**(1), 86–95 (2012)
42. TH Hildebrandt, WT Liu, Optical recognition of handwritten Chinese characters: advances since 1980. Pattern Recogn. **26**, 205–225 (1993)
43. RI Oka, Handwritten Chinese-Japanese character recognition by using cellular feature. Proc. Int. Conf. Pattern Recogn. **2**, 783–785 (1982)
44. OD Trier, AK Jain, T Taxt, Feature extraction methods for character recognition-a survey. Pattern Recogn. **29**(4), 641–662 (1996)
45. AW Senior, AJ Robinson, An off-line cursive handwriting recognition system. IEEE Trans. Pattern Anal. Mach. Intell. **20**, 309–321 (1998)
46. A Amin, Off-line Arabic character recognition: the state of the art. Pattern Recogn. **31**(5), 517–530 (1998)
47. CL Liu, F Yin, DH Wang, QF Wang, Online and offline handwritten Chinese character recognition: benchmarking on new databases. Pattern Recogn. **46**, 155–162 (2013)
48. S Basu, N Das, R Sarkar, M Kundu, M Nasipuri, DK Basu, A hierarchical approach to recognition of handwritten Bangla characters. Pattern Recogn. **42**, 1467–1484 (2009)
49. CW Liao, JS Huang, A transformation invariant matching algorithm for handwritten Chinese character recognition. Pattern Recogn. **23**(1), 1167–1188 (1990)
50. X Lin, X Ding, M Chen, R Zhang, W Youshou, Adaptive confidence transform based classier combination for Chinese character recognition. Pattern Recogn. Lett. **19**, 975–988 (1998)
51. RD Romero, DS Touretzky, RH Thibadeau, Optical Chinese character recognition using probabilistic neural networks. Pattern Recogn. **30**(8), 1279–1292 (1997)
52. D Shi, RI Damper, SR Gunn, Offline handwritten Chinese character recognition by radical decomposition. ACM Transac. Asian Lang. Info. Process. **2**(1), 27–48 (2003)
53. IJ Kim, JH Kim, Statistical character structure modeling and its application to handwritten Chinese character recognition. IEEE Trans. Pattern Anal. Mach. Intell. **25**(11), 1422–1436 (2003)
54. J Zeng, ZQ Liu, Markov random field-based statistical character structure modeling for handwritten Chinese character recognition. IEEE Trans. Pattern Anal. Mach. Intell. **30**(5), 767–780 (2008)
55. J Tan, X Xie, WS Zheng, JH Lai, Radical extraction using affine sparse matrix factorization for printed Chinese characters recognition. Int. J. Pattern Recognit. Artif. Intell. **26**(3) (2012). doi:10.1142/S021800141250005X
56. T Cover, P Hart, Nearest neighbor pattern classification. IEEE Trans. Inf. Theory **13**(1), 21–27 (1967)
57. R Duda, P Hart, *Pattern Classification and Scene Analysis* (Wiley, New York, 1973)
58. W Gong, Z Cai, L Jia, H Li, A generalized hybrid generation scheme of differential evolution for global numerical optimization, international journal of computational intelligence and applications. Int. J. Comp. Intel. Appl **35**(10), 560–567 (2011). doi: 10.1142/S1469026811002982

12

Audio bandwidth extension based on temporal smoothing cepstral coefficients

Xin Liu and Chang-Chun Bao*

Abstract

In this paper, we propose a wideband (WB) to super-wideband audio bandwidth extension (BWE) method based on temporal smoothing cepstral coefficients (TSCC). A temporal relationship of audio signals is included into feature extraction in the bandwidth extension frontend to make the temporal evolution of the extended spectra smoother. In the bandwidth extension scheme, a Gammatone auditory filter bank is used to decompose the audio signal, and the energy of each frequency band is long-term smoothed using minima controlled recursive averaging (MCRA) in order to suppress transient components. The resulting 'steady-state' spectrum is processed by frequency weighting, and the temporal smoothing cepstral coefficients are obtained by means of the power-law loudness function and cepstral normalization. The extracted temporal smoothing cepstral coefficients are fed into a Gaussian mixture model (GMM)-based Bayesian estimator to estimate the high-frequency (HF) spectral envelope, while the fine structure is restored by spectral translation. Evaluation results show that the temporal smoothing cepstral coefficients exploit the temporal relationship of audio signals and provide higher mutual information between the low- and high-frequency parameters, without increasing the dimension of input vectors in the frontend of bandwidth extension systems. In addition, the proposed bandwidth extension method is applied into the G.729.1 wideband codec and outperforms the Mel frequency cepstral coefficient (MFCC)-based method in terms of log spectral distortion (LSD), cosh measure, and differential log spectral distortion. Further, the proposed method improves the smoothness of the reconstructed spectrum over time and also gains a good performance in the subjective listening tests.

Keywords: Audio bandwidth extension; Temporal smoothing cepstral coefficients; Minima controlled recursive averaging; Gaussian mixture model

1 Introduction

In current mobile communication systems, the effective bandwidth of the transmitted wideband (WB) audio is limited to the frequency range of 50 ~ 7,000 Hz. Due to the loss of high-frequency (HF) information, the perceived quality of WB audio is significantly degraded compared to super wideband (SWB) audio, which is band-limited to 50 ~ 14,000 Hz in frequency. To improve the auditory quality of WB audio, bandwidth extension (BWE) techniques [1,2] have been developed to artificially restore the missing HF components at receiver only from the decoded WB audio signals and to reproduce the SWB audio signals without any modifications to the existing audio encoding and transmission components.

BWE methods for audio signals can be separated into two tasks: estimation of the spectral envelope and extension of the fine spectrum. The performance of spectral envelope estimation is crucial for the auditory quality of the extended audio [2]. Therefore, research on spectral envelope estimation is attractive in the BWE area. In general, a set of time-domain and frequency-domain features is extracted from WB audio signals, and the HF spectral envelope is estimated by using some statistical learning methods, e.g., codebook mapping [3], neural networks [4], or Bayesian estimation [1], on the basis of a priori knowledge between the WB features and the HF spectral envelope. In early studies, more concern was laid on the bandwidth extension of speech signals. The relationship between the high-frequency and low-frequency (LF) parameters was quantified through mutual information, discrete entropy and separability [5-7], and the upper bounds on

* Correspondence: baochch@bjut.edu.cn
Speech and Audio Signal Processing Lab, School of Electronic Information and Control Engineering, Beijing University of Technology, 100124 Beijing, China

the quality of bandwidth extension have been presented for both the clean and noisy speech [8,9]. Furthermore, Kleijn et al. suggested that BWE methods need to make use of perceptual properties to upgrade the subjective auditory quality of the extended signals [6]. Based on this view, Mel frequency cepstral coefficient (MFCC) has been applied into BWE methods [10] for parameterizing the WB spectrum. By using Mel-scale filters and cepstrum analysis, MFCC provides more certainty about the HF components, which is quantified as the ratio of mutual information between the HF and LF parameters to the discrete entropy of HF parameters. Then, the joint probability density function of the HF and LF feature vectors is approximated by a Gaussian mixture model (GMM), and the HF spectral envelope is estimated according to the minimum mean square error (MMSE) criterion [11,12]. The method based on MFCC and GMM can effectively reduce the spectral distortion of the extended speech compared to the method based on line spectral frequency parameters [10] and also achieves a good extension performance for audio signals.

The studies on audio perceptual quality [13-15] show that the temporal evolution of audio signals is as perceptually important as the reconstruction of spectral envelope. To some degree, artificially making the reconstructed spectra of audio signals temporally smooth is beneficial for the subjective quality and even can compensate some perceptual distortion caused by the error of the reconstructed spectra. Therefore, hidden Markov models (HMM) have been introduced into the BWE schemes to further capture the temporal relationship between the adjacent frames of audio signals [16-18]. In the HMM-based BWE methods, each state represents the characteristics of one particular sound. The state-specific probability density between the HF and LF features is approximated by GMM, and the envelope change of audio signals is modeled by means of the state transition process. Thus, HMMs contribute to a smoother temporal evolution of the reconstructed spectra which improves the listening quality. In addition, delta-MFCC features can be utilized to directly represent the difference between frames to include some temporal memory in the frontend of BWE schemes. The delta-feature-based BWE methods provide a higher certainty of the HF parameters and achieve better performance in terms of objective quality [19-21]. However, two abovementioned methods also have some drawbacks. HMM brings plenty of model parameters, and the storage space and computational complexity has to increase for model training and Bayesian estimation. Similarly, employing the delta-MFCC objectively adds the dimension of the input vectors for the statistical models and increases the burden on computational complexity.

1.1 Paper overview

In this paper, we attempt to embed the extraction of the temporal relationship of audio signals into cepstral coefficients and propose a novel temporal smoothing cepstral coefficient (TSCC)-based scheme for BWE of audio signals. It can improve the temporal smoothness of the extended HF spectrum, without increasing the dimension of input feature and without burdening the storage space and computational complexity for enhanced/more sophisticated statistical models. Firstly, a Gammatone filter bank is adopted to decompose the audio signals, and the audio energy of each frequency band is long-term smoothed by means of minima controlled recursive averaging (MCRA) to suppress transient signal components. Secondly, the resulting 'steady-state' spectrum is processed by frequency weighting, and TSCCs are extracted by means of the power-law loudness function and cepstral normalization. Finally, the extracted cepstral coefficients are applied into a GMM-based BWE scheme to restore the HF components. The proposed method suppresses the transient components existing in the WB audio signal in the BWE frontend and provides higher mutual information between the LF and HF parameters. Informal listening tests show that, for most audio signals, TSCC plays an important role on the improvement of spectral distortion and the temporal smoothness of the reconstructed audio signals, while for some rock music with the accompaniment of strong percussion music, temporal envelope modification needs to be applied in order to maintain a good extension performance.

In the next section, the TSCC extraction is described in detail, and then the application of the proposed feature to the GMM-based BWE scheme is briefly discussed. Section 3 gives the analysis of mutual information between the HF and LF parameters processed by different parameterizations. In Section 4, the proposed method and the reference BWE methods are evaluated in terms of objective quality measurements and subjective listening tests. Finally, conclusions are drawn in Section 5.

2 Bandwidth extension based on temporal smoothing cepstral coefficients

A block diagram of the proposed BWE method is shown in Figure 1. The input signal is the WB audio signals sampled at 16 kHz with a bandwidth of 7 kHz and is separated into 40-ms frames with 20-ms overlap. Then, the new TSCC features are extracted from the audio frame and are fed into a GMM-based Bayesian estimator to estimate the HF spectral envelope. Here, the HF spectral envelope $M_{HF}(r)$, $r = 0, 1, 2, 3$ is represented by the root mean square (RMS) values of four sub-bands in the HF band. These sub-bands are distributed on the perceptually correlated equivalent rectangular bandwidth (ERB) scale [22] without overlapping, and their central

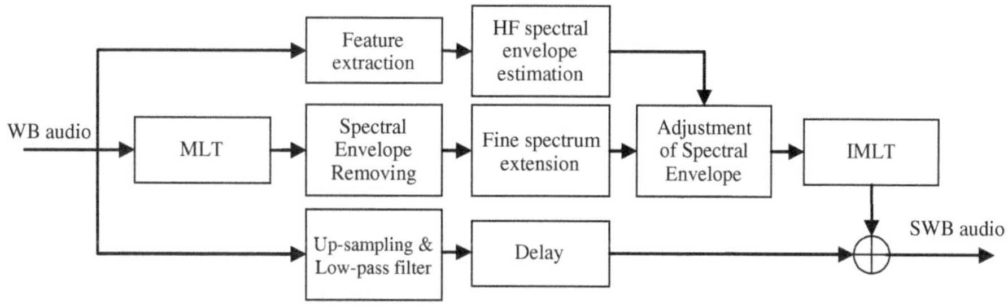

Figure 1 Block diagram of the proposed BWE method.

frequencies are arranged at 8,470, 9,338, 11,653, and 13,657 Hz.

In addition, the fine spectrum is extended by using the traditional spectral translation method [23,24]. Modulated lapped transform (MLT) [25] is performed to convert the input audio signals into the frequency domain, and 320 MLT coefficients $C_{mlt}(i)$, i =0,...,319 are obtained for describing the spectrum below 8 kHz. The fine spectrum of WB audio is obtained through the spectral envelope normalization. In order to keep the normalized spectrum flat, the LF spectral envelope is described as the RMS values of 14 sub-bands which are uniformly spaced in the range of 0 ~ 7 kHz,

$$M_{\mathrm{LF}}(r) = \sqrt{\frac{1}{h(r)-l(r)+1}\sum_{i=l(r)}^{h(r)} C_{\mathrm{mlt}}{}^{2}(i)}, 0 \le r < 14 \tag{1}$$

where $C_{mlt}(i)$, i =0,...,279 is the MLT coefficients in the range of 0 ~ 7 kHz, and $l(r)$ and $h(r)$ correspond to the lower and upper boundaries of the MLT coefficients in the rth sub-band. Thus, the LF normalized MLT coefficients $C_{norm\_mlt}(i)$, i =0,...,279 are given as,

$$C_{\mathrm{norm\_mlt}}(i) = \frac{C_{\mathrm{mlt}}(i)}{M_{\mathrm{LF}}(r)}, r = \left\lfloor \frac{i}{20} \right\rfloor \tag{2}$$

Next, the fine spectrum in the frequency band of 0 ~ 7 kHz is directly translated into the range of 7 ~ 14 kHz, and the HF spectral envelope is adjusted according to the estimated HF sub-band RMS value $\hat{M}_{HF}(r)$ [26]. Through inverse modulated lapped transform (IMLT), the HF components are transformed into the time domain. Finally, the original WB signals are up-sampled and low-pass filtered. The resulting signals are delayed and combined with the artificially restored HF signals to reconstruct the extended SWB audio. The total algorithmic delay of the proposed method is 40.75 ms, which is caused mainly by the time-frequency transform, low-pass filter, and filter bank used in feature extraction. A more detailed description of the algorithm is provided below.

2.1 Extraction of temporal smoothing cepstral coefficients

For the stationary audio signals, the HF spectrum changes at a relatively slow rate over time in comparison with the LF spectrum. By using the traditional BWE method, the envelope of the extended HF spectrum for the stationary audio signals is affected by the transient components contained in the LF spectrum, and some sudden changes of energy for the HF spectrum may occur between frames and degrade the auditory quality of the stationary audio signals according to the informal listening tests. Inspired by the studies on audio perceptual quality [13-15] that the temporal smoothness of audio spectrum is partially beneficial for the subjective quality, we attempt to eliminate the impact of transients from the features in order to improve the temporal smoothness of the extended HF spectrum. In the proposed method, the 'steady-state' spectrum of audio signals is first obtained by using MCRA for describing the spectrum components which are slowly evolved over time. Then, the resulting 'steady-state' spectrum is further used to remove the transients from the original spectrum of audio signals by the spectral weighting method. Finally, the cepstral features are computed from the weighted spectrum. Though some transients in the HF spectrum may be not well reconstructed by using the proposed method, the quality of the extended audio signals is partially improved on the whole according to the objective and subjective tests shown in Section 4.

Figure 2 shows the extraction principle of the TSCC feature proposed in this paper. Firstly, a Gammatone filter bank is used to decompose the input audio signals into 20 channels, whose central frequencies are non-uniformly distributed in the frequency range of 50 ~ 7,000 Hz [27]. As a standard model of cochlear filtering, the Gammatone filter bank can model the nonlinear response and frequency selectivity of basilar membranes in human ears by means of low-order causal filters with few parameters, and its impulse response function is represented as,

$$g(f,t) = \begin{cases} t^{a-1}e^{-2\pi bt}\cos(2\pi f t), & t \ge 0 \\ 0, & t < 0 \end{cases} \tag{3}$$

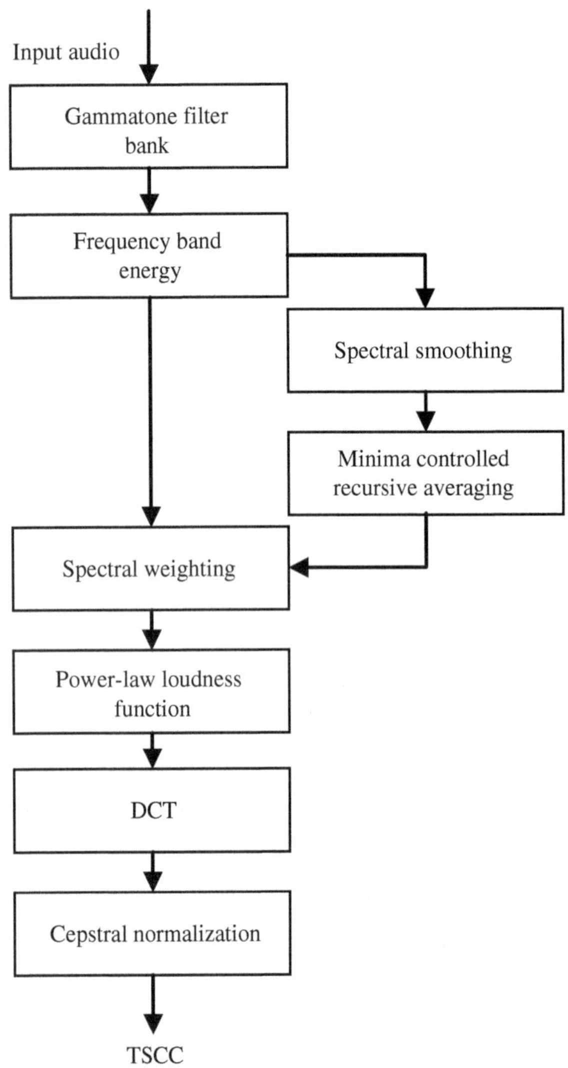

Figure 2 Principle of the TSCC feature extraction algorithm.

where t and f correspond to the time samples and central frequencies of the Gammatone filter in each channel, separately. In order to make the Gammatone filters better-matched to the filtering characteristics of basilar membranes, the filter order a is set to 4, the central frequencies of 20 Gammatone filters are spaced on the ERB scale, and the rectangular bandwidth b of each channel increases with the central frequency f.

The WB audio signal $x(i)$ is fed into the jth Gammatone filter, where i is the sample index. The resulting signal, $x_g(i,j)$, $j = 0,\ldots,19$ is down-sampled to the frame rate of 50 Hz along the time dimension to obtain the energy spectrum $S_f(m,j)$ in the current frame, where m is the frame index.

In order to verify the effects of the proposed cepstral coefficients, a segment of jazz music signal with a length of 11 s is used as an example. The resulting Gammatone spectrogram is shown in Figure 3. It can be clearly found that some transient components appear in the spectrum of the original audio signal and damage the continuity of each frequency band over time. This may degrade the extension performance for the audio signals.

2.1.1 Minima controlled recursive averaging

In the proposed method, MCRA [28] is first employed to extract the 'steady-state' components from the audio energy spectrum, and the adverse effects of BWE caused by the transient components are restrained in the frontend. Dependent on the presence 'probability' of the rapidly evolving components in each channel, the time smoothing factor is adaptively adjusted to estimate the 'steady-state' audio spectrum at the current frame. The detailed procedure of MCRA is as follows.

1) Temporal smoothing

The energy spectrum in the jth channel $S_f(m,j)$ at the current frame is temporally smoothed. The smoothed energy spectrum $S(m,j)$ is given as,

$$S(m,j) = \alpha_s S(m-1,j) + (1-\alpha_s) S_f(m,j) \tag{4}$$

where m is the frame index, $S(m-1,j)$ is the smoothed energy spectrum at the previous frame, and $\alpha_s = 0.7$ is the smoothing factor.

2) Asymmetric filtering

The spectrum of the transient components evolves more rapidly than the 'steady-state' components, so the asymmetric filtering is adopted to obtain the slowly evolving components in the jth channel $S_{slow}(m,j)$ as,

$$S_{\text{slow}}(m,j) = \begin{cases} \gamma S_{\text{slow}}(m-1,j) + \dfrac{(1-\gamma)}{(1-\beta)}\left(S(m,j) - \beta S(m-1,j)\right), & \text{if } S_{\text{slow}}(m-1,j) < S(m,j) \\ S(m,j), & \text{if } S_{\text{slow}}(m-1,j) \geq S(m,j) \end{cases} \tag{5}$$

where m and j are the frame index and the index of frequency bands. $\gamma = 0.92$ denotes the smoothing factor, and a 'look-ahead' factor $\beta = 0.7$ is introduced in the asymmetric filtering to adjust the adaptation time of the algorithm [29]. Through asymmetric filtering, the resulting $S_{slow}(m,j)$ approaches the lower envelope of the smoothed energy spectrum $S(m,j)$ to describe the slowly evolving components of the audio spectrum in the jth channel, and the rapidly evolving components are presented by $S(m,j) - S_{slow}(m,j)$. Figure 4 shows the slowly evolving components of the jazz music signal shown in Figure 3.

3) Calculating the presence 'probability' of the rapidly evolving components

The ratio of the smoothed energy spectrum $S(m,j)$ to the slowly evolving components $S_{slow}(m,j)$ can be expressed as,

$$R(m,j) = \frac{S(m,j)}{S_{\text{slow}}(m,j)} \tag{6}$$

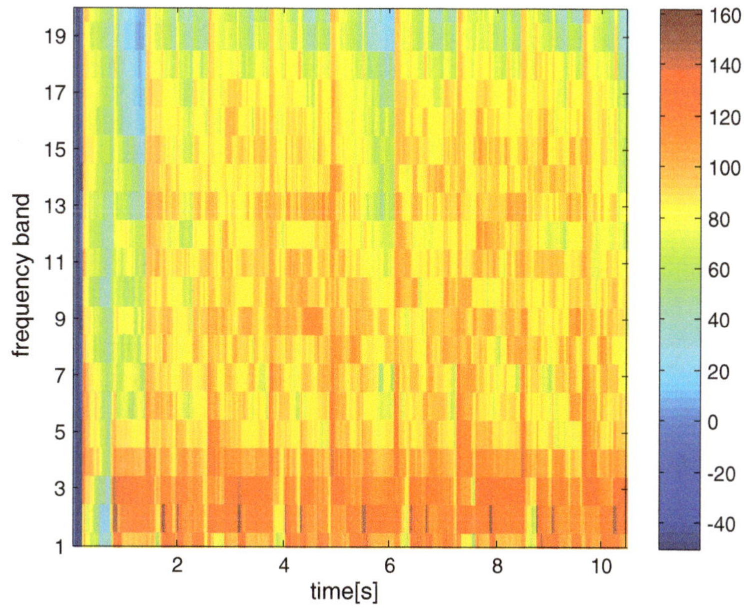

Figure 3 The resulting Gammatone spectrogram for a jazz music signal.

By using the hard threshold method, $R(m,j)$ can be exploited to determine whether the rapidly evolving components are present in the jth channel at the mth frame. The presence 'probability' of the rapidly evolving components is obtained as,

$$p(m,j) = \begin{cases} 1, if & R(m,j) > \delta \\ 0, if & R(m,j) \leq \delta \end{cases} \quad (7)$$

where δ =3, and if $R(m,j) > \delta$, then we assume that undesired rapidly evolving components are present, i.e., $p(m,j)$ =1, otherwise we assume that the slowly evolving components are dominant, i.e., $p(m,j)$ =0. Through recursive averaging, the presence 'probability' of the rapidly evolving components $p'(m,j)$ which are continuously evolved over frames can be further given as,

$$p'(m,j) = \alpha_p p'(m-1,j) + (1-\alpha_p)p(m,j) \quad (8)$$

where the smoothing factor is α_p =0.2.

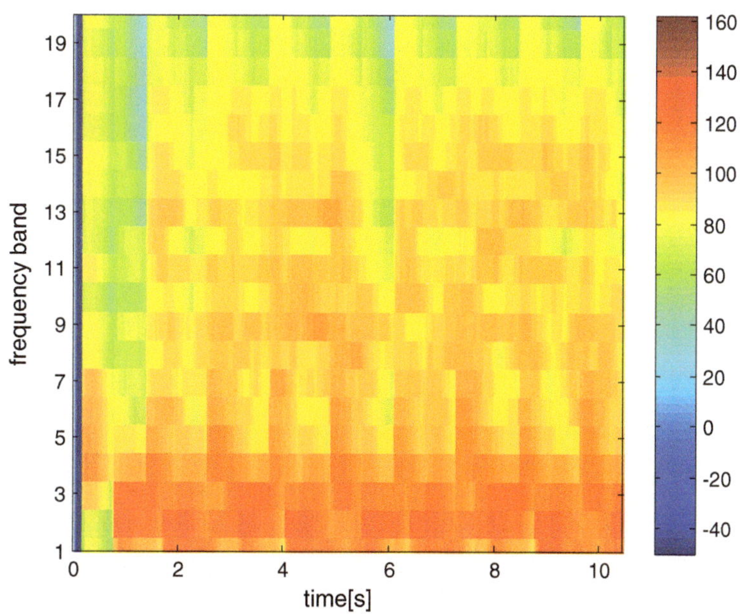

Figure 4 The spectrogram of the slowly evolving components for a jazz music signal.

4) Achieving the 'steady-state' spectrum of audio signals

The so-called 'steady-state' spectrum in the jth channel at the current frame $S_{steady}(m,j)$ can be estimated by recursive averaging as,

$$S_{\text{steady}}(m,j) = \alpha_d(m,j)S_{\text{steady}}(m-1,j) + (1-\alpha_d(m,j))S_f(m,j) \tag{9}$$

where $S_{steady}(m\text{-}1,j)$ is the 'steady-state' spectrum at the previous frame. The 'steady-state' smoothing factor $\alpha_d(m,j)$ is dynamically adjusted according to the presence 'probability' of the rapidly evolving components $p'(m,j)$ and is updated as,

$$\alpha_d(m,j) = \alpha + (1-\alpha)p'(m,j) \tag{10}$$

where α =0.85.

Figure 5 illustrates the 'steady-state' spectrum of the jazz music signal shown in Figure 3. It is obvious that, compared with the original spectrogram shown in Figure 3, the transient components are effectively restrained, but the fine structure of the audio spectrum is also blurred owing to long-term smoothing.

2.1.2 Spectral weighting

As shown in Figure 5, the 'steady-state' spectrum of audio signals is able to describe the spectrum components which are slowly evolved over time, but the fine structure of audio spectrum is blurred. If the 'steady-state' spectrum was directly adopted for extracting the cepstral coefficients, the mutual dependencies between the LF and HF parameters were obviously decreased according to the experimental observation, because much information contained in the fine spectrum was missing. In order to make a balance between temporal smoothness of the extended spectrum and mutual dependencies between the LF and HF parameters, the resulting 'steady-state' spectrum is further used to remove the transients from the original spectrum $S_f(m,j)$ of audio signals by the spectral weighting method.

First, the ratio of the 'steady-state' spectrum to the smoothed spectrum of audio signals is determined for each channel at the current frame. Next, the resulting ratio is multi-point averaged over frequency for weighting the audio spectrum. Then, the weighted spectrum can be computed as follows,

$$S_w(m,j) = w(m,j)S_f(m,j) \tag{11}$$

$$w(m,j) = \frac{1}{h(j)-l(j)+1}\sum_{k=l(j)}^{h(j)} \frac{S_{\text{steady}}(m,k)}{S(m,k)} \tag{12}$$

where $h(j)$ = max(j +2, 19) and $l(j)$ = min(j −2, 0).

Figure 6 shows an example for the weighted spectrum of the jazz music signal given in Figure 3. It can clearly be seen that spectral weighting decreases the weight value of transient components and retains the 'steady-state' components in the original audio spectrum to upgrade the smoothness of the processed audio spectrum over both time and frequency. Compared with the 'steady-state' spectrum shown in Figure 5, spectral weighting does not only repress the transient components but also preserves more fine structure which is beneficial to describe the time-frequency characteristics of audio signals.

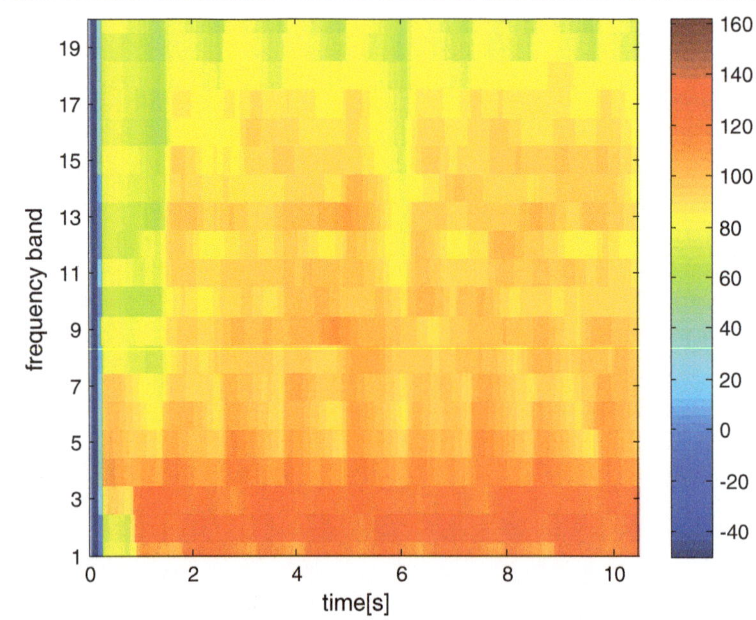

Figure 5 The 'steady-state' spectrum for a jazz music signal.

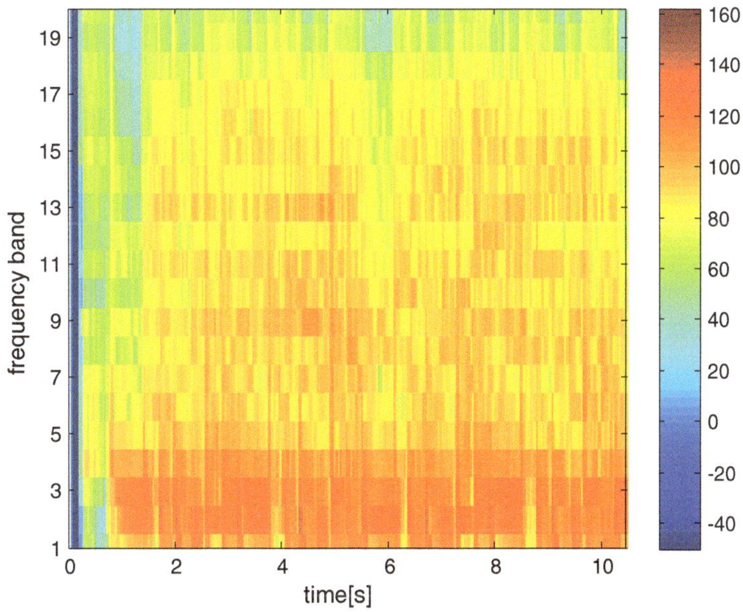

Figure 6 The weighted spectrum for a jazz music signal.

2.1.3 Extraction of cepstral coefficients

In consideration of the BWE performance, the cubic root loudness function [30] is selected to approximate the nonlinear relationship between the intensity of sound and its perceived loudness and can be expressed as,

$$S_c(m,j) = \sqrt[3]{S_w(m,j)} \tag{13}$$

Next, discrete cosine transform (DCT) is applied to de-correlate the coefficients, and the resulting cepstral coefficients are referred to as TSCCs,

$$C(m,j) = \sqrt{\frac{2}{N}}\sum_{i=0}^{19} S_c(m,i)\cos\left(\frac{j\pi}{2N}(2i+1)\right) \tag{14}$$

where j =0,...,19 is the TSCC index.

In order to compress the dynamic range, interval normalization is applied to the TSCCs. Since the silence frames in audio signals affect the value of TSCC, the pre-trained maximum and minimum values of the cepstral coefficients are searched only in the non-silence frames of the training data and are fixed in practical application. The normalized TSCC $C_{norm}(m,j)$ can be obtained as,

$$C_{\text{norm}}(m,j) = \frac{C(m,j) - C_{\min}(j)}{C_{\max}(j) - C_{\min}(j)} \tag{15}$$

where $C_{max}(j)$ and $C_{min}(j)$ correspond to the pre-trained maximum and the minimum of TSCC in the jth channel.

2.2 HF spectral envelope estimator based on Gaussian mixture model

A 20-dimensional TSCC vector is extracted as the WB audio feature vector $F_X = [C_{norm}(m,0), C_{norm}(m,1), \ldots, C_{norm}(m,19)]^T$ and is fed into the GMM-based Bayesian estimator [11] to estimate the RMS values of the HF subbands $F_Y = [M_{HF}(0), M_{HF}(1), \ldots, M_{HF}(3)]^T$ under the MMSE criterion. The joint vector of the HF and LF vectors is referred to as $F_Z = [F_X^T F_Y^T]^T$, and a GMM is utilized to model the joint probability density function $p(F_Z)$ as,

$$p(F_Z) = \sum_{i=1}^{N_G} w_i N(F_Z; \mu_i, C_i) \tag{16}$$

where N_G is the number of Gaussian mixtures, and $N(.)$ is the multi-variate Gaussian probability density function. w_i, μ_i, and C_i represent the weight, mean vector, and covariance matrix of the ith Gaussian components, respectively. These model parameters were off-line trained from the SWB audio dataset of the live concert recordings with the length of about 1 h. During the training phase, the parallel WB audio signals were generated by low-pass filtering, down-sampling, and time alignment. Then, the level of WB and SWB signals was normalized to −26 dBov, and the WB feature vectors F_X and the HF subband RSM values F_Y were extracted from the parallel dataset. Finally, the model parameters of GMM were estimated under the maximum likelihood criterion by using the standard expectation-maximization algorithm.

Given the joint density distribution function $p(F_Z)$, we use the MMSE-based Bayesian estimator and compute the conditional expectation $E[F_Y|F_X]$ of F_Y given the WB feature vector F_X as the estimated value of HF spectral envelope $\hat{F}_Y$ as follows,

$$\hat{F}_Y = \underset{\hat{F}_Y}{\operatorname{argmin}} \int \|F_Y - \hat{F}_Y\|^2 p(F_Y|F_X) dF_Y = E[F_Y|F_X] = \sum_{i=1}^{N_G} p(i|F_X) E[F_Y|i, F_X] \tag{17}$$

where $p(i|F_X)$ is the posterior probability of the ith Gaussian component given F_X,

$$p(i|F_X) = \frac{w_i p(F_X|i)}{\sum_{j=1}^{N_G} w_j p(F_X|j)} \tag{18}$$

$p(F_X|i)$ is the observed probability of F_X in the ith Gaussian component. It can be computed from the joint probability of F_Z in the ith Gaussian component $p(F_Z|i)$. $E[F_Y|i,F_X]$ is the conditional expectation of F_Y given F_X in the ith Gaussian component and is given as,

$$E[F_Y|i, F_X] = \mu_{Y,i} + C_{YX,i} C_{XX,i}^{-1} \left(F_X - \mu_{X,i}\right) \tag{19}$$

where $\mu_{X,i}$ and $\mu_{Y,i}$ are the mean vector of F_X and F_Y in the ith Gaussian component, $C_{XY,i}$ is the cross-correlation matrix, and $C_{XX,i}$ is the correlation matrix of F_X.

2.3 High-frequency component synthesis

In this paper, the fine spectrum of audio signals in the frequency range of 0 ~ 7 kHz is described as the normalized LF MLT coefficients $C_{norm\_mlt}(i)$, i =0,...,279. According to spectral translation [2], the normalized MLT coefficients in the LF band are directly copied to the HF band of audio spectrum to extend the fine spectrum. On the basis of the ERB scale, the HF fine spectrum is also divided into four non-overlapping sub-bands and is multiplied by the estimated RMS values of the HF sub-bands to reproduce the extended HF spectrum. At last, the HF spectrum is converted into the time domain through IMLT. The original WB audio signals are up-sampled, low-pass filtered, and temporally aligned, and then it is combined with the waveform of the HF components to reconstruct the extended SWB audio signals.

3 Mutual information analysis

The BWE performance depends on all mutual dependencies between the HF and LF parameters. On the basis of the information theory, all the linear and nonlinear dependencies between the LF feature vectors F_X and the HF spectral envelope F_Y can be described by their mutual information (MI) $I(F_X;F_Y)$. The larger the MI, the lower is the upper bound on the minimum achievable mean square error of the HF spectral envelope estimated from the LF features [8]. For this reason, the effects of mutual information caused by different WB audio features are studied in this paper.

The marginal probabilities $p_X(F_X)$, $p_Y(F_Y)$ and the joint probability density $p_{XY}(F_X,F_Y)$ of the WB audio features and the HF spectral envelope coefficients are modeled by a GMM with 128 mixtures and with full covariance matrices. The mutual information measure $I(F_X;F_Y)$ can be approximated in the light of the numerical integration method, in order to evaluate the dependencies between the HF and LF parameters as follows,

$$I(F_X; F_Y) = E\left[\log_2\left(\frac{p_{XY}(F_X, F_Y)}{p_X(F_X)p_Y(F_Y)}\right)\right] \approx \frac{1}{M}\sum_{i=0}^{M=1} \log_2\left(\frac{p_{XY}(F_X(m), F_Y(m))}{p_X(F_X(m))p_Y(F_Y(m))}\right) \tag{20}$$

where m is the frame index, M is the frame number. $F_X(m)$ and $F_Y(m)$ are referred to as the LF and HF feature vectors at the mth frame.

In this paper, the traditional MFCC [10] and Gammatone frequency cepstral coefficients (GFCC) [30] are adopted as the reference WB audio feature vectors to analyze the mutual information between the HF and LF parameters processed by different parameterizations, in comparison to the proposed TSCC. For the MFCC parameterization, Fourier transform is adopted to obtain the magnitude spectrum of audio signals, and Mel-scale triangular filters are applied to compute the log energy of 20 non-uniform sub-bands in the frequency range of 50 ~ 7,000 Hz. DCT is applied to the log energies to de-correlate the coefficients and to yield the cepstral coefficients. GFCC bring more auditory properties on the basis of MFCC. A Gammatone auditory filter bank takes the place of the Mel-scale triangular filter bank to mimic the properties of the basilar membrane in cochlear, and the log nonlinearity is replaced by cubic root loudness function for cepstral analysis. The proposed TSCC further adds the minima controlled recursive averaging-based long-term smoothing and spectral weighting to enhance the temporal smoothness of audio spectrum. In order to reduce the dynamic range, all the three versions of cepstral coefficients employ the same cepstral normalization methods. In addition, the dimension of the cepstral coefficients needs to be reduced by removing the trailing dimensions for MI analysis with different feature dimensions, as shown in Table 1.

It is shown that, for all the 20 cepstral coefficients, GFCC gains the MI improvement of 0.1 bit/frame in comparison with MFCC, and TSCC achieves a 1.4 bits/frame higher MI. Exploiting the Gammatone filter bank and loudness function is helpful for improving the MI

Table 1 Comparison of MI between the LF features and the HF spectral envelope

MFCC		GFCC		TSCC	
Dimension	MI	Dimension	MI	Dimension	MI
20	3.8537	20	3.9717	20	5.2553
18	3.4005	18	3.6244	18	4.6849
16	3.2117	16	3.3796	16	4.2116
14	3.0344	14	3.1797	14	3.7367
12	2.7832	12	2.9214	12	3.4479
10	2.6181	10	2.6686	10	3.1966
5	2.3627	5	2.3977	5	2.5690

between the HF and LF parameters. Long-term smoothing and spectral weighting are beneficial to the smoothness of audio spectrum and retain more MI which is positive about the extension performance without increasing the input feature dimension in the frontend of BWE. In addition, with the decreasing dimension of features, the MI measures of the three versions of cepstral coefficients are gradually reduced. Here, the MI about GFCC gradually approaches to that of traditional MFCC and TSCC still shows the superiority over the reference cepstral coefficients. When the dimension is reduced to 5, the MI value of the three versions of cepstral coefficients is down to 2.6 bits/frame.

Additionally, the selection of the smoothing factors in the TSCC extraction process, such as α_s, β, γ, δ, α_p, and α, may also affect the extension performance of the proposed BWE scheme. So, we experimented with several values of the smoothing factors and empirically optimized them under the MI measure. First, the smoothing factor α_s is analyzed when β, γ, δ, α_p, and α are respectively set to the fixed values, 0.7, 0.92, 3, 0.2, and 0.85, and the MI measure of the TSCC-based BWE method with different α_s is shown in Figure 7. The result demonstrates that the highest MI measure is obtained when α_s =0.7 and the value of the MI measure is rapidly diminished to about 4.9 bits/frame with the decrease of α_s. When α_s >0.7, the MI measure is marginally worse because the excessively smoothed spectrum may blur the fine structure and degrade the differentiation of the extracted cepstral coefficients. Also, we analyze the effect of the parameters for asymmetric filtering on the extension performance in the light of the MI measure with the fixed smoothing factors α_s =0.7, δ =3, α_p =0.2, and α =0.85. Figure 8 shows the MI measure between the WB audio features and HF spectral envelope coefficients as a function of the 'look-ahead' factor β with different γ. For a fixed γ, the value of β which ranges from 0.6 to 0.8 gives a higher MI, and when β >0.8 the MI measure would also decrease because a small value of β makes the audio spectrum excessively smooth and reduces the mutual

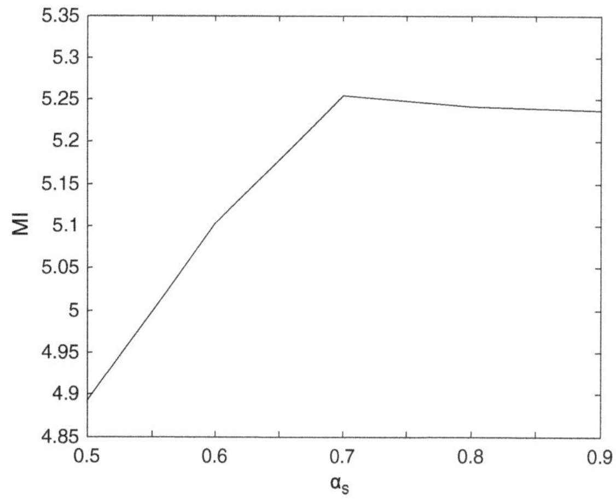

Figure 7 **MI measure of the TSCC-based BWE method with different α_s.**

dependencies between the LF and HF parameters. Besides, γ =0.92 provides a higher MI value in comparison with other values of γ. The threshold value δ is used to determine whether the rapidly evolving components are present in one channel. Figure 9 shows the MI measure of the TSCC-based BWE method with different threshold values δ. When δ =3, the MI measure reaches 5.2553 bits/frame. When the value of δ is larger or smaller, the MI measure would also decrease because the rapidly evolving components could not be differentiated from the slowly evolving components. Similarly, α_p and α are analyzed when other parameters are set to the fixed values. Figures 10 and 11 show that when α_p =0.2 and α =0.85, the maximum MI measure could be achieved. After all, according to the results of MI analysis, the smoothing factors could be experimentally determined as α_s =0.7, β =0.7, γ =0.92, δ =3, α_p =0.2, and α =0.85.

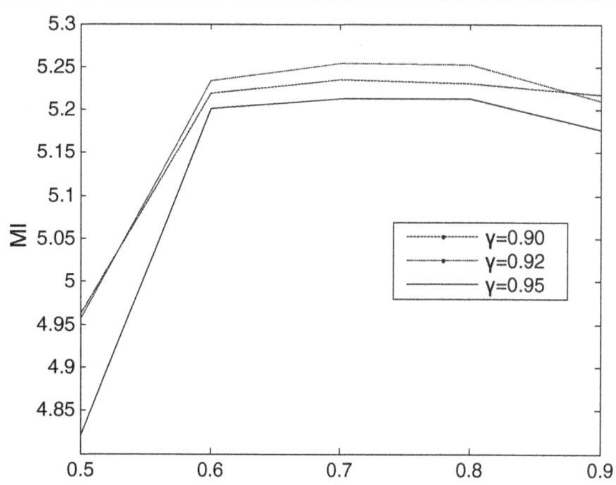

Figure 8 **MI measure of the TSCC-based BWE method with different β and γ.**

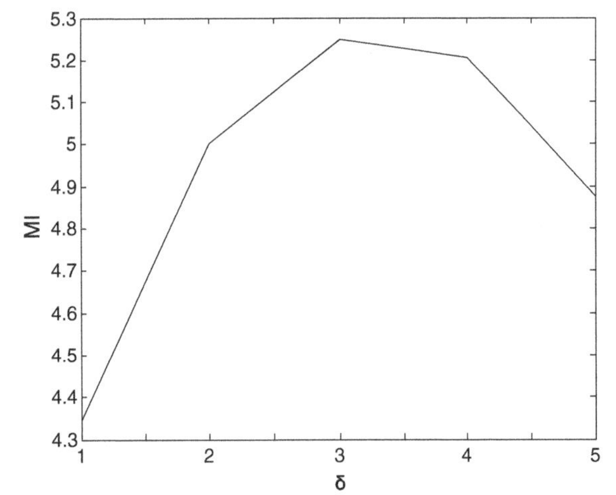

Figure 9 MI measure of the TSCC-based BWE method with different *δ*.

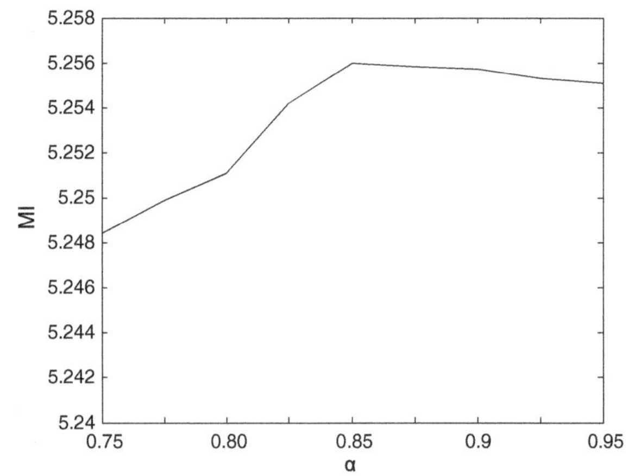

Figure 11 MI measure of the TSCC-based BWE method with different *α*.

4 Performance evaluation and test results

The goal of this work is to enhance the auditory quality of WB audio, without any side information for describing the missing HF components. In this section, we apply the proposed method and the MFCC-based BWE method to extend the bandwidth of the WB audio reproduced by the G.729.1 WB codec [26,31], in order to make a direct comparison with the SWB extension method with side information which is employed in G.729.1 Annex E [32] in terms of the objective and subjective quality tests. In addition, we make a subjective comparison between the original WB audio signals and the audio signals extended by the proposed method for further evaluating the improvement of the proposed method on the auditory quality.

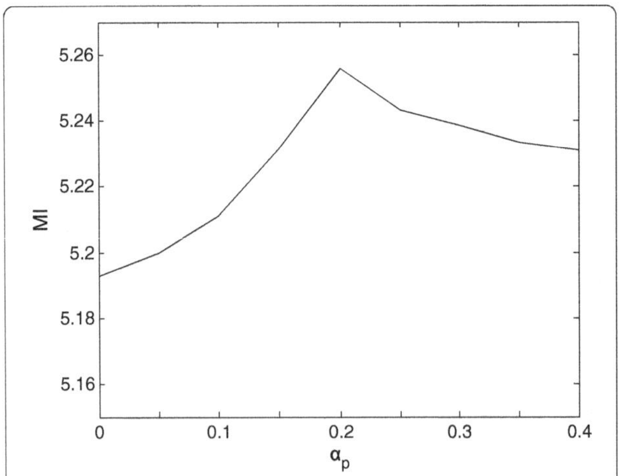

Figure 10 MI measure of the TSCC-based BWE method with different α_p.

4.1 Training

In the proposed method, the joint probability density between the features computed from the WB audio and the HF spectral envelope are modeled as a GMM. The training data comes from the lossless audio data of live concert recordings with the length of about 1 h. It contains different types of dialogues, music, singing, and live background sound. The audio signals were digitally stored by using 16-bit PCM with the sampling frequency of 32 kHz and the bandwidth of 14 kHz. The parallel WB audio signals were generated by low-pass filtering, down-sampling, and time alignment. The cepstral features extracted from the WB signals were used as the 20-dimensional input features F_X. The 4-dimensional HF spectral envelope F_Y was computed from the training samples in the parallel SWB database. The probability density of the joint vector $F_Z = \{F_X, F_Y\}$ were modeled by a GMM. The model parameters were trained using the standard expectation-maximization algorithm. The model has 128 mixtures and full covariance matrices. According to an informal listening test, no evident difference can be perceived with increase of the number of Gaussian mixtures.

4.2 Test data and reference methods

Fifteen SWB audio signals were selected from the MPEG auditory quality test database and were not part of the dataset for GMM training. The test signals contained pop music, instrumental solo music, symphonic pieces, and speech. Each signal had a length between 10 and 20 s. They were low-pass filtered with the cutoff frequency of 7 kHz and down-sampled to WB audio. Before further processing, the level of these WB audio signals [33] was required to be normalized to −26 dBov.

The test signals were processed with the G.729.1 WB codec at 32 kbit/s as WB references. SWB references were produced with the G.729.1 Annex E at 36 kbit/s. The core layer of G.729.1 Annex E with the bit rate of 32 kbit/s is fully identical to ITU-T G.729.1 to reproduce the WB audio signals, and the SWB extension layer with the additional bit rate of 4 kbit/s adopts a two-mode BWE method [34] in the modified discrete cosine transform (MDCT) domain to extend the bandwidth of the reproduced audio to 14 kHz. The mode selection is done by estimating the tonality of the input audio signals. For non-tonal signals, the 'generic mode' reconstructs the HF components through transposing the LF components with a gain adjustment. For tonal signals, the 'sinusoidal mode' adds a finite set of MDCT peaks to the HF spectrum. In addition, post-processing in time domain is used to improve the synthesized SWB audio at the decoder of G.729.1 Annex E.

Test items for BWE were obtained by applying the TSCC-based method and MFCC-based method to the G.729.1-coded WB signals. TSCC or MFCC is directly extracted from the G.729.1-coded WB signals and is fed into the GMM-based Bayesian estimator for obtaining the HF spectral envelope. Moreover, the reproduced WB signals are transformed into frequency domain via MLT, and the fine structure of the LF-MLT spectrum is translated into the HF band. After the energy adjustment, the regenerated HF components are converted into time domain by inverse MLT, to further form SWB audio signals with the WB audio decoded by G.729.1. In the interest of fairness, the central frequencies of the triangular filters adopted by MFCC should be the same as the central frequencies of the Gammatone filter bank in TSCC, and the resulting cepstral coefficients are also normalized to compress the dynamic range as shown in Equation 15.

4.3 Objective evaluation

Audio signals generated using the proposed TSCC-based method, MFCC-based method, G.729.1 at 32 kbit/s, and G.729.1 Annex E at 36 kbit/s were objectively evaluated in terms of the log spectral distortion (LSD) [35,36], cosh measure [20,35], and differential log spectral distortion (DLSD) [37] in comparison with the original SWB audio.

4.3.1 Log spectral distortion

The LSD [35,36] between the audio signals processed by different methods and the original SWB audio in the HF band of 7 ~ 14 kHz was computed directly from the FFT power spectra as,

$$d_{\text{LSD}}(i) = \sqrt{\frac{1}{N_{\text{high}} - N_{\text{low}} + 1} \sum_{n=N_{\text{low}}}^{N_{\text{high}}} \left[10 \log_{10} \frac{P_i(n)}{\hat{P}_i(n)} \right]^2} \quad (21)$$

where $d_{\text{LSD}}(i)$ is the LSD value at the ith frame and P_i and $\hat{P}_i$ are the FFT power spectra of the original SWB audio signals and the audio signals processed with different methods, respectively. N_{high} and N_{low} are the indices corresponding to the upper and lower bound of the HF band with the frequency range of 7 ~ 14 kHz. Before computing the LSD values, all the date needed to be re-sampled to 32 kHz and were temporally aligned with the original SWB audio. Especially for the up-sampled WB signals decoded by G.729.1, the spectral components above 7 kHz need to be removed by low filtering. Then, the LSD values were computed only for the HF band with the range of 7 ~ 14 kHz. The resulting LSD values were averaged over all the frames for each test signal, and the mean LSD was used as the distortion measure. For different types of test signals, the LSD measures of the reconstructed audio processed by different methods are shown in Table 2.

As shown in Table 2, G.729.1 only reproduces the WB audio signals and provides an LSD value of 11.3182 dB on average, and the LSD improvement can be achieved through all the BWE methods due to extending the bandwidth to 14 kHz. The reproduced SWB audio signals based on the proposed TSCC achieves an LSD improvement of about 1.2 dB in comparison with the traditional MFCC. Since the additional bit rate of 4 kbit/s is adopted to describe the real HF spectrum, G.729.1 Annex E can reconstruct the HF spectrum more accurately than the BWE methods without side information and further decreases the LSD value down to 7.4005 dB. Additionally, different types of audio signals show different extension

Table 2 LSD measures of the reconstructed audio based on different methods

Data type	TSCC-BWE	MFCC-BWE	G.729.1 Annex E 36 kbit/s	G.729.1 32 kbit/s
Country music	8.6300	10.1043	7.7542	11.8330
Jazz music	8.5145	10.6131	7.9227	12.2260
Rock music	9.4281	11.7701	8.1582	13.7525
Violin solo	6.1353	6.2611	5.9824	6.6624
Symphony	7.7741	8.3748	7.0836	9.0386
Speech	9.7205	10.1199	7.5021	14.3964
Average	8.3671	9.5406	7.4005	11.3182

Table 3 Cosh measures of the reconstructed audio based on different methods

Data type	TSCC-BWE	MFCC-BWE	G.729.1 Annex E 36 kbit/s	G.729.1 32 kbit/s
Country music	19.0821	44.1792	10.7046	183.6361
Jazz music	35.7912	118.3688	28.0020	227.3057
Rock music	57.7031	126.2545	12.9660	333.6178
Violin solo	7.0643	7.5069	6.9839	69.5002
Symphony	14.0920	24.5220	13.7074	51.7750
Speech	103.4225	176.1067	15.5153	377.6968
Average	39.5259	82.8230	14.6465	207.2553

performance in terms of LSD. For jazz music and rock music, the audio signals are accompanied with strong percussion and contain abundant transient component. The TSCC-based method suppresses the transient components of audio signals through MCRA and makes the spectrum of reconstructed audio temporally smooth. So, the TSCC-based method obtains the 2-dB LSD improvement in comparison with the MFCC-based method. In addition, the spectrum of symphony and violin solo presents few transients and maintains the continuity over time, thus the difference between the TSCC-based method and the MFCC-based method is not significant in terms of LSD. For speech, the results of LSD measure show that the TSCC-based method achieves a 0.4-dB improvement and the spectral distortion usually occurs in the starting frames of the unvoiced sound and the transition between the voiced and unvoiced frames.

4.3.2 Cosh measure

In contrast with LSD, the Itakura-Saito distortion provides more heavy weights on the peaks of an audio spectrum and is closely related to the subjective listening quality. It is defined as,

$$d_{\mathrm{IS}}\left(P_i, \hat{P}_i\right) = \frac{1}{N_{\mathrm{high}} - N_{\mathrm{low}} + 1} \sum_{n=N_{\mathrm{low}}}^{N_{\mathrm{high}}} \left[\frac{P_i(n)}{\hat{P}_i(n)} - \log_{10} \frac{P_i(n)}{\hat{P}_i(n)} - 1 \right] \tag{22}$$

Because the Itakura-Saito distortion is not symmetric as distance metric, the cosh measure [35] is used as the modified measure to describe the perceptual distortion of the reconstructed audio signals. Thus, we further introduced the cosh measure to evaluate the performance of the reconstructed audio signals and could obtain more subjectively correlated results [20]. The cosh measure is defined as,

$$d_{\mathrm{COSH}}(i) = \frac{1}{2} \left[d_{\mathrm{IS}}\left(P_i, \hat{P}_i\right) + d_{\mathrm{IS}}\left(\hat{P}_i, P_i\right) \right] \tag{23}$$

The cosh measure was computed only for the HF band and was also averaged over all the frames for test signals to obtain the final cosh measure. For different types of test signals, the cosh measures of the reconstructed audio processed by different methods are shown in Table 3.

As shown in Table 3, since the HF energy is large, the cosh measure of jazz music and rock music is a little higher. The unvoiced speech has abundant HF components, so its cosh measure is also large. For violin solo, the LF components are rich, and the energy of the HF spectrum is faint with the increase of frequency, thus its cosh measure value is lower and the difference between the BWE methods is not significant. Overall, the SWB audio reproduced by G.729.1 Annex E can achieve the best objective quality for all the types of audio. The cosh measure of G.729.1 is the largest because the HF spectrum is truncated. Compared with MFCC, the proposed cepstral coefficients also gain a better extension performance in terms of the subjectively correlated cosh measure, and the mean cosh measure is much closer to that of G.729.1 Annex E.

Table 4 DLSD measures of the reconstructed audio based on different BWE methods

Data type	TSCC-BWE	MFCC-BWE	G.729.1 Annex E 36 kbit/s	G.729.1 32 kbit/s
Country music	5.3129	6.0940	3.8121	7.1232
Jazz music	5.0191	6.5015	5.0955	7.4515
Rock music	5.7910	6.9800	5.1260	8.2231
Violin solo	4.3011	4.7076	3.8624	4.8863
Symphony	4.9078	5.0013	4.5332	5.5112
Speech	5.7345	6.0280	4.7101	8.0015
Average	5.1777	5.8854	4.5232	6.8661

4.3.3 Differential log spectral distortion

The smoothness of the audio spectrum between frames is important for the perceptual quality as well as the accurate reconstruction of the audio spectrum. Thus, DLSD is defined to evaluate how smoothly the spectral envelope of the extended audio signals is temporally evolved [37]. If the DLSD value of the reconstructed audio is low, the spectrum evolves smoothly over time. It is helpful for the overall subjective auditory quality of the reconstructed audio. DLSD is defined as,

$$d_{\mathrm{DLSD}}(i) = \sqrt{\frac{1}{2(N_{\mathrm{high}}-N_{\mathrm{low}}+1)}\sum_{n=N_{\mathrm{low}}}^{N_{\mathrm{high}}}\left[10\log_{10}\frac{P_i(n)}{P_{i-1}(n)}-10\log_{10}\frac{\hat{P}_i(n)}{\hat{P}_{i-1}(n)}\right]^2} \quad (24)$$

where $P_{i\ -1}$ and $\hat{P}_{i\ -1}$ are the power spectrum of the original SWB audio signals and extended SWB audio signals at the previous frame.

The DLSD values of the reconstructed SWB audio based on different BWE methods are shown in Table 4. Similar to the result of LSD and cosh measure, the transient components in country music, jazz music, and rock music increase the DLSD values of the reconstructed audio signals. For the temporally smooth symphony signals, different BWE methods have minor differences in terms of DLSD. For speech, the improvement of the proposed method in DLSD is not remarkable, because spectral distortion occurring in the unvoiced frames has a negative influence on the spectral smoothness of the extended audio. On the average, the proposed TSCC improves the smoothness of temporal evolution of the reconstructed audio spectrum, gains the DLSD improvement of 0.7 dB in comparison with MFCC, and is inferior to G.729.1 Annex E which partly provides the real HF spectrum.

4.4 Subjective listening tests

A CCR listening test which is similar to the subjective assessment method recommended by ITU-T P.800 [38] was used to pair-wise evaluate the differences of audio quality for the proposed TSCC-based method, MFCC-based method, and G.729.1 Annex E. In each test case, two differently processed versions of the same test signals were presented to the listeners. Listeners used the following seven-point comparison mean opinion scores (CMOS) to judge the quality of the second audio sample relative to that of the first: 3, much better; 2, better; 1, slightly better; 0, the same; −1, slightly worse; −2, worse; −3 much worse.

Fifteen male and five female listeners took part in the CCR test, and the age range was from 22 to 30 years old. The test was arranged in the quiet room, and only the test attendee was present in the room during the test. Six test signals including pop music, guitar, sax, drums, and speech were selected at random from the MPEG database, and the level of the original test signals and the processed signals was normalized to −26 dBov. The differently processed types of the six MPEG testing signals were played to both ears through AKG K271

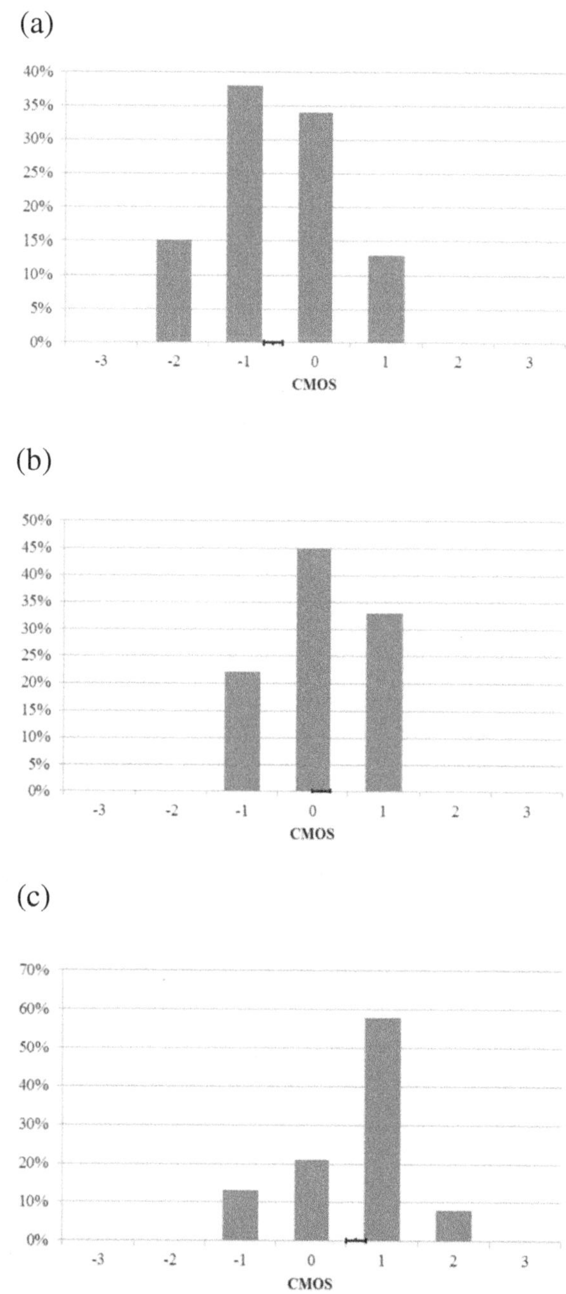

Figure 12 Distributions of listener rating in CCR tests. (a) Comparison between the TSCC-based method and the MFCC-based method. **(b)** Comparison between the TSCC-based method and G.729.1 Annex E. **(c)** Comparison between the MFCC-based method and G.729.1 Annex E.

MKII headphones for listening tests. Each listener had a short practice before actual tests to adjust the volume setting to a suitable level and was allowed to repeat each pair of testing data with no time limitation before giving their answers. After listeners gave their grade, their assessment of the quality for each pair of testing data was recorded in a previously prepared form, and the comments on the test signals for the listeners were collected and analyzed by the experimenters.

Three groups of tests were presented to each listener. They are the comparison between the TSCC-based method and MFCC-based method, comparison between the TSCC-based method and G.729.1 Annex E, and comparison between the MFCC-based method and G.729.1 Annex E. The distributions of listener rating for the test is shown in Figure 12. The bars indicate the relative frequencies of the scores given in the comparisons between the two processing methods. Bars on the positive side show preference for the latter method. The mean score for each group of tests is also shown on the horizontal axis with the 95% confidence interval. As shown in Figure 12, the proposed TSCC-based method and G.729.1 Annex E were considered substantially better than the MFCC-based method in terms of subjective listening tests, and the proposed method showed a little impairment compared to G.729.1 Annex E. According to the comments of the listeners, the proposed method is able to enhance the quality of WB audio signals decoded by G.729.1 and achieved a good performance on the whole, especially for the country music and jazz music. For speech signals, the proposed method effectively extends the bandwidth and improves the naturalness of the G.729.1-coded signals, but some spectral distortion occurred in the unvoiced sound may be perceived and slightly degrade the subjective quality of the audio signals extended by the TSCC-based method. For some rock music with the accompaniment of strong percussion signals, the HF transients of the reconstructed audio are not well restored due to the smoothing operations of TSCC for the WB audio. It might degrade the auditory quality in comparison with the SWB audio signals reproduced by G.729.1 Annex E which could partly restore the HF spectrum by using side information. So, the temporal envelope modification method could be further introduced in the future work in order to maintain a good extension performance for the percussion music signals.

Additionally, we made a comparison between the original WB audio signals and the extended audio signals in terms of the CCR test for further evaluating the performance of the proposed method. Here, the original WB audio signals were referred to as the WB reference, and the proposed TSCC-based method was adopted to extend the bandwidth of the original WB signals. The distribution of listener ratings for the test is shown in Figure 13. The proposed TSCC-based method achieved a better subjective quality than the WB audio signals on the whole. From the comments of the listeners, the quality of audio signals is effectively improved by using the BWE method, though some distortion can be perceived in the frames where the energy of the HF band is relatively high.

4.5 Computational complexity

We made a comparison of computational complexity between the proposed method and the MFCC-based method. The computational complexity was approximately measured in the number of multiplications per frame. Because the main modules of two BWE methods were the same except for feature extraction, only the computational complexity of feature extraction was compared. For computational requirements, the proposed TSCC needs about 53,380 multiplications per frame, and MFCC costs about

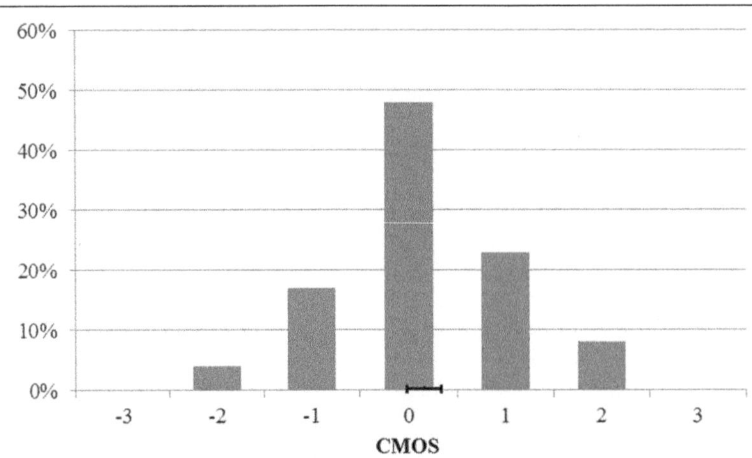

Figure 13 Distributions of listener rating in CCR tests. Comparison between the original WB audio and the audio signals extended by the TSCC-based method.

52,480 multiplications. For TSCC, the major contributions come from the Gammatone filter bank and DCT, and MCRA and spectral weighting only increase a very minor amount of computational complexity in comparison with MFCC.

5 Conclusions

This paper presents a novel BWE method based on TSCC. The audio signals are decomposed by using a Gammatone filter bank, and the audio spectrum is smoothed through MCRA and spectral weighting to suppress the transient components. Combining with the power-law nonlinear loudness function and cepstral normalization, TSCC is computed and then is applied into the GMM-based baseline BWE system for extending the bandwidth of the WB audio signals decoded by G.729.1. The extension performance of the proposed method is evaluated in comparison with the MFCC-based method and SWB extension method with side information in G.729.1 Annex E. Evaluation results show that the TSCC-based BWE method objectively gains the improvement to the MFCC-based method in terms of LSD, cosh measure, and DLSD but is inferior to G.729.1 Annex E. The subjective test results also indicate that the proposed method effectively enhances the auditory quality of the WB audio, makes the temporal evolution of the reconstructed audio signals smoother, and outperforms the MFCC-based reference method.

Competing interests

The authors declare that they have no competing interests.

Acknowledgements

This work was supported by the National Natural Science Foundation of China under Grants 61072089, 60872027, and 61471014.

References

1. P Vary, R Martin, *Digital Speech Transmission -Enhancement, Coding and Error Concealment* (John Wiley & Sons Ltd, UK, 2006)
2. E Larsen, RM Aarts, *Audio Bandwidth Extension-Application of Psychoacoustics, Signal Processing and Loudspeaker Design* (John Wiley & Sons Ltd, UK, 2004)
3. Y Qian, P Kabal, *Wideband Speech Recovery from Narrowband Speech using Classified Codebook Mapping, in Proceedings of the 9th Australian International Conference on Speech Science & Technology* (Australian Speech Science & Technology Association Inc., Melbourne, Australia, 2002)
4. H Pulakka, P Alku, Bandwidth extension of telephone speech using a neural network and a filter bank implementation for highband Mel spectrum. IEEE Trans. Audio Speech Lang. Process. **19**(7), 2170–2183 (2011)
5. P Jax, P Vary, *Feature Selection for Improved Bandwidth Extension of Speech Signals, in Proceedings of the International Conference on Acoustics, Speech, and Signal Processing* (IEEE, Montreal, Quebec, 2004)
6. M Nilsson, H Gustafsson, SV Andersen, WB Kleijn, *Gaussian Mixture Model Based Mutual Information Estimation between Frequency Bands in Speech, in Proceedings of the IEEE International Conference on Acoustics, Speech, and Signal Processing* (IEEE, Orlando, FL, USA, 2002)
7. M Nilsson, SV Andersen, WB Kleijn, *On the Mutual Information between Frequency Bands in Speech, in Proceedings of the IEEE International Conference on Acoustics, Speech, and Signal Processing* (IEEE, Istanbul, Turkey, 2000)
8. P Jax, P Vary, *An Upper Bound on the Quality of Artificial Bandwidth Extension of Narrowband Speech Signals, in Proceedings of the IEEE International Conference on Acoustics, Speech, and Signal Processing* (IEEE, Orlando, FL, USA, 2002)
9. T Esch, P Vary, *An Information Theoretic View on Artificial Bandwidth Extension in Noisy Environments, in Proceedings of the IEEE International Conference on Acoustics, Speech and Signal Processing* (IEEE, Kyoto, Japan, 2012)
10. AH Nour-Eldin, P Kabal, *Mel-frequency Cepstral Ccoefficient-based Bandwidth Extension of Narrowband Speech, in Proceedings of the INTERSPEECH (ISCA* (Brisbane, Australia, 2008)
11. KY Park, HS Kim, *Narrowband to Wideband Conversion of Speech using GMM Based Transformation, in Proceedings of the IEEE International Conference on Acoustics* (Speech and Signal Processing (IEEE, Istanbul, Turkey, 2000)
12. H Pulakka, U Remes, K Palomaki, M Kurimo, P Alku, *Speech Bandwidth Extension using Gaussian Mixture Model-based Estimation of the Highband Mel Spectrum, in Proceedings of the IEEE International Conference on Acoustics, Speech and Signal Processing* (Prague, Czech Republic, IEEE, 2011)
13. HP Knagenhjelm, WB Kleijn, *Spectral Dynamics is more Important than Spectral Distortion, in Proceedings of the IEEE International Conference on Acoustics* (Speech and Signal Processing (IEEE, Detroit, Michigan, USA, 1995)
14. F Norden, T Eriksson, *A Speech Spectrum Distortion measure with Interframe Memory, in Proceedings of the IEEE International Conference on Acoustics, Speech and Signal Processing* (IEEE, Salt Lake City, Utah, USA, 2011)
15. B Geiser, P Vary, *Beyond Wideband Telephony - Bandwidth Extension for Super-wideband Speech, in Proceedings of the German Annual Conference on Acoustics (DAGA)* (German Society for Audiology, Dresden, Germany, 2008)
16. P Jax, P Vary, *Wideband Extension of Telephone Speech using a Hidden Markov model, in Proceedings of the IEEE Workshop on Speech Coding* (IEEE, Delavan, WI, USA, 2000)
17. C Yağlı, T TTuran, E Erzin, Artificial Bandwidth Extension of Spectral Envelope along a Viterbi Path. Speech Comm. **55**(1), 111–118 (2013)
18. GB Song, P Martynovich, A study of HMM-based bandwidth extension of speech signals. Signal Process. **89**(10), 2036–2044 (2009)
19. AH Nour-Eldin, P Kabal, *Memory-based Approximation of the Gaussian Mixture Model Framework for Bandwidth Extension of Narrowband Speech, in Proceedings of the Interspeech (ISCA* (Florence, Italy, 2011)
20. AH Nour-Eldin, P Kabal, *Combining Frontend-based Memory with MFCC Features for Bandwidth Extension of Narrowband Speech, in Proceedings of the IEEE International Conference on Acoustics* (Speech and Signal (IEEE), Taipei, Taiwan, 2009)
21. AH Nour-Eldin, TZ Shabestary, P Kabal, *The Effect of Memory Inclusion on Mutual Information between Speech Frequency Bands, in Proceedings of the IEEE International Conference on Acoustics* (Speech and Signal Processing (IEEE, Toulouse, France, 2006)
22. BCJ Moore, BR Glasberg, Suggested formulae for calculating auditory-filter bandwidths and excitation patterns. J. Acoust. Soc. Am. **74**, 750–753 (1983)
23. J Makhoul, M Berouti, *High-frequency Regeneration in Speech Coding Systems, in Proceedings of the IEEE International Conference on Acoustics, Speech and Signal Processing* (IEEE, Washington, DC, USA, 1979)
24. U Chong, D Magill, The residual-excited linear prediction vocoder with transmission rate below 9.6 kbit/s. IEEE Trans. Comm **23**(12), 1466–1474 (1975)
25. S Shlien, The modulated lapped transform, its time-varying forms, and its applications to audio coding standards. IEEE Trans. Speech Audio Process. **5**(4), 359–366 (1997)
26. International Telecommunication Union, *ITU-T Recommendation G.729.1, G.729-based Embedded Variable Bit-Rate Coder: An 8–32 kbit/s Scalable Wideband Coder Bitstream Interoperable with G.729* (International Telecommunication Union, Geneva, 2006)
27. M Slaney, Auditory toolbox version 2. Interval Research Corporation. Tech. Rep. **10**, 1–52 (1998)
28. I Cohen, B Berdugo, Noise estimation by minima controlled recursive averaging for robust speech enhancement. IEEE Signal Process. Lett. **9**(1), 101–112 (2002)
29. G Doblinger, *Computationally Efficient Speech Enhancement by Spectral , Minima Tracking in Subbands, in Proceedings of EUROSPEECH* (ISCA, Madrid, Spain, 1995)
30. Y Shao, DL Wang, *Robust Speaker Identification using Auditory Features and Computational Auditory Scene Analysis, in Proceedings of the IEEE International Conference on Acoustics, Speech and Signal Processing* (IEEE, Las Vegas, USA, 2008)

31. S Ragot, B Kovesi, R Trilling, D Virette, N Duc, D Massaloux, S Proust, B Geiser, M Gartner, S Schandl, H Taddei, Y Gao, E Shlomot, H Ehara, K Yoshida, T Vaillancourt, R Salami, MS Lee, DY Kim, *ITU-T G.729.1: an 8–32 kbit/s Scalable Coder Interoperable with G.729 for Wideband Telephony and Voice over IP, in Proceedings of the IEEE International Conference on Acoustics, Speech and Signal Processing* ((IEEE), Honolulu, HI, 2007)
32. International Telecommunication Union, *ITU-T Recommendation G.729.1 Amendment 6, New Annex E on Superwideband Scalable Extension* ((International Telecommunication Union, Geneva, 2010)
33. International Telecommunication Union, *ITU-T Recommendation P.56, Objective Measurement of Active Speech Level* ((International Telecommunication Union, Geneva, 2011)
34. M Tammi, L Laaksonen, A Rämö, H Toukomaa, *Scalable Superwideband Extension for Wideband Coding, in Proceedings of the IEEE International Conference on Acoustics, Speech, and Signal Processing* (IEEE, Taiwan, 2009)
35. AH Gray, JD Markel, Distance measures for speech processing. IEEE Trans. Audio Speech Lang. Process. **24**(5), 380–391 (1976)
36. H Pulakka, L Laaksonen, M Vainio, J Pohjalainen, P Alku, Evaluation of an artificial speech bandwidth extension method in three languages. IEEE Trans. Audio Speech Lang. Process. **16**(6), 1124–1137 (2008)
37. F Norden, T Eriksson, Time evolution in LPC spectrum coding. IEEE Trans. Speech Audio Process. **12**(3), 290–301 (2004)
38. International Telecommunication Union, *ITU-T Recommendation P.800, Methods for Subjective Determination of Transmission Quality* ((International Telecommunication Union, Geneva, 1996)

13

A sub-band-based feature reconstruction approach for robust speaker recognition

Furong Yan*, Yanbin Zhang and Jiachang Yan

Abstract

Although the field of automatic speaker or speech recognition has been extensively studied over the past decades, the lack of robustness has remained a major challenge. The missing data technique (MDT) is a promising approach. However, its performance depends on the correlation across frequency bands. This paper presents a new reconstruction method for feature enhancement based on the trait. In this paper, the degree of concentration across frequency bands is measured with principal component analysis (PCA). Through theoretical analysis and experimental results, it is found that the correlation of the feature vector extracted from the sub-band (SB) is much stronger than the ones extracted from the full-band (FB). Thus, rather than dealing with the spectral features as a whole, this paper splits full-band into sub-bands and then individually reconstructs spectral features extracted from each SB based on MDT. At the end, those constructed features from all sub-bands will be recombined to yield the conventional mel-frequency cepstral coefficient (MFCC) for recognition experiments. The 2-sub-band reconstruction approach is evaluated in speaker recognition system. The results show that the proposed approach outperforms full-band reconstruction in terms of recognition performance in all noise conditions. Finally, we particularly discuss the optimal selection of frequency division ways for the recognition task. When FB is divided into much more sub-bands, some of the correlations across frequency channels are lost. Consequently, efficient division ways need to be investigated to perform further recognition performance.

Keywords: Robustness; Missing data technique (MDT); Reconstruction; Sub-band (SB); Full-band (FB); Principal component analysis (PCA)

1 Introduction

The performance of speaker or speech recognition systems degrades rapidly when they operate under conditions that differ from those used for training. Therefore, accomplishing noise robustness is a key issue to make these systems deployable in real world conditions. Solutions have been presented to solve this issue, such as feature-based [1-3], score-based [4,5], model-based [6-8], i-vectors [9], and the missing data technique (MDT) [10-12].

MDT can compensate for disturbances of the arbitrated type, so that this method which is based on the time-frequency representation is suitable to the problem of noise mismatch [12].

*Correspondence: yfrhappy@126.com
Beijing University of Posts and Telecommunications, No.10 Xitucheng Road, Beijing 100876, China

In MDT, two different methods have been considered to perform speech or speaker recognition with incomplete data: marginalization [13-15] and reconstruction [16,17]. In marginalization, the unreliable components are discarded or integrated up to the observed values. While the reconstruction method involves the estimation of the corrupted features using statistical methods, such as minimum mean square error (MMSE) [10], maximum *a posteriori* (MAP), and maximum likelihood (ML). Marginalization [11,14] and reconstruction [10] have been applied in speaker recognition system. However, marginalization suffers from two main drawbacks [17,18]. First, as known to us, utterance-level processing, such as mean and variance normalization, is capable of improving the recognition performance, but it cannot be performed with an incomplete spectrum [18]. Second, recognition has been carried out with spectral features. However, it is well known that cepstral features outperform spectral ones. Moreover, of all the methods, marginalization is assumed to have the most overhead. Consequently, if the complete reconstructed spectrogram is available, the recognizer is no longer constrained to perform recognition using spectral features. A more optimal set of parameters from the reconstructed spectrum will be derived.

In this paper, MAP reconstruction method [10] is used. Its efficiency significantly depends on the correlation between the spectral features. Conventional MAP reconstruction method is conducted on full-band [18,19]. According to our analysis, the spectral vectors extracted from the sub-band have more relevance than the ones extracted from the full-band. The conclusion will be illustrated in Section 2. Based on the above theory and the sub-band idea [20-22], a multi-sub-band reconstruction approach is proposed to improve on the recognition performance. The principle is to divide the full-band into multiple sub-bands and then independently reconstruct missing features extracted from every sub-band. After that, those features from all sub-bands will be recombined to yield the typical mel-frequency cepstral coefficient (MFCC) vector.

As one of many feature enhancement methods, the proposed reconstruction approach can be used in speaker and speech recognition system. To evaluate its validity, this paper will combine the new reconstruction method with speaker recognition system.

This paper is organized as follows. In the next section, the theory of the proposed reconstruction approach is analyzed. Section 3 is devoted to describing the proposed reconstruction approach. Section 4 describes the baseline experiment system and the experimental framework which is adopted to evaluate the proposed technique. Finally, Section 5 concludes this paper and discusses some future directions.

2 The analysis of concentration

As we know, the more concentrated the feature vector is, the higher its redundancy is, that is, the greater its correlation is [23]. It is measured by the degree of concentration with principal component analysis (PCA) .

In this paper, the P-dimensional mel log-spectral vector is used for reconstruction. Mel filters are used to represent a frame spectrum as a log-spectral vector of P-dimensional (termed as full-band feature vector). The frequency region $(0, f_s/2)$ is divided into C sub-bands. Let P_i denote the number of mel filters corresponding to the ith sub-band. Apparently,

$$\sum_{i=1}^{C} P_i = P \tag{1}$$

Corresponding to the tth frame and ith sub-band, the output of mel filters (termed as the ith sub-band feature vector) is represented as follows:

$$\overrightarrow{Y}_i^t = (y(1,t), \cdots y(P_i,t))^T \tag{2}$$

In order to analyze the degree of concentration of the feature vector $\overrightarrow{Y}_i^t$, the eigenvalues of associated covariance matrix Θ_i need to be calculated and then need to be arranged in descending order. It is represented as $[\lambda_{i,1}, \lambda_{i,2}, \cdots \lambda_{i,P_i}]$.

To learn how closely the ith sub-band feature vector $\overrightarrow{Y}_i^t$ is in the space of the P_i-dimension, the so-called concentration level $M_R^i(r)$ is introduced and computed as follows:

$$R_i(m) = \frac{\sum_{i=1}^{m} \lambda_{i,l}}{\sum_{i=1}^{P_i} \lambda_{i,l}}, m = 1, 2, \cdots P_i \tag{3}$$

$$M_R^i(r) = \arg\min_m (R_i(m) > r) \tag{4}$$

That is, $R_i(m)$ is the accumulative contribution rate of the first m principle components. Concentration level $M_R^i(r)$ is the minimum m that makes $R_i(m) > r$, where r is a predefined concentration coefficient.

For certain r, a smaller $M_R^i(r)$ implies that the ith sub-band feature vector is confined along a smaller number of principle directions, and therefore, the feature vector is much more closely related to each other according to the above definition.

In the same manner, the degree of concentration of the full-band feature vector could be analyzed.

The accumulative contribution rate of the first m principle components corresponding to the 4-sub-band and full-band is shown in Figure 1. The conclusion should be clear. The concentration level corresponding to each sub-band in the 4-sub-band is smaller than the one corresponding to the full-band.

The correlation between the redundancy and accuracy of the prediction is best visualized using 2-dimensional examples as shown in Figure 2. The 2-dimensional examples involve the feature vector extracted from clean and noisy utterances, together with MAP reconstruction obtained for the noisy utterance. Babble noise at 0 dB signal-to-noise ratio (SNR) has been added to obtain the noisy utterance. Panels (a) and (b), respectively, reflect a range of 2-dimensional feature vectors with different redundancies. The redundancy of data in panel (b) is lower than that in panel (a). The reconstruction data corresponding to the data with high redundancy and low redundancy is defined along the first principle direction and scattered. In short, the fatter the cloud is, the lower the prediction accuracy is in a 2-dimensional case.

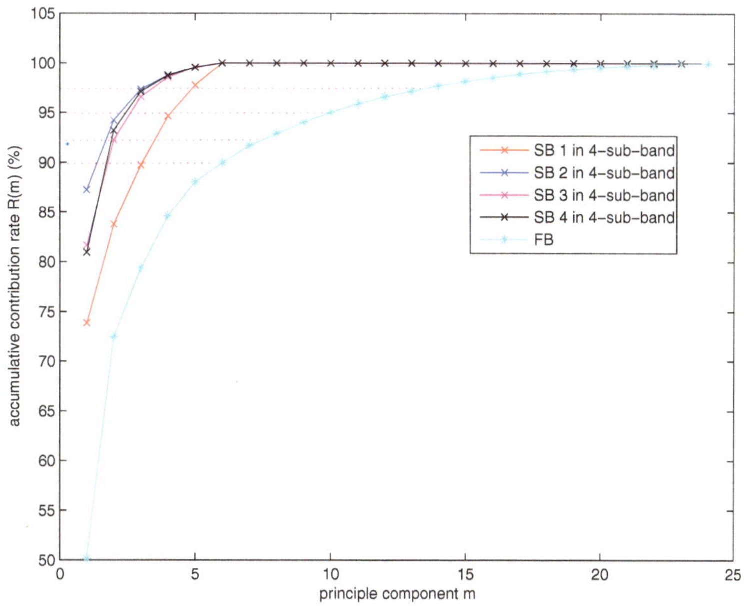

Figure 1 The accumulative contribution rate of the first *m* principle components corresponding to the 4-sub-band and full-band.

Figure 3 shows the contribution rate of two principle components which are obtained from the covariance matrix of the 2-dimensional feature vector. When the value of the predefined concentration coefficient r is 0.9, the concentration level which is corresponding to the data shown in Figure 2a,b is $M_R^{(high)}(r) = 1$ and $M_R^{(low)}(r) = 2$, respectively.

Considering the recorded positions of the 2-dimensional feature vector in Figure 2 and the corresponding contribution rate, together with our analysis, the following conclusion is obtained: the higher the redundancy of the data is, that is, the greater its correlation is, the smaller the corresponding concentration level is. As MAP reconstruction method is based on

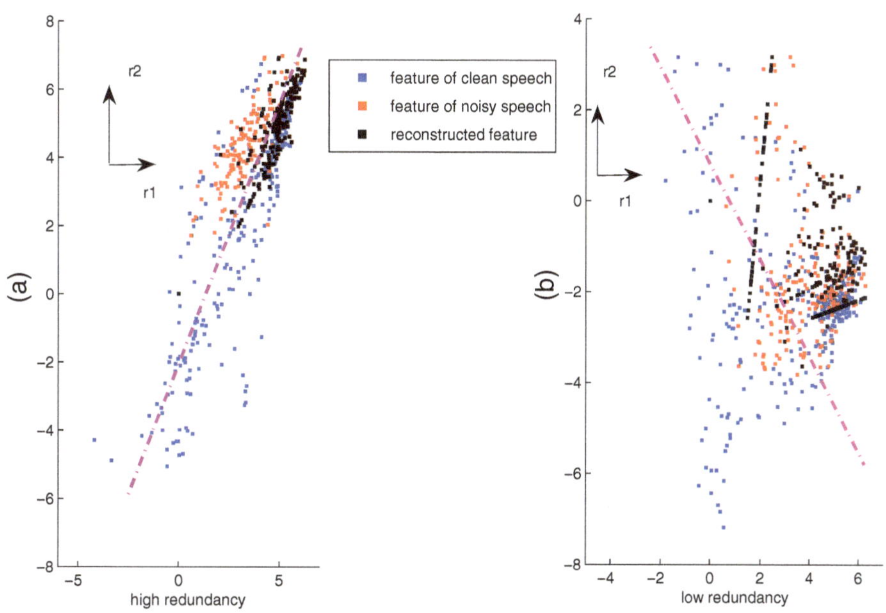

Figure 2 A spectrum of redundancies in data from the two separate recordings r1 and r2. The best-fit line is indicated by the dashed lines for **(a)** high redundancy and **(b)** low redundancy.

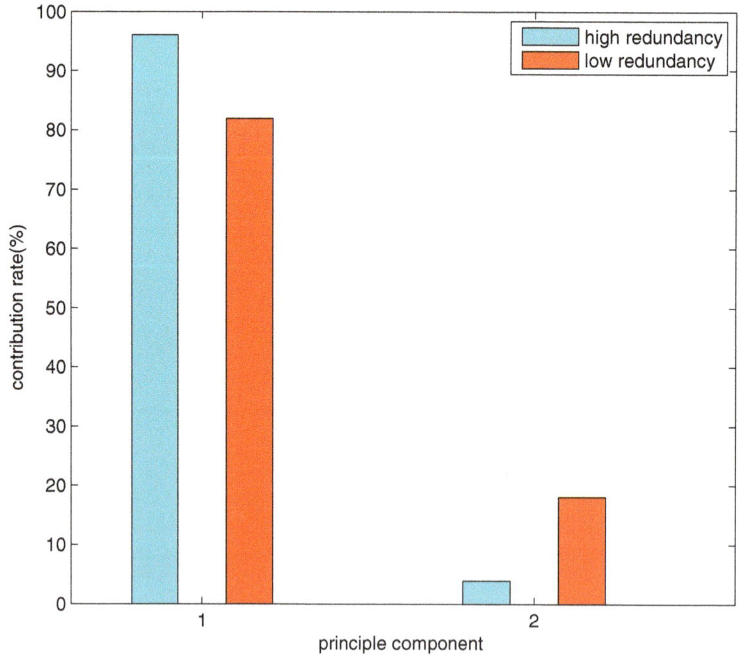

Figure 3 The contribution rate of two principle components in a 2-dimensional case.

the correlation between the feature vectors, the smaller the concentration level is, the higher the validity of the reconstruction is.

3 Multi-sub-band reconstruction for speaker recognition system

As one of many feature enhancement methods, the multi-sub-band reconstruction method in MDT can be applied in the Gaussian mixture model (GMM) [24], the SVM-GMM [25], and the universal background model (UBM)-GMM recognition system. Based on the validity of the UBM-GMM system shown in [11], the proposed reconstruction method is evaluated in a UBM-GMM speaker recognition system. In this section, the MDT-based speaker recognition system is described.

3.1 UBM-GMM model

In this paper, a speaker-independent UBM is used. A speaker-dependent model can be derived from UBM by adapting the UBM parameters to the speech material of the corresponding speaker using MAP estimation [11,26].

3.2 Feature vector

Mel log-spectral vector and MFCC are used in the reconstruction and recognition stage, respectively. The unreliable components are reconstructed based on the statistical relationship between the log-spectral vector.

3.3 Mask estimation

In order to perform MDT, a mask must be required which classifies the time-frequency (T-F) units into reliable and unreliable components. Various strategies have been proposed to estimate a mask, such as SNR-based estimation [27], auditory and perceptual estimation [14,28], classifier-based estimation [29], and DNN-based estimation [30]. It is, however, outside the scope of this paper to analyze and compare all existing approaches. Because the focus of this paper is to robustly identify speakers in the presence of noise, the mask $m(t,k)$ is determined by estimating the local SNR in individual T-F units. SNR-based mask estimation method is applied to decide whether a T-F unit is reliable.

$$m(t,k) = \begin{cases} 0, \text{ if } \left|\widehat{\vec{S}}(t,k)\right|^2 \leq \left|\widehat{\vec{N}}(t,k)\right|^2 \\ 1, \text{ otherwise} \end{cases} \tag{5}$$

where $\left|\widehat{\vec{S}}(t,k)\right|^2$ and $\left|\widehat{\vec{N}}(t,k)\right|^2$ represent the kth frequency bands of the power spectrum of speech and noise, respectively, in individual T-F units. What calls for special attention is that the estimation of speech and noise components is carried out in the spectral domain before applying mel filter.

The estimate of the noise spectrum is derived from the noisy signal spectrum. The estimation method is shown in [31]. The estimate of the speech spectrum $\left|\widehat{\vec{S}}(t,k)\right|^2$ can be derived by subtracting the estimated noise spectrum

$\left|\widehat{\vec{N}}(t,k)\right|^2$ from the corrupted signal spectrum. In this paper, the technique to accomplish this is to perform spectral subtraction by applying an SNR-dependent gain function MMSE log-STSA [32] in the frequency domain.

3.4 MAP estimation for unreliable components

In MAP estimation, the unreliable components are estimated by making their likelihood condition on the reliable components [18] be maximum.

$$\hat{\vec{x}}_u = \arg\max_{\vec{x}_u} p\left(\vec{x}_u | \vec{x}_r, \vec{\mu}, \Theta\right) \tag{6}$$

A feature vector $\vec{x} \in \Re^{P_j * 1}$ is divided into reliable and unreliable components based on SNR-based mask estimation method.

$$\vec{x_r} \in \Re^{D_1 * 1} \quad \vec{x_u} \in \Re^{D_2 * 1}, D_1 + D_2 = P_j \tag{7}$$

$$\vec{x} = \left[\vec{x}_r, \vec{x}_u\right] \tag{8}$$

assuming that $p\left(\vec{x}; \vec{\mu}, \Theta\right)$ is the probability distribution function (pdf) of a Gaussian distribution with mean vector μ and covariance matrix Θ. According to the nature of Gaussian distribution, $p\left(\vec{x}_r; \vec{\mu}, \Theta\right)$ and $p\left(\vec{x}_u; \vec{\mu}, \Theta\right)$ would therefore also be Gaussian [33]. Consequently,

$$\vec{\mu} = \left[\vec{\mu}_r, \vec{\mu}_u\right] \tag{9}$$

$$\Theta = \begin{bmatrix} \Theta_{rr} & \Theta_{ru} \\ \Theta_{ur}, & \Theta_{uu} \end{bmatrix} \tag{10}$$

$$p\left(\vec{x}_r, \vec{\mu}_r, \Theta_{rr}\right) = C_1 \exp\left[-0.5\left(\vec{x}_r - \vec{\mu}_r\right)^T \Theta_{rr}^{-1}\left(\vec{x}_r - \vec{\mu}_r\right)\right] \tag{11}$$

$$p\left(\vec{x}_u, \vec{x}_r, \vec{\mu}, \Theta\right) = C_2 \exp\left[-0.5\left(\vec{x} - \vec{\mu}\right)^T \Theta^{-1}\left(\vec{x} - \vec{\mu}\right)\right] \tag{12}$$

where Θ_{ru} is the cross covariance between $\vec{x}_r$ and $\vec{x}_u$ and $\Theta_{ru} = \Theta_{ur}^T$.

It can now be shown that $p\left(\vec{x}_u | \vec{x}_r, \vec{\mu}, \Theta\right)$ is given by

$$p\left(\vec{x}_u | \vec{x}_r, \vec{\mu}, \Theta\right) = \frac{p\left(\vec{x}_u, \vec{x}_r, \vec{\mu}, \Theta\right)}{p\left(\vec{x}_r, \vec{\mu}_r, \Theta_{rr}\right)}$$

$$= C \exp\left[-0.5\left(\vec{x}_u - \vec{\mu}_u\right) - \Theta_{ur}\Theta_{rr}^{-1}\left(\vec{x}_r - \vec{\mu}_r\right)\right] \tag{13}$$

where C is a normalizing constant. The following equation can be obtained from Equations 11, 12, and 13.

$$\widehat{\vec{x}}_u = \arg\max_{\vec{x}_u}\left[p\left(\vec{x}_u | \vec{x}_r, \vec{\mu}, \Theta\right)\right] = \vec{\mu}_u + \Theta_{ur}\Theta_{rr}^{-1}\left(\vec{x}_r - \vec{\mu}_r\right) \tag{14}$$

Figure 4 shows the process of reconstruction. The values of the statistical parameters such as $\vec{\mu}_r$, $\vec{\mu}_u$, Θ_{ur}, and Θ_{rr} must be learned from the training corpus. A vector is said to belong to the cluster that is most likely to have generated it. As the distribution of the vector is assumed to be Gaussian, the cluster membership $\widehat{m}_{\vec{x}(t)}$ of a vector $\vec{x}(t)$ is defined as

$$\widehat{m}_{\vec{x}(t)} = \arg\max_m\left[p\left(m | \vec{x}(t)\right)\right] = \arg\max_m\left[p\left(\vec{x}(t) | m\right) p(t)\right] \tag{15}$$

and then the unreliable components of the vector are reconstructed using MAP estimation method.

3.5 The proposed multi-sub-band reconstruction approach

Assuming that utilizing P mel filter to smooth the N FFT magnitude coefficients. The reconstruction is individually conducted on 2 sub-bands consisting of consecutive channels ($P/2$-dimensional channels) with no band overlap (sub-band 1: channel 1 to $P/2$, sub-band 2: channel $P/2+1$ to P). The reconstruction method falls neatly into two parts as shown in Figure 5. In the first part, the statistical parameters (SP) used in construction are individually trained for different sub-bands. The steps of the second part are as follows:

(a) The estimation of speech and noise components is carried out in the spectral domain.
(b) A mask will be obtained which classifies the T-F representation into reliable and unreliable components corresponding to the frequency range of P mel filters. The above two steps are carried out before applying the mel filter.
(c) P mel filters are used to smooth the power spectrum and then its logarithm is taken.
(d) The mel log-spectral vector is multiplied by the mask estimated in step (b).
(e) The feature vector corresponding to full-band is divided into ones corresponding to 2 sub-bands.
(f) Based on SP trained in the first part, the feature vectors corresponding to every sub-band are reconstructed, individually.
(g) The reconstructed vector of 2-sub-band is recombined to yield the typical MFCC vector.

3.6 Baseline system

The system described in [11] assumes that the unreliable components are bounded between zero and the observed mel log-spectrum and the mel log-spectrum is independent, and marginalization is applied to process the corrupted vector. The feature vector used in recognition is a P-dimensional mel log-spectrum. We compare the performance of the proposed system with the baseline system.

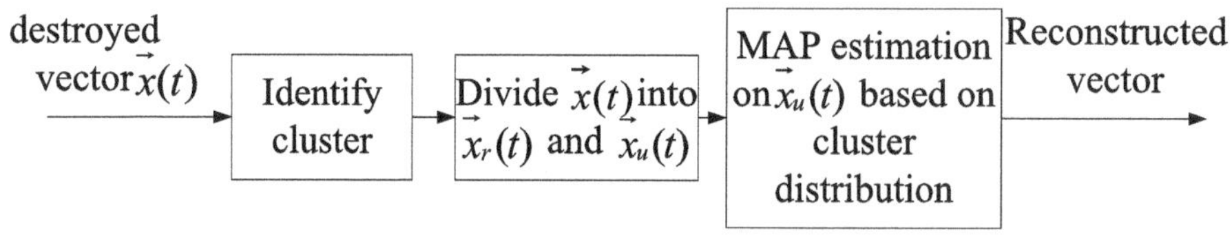

Figure 4 Block diagram presenting estimation of missing components in a vector using MAP estimation method.

4 Experiments

New reconstruction method is evaluated on a closed set of 30 speakers and 140 utterances per speaker. The sampling frequency is 16 KHz. For each speaker, 70% of the available speech material is randomly selected to train the corresponding speaker model, 7% is used for training SP for reconstruction stage, and the remaining 23% is used for test.

In the training stage, we use a voice activity detector (VAD) based on power to ensure that silence frames would not impact on the establishing model.

Speaker recognition performance is evaluated on a subset of ten randomly selected speakers involving a total of 30 sentences per speaker (20 sentences for training speaker-dependent GMM and 10 sentences for testing). In the test phase, utterances are mixed at various SNRs with noise signals drawn from the NOISEX database [34].

Figure 6 describes evaluation system in which 24 mel filters are used to smooth the spectrum and the full-band is divided into 2 sub-bands (SB1: channel 1 to 12, SB2: channel 13 to 24), and 34-demensional MFCC consisting of 16 static MFCC coefficients including the 0th order coefficient and first order temporal derivatives is used for recognition. At the end, cepstral mean normalization (CMN) is applied to improve robustness.

4.1 Experiment 1: performance comparison between marginalization and reconstruction including full-band and 2-sub-band reconstruction

In the first experiment, we compare the performances of two systems which use the marginalization and reconstruction methods to process the corrupted features and then evaluate the validity of the proposed reconstruction method. The point is that recognition has to be carried out with spectral features in the former system. While in the latter system, MFCC are extracted for recognition.

The DET curves visualize the trade-off between missed detections and false alarms [35]. Figure 7 gives the recognition performance of two systems in destroyer-engine noise at a SNR of 0 dB. The results in the figure show that cepstral features outperform spectral ones for speaker recognition.

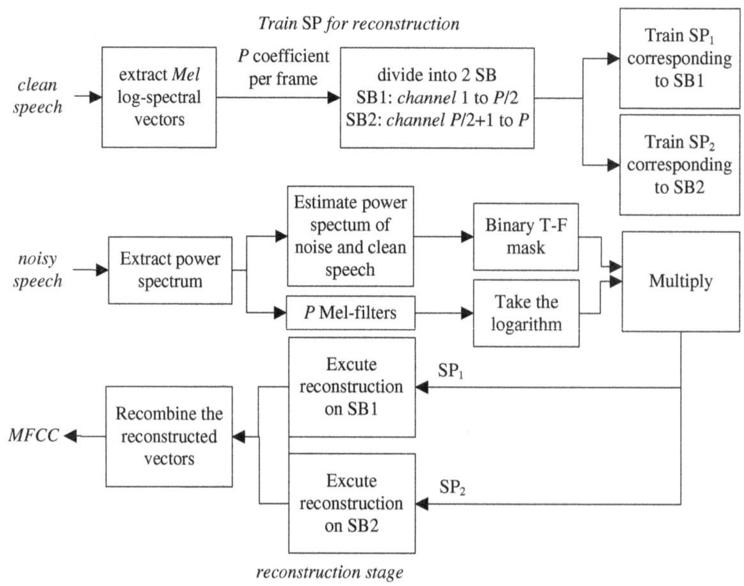

Figure 5 Diagram of the proposed reconstruction method.

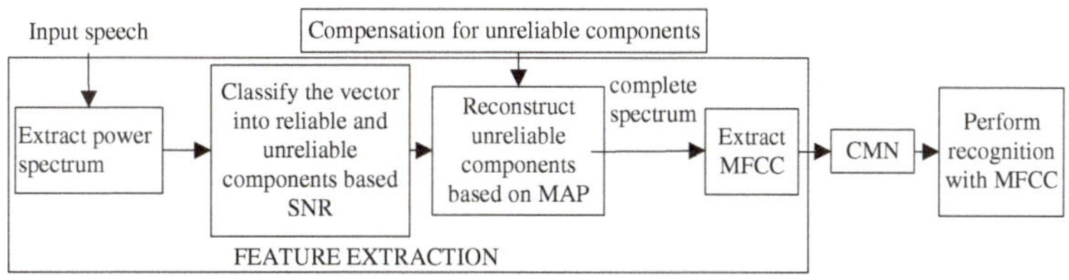

Figure 6 Schematic diagram of the evaluation system.

Figure 8 shows that the recognition performance of the latter system improves when reconstruction is applied to process the corrupted features, and the recognition accuracy of the latter system using 2-sub-band reconstruction method is improved 5.65% more than full-band reconstruction method.

In order to evaluate the validity of the proposed reconstruction method in various noise types, this paper conducts recognition experiments in babble, factory1, pink, white, and destroyer-engine noise. The SNR-dependent recognition accuracy for recognition system is presented in Table 1. The last table depicts the average performance over all noise conditions.

Based on the experimental results reported in Table 1, the corresponding SNR-dependent curves are shown for all noise types in Figures 9, 10, and 11.

The following observations can be made:

(a) It can be observed in Table 1 that the performance obtained from both reconstruction methods clearly outperforms the baseline system.
(b) The results show that 2-sub-band reconstruction method performs better than full-band for all noise types. The recognition performance is higher at a larger SNR.
(c) The recognition performance in babble noise is higher than the other four noise types in most cases for two kinds of reconstruction methods.
(d) The corresponding relative improvements regarding full-band reconstruction are 2.55%, 1.49%, 1.10%, 1.63%, and 1.03% at a SNR of 0, 5, 10, 15, and 20 dB, respectively. Recognition performance improves the most at a SNR of 0 dB.

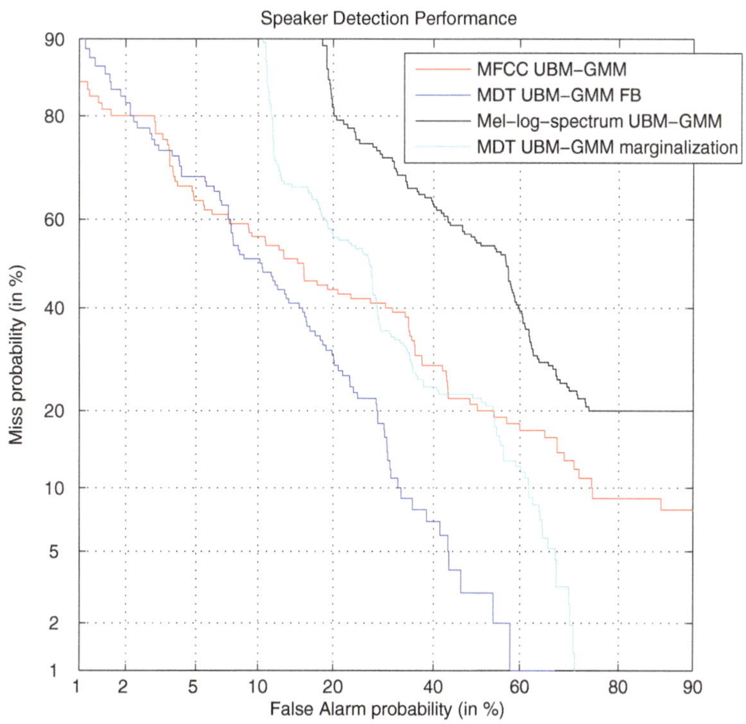

Figure 7 The DET curves for two recognition systems in destroyer-engine noise at a SNR of 0 dB.

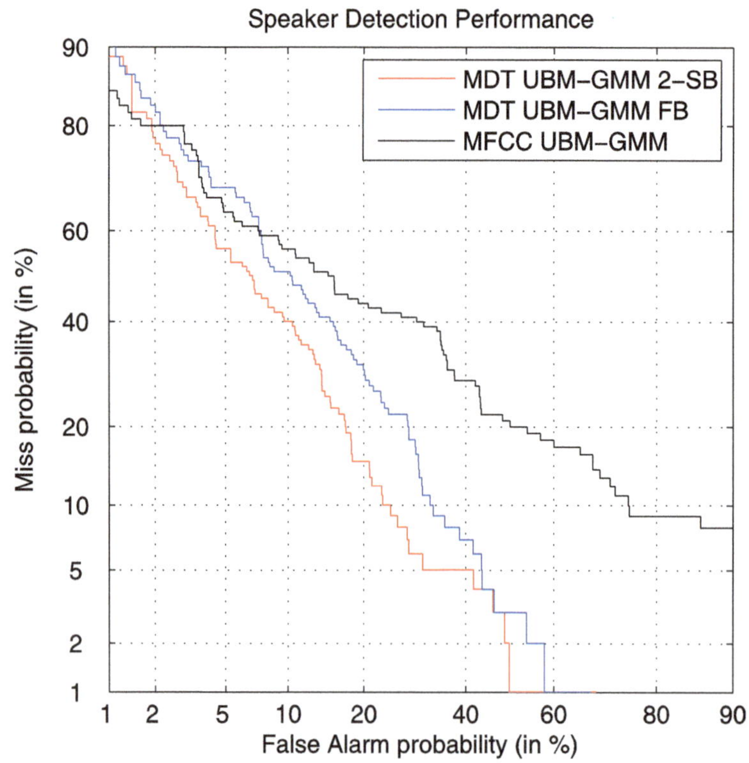

Figure 8 The DET curves for two kinds of reconstruction methods in destroyer-engine noise at a SNR of 0 dB.

Table 1 Recognition performance of FB, 2-SB reconstruction, and marginalization in the presence of different types of noise (unit: %)

		0 dB	5 dB	10 dB	15 dB	20 dB
Babble	FB	82.60	84.02	87.01	87.49	89.49
	2-SB	82.96	85.30	87.84	89.30	91.25
	Marginalization	63.71	64.88	66.44	67.69	69.75
Factory1	FB	76.52	82.10	87.40	88.00	88.33
	2-SB	80.01	82.25	87.70	89.13	90.23
	Marginalization	67.54	68.10	68.33	68.51	69.40
Pink	FB	75.12	80.78	84.83	87.55	89.11
	2-SB	76.66	82.43	87.40	89.62	90.27
	Marginalization	67.79	68.54	69.00	69.06	69.91
White	FB	77.10	83.02	84.40	87.16	89.71
	2-SB	78.81	83.60	86.09	88.70	89.82
	Marginalization	68.00	68.92	70.21	70.77	71.21
Destroyer-engine	FB	76.51	82.81	86.43	86.29	88.47
	2-SB	82.16	86.60	86.52	87.92	88.68
	Marginalization	66.64	67.26	68.30	68.48	69.97
Average	FB	77.57	82.55	86.01	87.30	89.02
	2-SB	80.12	84.04	87.11	88.93	90.05
	Marginalization	66.74	67.54	68.46	68.90	70.05

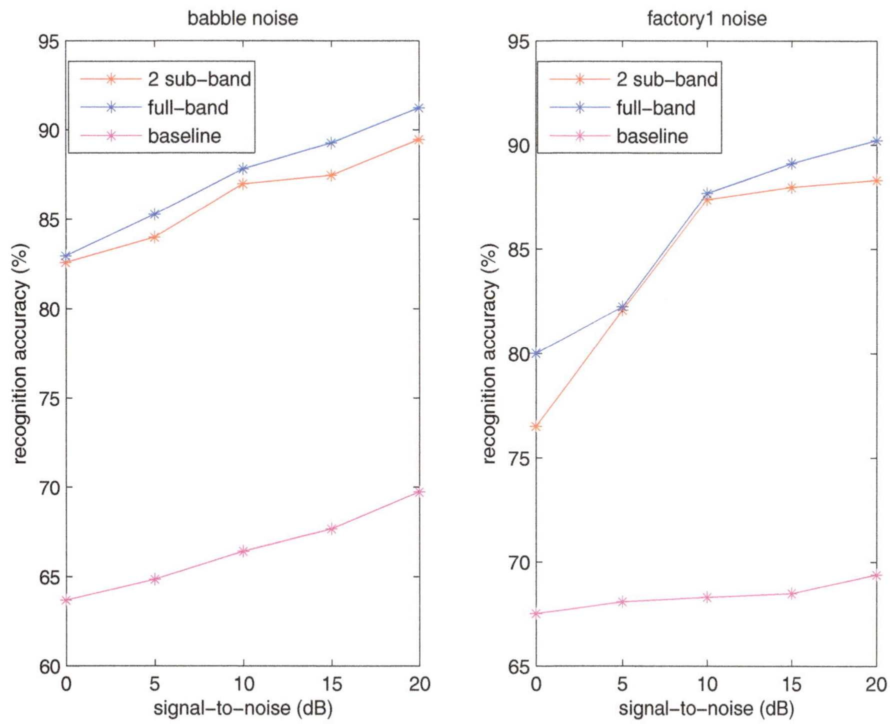

Figure 9 SNR-dependent speaker recognition performance in babble and factory1 noise.

(e) The improved recognition performance is 6.04%, 6.97%, 8.99%, 5.63%, and 11.37% in babble, factory1, pink, white, and destroyer-engine noise, respectively. Recognition performance improves the most in destroyer-engine noise.

We analyze the relationship between reconstruction performance and the correlation of the feature vector with PCA. Table 2 shows the contribution rate of every principle component. When the value of concentration coefficient r is 0.95, the corresponding concentration

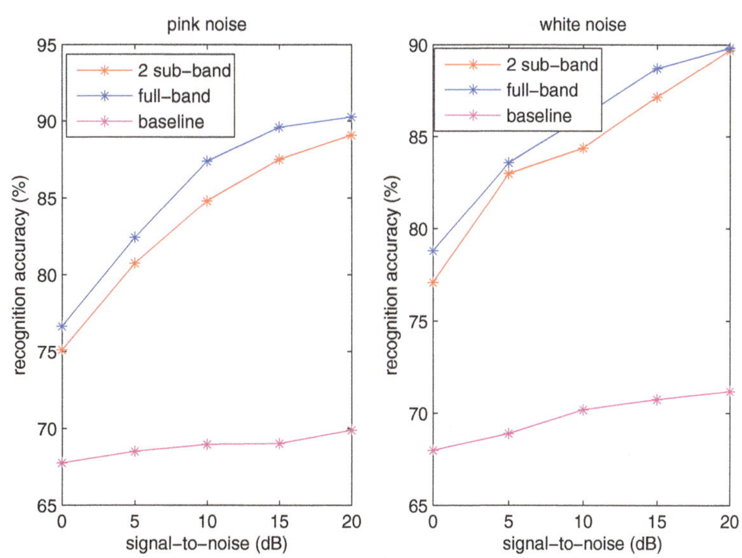

Figure 10 SNR-dependent speaker recognition performance in pink and white noise.

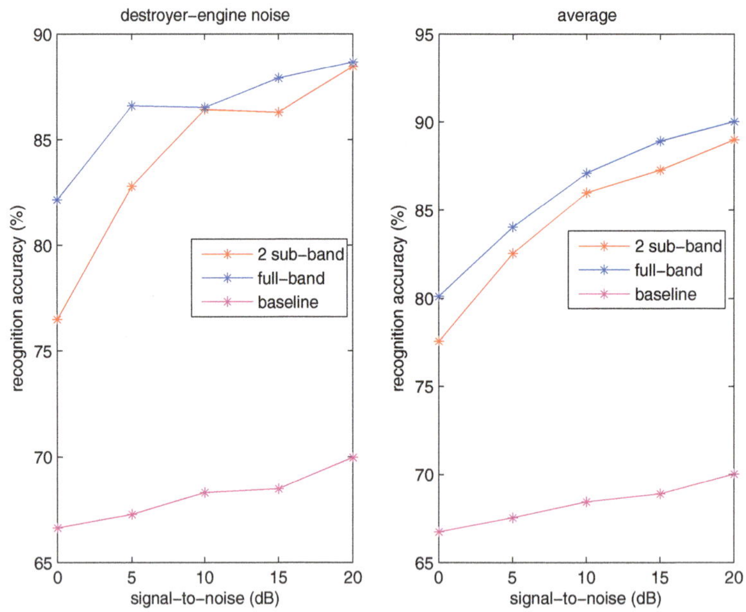

Figure 11 SNR-dependent speaker recognition performance in destroyer-engine noise and average condition.

levels are $M_R^{FB}(r) = 10$, $M_R^1(r) = 6$, and $M_R^2(r) = 5$. Based on the conclusion shown in Section 2, a smaller $M_R^i(r)$ implies a stronger concentration for the feature vector. Consequently, since the correlation of every sub-band is stronger than the full-band, the performance of the 2-sub-band reconstruction approach is better.

The result of PCA will be obtained by decomposing eigenvalues of the covariance matrix, which is relevant to the reconstruction. The accumulative contribution rate of the principle components is shown in Figure 12.

Table 2 The contribution rate (%) of every principle component

	FB	SB1	SB2
1	50.161	69.455	66.811
2	22.291	13.475	17.025
3	6.935	3.292	6.812
4	5.236	2.315	3.084
5	3.424	2.315	2.051
6	1.945	2.139	1.345
7	1.708	1.514	0.951
8	1.218	1.185	0.715
9	1.125	0.898	0.490
10	1.002	0.651	0.327
11	0.869	0.445	0.233
12	0.698	0.240	0.157

4.2 Experiment 2: influence of different division ways of full-band

The conclusion that the recognition performance obtained by the proposed reconstruction method is superior to full-band reconstruction has been obtained in Experiment 1. The choice of an optimal division of full-band seems to be crucial for sub-band reconstruction method. In order to find the optimal division, this paper conducts a series of recognition experiments. The division ways and the corresponding recognition performance are shown in Table 3.

These experiments are conducted in babble noise which is highly non-stationary and a SNR of 0 dB. The results are shown in Figure 13.

The recognition performance is ranked corresponding to different division ways starting with the highest performance: 4 sub-bands, 2 sub-bands, 3 sub-bands, 8 sub-bands, 6 sub-bands, and 12 sub-bands. The relationship between channel number and recognition performance is not obvious. In order to explain the relationship between the recognition performance and the division ways, we analyze the case of 4 sub-band and 2 sub-band. The results are shown in Figure 14. Assume that the amount of information presented by the original data is 100%. When full-band is divided into 4 sub-bands, the amount of information presented by the first four principle components

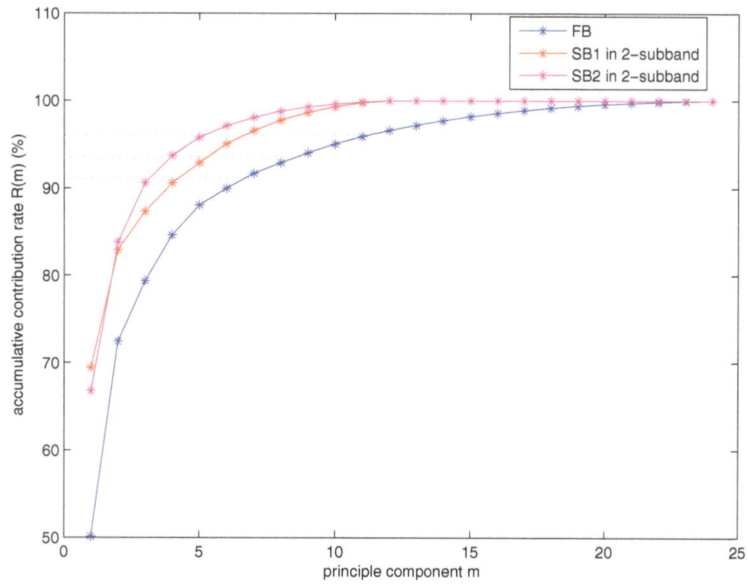

Figure 12 Accumulative contribution rate of principle components derived from feature vectors corresponding to 2 sub-bands and full-band.

derived from sub-band 1, sub-band 2, sub-band 3, and sub-band 4 is 94.67%, 98.79%, 98.58%, and 98.73%, respectively. However, if the full-band is divided into 2 sub-bands, the amount of information presented by the first four principle components derived from both sub-bands is 90.61% and 93.73%. That is, the redundancy of the feature vector extracted from each sub-band is higher on the condition that the full-band is divided into 4 sub-bands.

When the full-band is divided into 12 sub-bands, the recognition performance is inferior. The observation shows that the correlations between the feature vector are lost when the number of sub-bands is more numerous.

5 Conclusions

This paper presents a new feature enhancement method, which is evaluated in a UBM-GMM speaker recognition system. In the proposed method, the reconstruction is executed on a partial sub-band independently and then the reconstructed spectrum is recombined into a complete spectrum to yield the conventional MFCC for recognition. Compared to full-band reconstruction method, recognition performance obtained by the proposed reconstruction approach has been shown to be higher in five noise types. The experiment has also reflected that the recognition performance depends on the frequency division ways, thus the optimal division ways need to be developed.

The first experiment has revealed the following results. First, MFCC features outperform spectral ones for speaker recognition. Second, the recognition performance obtained by reconstruction is higher than marginalization. Third, the recognition performance obtained by the 2-sub-band reconstruction method is superior to the full-band reconstruction in five noise types and at all SNRs. The second experiment has shown that different frequency division ways could influence on the recognition performance.

In order to achieve further recognition performance improvements, on the one hand, an optimal frequency division way will be very important. On the other hand, analyzing the distribution property of various noise types and then accurately identifying destroyed components are also research hot spots. In the end, research on mask estimation algorithms is required to precisely separate reliable from unreliable components.

Table 3 Different division ways of full-band and the corresponding recognition performance

Channel number	Sub-bands	Recognition (%)
12	1-2, 3-4, 5-6, 7-8, 9-10, 11-12,13-14,15-16, 17-18, 19-20, 21-22, 23-24	74.18
8	1-3, 4-6, 7-9, 10-12, 13-15, 16-18, 19-21, 22-24	78.15
6	1-4, 5-8, 9-12, 13-16, 17-20, 21-24	77.83
4	1-6, 7-12, 13-18, 19-24	84.86
3	1-8, 9-16, 17-24	81.78
2	1-12, 13-24	82.96

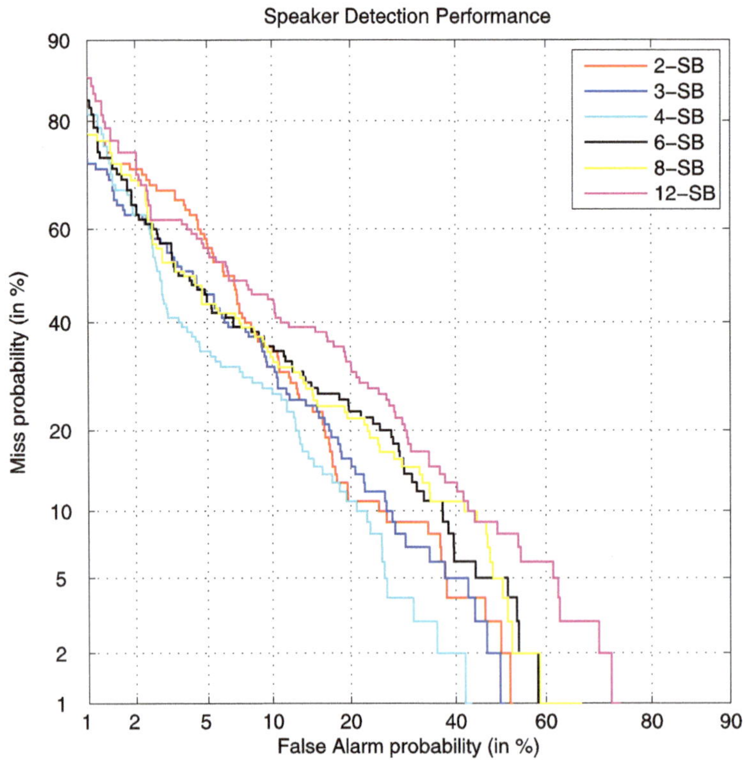

Figure 13 The recognition performance derived from different division ways.

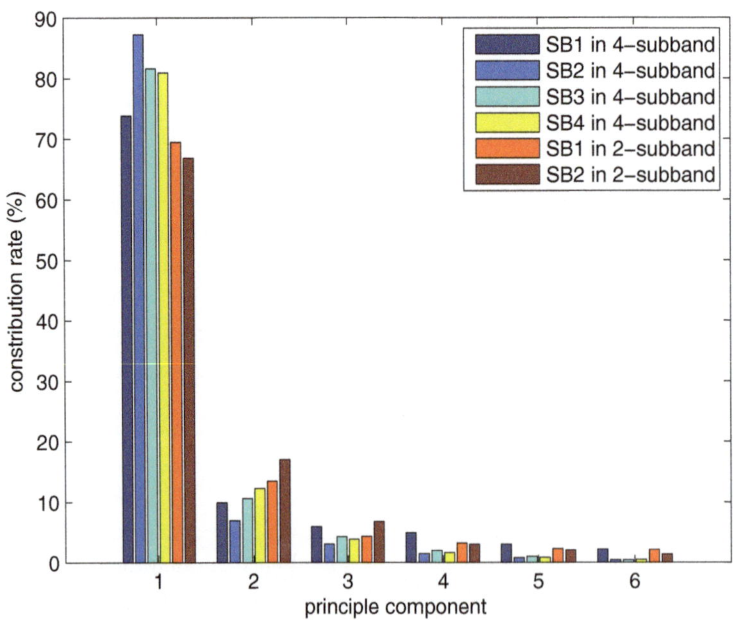

Figure 14 The comparison in contribution rate derived from 4 sub-bands and 2 sub-bands.

Competing interests
The authors declare that they have no competing interests.

References

1. J Pelecanos, S Sridharan, Feature warping for robust speaker verification. ISCA Workshop Speaker Recognition, June 213–218 (2001)
2. DA Reynolds, Channel robust speaker verification via feature mapping. ICASSP. **2**, 53–56 (2003)
3. V Chandran, D Ning, S Sridharan, in *Biometric Authentication*. Speaker identification using higher order spectral phase features and their effectiveness vis-avis mel-cepstral features, vol. 3072 (Springer Verlag Berlin, 2004), pp. 1–20
4. DA Reynolds, TF Quatieri, RB Dunn, Speaker verification using adapted Gaussian mixture models. Digital Signal Process. **10**(1–3), 19–41 (2000)
5. R Auckenthaler, M Carey, H Lloyd-Thomas, Score normalization for text-independent speaker verification systems. Digital Signal Process. **10**(1–3), 42–54 (2000)
6. P Kenny, P Dumouchel, in *Proc. ODYSSEY 2004-The Speaker and Language Recognition Workshop*. Experiments in speaker verification using factor analysis likelihood ratios (Toledo, Spain, May 31–June 3 2004), pp. 219–226
7. P Kenny, G Boulianne, P Ouellet, P Dumouchel, Factor analysis simplified. ICASSP. **1**, 637–640 (2005)
8. P Jančovič, M Köküer, Estimation of voicing-character of speech spectra based on spectral shape. IEEE Signal Process. Lett. **14**(1), 66–69 (2007)
9. M McLaren, D van Leeuwen, in *ICASSP*. Improved speaker recognition when using i-vectors from multiple speech sources (Prague, 2011), pp. 5460–5463
10. JA González, AM Peinado, N Ma, AM Gómez, J Barker, MMSE-based missing-feature reconstruction with temporal modeling for robust speech recognition. IEEE Trans. Audio, Speech, Lang. Process. **21**(3), 624–635 (2013)
11. T May, S van de Par, A Kohlrausch, Noise-robust speaker recognition combining missing data techniques and universal background modeling. IEEE Trans. Audio, Speech, Lang. Process. **20**(1), 108–121 (2012)
12. R Togneri, D Pullella, An overview of speaker identification: accuracy and robustness issues. IEEE Circuits Syst. Mag. **Second Quarter**, 23–61 (2011)
13. M Cooke, P Green, L Josifovski, A Vizinho, Robust automatic speech recognition with missing and unreliable acoustic data. Speech Commun. **34**, 267–285 (2001)
14. X Zhao, Y Shao, D Wang, CASA-Based Robust Speaker Identification. IEEE Trans. Audio, Speech Lang. Process. **20**, 1608–1616 (2012)
15. N Ma, J Barker, H Christensen, P Green, Combining speech fragment decoding and adaptive noise floor modelling. IEEE Trans. Audio Speech Lang. Process. **20**(3), 818–827 (2012)
16. JF Gemmeke, VanH Hamme, B Cranen, L Boves, Compressive sensing for missing data imputation in noise robust speech recognition. IEEE J. Sel. Topics Signal Process. **4**(2), 272–287 (2010)
17. JA González, AM Peinado, AM Gómez, N Ma, J Barker, in *IEEE Trans. Audio Speech Lang. Process*. Combining missing-data reconstruction and uncertainty decoding for robust speech recognition (Kyoto, 2012), pp. 4693–4696
18. B Raj, ML Seltzer, RM Stern, Reconstruction of missing features for robust speech recognition. Speech Commun. **43**, 195–202 (1997)
19. B Raj, *Reconstruction of incomplete spectrograms for robust speech recognition*. PhD dissertation, Pittsburgh, PA, Carnegie Mellon Univ, 2000
20. L Raj, JF Bonastre, in *Proc. Audio and Video based Biometric Person Authentication*. LNCS, ed. by J Bigün, G Chollet, and G Borgefors. Subband approach for automatic speaker recognition: optimal division of the frequency domain (Springer Heidelberg, 1997), pp. 195–202
21. L Besacier, JF Bonastre, C Fredouille, Localization and selection of speaker-specific information with statistical modeling. Speech Comm. **31**, 89–106 (2000)
22. H Bourlard, S Dupont, A new ASR approach based on independent processing and recombination of partial frequency bands. ICSLP. **1**, 426–429 (1996)
23. J Shlens, A tutorial on principal component analysis. Systems Neurobiology Laboratory, Salk Institute for Biological Studies, version 2, 1–13 (2005)
24. D Reynolds, R Rose, Robust text-independent speaker identification using Gaussian mixture speaker models. IEEE Trans. Speech Audio Process. **3**(1), 72–83 (1995)
25. W Campbell, D Sturim, D Reynolds, Support vector machines using GMM supervectors for speaker verification. IEEE Signal Process. Lett. **13**(5), 308–311 (2006)
26. DA Reynolds, TF Quatieri, RB Dunn, Speaker verification using adapted Gaussian mixture models. Digital Signal Process. **10**, 19–41 (2000)
27. B Raj, RM Stern, Missing-feature approaches in speech recognition. IEEE Signal Process. Mag. **22**(5), 101–116 (2005)
28. GJ Brown, D Wang, *Separation of Speech by Computational Auditory Scene Analysis*. (Springer Verlag, New York, 2005), pp. 371–402
29. ML Seltzer, B Raj, RM Stern, A Bayesian classifier for spectrographic mask estimation for missing feature speech recognition. Speech Commun. **43**(4), 379–393 (2004)
30. X Zhao, Y Wang, D Wang, Robust speaker identification in noisy and reverberant conditions. IEEE Trans. Audio, Speech Lang. Process. **22**, 836–845 (2014, in press)
31. Martin Rainer, Noise power spectral density estimation based on optimal smoothing and minimum statistics. IEEE Trans. Speech Audio Process. **9**, 504–512 (2001)
32. M Brookes, VOICEBOX: Speech Processing Toolbox for MATLAB (2009). [Online] Available: http://www.ee.ic.ac.uk/hp/staff/dmb/voicebox/voicebox.html
33. A Papoulis, *Probability, Random Variables, and Stochastic Processes*. (Academic Press, New York, 1991). Third Edition, McGraw Hill, Inc
34. A Varga, H Steeneken, M Tomlinson, D Jones, in *Tech. Rep., Speech Res. Unit, Defense Res. Agency*. The NOISEX-92 study on the effect of additive noise on automatic speech recognition (Malvern, U.K., 1992). (Available from NOISEX-92 CD-ROMS)
35. A Martin, G Doddington, T Kamm, M Ordowski, in *Proceedings of the European Conference on Speech communication and Technology*. The DET curve in assessment of detection task performance, (1997), pp. 1895–1898

14

An imperceptible and robust audio watermarking algorithm

Ali Al-Haj

Abstract

In this paper, we propose a semi-blind, imperceptible, and robust digital audio watermarking algorithm. The proposed algorithm is based on cascading two well-known transforms: the discrete wavelet transform and the singular value decomposition. The two transforms provide different, but complementary, levels of robustness against watermarking attacks. The uniqueness of the proposed algorithm is twofold: the distributed formation of the wavelet coefficient matrix and the selection of the off-diagonal positions of the singular value matrix for embedding watermark bits. Imperceptibility, robustness, and high data payload of the proposed algorithm are demonstrated using different musical clips.

Keywords: Audio watermarking; Copyright protection; Discrete wavelet transform; Singular value decomposition; Imperceptibility; Robustness; Data payload

1 Introduction

The recent advancements of digital audio technology have increased the ease with which audio files are stored, transmitted, and reproduced. However, along with such conveniences come new risks such as copyright violation. Conventional encryption algorithms permit only authorized users to access encrypted digital data; however, once decrypted, there is no way to prohibit illegal copying and distribution of the data [1]. A promising solution to the copyright violation problem is to apply audio watermarking in which audio files are marked with secret, robust, and imperceptible watermarks to achieve copyright protection [2-5]. Indeed, a digital watermark is a good deterrent to illicit copying and dissemination of copyrighted audio since it can provide evidence of copyright infringements after the copyright violation has occurred.

Audio watermarking techniques which are used for copyright protection of digital audio signals must satisfy two main requirements: imperceptibility and robustness [6]. Imperceptibility refers to the condition that the embedded watermark should not produce audible distortion to the sound quality of the original audio. That is, the watermarked version of the audio signal must be indistinguishable from the original audio signal. On the other hand, robustness ensures the resistance of the watermark against removal or degradation. The watermark should survive malicious attacks such as random cropping and noise adding. Some watermarking applications may demand additional requirements such as high data payload and low computational time of the watermarking algorithm [3]. In practice, there exists a fundamental trade-off between the different watermarking requirements.

Audio watermarking can be carried out in the time domain or the transform domain of the audio signal. Time-domain techniques based on least significant bit substitution and echo hiding are found extensively in literature [7-12]. In general, time-domain audio watermarking techniques are relatively easy to implement and require few computing resources. However, they are less robust than transform-domain techniques which employ the human perceptual properties and frequency masking characteristics of the human auditory system [13]. Popular transforms that have been widely used in digital watermarking include the discrete Fourier transform (DFT), the discrete cosine transform (DCT), the discrete wavelet transform (DWT), and the singular value decomposition (SVD) [14-20].

It has been reported recently that imperceptible and robust audio watermarking can be achieved by applying

Correspondence: ali@psut.edu.jo
Department of Computer Engineering, King Abdullah II Faculty of Engineering, Princess Sumaya University for Technology, Al-Jubeiha, PO Box 1438, Amman 11941, Jordan

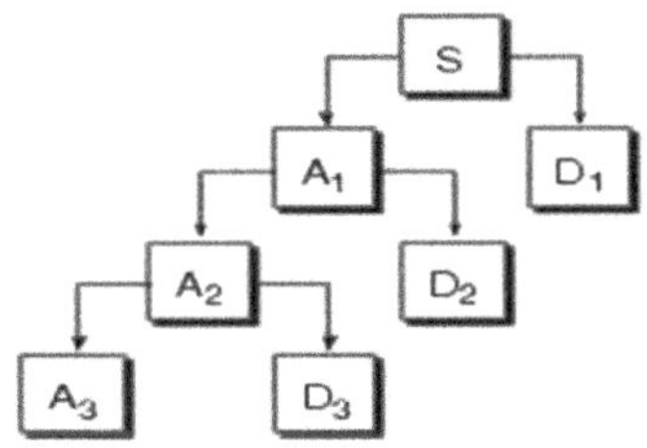

$$S = A_1 + D_1$$
$$= A_2 + D_2 + D_1$$
$$= A_3 + D_3 + D_2 + D_1$$

Figure 1 Three-level DWT decomposition of signal *S*.

a cascade of two different transforms on the original audio signal. Being different, the cascaded transforms may provide different, but complementary, levels of robustness against the same attack. Many audio watermarking techniques based on hybrid transforms have been proposed in literature. These techniques include but are not limited to DWT-DCT [21], DWT-SVD [22], and SVD-STFT [23].

Several hybrid algorithms based on the SVD transform have been recently proposed in literature. In the algorithm proposed by [23], the audio signal is first converted into a matrix form using the short-time Fourier transform (STFT), the SVD transform is then applied on the matrix, and finally embedding is carried out by adaptively modifying the SVD coefficients with watermark bits. In the hybrid algorithm proposed by [24], the audio signal is partitioned into blocks, and the watermark bits are embedded using dither modulation quantization of the singular values of the blocks. In [23], an audio watermarking algorithm is proposed in which watermark embedding and extraction procedures are based on the quantization of the norms of the singular values of audio blocks. The same authors proposed in [25] a hybrid algorithm in which watermark bits are embedded by applying quantization index modulation (QIM) on the singular values of wavelet-domain blocks. All of the abovementioned SVD-based hybrid algorithms employ some sort of quantization to embed watermark bits. Although quantization is simple, an acceptable level of robustness against noise and filtering attack may not always be achieved.

In this paper, we propose a semi-blind hybrid audio watermarking algorithm based on the DWT and SVD transforms. In the proposed algorithm, the audio signal is sampled, partitioned into short audio segments called frames, and a four-level DWT decomposition is applied on each frame. A matrix is then formed by arranging the wavelet coefficients of all detail sub-bands in a unique distributed pattern which scatters the watermark bits throughout the transformed frame to provide a high degree of robustness. The SVD operator is then applied on the matrix, and the watermark bits are embedded onto the off-diagonal zero elements of the ***S*** matrix produced by the SVD transform. Unlike the other SVD-based algorithms, the proposed algorithm leaves the non-zero singular values of the ***S*** matrix unchanged to ensure high watermarking imperceptibility.

The rest of the paper is organized as follows. In the next section, the DWT and SVD transforms are described, and their unique utilization in the proposed algorithm is outlined. The proposed audio DWT-SVD watermarking algorithm is described in detail in Section 3, and evaluated with respect to imperceptibly, robustness, and data payload in Section 4. Concluding remarks are given in Section 5.

2 Related work and contribution

The proposed algorithm is based on cascading the two transforms: DWT and SVD. The uniqueness of the proposed algorithm is twofold: the distributed formation of the DWT coefficient matrix and the selection of the off-diagonal positions of SVD's singular value matrix for embedding watermark bits. Description of the two transforms and their exact utilization in the proposed algorithm is given in this section.

2.1 DWT-based audio watermarking

DWT is a frequency transform capable of giving a time-frequency representation of any given signal [26]. Starting from an audio signal ***S***, DWT produces two sets of coefficients: the approximation coefficients A_1 produced by passing ***S*** through a low-pass filter and the detail coefficients D_1 produced by passing ***S*** through a high-pass filter. Depending on the application and the length of ***S***, A_1 can be further decomposed into more levels. Figure 1

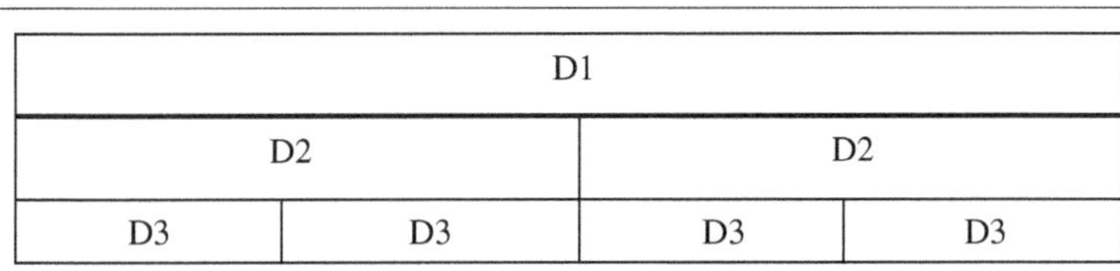

Figure 2 Matrix formation of the of detailed coefficient sub-bands.

$$\begin{bmatrix} U_{1,1} & . & . & U_{1,n} \\ U_{2,1} & . & . & U_{2,n} \\ . & . & . & . \\ U_{n,1} & . & . & U_{n,n} \end{bmatrix} \begin{bmatrix} \sigma_{11} & 0 & 0. & 0 \\ 0 & \sigma_{22} & 0 & 0 \\ . & . & . & . \\ 0 & 0 & 0 & \sigma_{nn} \end{bmatrix} \begin{bmatrix} V_{1,1} & . & . & V_{1,n} \\ V_{2,1} & . & . & V_{2,n} \\ . & . & . & . \\ V_{n,1} & . & . & V_{n,n} \end{bmatrix}^T$$

Figure 3 The SVD operation SVD $(A) = U\,\Sigma\,V^T$.

illustrates a three-level DWT decomposition of the audio signal $\boldsymbol{S}$.

Many DWT-based audio watermarking algorithms can be found in literature. Many variations among the different algorithms exit; however, the main variation is in the sub-band chosen for embedding the watermark bits. In [27-29], the approximation sub-band is used for embedding the watermark bits, while in most algorithms, only one detail sub-band is used to embed the watermark bits [30-36]. Claims of good imperceptibility and robustness have been reported using the two embedding approaches.

In this paper, watermark bits are not embedded in one sub-band only, rather the bits are distributed among all multi-resolution detail sub-bands. For a three-level DWT decomposition, this is done by forming a matrix of the detail sub-bands (D_1, D_2, and D_3) as shown in Figure 2. This matrix formation allows for better scattering of the watermark bits throughout the sub-bands, leading to a higher degree of robustness. The resultant DWT matrix is processed by the SVD transform to embed the watermark bits, as will be explained in the next subsection.

2.2 SVD-based audio watermarking

The SVD of matrix $\boldsymbol{A}$ is defined by the operation $\boldsymbol{A} = \boldsymbol{U}\,\boldsymbol{\Sigma}\,\boldsymbol{V}^T$, as shown in Figure 3. The non-zero diagonal entries of $\boldsymbol{\Sigma}$ are called the singular values of $\boldsymbol{A}$ and are assumed to be arranged in decreasing order $\sigma_i > \sigma_{i+1}$. The columns of the $\boldsymbol{U}$ matrix are called the left singular vectors, while the columns of the $\boldsymbol{V}$ matrix are called the right singular vectors of $\boldsymbol{A}$.

The SVD transform has been used in several audio watermarking algorithms [22-25,37-39]. The algorithms varied in the way the singular values were used in the watermarking process. For example, in [37], the single largest singular value, σ_{11}, was quantized and used to embed the watermark, whereas in [38], the encrypted watermark signal was added to all singular values of matrix $\boldsymbol{\Sigma}$. In [22,24,25], the norms of all singular values were quantized and used in the watermark embedding process.

In our proposed algorithm, matrix $\boldsymbol{A}$ represents the detail sub-bands matrix shown in Figure 2, which is produced after applying DWT on the original audio signal. After applying the SVD operator on the DWT matrix, watermark bits are embedded onto the off-diagonal zero elements of the $\boldsymbol{S}$ matrix, while the diagonal singular values of the matrix remain unchanged. This embedding procedure will eliminate the possibility of any distortion caused to the singular values which may affect imperceptibility and watermarking quality. Related preliminary works have been published by the author and others in [40,41]. The algorithms reported in those papers have low capacity as they embed the watermark bits in the single largest singular value, σ_{11}, and not in the off-diagonal zero elements of the $\boldsymbol{\Sigma}$ matrix, as it is the case in the proposed algorithm.

3 Proposed DWT-SVD audio watermarking algorithm

In this section, we describe the proposed DWT-SVD algorithm. The algorithm consists of two procedures: watermark embedding and watermark extraction procedures.

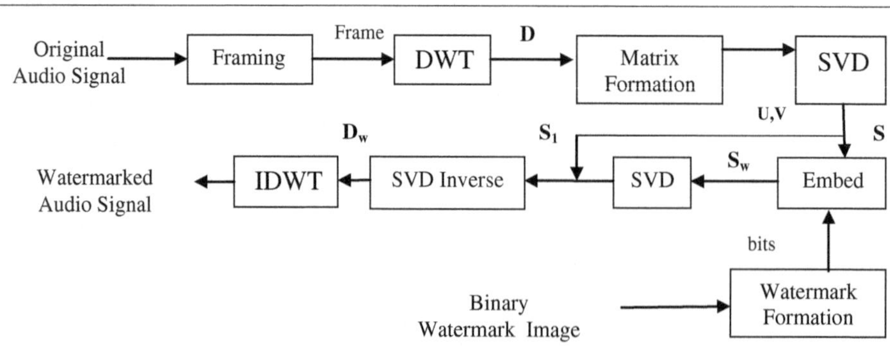

Figure 4 The watermark embedding procedure.

A4	D4	D3	D2	D1

Figure 5 A vector representing the five DWT multi-resolution sub-bands.

3.1 Watermark embedding procedure

The watermark embedding procedure transforms the audio signal using DWT and SVD, embeds the bits of a binary image watermark in appropriate locations in the transformed signal, and finally produces a watermarked audio signal by performing inverse SVD and DWT operations. The procedure is illustrated in the block diagram shown in Figure 4 and described thereafter.

Step 1: Convert the binary image watermark into a one-dimensional vector $\boldsymbol{b}$ of length $M \times N$. A watermark bit $\boldsymbol{b_i}$ may take one of two values: 0 or 1.

$$b_i = \{[0, 1], 1 \le i \le (M \times N)\} \tag{1}$$

Step 2: Sample the original audio signal at a sampling rate of 44,100 samples per second and partition the sampled file into N frames. The optimal frame length will be determined experimentally in such a way to increase data payload.

Step 3: Perform a four-level DWT transformation on each frame. This operation produces five multi-resolution sub-bands: $\boldsymbol{D_1}$, $\boldsymbol{D_2}$, $\boldsymbol{D_3}$, $\boldsymbol{D_4}$, and $\boldsymbol{A_4}$. The $\boldsymbol{D}$ sub-bands are called 'detail sub-bands' and the $\boldsymbol{A_4}$ sub-band is called 'approximation sub-band'. The five sub-bands are arranged in the vector shown in Figure 5.

Step 4: Arrange the four detail sub-bands $\boldsymbol{D_1}$, $\boldsymbol{D_2}$, $\boldsymbol{D_3}$, and $\boldsymbol{D_4}$ in a matrix $\boldsymbol{D}$ as shown in Figure 6. The matrix formation is done this way to distribute the watermark bits throughout the multi-resolution sub-bands $\boldsymbol{D_1}$, $\boldsymbol{D_2}$, $\boldsymbol{D_3}$, and $\boldsymbol{D_4}$. Forming the matrix with the $\boldsymbol{D}$s, rather than using $\boldsymbol{A}$ alone, is done to allow for matrix formation and subsequent application of the matrix-based SVD operator. The size of matrix $\boldsymbol{D}$ is $4 \times (L/2)$, where L refers to the length of the frame.

Step 5: Decompose matrix $\boldsymbol{D}$ using the SVD operator. This operation produces the three orthonormal matrices $\boldsymbol{\Sigma}$, $\boldsymbol{U}$, and $\boldsymbol{V^T}$ as follows:

$$D = U * \Sigma * V^T \tag{2}$$

where the diagonal matrix $\boldsymbol{\Sigma}$ has the same size of the $\boldsymbol{D}$ matrix. The diagonal $\boldsymbol{\sigma_{ii}}$ entries correspond to the singular values of the $\boldsymbol{D}$ matrix. However, for embedding purposes, only a 4×4 subset of matrix $\boldsymbol{\Sigma}$, assigned the name $\boldsymbol{S}$ hereafter, is used as shown below. This is a trade-off between imperceptibility (inaudibility) and payload (embedding capacity). That is, using the whole $\boldsymbol{\Sigma}$ matrix for embedding will increase embedding capacity but will lead to severe distortion in imperceptibility (inaudibility) of the watermarked audio signal.

$$S = \begin{bmatrix} S_{11} & 0 & 0 & 0 \\ 0 & S_{22} & 0 & 0 \\ 0 & 0 & S_{33} & 0 \\ 0 & 0 & 0 & S_{44} \end{bmatrix} \tag{3}$$

Step 6: Arrange 12 bits of the original watermark bit vector $\boldsymbol{b}$ into a scaled 4×4 watermark matrix $\boldsymbol{W}$. The watermark bits must be located in the non-diagonal positions within the matrix, as shown below.

$$W = \begin{bmatrix} 0 & \text{bit1} & \text{bit2} & \text{bit3} \\ \text{bit4} & 0 & \text{bit5} & \text{bit6} \\ \text{bit7} & \text{bit8} & 0 & \text{bit9} \\ \text{bit10} & \text{bit11} & \text{bit12} & 0 \end{bmatrix} \tag{4}$$

As an example, the watermark 12-bit watermark pattern 1010 0011 0101 must be converted to the following matrix form before the actual embedding is carried out.

$$W = \begin{bmatrix} 0 & 1 & 0 & 1 \\ 0 & 0 & 0 & 0 \\ 1 & 1 & 0 & 0 \\ 1 & 0 & 1 & 0 \end{bmatrix} \tag{5}$$

Step 7: Embed watermark matrix $\boldsymbol{W}$ bits into matrix $\boldsymbol{S}$ according to the following 'additive-embedding' formula:

<table>
<tr><td colspan="8">D1</td></tr>
<tr><td colspan="4">D2</td><td colspan="4">D2</td></tr>
<tr><td colspan="2">D3</td><td colspan="2">D3</td><td colspan="2">D3</td><td colspan="2">D3</td></tr>
<tr><td>D4</td><td>D4</td><td>D4</td><td>D4</td><td>D4</td><td>D4</td><td>D4</td><td>D4</td></tr>
</table>

Figure 6 Matrix formation of the detail coefficient sub-bands (*D* matrix).

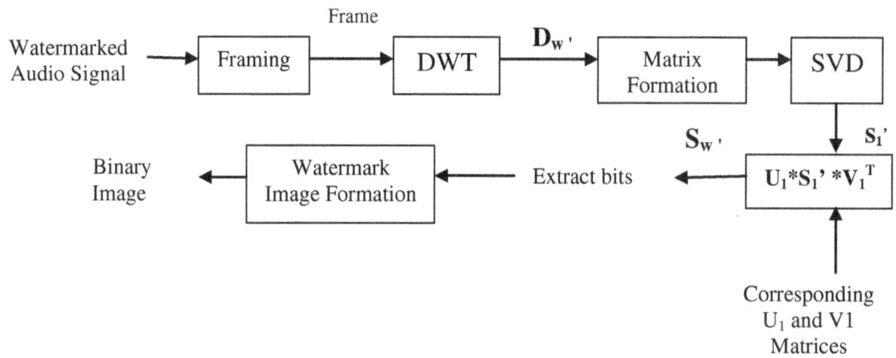

Figure 7 The watermark extraction procedure.

$$S_w = S + \alpha * W \tag{6}$$

where S_w is the watermarked S matrix, and α is the watermark intensity which should be chosen to tune the trade-off between robustness and imperceptibility. With this type of embedding, the singular values of D remain unchanged, and thus, audible distortion caused by modifying the singular values is avoided.

Step 8: Decompose the new watermarked matrix S_w using the SVD operator. This operation produces three new orthonormal matrices as follows:

$$S_w = U_1 * S_1 * V_1^T \tag{7}$$

The matrices U_1 and V_1^T are stored for later use in the extraction process. This makes the proposed watermarking algorithm semi-blind, as the whole original audio frame is not required in the extraction process.

Step 9: Apply the inverse SVD operation using the U and V^T matrices, which were unchanged, and the S_1 matrix, which has been modified according to Equation (6). The D_w matrix given below is the watermarked D matrix given in Equation (2).

Figure 8 The watermark image.

$$D_w = U * \Sigma' * V^T \tag{8}$$

where matrix Σ' is the original Σ matrix with the S sub-matrix replaced by the S_1 sub-matrix.

Step 10: Apply the inverse DWT operation on the D_w matrix to obtain the watermarked audio frame.

Step 11: Repeat all previous steps on each frame. The overall watermarked audio signal is obtained by concatenating the watermarked frames obtained in the previous steps.

3.2 Watermark extraction procedure

Given the watermarked audio signal and the corresponding U_1 and V_1 matrices that were computed in Equation (7) and stored for each frame, the embedded watermark can be extracted according to the procedure outlined in Figure 7 and described in detail in the following steps:

Step 1: Obtain the matrix S_1' from each frame of the watermarked audio signal following the general steps presented in Figure 7.

Step 2: Multiply matrix S_1' by U_1 and V_1 which were computed in the watermark embedding procedure and stored for use in the extraction process. This results in the following matrix.

Table 1 SNR values for different audio signals

Audio type	SNR_{dB}
Pop audio	38.75
Instrumental audio	39.02
Speech audio	37.50
Average	38.17

Table 2 Subjective and objective grades for audio quality measurement

Audio quality	Subjective difference grade (SDG)	Objective difference grade (ODG)
Imperceptible	5	0
Perceptible, but not annoying	4	−1.0
Slightly annoying	3	−2.0
Annoying	2	−3.0
Very annoying	1	−4.0

$$S_w{}' = U_1 * S_1{}' * V_1^T \quad (9)$$

Step 3: Extract the 12 watermark bits from each frame by examining the non-diagonal values of matrix $S_w{}'$. It has been experimentally noticed that there are two groups of non-diagonal values that are extremely distinct. The values at the positions where a 0 bit has been embedded tend to be much smaller than those values at the positions where a 1 bit has been embedded. Thus, to determine the watermark bit *W(n)*, the average of non-diagonal values is first computed, name it avg, then for each non-diagonal value $S_w{}'_{ij}$, *W(n)* is extracted according to the following formula:

$$W(n) = \left\{ \begin{matrix} 0 & S_{w_{ij}}' \leq \text{avg} \\ 1 & \text{otherwise} \end{matrix} \right\} \quad (10)$$

Step 4: Construct the original watermark image by assembling the bits extracted from all frames.

4 Experimental results

Different types of audio signals have different perceptual properties, and therefore, watermarking performance may vary from type to another. Accordingly, we evaluated the performance of the proposed algorithm using three mono audio signals representing pop music, instrumental music, and speech. Each signal has a duration of 11 s and was sampled at 44.1 kHz and quantized to 16 bits per sample. The watermark used for experimentation is the 12 × 10 binary image shown in Figure 8. The watermark is embedded repeatedly throughout the sampled signal, such that one single watermark image is embedded in a sequence of ten frames.

Table 3 SDG and ODG values for different audio signals

Audio type	SDG	ODG
Pop audio	4.67	−0.67
Instrumental audio	4.72	−0.71
Speech audio	4.81	−0.91
Average	4.73	−0.76

Table 4 Imperceptibility results for different algorithms

Reference	Algorithm	SNR (average)	SDG (average)	ODG (average)
Wang and Zhao [21]	DWT-DCT based	43.11	-	-
Xiang [27]	DWT based	23.98	-	−1.98
Fallahpour and Megias [30]	DWT based	30.65	-	−0.7
Bhat et al. [24]	SVD-DM based	-	4.64	−0.73
Bhat et al. [25]	SVD-DWT based	24.37	4.46	-
Proposed algorithm	SVD-DWT based	38.17	4.73	−0.76

Four-level DWT decomposition is applied on each frame using the Daubechies wavelet (***db1***). Using other wavelet types has a little effect on the performance, as it was observed experimentally. Values ranging from 1 to 5 were used for the watermark intensity $\boldsymbol{\alpha}$. However, the results reported in this paper were obtained when the intensity value was set to 3. In what follows, we present performance results of the proposed algorithm with respect

Table 5 BER values for common signal processing operations

Attack type	Pop audio	Instrumental audio	Speech audio	Average BER
Gaussian noise (15_{dB})	0	0	0	0
Gaussian noise (20_{dB})	0	0	0	0
Re-sampling 22.05	0.0021	0.000	0.0363	0.0128
Re-sampling 11.025	0.0061	0.0011	0.0448	0.0173
Re-sampling 6 kHz	0.0901	0.0330	0.0543	0.0591
Re-quantization 24	0	0	0	0
Re-quantization 8	0	0	0	0
MP3 compression 128 kbps	0	0	0	0
MP3 compression 96 kbps	0.0301	0.0541	0.0721	0.0430
MP3 compression 64 kbps	0.0521	0.0841	0.0820	0.0727
MP3 compression 32 kbps	0.0810	0.1410	0.2901	0.1707
Echo (delay 100 ms, decay 50%)	1.1264	1.5932	1.878	1.5325
Echo (delay 100 ms, decay 40%)	1.0536	1.5641	1.7330	1.4500
Low-pass filtering 8 kHz	0.0972	0.1540	0.3168	0.1893
High-pass filtering 50 Hz	0.2701	0.2810	0.5231	0.3580
Band-pass filtering (100 to 4,000 Hz)	0.1080	0.132	0.2130	0.1510

Table 6 Comparison between BER values of different transform-based algorithms

	Algorithm					
	DWT based [29]	DWT-DCT based [21]	DWT based [50]	SVD-QIM based [22]	SVD-DWT based [25]	SVD-DWT based Proposed algorithm
Gaussian noise (20_{dB})	7.525	0.0115	0	0	0	0
Re-quantization 8	0	0	0	0	0	0
Re-sampling 22.05	0	0	0	0	2	0.0128
MP3 compression 64 kbps	4.34	0	0.08	0.5615	0	0.0727
MP3 compression 32 kbps	17.22	0.03525	0.67	2.2094	1	0.1707
Echo (delay 100 ms, decay 40%)	-	-	5.83	3.955 (98, 41)	2 (98, 41)	1.450
Low-pass filtering 8 kHz	-	-	0.97	0.3540 (11,025 Hz)	0 (11,025 Hz)	0.1893

to three metrics: imperceptibility, robustness, and data payload [42,43].

4.1 Imperceptibility results

Imperceptibility ensures that the quality of the signal is not perceivably distorted and the watermark is imperceptible to listeners. To measure imperceptibility, different authors use different metrics; however, the most commonly used metrics are signal-to-noise ratio (SNR) and listening tests.

4.1.1 Signal-to-noise ratio

SNR is a statistical difference metric which is used to measure the similitude between the undistorted original audio signal and the distorted watermarked audio signal. The SNR computation is done according to Equation (11), where $\boldsymbol{A}$ corresponds to the original signal, and $\boldsymbol{A}'$ corresponds to the watermarked signal.

$$\mathrm{SNR}(dB) = 10\log_{10}\frac{\Sigma_n A_n^2}{\Sigma_n\left(A_n - A_n'\right)^2} \qquad (11)$$

We obtained the $\mathrm{SNR}_{\mathrm{dB}}$ values given in Table 1. As shown in the table, the values are much higher than the 20_{dB} minimum requirement set by the International Federation of Phonographic Industry [13]. Although SNR is a simple metric to measure the noise introduced by the embedded watermark and can give a general idea of imperceptibility, it does not take into account the specific characteristics of the human auditory system.

4.1.2 Listening tests

For better evaluation of imperceptibility, subjective and objective listening tests are used. Subjective difference grade (SDG) listening tests are implemented by human listeners, and objective difference grade (ODG) listening tests are implemented by software packages incorporating the human auditory system. The two listening tests use the 5-grade scale shown in Table 2.

We employed a blind subjective listening test to estimate the audio quality of the watermarked signals. The listening test was performed repeatedly with five adults in a listening room equipped with audio testing and recording devices. A computer system running a special software was also used for computer-controlled presentation of the watermarked signals to the listeners and for recording their responses. Each person was presented with ten pairs of signals (original and watermarked) and then asked to give performance scores using the 5-grade impairment scale given in Table 1. The five persons listened to each pair of signals ten times and gave an average SDG value for each pair. The average grade for each pair submitted by all persons is considered the final grade for that particular pair of signals. The SDG averages obtained for the subjective listening tests are 4.67, 4.72, and 4.81 for the pop, instrumental, and speech signals, respectively. These values clearly indicate that imperceptibility has been achieved by the proposed audio watermarking algorithm.

The ODG scores were also computed using the Perceptual Evaluation of Audio Quality (PEAQ) standard. The standard is specified in ITU-R BS.1387 [44] and implemented by the software tool EAQUAL [45]. The ODG values we obtained are −0.67, −0.71, and −0.91 for the pop, instrumental, and speech signals, respectively. These results confirm with those obtained by subjective listening tests. The measured SDG and ODG values are given in Table 3.

Comparing imperceptibility results with results achieved by other algorithms is not straightforward, since different authors use different evaluation metrics. Moreover, subjective evaluation is relative and may differ from one listener to another. This may explain why imperceptibly results are hardly compared in literature. Nonetheless, and for the sake of completion, we present in Table 4 some imperceptibility results achieved by recently proposed algorithms. It is important to note that the values in table are average values taken over different audio types.

Table 7 BER values due to Stirmark® attacks

Stirmark attack	Extracted watermark (pop)	Pop audio	Instrumental audio	Speech audio	Average BER
AddBrumm (55 Hz Sinus)		0	0	0	0
AddSinus (3000 Hz sinus)		0	0	0	0
AddNoise (20 dB level)		0	0	0	0
Stat1 (statistical distortion)		0	0	0	0
Stat2 (statistical distortion)		0	0	0	0
Smooth1 (simple smoothing)		0.80	1.40	0.36	0.853
Smooth2 (simple smoothing)		0.65	1.34	0.29	0.760
Amplify (increases amplitude)		0	0	0	0
Invert (phase shift 180°)		0	0	0	0
Exchange (swaps samples)		5.01	5.54	3.68	4.74

Table 7 BER values due to Stirmark® attacks *(Continued)*

CutSamples (7 samples per 1,000)		2.41	3.11	1.08	2.20
LSBZero (reset LSBs)		0	0	0	0
ZeroCross (reset samples)		0	0	0	0
ZeroRemove (removes 0 samples)		0	0	0	0

4.2 Robustness results

Watermarked audio signals may undergo signal processing operations such as linear filtering, lossy compression, among many other operations [46,47]. Although these operations may not affect the perceived quality of the host signal, they may corrupt the watermark embedded within the signal. Two sets of attacks were performed to test the robustness of our proposed algorithm. The first set includes the following set of common signal processing operations: Gaussian noise addition, re-quantization, re-sampling, MP3 compression, low-pass filtering, and echo addition. The other set is the Stirmark® audio watermarking benchmark which includes a whole set of add, modify, and filter attacks [48,49].

Robustness is measured using the bit error rate (BER) metric since the watermark used in the simulation is a binary image. BER is defined as the ratio of incorrect extracted bits to the total amount of embedded bits, as expressed in Equation (12).

Table 8 Comparison between BER values due to Stirmark® attacks

	Algorithm					
	DCT based [51]	SVD-STFT based [23]	DWT based [27]	DWT based [30]	SVD-QIM based [22]	SVD-DWT based Proposed algorithm
AddBrumm (55 Hz Sinus)	1.25	0	15.79	0	-	0
AddSinus (3,000 Hz sinus)	0.77	0	0	0	-	0
AddNoise (20 dB level)	0.78	0	5.875	0	0	0
Stat1 (statistical distortion)	0	0	0	9	0.1831	0
Stat2 (statistical distortion)	0	0	-	-	0.7324	0
Smooth1 (simple smoothing)	0	0	0	14	2.0874	0.853
Smooth2 (simple smoothing)	0	0	0	-	1.0986	0.760
Amplify (increases amplitude by 50%)	0	0.375	0	0	-	0
Invert (phase shift 180°)	52.42	0	0	0	0	0
Exchange (swaps samples)	0	0	0	-	0	4.74
CutSamples (7 samples per 1,000)	100	1.5	0	-	-	2.20
LSBZero (sets LSBs to 0)	0	0	-	0	-	0
ZeroCross (reset samples)	0	0	0	-	0	0
ZeroRemove (removes 0 samples)	100	0	0	-	-	0

Table 9 Effect of frame length on data payload

	Frame length (samples)							
	512	**1,024**	**2,048**	**4,096**	**8,192**	**16,384**	**32,768**	**65,536**
Data payload (bps)	1,032	516	258	129	64	32	16	8

$$\text{BER} = \frac{100}{l}\sum_{n=0}^{i-1}\left\{\begin{matrix} 1, & W'_n = Wn \\ 0, & W'_n \neq Wn \end{matrix}\right\} \quad (12)$$

where l is the watermark length, $\boldsymbol{W_n}$ is the nth bit of the embedded watermark, and $\boldsymbol{W'_n}$ is the nth bit of the extracted watermark.

4.2.1 Common signal processing operations

The following common signal processing attacks were applied to test the robustness of the proposed algorithm:

1. *Additive white Gaussian noise*: White Gaussian noise is added to corrupt the watermarked signal to SNR levels of 15_{dB} and 20_{dB}.
2. *Re-quantization*: The 16-bit watermarked audio signal is re-quantized to 8 bits per sample and 24 bits per sample.
3. *Re-sampling*: The watermarked signal, originally sampled at 44.1 kHz, is down-sampled to 22.05, 11.025, and 6 kHz.
4. *MP3 compression*: The watermarked audio signal is compressed at different bit rates: 128, 96, 64, and 32 kbps.
5. *Low, high, and band-pass filtering*: Filtering at different cutoff frequencies is applied to the watermarked signal.
6. *Echo addition*: An echo signal with a delay of 100 ms and different decay rates are added the watermarked signal.

The BER values we obtained after applying the common signal processing operations are listed in Table 5. As shown in the table, the BER values, which have been computed over the whole period of the test signals, are very small in magnitude and thus reflect the robustness of the proposed algorithm against common signal operations. Maximum robustness has been achieved against the Gaussian noise attacks, re-quantization, and MP3 compression at 128 kbps. BER values due to re-sampling increased as the watermarked signal was down-sampled to lower frequencies. The same observation is also seen for the MP3 compression attack, where higher BER values were obtained as the compression rate of the watermarked signal was increased. The watermarked signal is also robust against filtering operations as shown in the corresponding small BER values. The least robustness is seen against the echo addition operation as indicated by the relatively higher BER values.

Finally, we compared the robustness of the proposed algorithm with the robustness of recently published transform-based algorithms. Its clear from Table 6 that the proposed algorithm performs better when compared with the other algorithms. It is important to note that the values in Table 6 represent average values taken over different audio types.

4.2.2 Stirmark© attacks

To evaluate robustness of the proposed algorithm furthermore, we implemented a set of attacks defined by *Stirmark® benchmark for audio* [48,49]. The attacks are comprehensive as they include add, filter, and modification attacks. The results are recorded in Table 7 alongside with snapshots of extracted watermarks from the watermarked signals. It is noted in Table 7 that BER values due to most of the attacks are zero. It is also noted that the proposed algorithm performs comparably well with regard to the three audio signal types.

The *Stirmark®* attacks have been used by several transform-based algorithms. Table 8 compares the BER results we obtained and the BER results reported in four relevant references. As shown in the table, the results are comparable among the different transform-based references with regard to most of the *Stirmark®* attacks. It is instructive to note here that *Stirmark®* package can be used to simulate composite attacks, where two or more attacks are tested in one run. Such composite attacks may give better comparison between the different algorithms; however, they are rarely reported in literature.

4.3 Data payload results

Data payload is defined as the data embedding capacity of the algorithm and is measured as the number of bits

Table 10 Data payload results for different algorithms

	Algorithm					
	DWT based [29]	**SVD-STFT based [23]**	**DWT based [50]**	**DWT based [27]**	**SVD-DWT based [25]**	**SVD-DWT based Proposed**
Payload (bits per second)	172	32	25	28.71	45.9	258

embedded within one second of the audio signal (bps). In the proposed algorithm, the audio signal is segmented into frames, with each frame having a fixed embedding capacity of 12 watermark bits, as shown in matrix $\boldsymbol{W}$ given in (5). Therefore, the payload is computed by multiplying number of frames per second by the bit capacity of the frame. The number of frames per second depends on the frame length and is computed by dividing the 44.1 KHz sampling rate by the frame length. Table 9 shows the data payload as a function of the frame length.

As shown in the table, the payload increases as the frame length decreases. However, short-length frames degrade performance and result in unacceptable imperceptibility and robustness results. A frame length of 2,048 samples has been fixed and used to evaluate imperceptibly and robustness of the proposed algorithm.

The data payload we obtained is higher than payload rates obtained by other recently proposed algorithms. Table 10 lists the payload of different transform-based audio watermarking algorithms.

5 Conclusions

In this paper, we proposed an imperceptible and a robust audio watermarking technique based on cascading two well-known transforms: the discrete wavelet transform and the singular value decomposition. The two transforms were used in a unique way that scatters the watermark bits throughout the transformed frame in order to achieve high degrees of imperceptibility and robustness. High data payloads were also achieved. The simulation results obtained were in total agreement with the requirements set by IFPI for audio watermarking, thus proving the effectiveness of the proposed algorithm.

Future research will focus on enhancing the proposed algorithm to resist de-synchronization attacks such as random cropping, pitch shifting, amplitude variation, time-scale modification, and jittering. Methods proposed in the literature that counter de-synchronization attacks include the all-list-search method, the combination of spread spectrum and spread spectrum code method, the self-synchronization strategy method, and the synchronization code method. Our approach will be based on embedding synchronization codes with the watermark bits so that the hidden data have the self-synchronization capability.

Competing interests

The author has no competing interests.

References

1. B Furht, D Kirovski, *Encryption and Authentications: Techniques and Applications* (Auerbach, USA, 2006)
2. M Arnold, S Wolthusen, M Schmucker, Techniques and applications of digital watermarking and content protection. Artech House, in *Psychoacoustics: Facts and Models*, ed. by E Zwicker, H Fastl (Springer-Verlag, Massachusetts, USA, 2003)
3. A Acevedo, *Digital Watermarking for Audio Data in Techniques and Applications of Digital Watermarking and Content Protection* (Artech House, USA, 2003), pp. 75–114
4. C Xu, J Wu, Q Sun, K Xin, Applications of watermarking technology in audio signals. J. Audio Eng. Soc. **47**(10), 805–812 (1999)
5. M Swanson, B Zhu, A Tewfic, L Boney, Current state of the art, challenges and future direction for audio watermarking, in *Proceeding of the IEEE International Conference on Multimedia Computing and Systems* (1999), pp. 19–24
6. M Arnold, Audio watermarking: Features, applications and algorithms, in *Proceeding of the IEEE International Conference on Multimedia and Expo* (2000), pp. 1013–1016
7. P Bassia, I Pitas, Robust audio watermarking in the time domain. IEEE Trans. Multimed. **3**(2), 232–241 (2001)
8. WN Lie, LC Chang, Robust and high-quality time-domain audio watermarking based on low-frequency amplitude modification. IEEE Trans. Multimed. **8**(1), 46–59 (2006)
9. S Dumitrescu, W Wu, Z Wang, Detection of LSB steganography via sample pair analysis. IEEE Trans. Signal Process. **51**(7), 1995–2007 (2003)
10. O Chen, W Wu, Highly robust, secure, and perceptual-quality echo hiding scheme. IEEE Trans. Speech Audio Process. **16**(3), 629–638 (2008)
11. BS Ko, R Nishimura, Y Suzuki, Time-spread echo method for digital audio watermarking. IEEE Trans. Multimed. **7**(2), 212–221 (2005)
12. H Kim, Y Choi, A novel echo-hiding scheme with backward and forward kernels. IEEE Trans. Circ. Syst. Video Tech. **13**(8), 885–889 (2003)
13. S Katzenbeisser, F Petitcloas, *Information Hiding Techniques for Steganography and Digital Watermarking* (Artech House, USA, 2000)
14. M Fallahpour, D Perez-Megias, High capacity audio watermarking using FFT amplitude interpolation. IEICE Electron. Express **6**(14), 1057–1063 (2009)
15. M Fan, H Wang, Chaos-based discrete fractional sine transform domain audio watermarking scheme. Comput. Electr. Eng. **35**(3), 506–516 (2009)
16. I Yeo, H Kim, Modified patchwork algorithm: a novel audio watermarking scheme. IEEE Trans. Speech Audio Process. **11**(4), 381–386 (2003)
17. M Hsieh, D Tseng, Y Huang, Hiding digital watermarks using multiresolution wavelet transform. IEEE Trans. Ind. Electron. **48**(5), 875–882 (2001)
18. C Chang, W Shen, H Wang, Using counter-propagation neural network for robust digital audio watermarking in DWT domain. Proc. IEEE Int. Conf. Syst. Man. Cybern. **2**, 1214–1219 (2006)
19. R Liu, T Tan, An SVD-based watermarking scheme for protecting rightful ownership. IEEE Trans. Multimed. **4**(1), 121–128 (2002)
20. A Mohammad, A Al-Haj, S Shaltaf, An improved SVD-based watermarking scheme for protecting rightful ownership. Signal Process. J. **88**(9), 2158–2180 (2008)
21. X Wang, H Zhao, A novel synchronization invariant audio watermarking scheme based on DWT and DCT. IEEE Trans. Signal Process. **54**(12), 4835–4840 (2006)
22. KV Bhat, I Sengupta, A Das, A new audio watermarking scheme based on singular value decomposition and quantization. Circ. Syst. Signal Process. **30**, 915–927 (2011)
23. H Ozer, B Sankur, N Memon, An SVD-based audio watermarking technique, in *ACM Workshop on Multimedia and Security* (2005), pp. 51–56
24. KV Bhat, I Sengupta, A Das, An audio watermarking scheme using singular value decomposition and dither-modulation quantization. Multimed. Tool. Appl. **52**, 369–383 (2011)
25. K Bhat, I Sengupta, A Das, An adaptive audio watermarking based on the singular value decomposition in the wavelet domain. Digit. Signal Process. **20**, 1547–1558 (2010)
26. G Strang, T Nguyen, *Wavelets and Filter Banks* (Wellesley-Cambridge Press, Wellesley, MA, 1996)
27. S Xiang, Audio watermarking robust against D/A and A/D conversions. EURASIP J Adv. Signal Process. **3**, 1–14 (2011)
28. H Peng, J Wang, Z Zhang, Audio watermarking scheme robust against desynchronization attacks based on kernel clustering. Multimed. Tool. Appl. **3**, 1–14 (2011)
29. S Wu, J Huang, D Huang, Y Shi, Efficiently self-synchronized audio watermarking for assured audio data transmission. IEEE Trans. Broadcast. **51**(1), 69–76 (2005)
30. M Fallahpour, D Megias, High capacity audio watermarking using the high frequency band of the wavelet domain. Multimed. Tool. Appl. **52**, 485–498 (2011)

31. M Swanson, B Zhu, A Tewfic, L Boney, Robust audio watermarking using perceptual masking. Signal Process. **66**(3), 337–355 (1998)
32. X Li, M Zhang, L Sun, Adaptive audio watermarking algorithm based on SNR in wavelet domain, in *International Conference on Natural Language Processing and Knowledge Engineering* (2003), pp. 287–292
33. Y Wu, S Shimamoto, A study on DWT-based digital audio watermarking for mobile ad hoc networks, in *International Conference on Sensor Networks, Ubiquitous, and Trustworthy Computing*, 2006. 5–7 June
34. E Erelebi, L Bataki, Audio watermarking scheme based on embedding strategy in low frequency components with a binary image. Digit. Signal Process. **19**(2), 265–277 (2009)
35. L Wei, X Xue, An audio watermarking technique that is robust against random cropping. Comput. Music. J. **27**(4), 58–68 (2003)
36. X Li, H Yu, Transparent and robust audio data hiding in sub-band domain, in *Proceedings of the International Conference on Information Technology: Coding and Computing* (2000), pp. 74–79
37. C Chang, P Tsai, C Lin, SVD-based digital image watermarking scheme. Pattern Recogn. Lett. **26**(10), 1577–1586 (2005)
38. F Abd El-Samie, An efficient singular value decomposition algorithm for digital audio watermarking. Int. J. Speech Tech. **21**, 27–45 (2009)
39. A Basso, F Bergadano, D Cavagnino, V Pomponiu, A Vernone, A novel block-based watermarking scheme using the SVD transform. Algorithms **2**(1), 46–75 (2009)
40. A Al-Haj, A Mohammad, Digital audio watermarking based on the discrete wavelets transform and singular value decomposition. Eur. J. Sci. Res. **39**(1), 6–21 (2010)
41. A Al-Haj, C Twal, A Mohammad, Hybrid DWT-SVD audio watermarking, in *Proceedings of the International Conference on Digital Information Management* (2010), pp. 525–529
42. M Sehirli, F Gurgen, S Ikizoglu, Performance evaluation of digital audio watermarking techniques designed in time, frequency and cepstrum domains, in *International Conference on Advances in Information Systems* (2004), pp. 430–440
43. J Grody, L Brutun, Performance evaluation of digital audio watermarking algorithms, in *The 43rd IEEE Midwest Symposium on Circuits and Systems* (2000), pp. 456–459
44. T Thielde, WC Treurniet, R Bitto, C Schmidmer, T Sporer, JG Beerends, C Colomes, M Keyhl, G Stoll, K Brandenburg, B Feiten, PEAQ – the ITU standard for objective measurement of perceived audio quality. J. Audio Eng. **48**(1/2/3), 3–29 (2000)
45. A Lerch, Zplane development, EAQUAL- Evaluate Audio QUALity, version:0.1.3alpha, 2002. http://www.mp3-tech.org/programmer/misc.html
46. S Voloshynovskiy, S Pereira, T Pun, Attacks on digital watermarks: classification, estimation-based attacks, and benchmarks. Comm. Mag. **39**(8), 118–126 (2001)
47. M Arnold, Attacks on digital audio watermarks and countermeasures, in *Proceedings of the IEEE International Conference on WEB Delivering of Music* (2003), pp. 1–8
48. M Steinebach, F Petitcolas, F Raynal, J Dittmann, C Fontaine, S Seibel, N Fates, LC Ferri, Stirmark benchmark: audio watermarking attacks, in *Proceedings of the International Conference on Information Technology: Coding and Computing* (2001), pp. 49–54
49. A Lang, Stirmark benchmark for audio (SMBA): evaluation of watermarking schemes for audio [Online], 2006. Available: http://omen.cs.uni-magdeburg.de/alang/smba.php#smba_LA
50. N Kalantari, M Akhaee, S Ahadi, S Feizi, H Amindavar, Robust multiplicative patchwork method for audio watermarking. IEEE Trans. Audio Speech Lang. Process. **17**(6), 1133–1141 (2009)
51. I Cox, J Kilian, T Leighton, T Shamoon, Secure spread spectrum watermarking for multimedia. IEEE Trans. Image Process. **6**(12), 1673–1687 (1997)

15

Audio segmentation-by-classification approach based on factor analysis in broadcast news domain

Diego Castán*, Alfonso Ortega, Antonio Miguel and Eduardo Lleida

Abstract

This paper studies a novel audio segmentation-by-classification approach based on factor analysis. The proposed technique compensates the within-class variability by using class-dependent factor loading matrices and obtains the scores by computing the log-likelihood ratio for the class model to a non-class model over fixed-length windows. Afterwards, these scores are smoothed to yield longer contiguous segments of the same class by means of different back-end systems. Unlike previous solutions, our proposal does not make use of specific acoustic features and does not need a hierarchical structure. The proposed method is applied to segment and classify audios coming from TV shows into five different acoustic classes: speech, music, speech with music, speech with noise, and others. The technique is compared to a hierarchical system with specific acoustic features achieving a significant error reduction.

Keywords: Audio segmentation; Factor analysis; Within-class variability compensation; Broadcast news; Albayzin 2010 evaluation

Introduction

Due to the increase in audiovisual content, it becomes necessary to use automatic tools for different tasks such as analysis, indexation, search, and information retrieval. Given an audio document, the first step is audio segmentation to obtain the delineation of a continuous audio stream into acoustically homogeneous regions. Secondly, each homogeneous region can be classified into predefined classes to provide labels that can be used as context for searching metadata or as identifiers for speaker adaptation techniques in speech recognition systems.

Segmentation of broadcast news (BN) recordings into audio events (like speech, music, speech with music) is very challenging because such documents contain different kinds of sequences with a very heterogeneous style. Several international evaluation campaigns, like the TREC NIST evaluations for Spoken Document Retrieval (SDR) [1], the ESTER evaluations campaign for Rich Transcription (RT) in French [2], and the COST278 evaluation for segmentation and speaker clustering in multi-lingual domain [3], have already been proposed to face this task in the past. Some examples of the audio sequences in BN (not pretending to be general definitions) in different conditions are as follows:

- *News anchor speech*: Traditional news anchor usually reading text in clean (or low-noise) conditions.
- *Interviews*: Conversations between two people with spontaneous speech or by following a script.
- *Debates*: Conversations between two or more people. They may contain overlapped speech in some parts.
- *Reporter in the field*: The audio comes from a wide range of noises generally overlapped with speech.
- *Advertising*: Speech with music in background and a variety of acoustic noise effects (slams, explosions, cars, screams, etc.).
- *Jingles*: Jingles are commonly used as a short tune to introduce different topics during the news.
- *Broadcasting of sports events*: Speech with a strong background noise and diegetic music and sounds.
- *Telephone connections*: Used when the reporters do not have camera and microphone.

The studies that can be found in the literature focus on either the feature extraction method or the segmentation/classification strategies. A good review of the features

*Correspondence: dcastan@unizar.es
Departamento Ingeniería Electrónica y Comunicaciones, Universidad de Zaragoza, María de Luna, 1, Zaragoza 50018, Spain

and the classification methods used in several solutions can be found in [4].

The proper selection of a set of acoustic features may help to describe the behavior of the acoustic classes (speech, music, environmental sounds, etc.) both in the time and frequency domains. Mel-frequency cepstrum coefficients (MFCC) or perceptual linear prediction (PLP) have been widely used throughout history in the context of audio and speech technologies [5-9] and more recently in [10]. More precisely, these features with other extended sets of features have been proposed for segmenting and classifying BN audio into broad classes. Among others, two pitch-density-based features are proposed in [11], short-time energy (STE) is used in [12-14], and harmonic features are used in [15-17]. The previously mentioned features are short-term characteristics because they are extracted within short periods of time (between 10 and 30 ms), usually known in the literature as frame-based features. These features are commonly used in speech-related tasks where the signal can be considered stationary over that short period. The frame-based features can be used directly in the classifier. However, some classes are better described by the statistics computed over consecutive frames (from 0.5 to 5 s long). These characteristics are referred in the literature as segment-based features [18,19]. For example, in [20], a content-based speech discrimination algorithm is designed to exploit the long-term information inherent in the modulation spectrum. In [21], the authors propose two segment-based features: the variance of the spectrum flux (VSF) and the variance of the zero crossing rate (VZCR).

Audio segmentation/classification systems can be divided into two different groups depending on how the segmentation is performed. The first group detects the boundaries in a first step and then classifies each delimited segment in a second step. We refer to them as segmentation-and-classification approaches. For example, in [22], an approach using a temporally weighted fuzzy C-means algorithm was proposed. The Bayesian information criterion (BIC) is widely employed in many studies as [23] to generate a break-point for every speaker change and environment/channel condition change in the BN domain. Also, [24] and [25] utilize BIC to identify mixed-language speech and speaker change, respectively. However, BIC has several shortcomings to be considered. It can only set one break-point for each analysis window, so a small window involves more precision but the Gaussian estimation may be inaccurate due to the scarcity of data. In [26], the authors propose a minimum description length (MDL) approach that allows multiple break-points for any generic data. The second group is known as segmentation-by-classification and consists of classifying consecutive fixed-length audio segments. The segmentation is produced directly by the classifier as a sequence of decisions. This sequence is usually smoothed to improve the segmentation performance. An example of this procedure can be found in [27] where the author combines different features with a Gaussian mixture model (GMM) and a maximum entropy classifier. The final decisions were smoothed with a hidden Markov model (HMM) to avoid sudden changes. In [28], an audio stream is segmented by classifying each window into five broad classes. The solution is a combination of different support vector machines (SVM) and evaluates the classification over some new proposed features. Three different smoothing rules were applied to avoid sudden changes in the decisions. Aronowitz suggests a framework in which the classification and the smoothing are unified [29]. The author models the audio segments as supervectors, and each class (speech, silence, music) is modeled by a distribution over the supervector space. The supervectors are classified with SVM or GMM.

Segment-based features are not suitable for training statistical models [21], and it is difficult to determine *a priori* the appropriate statistics for each class. However, they provide great discriminative power for audio classification if the segment is well-delimited using any segmentation-and-classification system [14]. On the other hand, frame-based features allow statistical models to make decisions over short-duration windows in segmentation-by-classification strategies [30], but they are usually less discriminative for audio classification since they were mainly designed for speech-related tasks such as automatic speech recognition (ASR) [21]. The most common solution to avoid the shortcomings and enjoy the benefits of each strategy is to create hierarchical systems with multiple steps where each level is designed with class-specific features and segmentation systems as in [31] and [32]. Nevertheless, these systems become very specific for the intended task and are quite difficult to adapt for other databases.

Recently, an audio segmentation task in the BN domain was proposed in [33] in the context of the Albayzin 2010 evaluation campaigns. The proposed evaluation task consisted of segmenting a broadcast news audio document into five acoustic classes: speech (SP), speech with noise (SN), speech with music (SM), music (MU), and others (OT). The main difficulty in this database is the classification between the classes with speech because these classes have some segments very homogeneous (specially, between SP and SN). In this context, we introduce a novel and generic segmentation-by-classification system based on factor analysis (FA) with two clear advantages: (1) the system does not need class-dependent features with hierarchical structure to classify different classes and (2) the algorithm compensates the within-class variability with high accuracy being able to classify well-defined classes in generic tasks. The FA technique has been successfully applied in speaker ID (recognition/verification) [34-38] and language recognition [39] with significant improvements with respect to GMM or SVM. However, the system proposed in this article has several differences from those systems. In contrast to a segmentation task, the speaker ID or language recognition has well-delimited segments (usually in separate files) and, therefore, FA is applied over the whole file. Unlike in the speaker ID, speaker diarization, or language recognition tasks, we can find here the same speaker in two different acoustic classes, for example, the situation where an anchor is in the studio with clean conditions (SP) and outside of the studio with noise in the background (SN). Due to all these factors, we propose an extension of a FA segmentation system proposed in [40] and [41] with a new and more discriminative scoring using class/non-class parameters and with a set of back-end systems that perform a better segmentation than the traditional FA systems for language recognition or speaker ID.

The remainder of the paper is organized as follows: the database and metric of the Albayzin 2010 evaluation is presented in the 'Albayzin audio segmentation evaluations and database description' section. The 'Novel factor analysis audio segmentation system' section shows the theoretical approach based on FA and a set of back-end subsystems. The experiments are presented in the 'Experimental results' section. Finally, the summary and the conclusions are presented in the 'Conclusions' section.

Albayzin audio segmentation evaluations and database description

The Albayzin campaigns are internationally open evaluations organized by the RTTH[a] every 2 years. A complete description of the Albayzin 2010 audio segmentation and classification evaluation can be found in [19] where the participant's approaches and the results are presented. We describe the database and the metric used in the evaluation in the next subsections.

Database

The database consists of BN audio in Catalan recorded by the TALP[b] Research Center. It includes approximately 87 h of annotated audio divided into 24 files. Five audio classes were defined for the evaluation. The classes are distributed as follows: clean speech, 37%; music, 5%; speech over music, 15%; speech over noise, 40%; others, 3%. The class 'others' is not evaluated in the final test. The database for the evaluation was split into two parts: for training (two thirds of the total amount of data divided into 16 files) and testing (the remaining one third divided into 8 files).

Each segment is labeled with one class previously described. Most of the segments are between 10 and 20 s long. However, there is an important amount of long segments (longer than 60 s). More details about the database and the labeling process can be found in [19].

Metric

The metric that was proposed for the evaluation represents the relative error averaged over all acoustic classes (ACs):

$$\text{Error} = \text{average}_i \frac{\text{dur}(\text{miss}_i) + \text{dur}(\text{fa}_i)}{\text{dur}(\text{ref}_i)}, \quad (1)$$

where $\text{dur}(\text{miss}_i)$ is the total duration of all deletion errors (misses) for the ith acoustic classes (AC), $\text{dur}(\text{fa}_i)$ is the total duration of all insertion errors (false alarms) for the ith AC, and $\text{dur}(\text{ref}_i)$ is the total duration of all the ith AC instances according to the reference file. The incorrectly classified audio segment (a substitution) is computed both as a deletion error for one AC and an insertion error for another. A collar of 1 s is not scored around each reference boundary to avoid the uncertainty about when an AC begins or ends.

Since the distribution of the classes in the database is not uniform, the errors from different classes are weighed differently (depending on the total duration of the class in the database). Therefore, the system has to detect correctly not only the best represented classes ('speech' and 'speech over noise,' 77% of total duration) but also the minor classes (like 'music,' 5%). This metric is different from the conventional NIST metric [42] for speaker diarization, where the score is defined as the ratio of the overall segmentation error time to the sum of the durations of the segments that are assigned to each class in the file. In this work, we will present the final results with both metrics.

Novel factor analysis audio segmentation system

We propose a framework for automatic audio segmentation-by-classification. The system deals with the problem of assigning a class label to each fixed-length window using factor analysis (FA) models. The FA approach has been successfully used in speaker recognition/verification [34-37], speaker diarization [38], and language recognition [39]. In these tasks, the systems have to face several sources of variability such as speaker, channel, and environment. The variability of the same class segments is known as within-class variability. The goal of these systems is to model (in the case of [38]) or compensate the within-class variability to reduce the mismatch between training and test. As we presented in the first section, there are some differences between those systems and the segmentation-by-classification system proposed in this work. The main difference is that in this task, the classes can have the same speaker. However, in speaker ID, speaker diarization, or language recognition, the speakers define an independent class. As a result, the within-class variability is more difficult to compensate in our task. Therefore, we introduce a novel approach with class/non-class parameters that compensate the within-class variability more accurately. Figure 1 illustrates the proposed framework where each block is described in the next subsections.

Acoustic feature extraction

In this work, we extract 16 MFCCs (including the zeroth-order cepstrum) computed in 25-ms frames with a 10-ms frame step and their first and second derivatives. The audio features are packed in windows of 3 s long with 0.1- or 0.5-s window steps depending on the desired computational load and resolution.

Statistics computation

The fixed-length windows are mapped to sufficient statistics by using a universal background model (UBM) [43] which is a class-independent GMM with C Gaussians estimated with the expectation-maximization (EM) algorithm [44] on the training data set. The UBM parameters are the mean vectors, $\boldsymbol{\mu}_k$, and the diagonal covariances matrices, $\boldsymbol{\Sigma}_k$, where k is the Gaussian component index. Let $P_{ksi} = P(k|\boldsymbol{\phi}_{si})$ represent the posterior probability of the kth UBM component, given the feature vector $\boldsymbol{\phi}_{si}$ and assuming frame independence [45]. For a window s, with feature vectors indexed $i = 1, 2, \ldots, N_s$, we define the zeroth- and first-order statistics, respectively, as

$$n_{sk} = \sum_{i=1}^{N_s} P_{ksi}, \tag{2}$$

$$\boldsymbol{f}_{sk} = \sum_{i=1}^{N_s} P_{ksi} \boldsymbol{\Sigma}_k^{-1/2} \left(\boldsymbol{\phi}_{si} - \boldsymbol{\mu}_k \right), \tag{3}$$

where these statistics are normalized to the UBM.

Theoretical background

Data from a particular class are modeled by a GMM defined by a set of mean vectors $\boldsymbol{m}_1, \boldsymbol{m}_2, \ldots, \boldsymbol{m}_C$, weights $w_1, w_2, \ldots, w_C$, and covariance matrices $\boldsymbol{\Sigma}_1, \boldsymbol{\Sigma}_2, \ldots, \boldsymbol{\Sigma}_C$, where C is the number of Gaussians. We can concatenate all GMM mean vectors to one mean supervector $\boldsymbol{m}$ of dimension $CF \times 1$ where F is the feature vector length:

$$\boldsymbol{m} = \left[\boldsymbol{m}_1^T, \boldsymbol{m}_2^T, \ldots, \boldsymbol{m}_C^T \right]^T. \tag{4}$$

The factor analysis model is the adaptation of the UBM model where the supervector of means is not fixed and it can vary from segment to segment due to several sources that increase the within-class variability [36]. We assume that these GMMs have segment- and class-dependent means but fixed weights and covariances chosen to be equal to the UBM weights and covariances. Specifically, we use a factor analysis model for the mean vector of the kth component of the GMM for segment s:

$$\boldsymbol{m}_k^s = \boldsymbol{t}_k^{c(s)} + \boldsymbol{U}_k \boldsymbol{x}_s, \tag{5}$$

where $c(s)$ denotes the class of segment s. The class location vector $\boldsymbol{t}_k^{c(s)}$ is obtained by using a single iteration of relevance MAP adaptation from the UBM [43]. This adaptation is expressed, in terms of statistics, as

$$\boldsymbol{t}_k^{c(s)} = \frac{\sum_s \boldsymbol{f}_{sk}}{r + \sum_s n_{sk}}, \tag{6}$$

where r is the relevance factor. $\boldsymbol{U}_k$ is the factor loading matrix and $\boldsymbol{x}_s$ is a vector of L segment-dependent-within-class-variability factors assumed to follow a normal distribution ($N(0, I_L)$).

We stack the component-dependent vectors into supervectors $\boldsymbol{m}_s$ and $\boldsymbol{t}^{c(s)}$ and the component-dependent $\boldsymbol{U}_k$

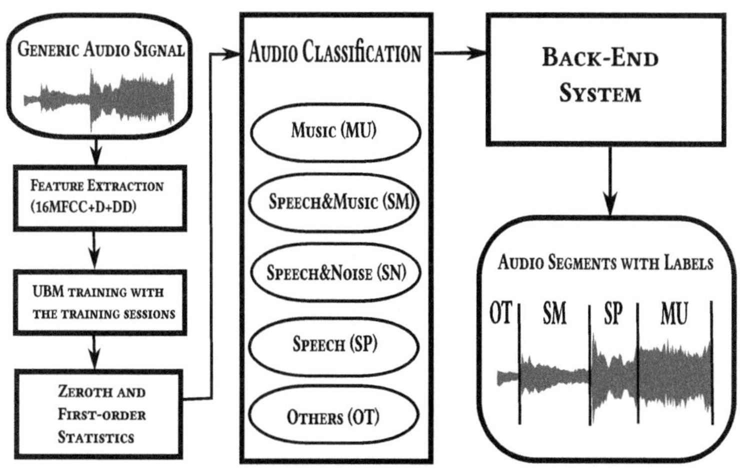

Figure 1 Block diagram of factor analysis segmentation-by-classification system for broadcast news classes.

matrices into a single tall matrix $\boldsymbol{U}$, so Equation 5 can be expressed as

$$\boldsymbol{m}^s = \boldsymbol{t}^{c(s)} + \boldsymbol{U}\boldsymbol{x}_s, \tag{7}$$

where $\boldsymbol{U}$ is known as the within-class variability matrix that we use to compensate that variability. The columns of the $\boldsymbol{U}$ matrix are the basis spanning the subspace of the within-class variability, and the within-class variability factors are the coordinates defining the position of the supervector in the subspace. The within-class variability factor dimension (L) is smaller than CF, so $\boldsymbol{U}$ has low rank ($CF \times L$ dimensions). Depending on the application, the value of L is between 50 and 200 and CF can be 98,304 if we have 2,048 Gaussians and 48-dim feature vector (with the MFCC-UBM).

Estimation of the within-class variability matrices

$\boldsymbol{U}$ can be estimated using the EM algorithm, where the $\boldsymbol{x}$ factors of each window are treated as hidden variables. In the E step, the expected value of $\boldsymbol{x}$ (denoted by $\hat{\boldsymbol{x}}$) are estimated for each window, using the current parameters as

$$\hat{\boldsymbol{x}}_s = \left(\boldsymbol{I} + \sum_k n_{sk}\boldsymbol{U}_k^T\boldsymbol{U}_k\right)^{-1} \boldsymbol{U}^T\boldsymbol{f}_s. \tag{8}$$

In the M step, we obtain $\boldsymbol{U}$ that maximize an auxiliary function involving the old and new parameters as

$$\boldsymbol{U}_k = \left[\sum_c\sum_s \left(\boldsymbol{f}_{sk} - \boldsymbol{t}_k^{c(s)} n_{sk}\right)\hat{\boldsymbol{x}}_s^T\right]\boldsymbol{A}_k^{-1}, \tag{9}$$

where

$$\boldsymbol{A}_k = \sum_s \left[\hat{\boldsymbol{x}}_s\hat{\boldsymbol{x}}_s^T\right]^T n_{sk}. \tag{10}$$

This paper does not aim to deepen into the training process of $\boldsymbol{U}$; more details and an exhaustive description can be found in [36].

Class model vs alternative model U matrices

The approach proposed in this paper has several differences with language recognition in the way within-class variability is compensated. Most of the approaches based on FA for language recognition are implemented with a single $\boldsymbol{U}$ matrix because the segments are well-delimited (typically in separated files) and the nature of the within-class variability is similar for all the languages as it can be seen in [36,46-48]. In [49], a segmentation system was proposed with five class location vectors (one vector per class) and a single compensation matrix $\boldsymbol{U}$ for all the classes. The paper compared the FA system with the winner of the Albayzin 2010 evaluation, and the conclusion was that the FA system is better as a classification system with oracle segments. On the other hand, the compensation matrix had a bad behavior in a segmentation-by-classification system for the music class due to the different nature of the rest of the classes. In [32], a hierarchical system was proposed with different features and different techniques in each level depending on the class. First, the system decides among MU, SM, or the rest of the classes by using HMM/GMM and a smoothed combination of MFCC and Chroma as feature vectors. Next, the system classifies SP and SN by using FA and MFCC as acoustic features to improve the performance of the speech classes because the confusion between these two classes is very high. The error rate achieved was lower than the one obtained by the best system presented in the Albayzin 2010 evaluation showing a clear advantage when the classes are homogeneous (like SN and SP), since $\boldsymbol{U}$ models the variability across speakers and phonemes. The background noise is, then, the discriminative information for the classification and segmentation. Nevertheless, hierarchical systems can be very specific for an intended task and are difficult to adapt to other databases with new classes.

Therefore, we propose here a non-hierarchical segmentation-by-classification system with ten class-specific vectors (one class vector and one non-class vector for each class) and five matrices modeling the within-class variability of each pair class/non-class. Let

$$\boldsymbol{T} = \left[\boldsymbol{t}^{\mathrm{MU}}, \boldsymbol{t}^{\overline{\mathrm{MU}}}, \boldsymbol{t}^{\mathrm{OT}}, \boldsymbol{t}^{\overline{\mathrm{OT}}}, \boldsymbol{t}^{\mathrm{SM}}, \boldsymbol{t}^{\overline{\mathrm{SM}}}, \boldsymbol{t}^{\mathrm{SN}}, \boldsymbol{t}^{\overline{\mathrm{SN}}}, \boldsymbol{t}^{\mathrm{SP}}, \boldsymbol{t}^{\overline{\mathrm{SP}}}\right] \tag{11}$$

and

$$\Xi = \left[\boldsymbol{U}^{\mathrm{MU}-\overline{\mathrm{MU}}}, \boldsymbol{U}^{\mathrm{OT}-\overline{\mathrm{OT}}}, \boldsymbol{U}^{\mathrm{SM}-\overline{\mathrm{SM}}}, \boldsymbol{U}^{\mathrm{SN}-\overline{\mathrm{SN}}}, \boldsymbol{U}^{\mathrm{SP}-\overline{\mathrm{SP}}}\right], \tag{12}$$

where $\boldsymbol{T}$ represents the locations of classes ($\boldsymbol{t}^C$) and non-classes ($\boldsymbol{t}^{\overline{C}}$) in the GMM space and Ξ the within-class variability matrices. This approach will be compared to the classic formulation with a single $\boldsymbol{U}$ matrix in 'Experimental results' section for the classification over the oracle segments and the final segmentation system.

Scoring

We study here the two scoring approaches most commonly used: the integration over the $\boldsymbol{x}$ factors distributions and the linear scoring, both of which are summarized in [50].

Score 1: The score based on the integration over the $\boldsymbol{x}$ factors distributions is a marginalization using a point estimate of the class $\boldsymbol{m}^s$, integrated only over the $\boldsymbol{x}$ factors, when the statistics are centered around the point estimate $\boldsymbol{m}^s$ as defined in [36].

Score 2: The linear scoring, which is faster than the previous one, is an approximation that makes use of the first-order Taylor expansion [50].

In [50,51] and [36], the score employed to detect the speaker is the log-likelihood ratio test (LLRT)

$$\mathrm{LLRT}_{\mathrm{class}} = \log \frac{P(\chi/\mathrm{class})}{P(\chi/\mathrm{UBM})}, \tag{13}$$

where the numerator is the likelihood for the class model and the denominator the likelihood for the UBM. Note that the UBM is used as a general model to describe the alternative hypothesis which is appropriated for speaker identification where the hypothesized speaker is not in the UBM. However, our problem has a small number of classes, and therefore, each class is highly represented by the UBM and may corrupt the test statistics.

We propose here a compensated log-likelihood ratio test (CLLRT) scoring:

$$\mathrm{CLLRT}_{\mathrm{class}} = \log \frac{P(\chi/\mathrm{class})}{P(\chi/\overline{\mathrm{class}})}, \tag{14}$$

where the alternative hypothesis is the likelihood for the non-class model which is compensated also with the within class variability matrix. The CLLRT is expected to be more discriminative than the LLRT for a segmentation task because the hypothesized class is not present in the denominator and, also, because the non-class model is compensated in the same way as the class model.

Back-end systems

We propose here three different back-end systems to combine, smooth, and improve the classification performance of the FA:

1. *Maximum a posteriori (MAP)*: This well-known method has been widely used in the literature [40,41]. To increase the detection performance, we optimize the prior probabilities in a Viterbi algorithm over the training files. Later, these priors are employed in the Viterbi over the test files.
2. *Derivative HMM/GBE*: There is an apparent correlation among the likelihood ratios of different classes. For example, if a segment is a jingle, the likelihood ratio for the MU class (music) should be the biggest, but it is very likely that the second one is the SM (speech with music). Also, SN (speech and noise) and SM (speech with music) are more correlated between them than to the SP class (speech) because both classes have background audio. The classification and, therefore, the segmentation can be improved by combining the outputs of each class-dependent subsystem [52]. Figure 2 shows the combination and smoothing back-end system proposed here. On a first step, a calibration of scores is made by a multi-class logistic regression [53] estimated using the training partition of the database. In order to benefit from the use of the dynamic behavior of the scores, we compute the first- and second-order time derivatives of the scores. To smooth the decisions after the calibration and the dynamic description, one Gaussian/HMM back-end is used for each class. A left-to-right topology was selected with a full-covariance Gaussian per state estimated with the scores from the training files. The mean vectors and the covariance matrices are estimated with the samples of the scores based on the class labels with the ML criterion. The number of states for the HMM depends on the desired level of smoothing. The Viterbi algorithm was chosen to determine the maximum likelihood transitions between the classes.
3. *Stacking HMM/GBE*: This back-end system can be considered as a modification of the previous back-end system. The main idea is to provide contextual information through longer term temporal scoring. Instead of the derivation of the scores, this back-end system proposes a stacking of past and future scores with the present score to model the dynamic behavior in a different way. Figure 3 shows this combination process where several score frames from the past and several score frames from the future are stacked with the present frame. The experiments are carried out with one, two, and three frames from the past and future and different numbers of states.

In an HMM segmentation system, it is usual to optimize the transition penalties on a development set since this can have a significant impact on performance. However, we do not optimize any transition penalty because our goal was to create a general approach to segment audio that could be used in other databases with different distributions or with other classes.

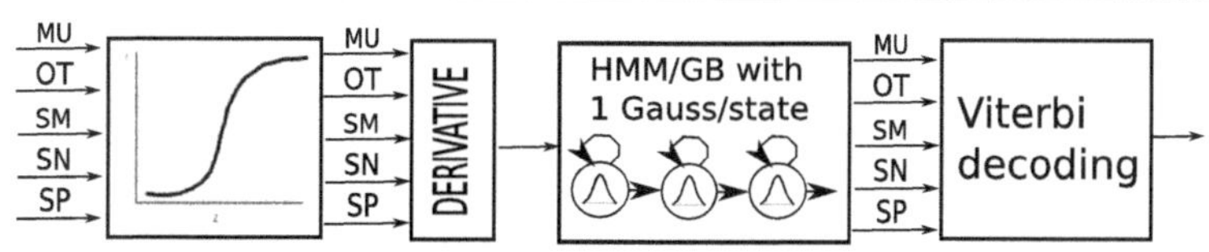

Figure 2 Back-end system 2 - derivative HMM/GBE block diagram.

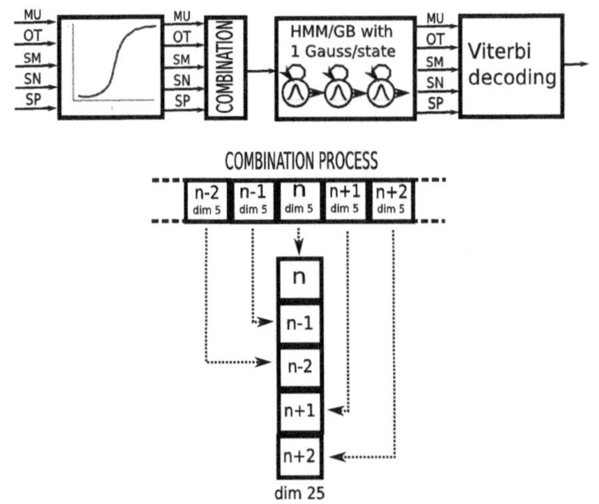

Figure 3 Back-end system 3 - stacking HMM/GBE block diagram.

Table 1 Baseline for classification experiments

GMM	MU	SP	SM	SN	Total
64G	10.68	45.74	36.68	45.44	34.63
128G	9.81	41.79	32.02	40.75	31.09
256G	10.4	37.6	31.8	37.6	29.3
512G	9.5	35.9	29.3	35.9	27.7
1,024G	9.3	34.9	27.0	34.3	26.4
2,048G	9.6	33.3	28.0	34.0	*26.2*

Classification error per class and total error for GMM systems with different numbers of Gaussians over the test files with perfect segmentation. The italicized number represents the best performance of the GMM system.

Experimental results

The errors can be produced in two ways: first, a classification error due to a bad labeled frame, and a segmentation error due to a temporal mismatch between the reference boundaries and the hypothesis boundaries. This section shows the experiments for the evaluation data described in the 'Albayzin audio segmentation evaluations and database description' section divided into two sets. In the first set, the boundaries between segments are given by the ground truth and the system decides the class of each segment with no segmentation error to evaluate the classification accuracy of the classical FA system versus GMMs.

The second set of experiments shows the segmentation and the classification error when the boundaries are not given. A final segmentation-by-classification system based on FA with a class/non-class parameters is proposed. The three back-end systems previously described are tested over this system. The back-end systems show that a combination and smoothing of the scores improve the previous results. Also, the systems are compared to the winner system of the 2010 Albayzin evaluation that has a hierarchical structure with specific features for each class.

Classification experiments with oracle segmentation

The classification is made over the segments extracted with the ground truth boundaries to evaluate the classification accuracy over the whole segment. Since the system decides the class that the whole segment belongs to, the smoothing is not needed.

We propose GMM systems as a baseline for classification experiments using the acoustic features described in the 'Acoustic feature extraction' section. Table 1 shows the results for these systems. We have evaluated different numbers of Gaussians (from 64 to 2,048). The classification is based on the highest accumulated likelihood over the whole segment. As it can be seen in Table 1, increasing the number of Gaussians improves the final result. The highest number of Gaussians evaluated was 2,048 because the error for MU and SM classes began to increase although the total result improved slightly compared to the 1024G model.

In the experiments with FA for classification with oracle segmentation, we assess different configurations for the number of $\boldsymbol{x}$ factors and the scoring methods described previously. The UBM employed to compute the statistics has a fixed number of 2,048 Gaussians to be able to compare the results of the FA systems with the best GMM baseline configuration. Because the boundaries are known, the statistics are calculated over the whole segment without merging underlying partitions. We compute the result using linear scoring and the integration trough the $\boldsymbol{x}$ factors distributions scoring (called as IoChD in this section). The linear scoring needs a final calibration because the scoring is scaled by the module of the target model. A Gaussian back-end (GBE) ([54,55]) provides benefits in two ways: calibration and score combination. The calibration for the IoChD scoring does not provide substantial benefits because the score is based on a likelihood ratio over a MAP adaptation using the same UBM and the marginal improvement comes from the combination of scores. The experiments are carried out with a single $\boldsymbol{U}$ matrix to compensate all the within-class variability and different numbers of $\boldsymbol{x}$ factors (50, 100, 150, 200, 250, and 300) providing the error for each class. Note that the increment of $\boldsymbol{x}$ factors involves an exponential increment of the computational cost.

Table 2 compares all the experiments with FA over a perfect segmentation. According to these results, the IoChD scoring is more accurate than the linear scoring for all the configurations and all the classes. Comparing Tables 1 and 2, a significant improvement can be seen using FA versus GMM. Using the best GMM configuration (2,048 Gaussians) as reference, the worst FA system improves

Table 2 FA systems for classification experiments

	One U for all classes									
	Linear GBE					IoChnf				
Number of chnf	MU	SP	SM	SN	Total	MU	SP	SM	SN	Total
50	21.6	16.9	23.6	23.4	21.4	10.2	15.9	24.2	21.4	17.9
100	21.8	17.4	21.0	22.9	20.8	9.1	16.0	20.2	20.0	16.3
150	20.8	17.7	20.5	23.5	20.6	9.4	15.5	18.0	18.9	15.4
200	20.7	17.8	20.5	22.4	*20.4*	9.0	15.7	17.3	19.1	15.3
250	20.0	19.2	20.2	23.1	20.6	8.5	16.7	16.0	19.4	*15.1*
300	21.3	19.5	20.5	21.7	20.8	9.8	15.0	18.9	18.9	15.6

Classification error per class and total error for linear and IoChD scoring with perfect segmentation and a single U for all the classes. The italicized numbers represent the best performance of the FA system for a given configuration.

the total result (18.3% relative error reduction with linear scoring and 50 $\boldsymbol{x}$ factors) and also compared to the best FA configuration (43.1% relative error reduction with IoChD scoring and 150 $\boldsymbol{x}$ factors). Note that the music has been better classified with GMMs than with linear GBE. However, the rest of the classes presents a high classification error with GMMs (as we knew from the results presented in the Albayzin evaluation).

An important fact about the distribution of the errors is shown in Table 3. The table shows the percentages of the segments that have been correctly classified for GMM with 2,048 Gaussians, FA with linear GBE scoring and FA with IoChD scoring both with 100 channel factors. The table is divided into two columns: the first column shows the percentage of the correctly classified segments between 0 and 3 s long. It clearly shows that, while the classification is better with FA systems as we have shown in Table 2, segments shorter than 3 s are better classified with GMMs. The second column shows the percentage of the segments longer than 3 s. It can be seen that the best classification system is based on FA with IoChD scoring. As a conclusion, the FA is a better classifier if the segments are longer than 3 s which is a common fact because most of the segments are between 10 and 20 s long and a collar of 1 s is not scored around each reference boundary.

Segmentation-by-classification experiments

In this subsection, no oracle segment boundaries are considered, so the audio stream is segmented by classifying each window into one of the five classes.

Table 3 Percentage of correctly classified segments shorter than 3 s and longer than 3 s

	Segment <3 s	Segment ≥3 s
GMM - 2,048G	*25.4*	56.9
Linear GBE - 100 chnf	19.2	57.3
IoChnf - 100 chnf	23.4	*60.8*

The total number of segments is 7,754. The italicized numbers represent the best performance for segments shorter than 3 s and longer than 3 s.

Table 4 shows the baseline results for this segmentation task. To be able to compare the results with the best baseline classification system of Table 1, the baseline segmentation systems in Table 4 are based on GMM with 2,048 Gaussians. The first row in this table shows the results of a basic GMM - 2,048G system. The segments in this system are delimited by the transition of the frame-by-frame classification and no smoothing is applied. Note the degradation of the GMM - 2,048G (54.6% of total error) compared to the GMM - 2,048G with perfect segmentation in Table 1 (26.2% of total error) where the decision of each class was based on the accumulated likelihood of the whole given segment. These results clearly show that a smoothing stage to avoid sudden changes in the decision sequence is needed in a segmentation task. A widely used technique to smooth the transitions between classes is the left-to-right HMM topologies. Table 4 shows different left-to-right HMM configurations where the 2,048G are divided by the number of states to maintain the same number of Gaussians in every configuration. The best baseline system for the segmentation task (33.3% of total error) has 32 states with 64G per state (keeping a total of 2,048G). This result proves the dramatic improvement when a temporal smoothing is applied to segmentation-by-classification system.

Classification experiments in the last subsection indicate that the IoChD scoring is more accurate than the

Table 4 Baseline for segmentation experiments

GMM/HMM LeftToRight	MU	SP	SM	SN	Total
GMM - 2,048G	35.5	59.2	65.0	58.6	54.6
2 ST - 1,024G	29.9	59.2	54.7	56.8	50.2
4 ST - 512G	26.0	49.8	45.9	50.2	43.0
8 ST - 256G	24.3	49.3	41.6	50.1	41.3
16 ST - 128G	17.8	40.2	36.0	43.0	34.2
32 ST - 64G	17.3	39.5	33.9	41.5	*33.3*

The table shows an error per class and total error for GMM-HMM systems over the test files with non-oracle segment boundaries. The italicized number represents the best performance of the GMM/HMM system.

Table 5 Error per class and total error for FA segmentation-by-classification systems

	IoChD scoring: step - 0.5 s, 100 chnf				
	MU	SP	SM	SN	Total
One single U	40.3	76.9	60.5	64.3	*60.5*
One U per class	33.3	45.6	36.2	47.4	*40.6*

The experiments are computed with one single U for all the classes and one U matrix for each class/non-class using IoChD scoring. No score combination or smoothing was carried out. The italicized numbers show the total error of each system.

linear scoring as we asserted in the 'Scoring' section. For the sake of clarity, results with the linear scoring are not presented in this subsection.

Unlike the oracle segmentation where the $\boldsymbol{x}$ factors were computed for each segment, in this subsection the $\boldsymbol{x}$ factors are computed for each window, so an increment in the number of $\boldsymbol{x}$ factors or a reduction of the window step increases the memory and the time needed to train the models dramatically. As a preliminary experiment, the FA segmentation-by-classification system computes the statistics every 0.5 s and 100 $\boldsymbol{x}$ factors. Because the windows (3 s long in our experiments) are smaller than the oracle segments, the useful information which describes the class of the window is scarcer. Therefore, a more discriminative scoring is needed and it is provided by the models with one $\boldsymbol{U}$ matrix for each class. Results with a single $\boldsymbol{U}$ matrix for all the classes and one $\boldsymbol{U}$ matrix for each class are presented in Table 5. There is a significant improvement in the classes with more data using one $\boldsymbol{U}$ matrix for each class because the CLLRT removes the information of the target class in the denominator as we pointed out in the 'Scoring' section. Figure 4 displays the confusion matrices for the experiments of Table 5. The percentages have been computed with the frames scored (affected by the collar) divided by all the frames of each class in the reference (dur(ref_i)). The table clearly shows less confusion between classes using one $\boldsymbol{U}$ matrix for each class. Specially, there is a significant reduction in the confusion between SP and SN and a slight reduction in the confusion between MU and SM. The more frequent the class in the data, the more significant the error reduction compared to a single $\boldsymbol{U}$ matrix for all the classes. Accordingly, the total error is reduced around 20%.

Once the benefits of the FA system with one $\boldsymbol{U}$ matrix for each class with IoChD scoring are determined, the window-step can be reduced to increase the resolution (0.1-s window step) at the expense of increasing the computational cost. The number of $\boldsymbol{x}$ factors is not increased because the computation time and the memory grow exponentially. Figure 5 shows the scores for each class over a chunk of a test file. The ground truth is plotted in the same figure, and it is represented with a square wave of amplitude 1. The green bars represent the forgiveness collar around each boundary. The color of each score class and the corresponding ground truth is the same. The figure clearly shows that the ratio of the winner class is bigger than zero and corresponds to the ground truth class for most of the frames. The results in Table 5 can be compared to the results in Table 6 showing a significant error reduction achieved by decreasing the window step because of the resolution increase.

To avoid sudden changes in the segmentation process, three back-end subsystems are evaluated here. The first back-end system is based on a MAP approach, and the two following systems are very similar but they model the temporal behavior in different ways: on a first step, the scores of both systems are conditioned using a multi-class logistic regression. The dynamic behavior of the scores is extracted with the first- and second-order time derivatives or by stacking the past and future frames of the scores (we know these systems as derivative HMM/GBE and stacking HMM/GBE, respectively). Finally, a left-to-right HMM/GBE with full covariance matrices is used to smooth the scores and improve the results with a scoring

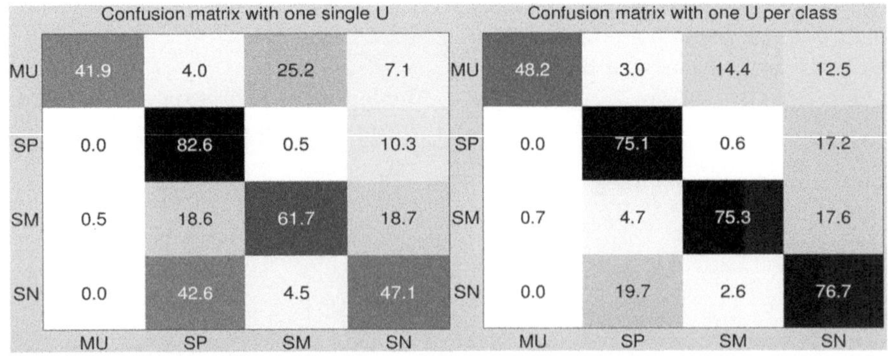

Figure 4 Confusion matrices for the experiments of Table 5. One single U for all the classes and one U matrix for each class/non-class are displayed. No score combination or smoothing was carried out.

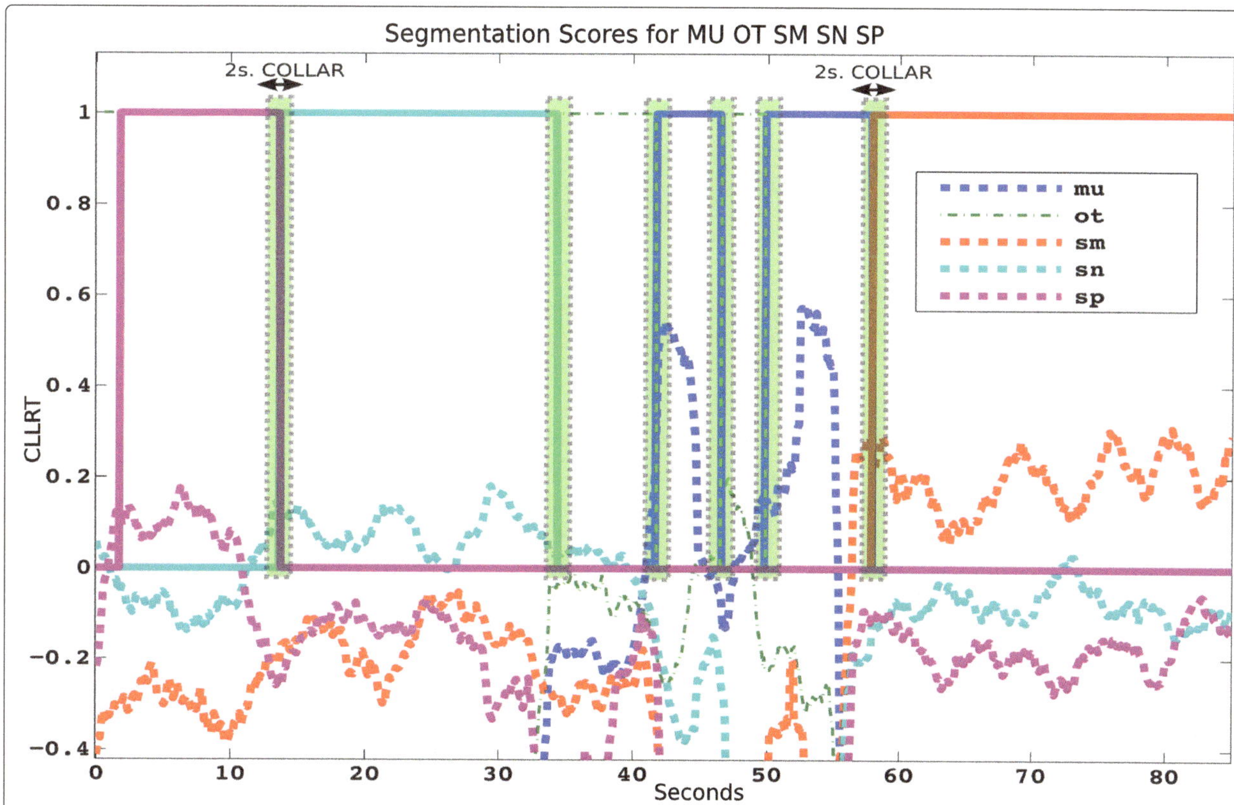

Figure 5 Scores and ground truth of each class over a chunk of a test file.

combination for both systems. The number of states in the HMM determines the minimum length of the segment.

We compare the error of the system proposed in this work with the winner system of the Albayzin 2010 evaluation [31] where 15 MFCCs, frame energy, and their corresponding first and second derivatives are extracted. In addition, the spectral entropy and the Chroma coefficients are calculated. The mean and variance of these features are computed over 1-s intervals creating 122 dimension feature vectors. The segmentation approach chosen is HMM-based. The acoustic modeling is performed using five HMMs with three emitting states and 256 Gaussians per state. Each HMM corresponds to one acoustic class. A hierarchical organization of binary HMM detectors is used. First, audio is segmented into music/non-music portions. Second, the non-music portions are further segmented into speech-over-music/non-speech-over-music portions. Finally, the non-speech-over-music portions are segmented into speech/speech over noise.

Table 6 Error per class and total error for FA segmentation-by-classification systems

	loChD scoring: step - 0.1 s, 100 chnf				
	MU	**SP**	**SM**	**SN**	**Total**
One U per class	27.9	37.9	32.4	40.9	*34.8*

The experiments are computed with one U matrix for each class/non-class using loChD scoring. No score combination or smoothing was carried out. The italicized number shows the total error of the system.

Figure 6 shows the results of the systems described in the previous two paragraphs. First, two straight lines represent the results for the winner system of the Albayzin 2010 evaluation [31] (30.2% of total error rate) and the FA system with MAP back-end [40] (28.8% of total error rate). The behaviors of the derivative HMM/GBE and the stacking HMM/GBE back-end systems are plotted in the same figure with a different number of states. The stacking HMM/GBE combines the present frame with one, two, and three frames from the past and the future to provide different levels of contextual information. The figure shows a slight improvement in derivative HMM/GBE for almost any number of states. However, the result are quite similar for derivative HMM/GBE and stacking HMM/GBE. Both systems for every configuration improve the results of the winner hierarchical-HMM [31] and the MAP-FA system [40]. Note that the final number of states is not critical because the difference among errors is less than 3%. The best result obtained was an error of 23.8% using 13 states, and the worst result was 26.6% with 25 states.

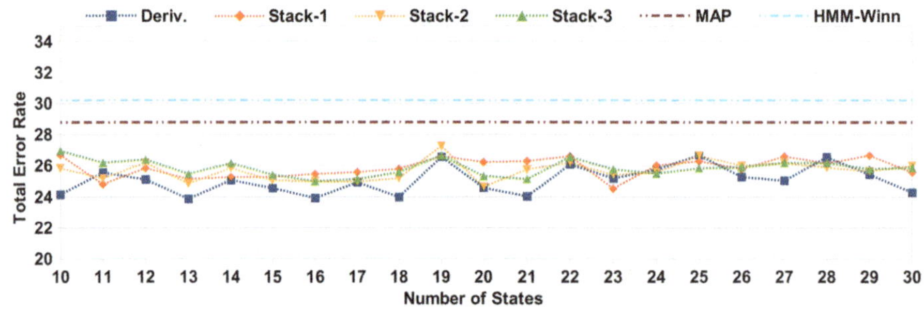

Figure 6 HMM/GBE-FA segmentation-by-classification system with different numbers of states.

Table 7 is divided into two parts: the first part shows the error for each class and the average error for the winner hierarchical-HMM system of the evaluation (HMM-Winn). The last column shows the NIST metric used in the NIST RT Diarization evaluations [42] to compare the systems with a well-known metric. To be able to compute the NIST error with the hierarchical-HMM system, we replicated the winner system according to [31] (HMM-Rep). The second part of the table shows the FA segmentation-by-classification system (FA-Segm) after the combination and the smoothing with the derivative HMM/GBE back-end subsystem because this subsystem is slightly better than the other back-end subsystems. We choose the best configuration (Best FA-Segm) and the worst configuration (Worst FA-Segm). Note that the final number of states is not very critical because the difference between errors is less than 3%. The best result obtained was an error of 23.8% using 13 states, and the worst result was 26.6% with 25 states. The hierarchical-HMM systems perform better than the worst FA system for the MU and SM classes, but their behavior is worse for SN and SP. Also, there is not a substantial benefit classifying the MU with the best FA system compared to the hierarchical-HMM system. This is due to the use of specific features to detect the music like the Chroma features. The worst FA system achieves a relative error reduction of 11.3% with respect to the hierarchical-HMM system. Finally, the best FA configuration improves the performance for all the classes and achieves a relative error reduction of 29.2% with respect to the hierarchical-HMM system.

Conclusions

This paper presents a novel system to segment and classify audios coming from broadcast TV news into five broad classes. The proposed system is based on a factor analysis (FA) approach to compensate the within-class variability with one factor loading matrix per class. Unlike other FA systems (like speaker ID and language recognition), the system proposed in this work does not have well-delimited segments, the same speaker can be found in different classes, and the nature of the classes can be very different (music, speech, or noise). The relevance of this approach can be summarized in two major aspects: it does not need specific features or hierarchical structure and it performs a very accurate segmentation and classification for all the classes. Therefore, the system is general enough to be used for different tasks and scenarios. The classification experiments with oracle segmentation ('Classification experiments with oracle segmentation' section) show a clear improvement compared to the baseline GMM system. A class/non-class FA system is proposed for the segmentation-by-classification experiments in the 'Segmentation-by-classification experiments' section. Different back-end systems have been evaluated in order to exploit the correlation among classes and avoid sudden changes in the decisions. This system is compared to a hierarchical solution with specific features for each level. The results show a significant improvement for all classes, metrics, and configurations achieving a 29.2% relative error reduction with respect to the hierarchical-HMM system for the best configuration.

Table 7 Results for the Albayzin evaluation winner system and factor analysis segmentation system over the test files

	Error for each class					
	MU	SP	SM	SN	Total	NIST
HMM-Winn [31]	19.2	39.5	25.0	37.2	30.2	-
HMM-Rep	16.3	40.8	24.0	38.8	30.0	19.3
Worst FA-Segm	19.3	29.5	24.6	33.1	26.6	16.7
Best FA-Segm	18.8	23.7	23.6	29.1	*23.8*	*14.7*

The table shows the error per class and the total error with the metric of the evaluation and the NIST metric. The italicized numbers show the total error and the NIST metric of the best FA system.

Endnotes

[a]Spanish Thematic Network on Speech Technologies: http://www.rthabla.es.

[b]The Center for Language and Speech Technologies and Applications (TALP) is a specific interdepartmental research center at the Technical University of Catalonia.

Abbreviations

AC, acoustic classes; BN, broadcast news; HMM, hidden Markov model; GMM, Gaussian mixture model; FA, factor analysis; UBM, universal background model.

Competing interests

The authors declare that they have no competing interests.

Acknowledgements

This work has been funded by the Spanish Government and the European Union (FEDER) under the project TIN2011-28169-C05-02.

References

1. NIST, TREC NIST Evaluations. http://www.itl.nist.gov/iad/mig//tests/sdr/. Accessed 6 Aug 2014
2. S Galliano, E Geoffrois, D Mostefa, The ESTER phase II evaluation campaign for the rich transcription of French broadcast news, in *Interspeech*, Lisbon, 4–8 Sept 2005, pp. 3–6
3. J Zibert, F Mihelic, J Martens, H Meinedo, J Neto, L Docio, C Garcia-Mateo, P David, E Al, The COST278 broadcast news segmentation and speaker clustering evaluation-overview, methodology, systems, results, in *Interspeech*, Lisbon, 4–8 Sept 2005
4. Y Lavner, D Ruinskiy, A decision-tree-based algorithm for speech/music classification and segmentation. EURASIP J. Audio Speech Music Process. **2009**, 1–15 (2009)
5. S Imai, Cepstral analysis synthesis on the mel frequency scale, in *IEEE International Conference on Acoustics, Speech, and Signal Processing Conference Proceedings*, Boston, 14–16 Apr 1983, pp. 93–96
6. R Vergin, D O'Shaughnessy, V Gupta, Compensated mel frequency cepstrum coefficients, in *IEEE International Conference on Acoustics, Speech, and Signal Processing Conference Proceedings*, vol. 1, Atlanta, 7–10 May 1996, pp. 323–326
7. R Vergin, Generalized mel frequency cepstral coefficients for large-vocabulary speaker-independent continuous-speech recognition. IEEE Trans. Speech Audio Process. **7**(5), 525–532 (1999)
8. E Wong, S Sridharan, Comparison of linear prediction cepstrum coefficients and mel-frequency cepstrum coefficients for language identification, in *International Symposium on Intelligent Multimedia, Video and Speech Processing*, Kowloon Shangri-La, Hong Kong, 2–4 May 2001, pp. 95–98
9. M Hasan, M Jamil, M Rahman, Speaker identification using mel frequency cepstral coefficients, in *International Conference on Computer and Electrical Engineering*, Dhaka, 28–30 Dec 2004
10. P Dhanalakshmi, S Palanivel, V Ramalingam, Classification of audio signals using AANN and GMM. Appl. Soft Comput. **11**(1), 716–723 (2011)
11. L Xie, Z-H Fu, W Feng, Y Luo, Pitch-density-based features and an SVM binary tree approach for multi-class audio classification in broadcast news. Multimed. Syst. **17**(2), 101–112 (2011)
12. J Saunders, Real-time discrimination of broadcast speech/music, in *IEEE International Conference on Acoustics, Speech, and Signal Processing Conference Proceedings*, Atlanta, 7–10 May 1996, pp. 993–996
13. D Li, I Sethi, N Dimitrova, T McGee, Classification of general audio data for content-based retrieval. Pattern Recogn. Lett. **22**, 533–544 (2001)
14. L Lu, H Zhang, H Jiang, Content analysis for audio classification and segmentation. IEEE Trans. Speech Audio Process. **10**(7), 504–516 (2002)
15. TL Nwe, H Li, Broadcast news segmentation by audio type analysis, in *IEEE International Conference on Acoustics, Speech and Signal Processing*, vol. 2, Philadelphia, 18–23 Mar 2005, p. 1065
16. A Hauptmann, R Baron, M Chen, Informedia at TRECVID 2003: analyzing and searching broadcast news video, in *Proc. TRECVID*, Gaithersburg, 17–18 Nov 2003
17. S Dharanipragada, M Franz, Story segmentation and topic detection in the broadcast news domain, in *DARPA Broadcast News Workshop*, Herndon, 28 Feb–3 Mar 1999, pp. 1–4
18. A Gallardo-Antolín, J Montero, Histogram equalization-based features for speech, music, and song discrimination. IEEE Signal Process. Lett. **17**(7), 659–662 (2010)
19. T Butko, C Nadeu, Audio segmentation of broadcast news in the Albayzin-2010 evaluation: overview, results, and discussion. EURASIP J. Audio Speech Music Process. **2011**(1), 1 (2011)
20. M Markaki, Y Stylianou, Discrimination of speech from nonspeeech in broadcast news based on modulation frequency features. Speech Commun. **53**(5), 726–735 (2011)
21. R Huang, J Hansen, Advances in unsupervised audio classification and segmentation for the broadcast news and NGSW corpora. IEEE Trans. Audio Speech Lang. Process. **14**(3), 907–919 (2006)
22. N Nguyen, M Haque, C-h Kim, J Kim, Audio segmentation and classification using a temporally weighted fuzzy C-means algorithm. Adv. Neural Netw. **6676**, 447–456 (2011)
23. SS Chen, PS Gopalakrishnan, Speaker, environment and channel change detection and clustering via the Bayesian information criterion, in *Proc. DARPA Broadcast News Workshop*, Lansdowne, 8–11 Feb 1998
24. C-h Wu, Y Chiu, Automatic segmentation and identification of mixed-language speech using delta-BIC and LSA-based GMMs. IEEE Trans. Audio Speech Lang. Process. **14**(1), 266–276 (2006)
25. M Kotti, E Benetos, C Kotropoulos, Computationally efficient and robust BIC-based speaker segmentation. IEEE Trans. Audio Speech Lang. Process. **16**(5), 920–933 (2008)
26. C-h Wu, C-h Hsieh, Multiple change-point audio segmentation and classification using an MDL-based Gaussian model. IEEE Trans. Audio Speech Lang. Process. **14**(2), 647–657 (2006)
27. A Misra, Speech/nonspeech segmentation in web videos, in *Proc. Interspeech*, Portland, 9–13 Sept 2012
28. L Lu, H-J Zhang, SZ Li, Content-based audio classification and segmentation by using support vector machines. Multimed. Syst. **8**(6), 482–492 (2003)
29. H Aronowitz, Segmental modeling for audio segmentation, in *IEEE International Conference on Acoustics, Speech, and Signal Processing Conference Proceedings*, Honolulu, 15–20 Apr 2007, pp. 393–396
30. J Foote, A similarity measure for automatic audio classification, in *American Association for Artificial Intelligence: Intelligence Integration and Use of Text, Image, Video, and Audio Corpora* (Stanford, March 1997)
31. A Gallardo, R San Segundo, UPM-UC3M system for music and speech segmentation, in *II Iberian SLTech*, Vigo, 10–12 Nov 2010, pp. 421–424
32. D Castan, C Vaquero, A Ortega, D Martínez, E Lleida, Hierarchical audio segmentation with HMM and factor analysis in broadcast news domain, in *Proc. Interspeech*, Florence, 15 Aug 2011
33. T Butko, CN Camprubí, H Schulz, Albayzin-2010 audio segmentation evaluation: evaluation setup and results, in *II Iberian SLTech*, Vigo, 10–12 Nov 2010, pp. 305–308
34. P Kenny, G Boulianne, P Dumouchel, Eigenvoice modeling with sparse training data. IEEE Trans. Speech Audio Process. **13**(3), 345–354 (2005)
35. P Kenny, Joint factor analysis of speaker and session variability: theory and algorithms, 1–17 (2006). http://www.crim.ca/perso/patrick.kenny. Accessed 6 Aug 2014
36. P Kenny, G Boulianne, P Ouellet, P Dumouchel, Joint factor analysis versus eigenchannels in speaker recognition. IEEE Trans. Audio Speech Lang. **15**(4), 1435–1447 (2007)
37. C Vaquero, A Ortega, J Villalba, A Miguel, E Lleida, Confidence measures for speaker segmentation and their relation to speaker verification, in *Proc Interspeech 2010*, vol. 2010, Makuhari, 26–30 Sept 2010, pp. 2310–2313
38. C Vaquero, A Ortega, E Lleida, Intra-session variability compensation and a hypothesis generation and selection strategy for speaker segmentation, in *IEEE International Conference on Acoustics, Speech and Signal Processing*, Prague, 22–27 May 2011, pp. 3–6
39. N Brummer, A Strasheim, V Hubeika, P Matějka, L Burget, O Glembek, Discriminative acoustic language recognition via channel-compensated GMM statistics, in *Proc Interspeech*, Brighton, 6–10 Sept 2009, pp. 2187–2190
40. D Castan, A Ortega, A Miguel, E Lleida, Broadcast news segmentation with factor analysis system, in *Proc. SLAM Workshop*, Marseille, 22–23 Aug 2012

41. D Castan, A Ortega, J Villalba, E Lleida, Segmentation-by-classification system based on factor analysis, in *IEEE International Conference on Acoustics, Speech, and Signal Processing Conference Proceedings*, Vancouver, 26–31 May 2013
42. NIST, *The 2009 (RT-09) Rich Transcription Meeting Recognition Evaluation Plan*, (Melbourne, 28–29 May 2009
43. D Reynolds, TF Quatieri, RB Dunn, Speaker verification using adapted gaussian mixture models. Digit. Signal Process. **10**(1–3), 19–41 (2000)
44. CM Bishop, *Pattern Recognition and Machine Learning*, vol. 4. Computers - Springer, Aug 17, 2006, p. 738
45. P Kenny, D Reynolds, F Castaldo, Diarization of telephone conversations using factor analysis. IEEE J. Selected Topics Signal Process. **4**(6), 1059–1070 (2010)
46. H Li, B Ma, K Lee, Spoken language recognition: from fundamentals to practice. Proceedings of IEEE. **101**(5), 1136–1159 (2013)
47. F Castaldo, D Colibro, E Dalmasso, P Laface, C Vair, Compensation of nuisance factors for speaker and language recognition. IEEE Trans. Audio Speech Lang. Process. **15**(7), 1969–1978 (2007)
48. R Vogt, S Sridharan, Explicit modelling of session variability for speaker verification. Comput. Speech Lang. **22**(1), 17–38 (2008)
49. D Castan, A Ortega, E Lleida, Factor analysis segmentation and classification in broadcast news domain, in *Proc. III Iberian SLTech*, Madrid, 21–23 Nov 2012
50. O Glembek, L Burget, N Dehak, N Brummer, P Kenny, Comparison of scoring methods used in speaker recognition with joint factor analysis, in *IEEE International Conference on Acoustics, Speech and Signal Processing*, Taipei, 19–24 Apr 2009, pp. 4057–4060
51. P Kenny, G Boulianne, P Ouellet, P Dumouchel, Factor analysis simplified, in *IEEE International Conference on Acoustics, Speech and Signal Processing*, vol. 1, Philadelphia, 18–23 Mar 2005, pp. 637–640
52. J Kittler, Combining classifiers: a theoretical framework. Pattern Anal. Appl. **1**(1), 18–27 (1998)
53. N Brummer, Measuring, refining and calibrating speaker and language information extracted from speech PhD thesis, University of Stellenbosch, (2010)
54. V Hubeika, A Strasheim, Data selection and calibration issues in automatic language recognition - investigation with BUT-AGNITIO NIST LRE 2009 system, in *Odyssey*, Brno, 28 June–1 July 2010, pp. 215–221
55. D Martínez, A Miguel, A Ortega, E Lleida, I3A language recognition system for Albayzin 2010 LRE, in *Proc. Interspeech*, Florence, 15 Aug 2011

16

Music selection interface for car audio system using SOM with personal distance function

Ning-Han Liu

Abstract

Devices such as smart phones and tablet PCs of various sizes have become increasingly popular, finding new applications, including in-car audio systems. This paper proposes a new car audio system. In the architecture, music data is stored in an online database, which users are then able to select a genre of music or a playlist from through a 2D interface. Self-organizing map, depending on a personalized distance function and music contents, is utilized to map music tracks to the interface. With this data model and interface, drivers can easily select the type of music they want to listen to without interfering with their driving. Artificial neural networks record and analyze user preference, allowing the system to select and order the music tracks to be played automatically. Experiments have shown that the system satisfies user requirements.

Keywords: Car audio system; Music database; Self-organizing map; User interface

1. Introduction

New vehicle telecommunication systems are constantly being developed. The problem of combining multimedia player technology with in-car systems has recently attracted significant attention. Among car-based multimedia systems, personal music and radio are less distracting for drivers and are thus the oldest and most popular types of in-car entertainment. The wide availability of third generation (3G) mobile networks has also enhanced car audio systems. A new type of Internet radio for vehicles has recently been released. It offers advanced radio functionalities including: AM/FM radio, CD, and MP3 playback, along with Internet radio. In the not-too-distant future, it is predicted that music will be stored in the cloud [1], in contrast to how it is currently stored locally, and will be accessed over wireless networks. However, this creates a new problem. The driver's primary focus needs to be on driving. If the driver spends too much time and attention on selecting music, there is an increased risk of accidents. Therefore, designing a new user interface to facilitate the selection of music from large databases is an important research topic.

Correspondence: gregliu@mail.npust.edu.tw
Department of Management Information Systems, National Pingtung University of Science and Technology, 1, Shuefu Road, Neipu, Pingtung 912, Taiwan

The original method to select a track on a car audio system is achieved using the 'Next,' 'Previous,' and 'Random' buttons. Newer music players often have a touch screen to let users look through albums or folders and find the desired songs. The former method is time-consuming, and the latter involves a number of interactions, distracting the driver and creating a safety issue, especially when the number of tracks is large. In this paper, a novel music selection interface is proposed, which utilizes the self-organizing map (SOM) to map tracks to one or two discrete grids. Users only need to press the appropriate spot on a touch screen corresponding to a particular type of music or slide their finger on a curve to create a playlist. A content-based method is adopted that uses musical content as the basis for the SOM clustering. Since each individual perceives the variation between certain music tracks differently, a dynamically adjustable measure is proposed. From here on the variation in music tracks perceived by the user is referred to as a distance function. The proposed method can calculate the users' implicit distance functions based on the ratings they give for the distances between music tracks. The distance function is then used to quantify the music similarity and is needed for the music clustering process. The superior clustering capability provided by the SOM is used to create a personalized index and interface for each user. In addition, an artificial intelligence technique is adopted to select tracks automatically to recommend to users. This car audio system and interface is implemented in experiments, to test the influence of the personalized distance function on the clustering performance, and the tracks retrieved. The clustering method and interface design are suitable for the next generation of car audio systems according to the experimental results.

The proposed car audio system is illustrated in Figure 1. All music data are stored in the music server. The system first utilizes the Internet to estimate the user's perception of 'music distance', which will then be used to generate a personalized distance function. The SOM clustering method is applied, with this distance function, to generate a personalized interface that is transferred to the in-car system through a 3G connection. The user will select the type of songs to play on the car system. This choice is sent to the playlist generator on the server side, which finds appropriate songs to transmit to the car.

The rest of this paper is organized as follows: Section 2 discusses related work. Section 3 discusses feature extraction from the music content and the method used to generate the distance function. Section 4 describes the design of the music clustering method and the user interface. Experimental results are presented and discussed in Section 5, while Section 6 concludes the work.

2. Related works

As more and more electronic devices are installed in modern vehicles, the I/O systems become increasingly important. The use of car navigation systems is an example of this. In Kern et al. [2], it was found that input and output device controls should be in the middle of the steering wheel for safety and suggested that multi-touch screens be used. However, it is not clear how the content of the screen was to be kept horizontal. Although a round screen that adjusts content along with the rotation of the steering wheel can be used [3], the cost is too high to commercialize and also prevents the installation of airbags. A study of car audio system interfaces [4] proposed mapping the control commands to a similar system as the fingerings of chords. The buttons on the steering wheel when pressed in certain patterns could then implement complex controls. In Schnitzer et al. [5], the design of a music player interface using the travelling salesman problem algorithm to sort tracks from most similar to least similar, circling round to more similar again, was demonstrated. In this way, tracks are arranged in an optimal and non-repetitive circular order, forming the interface. In Prescher et al. [6], a method was proposed to cleanly partition the touch-screen area. The interface needs to be convenient and easy to use and the touch-screen area should display information clearly. In Alonso et al. [7], the complexity of touch-screen interfaces was discussed, where it was suggested that it is better to use fewer steps for interaction. Rauber et al. [8] used 'islands of music' to visualize music based on a distribution of features. However, they did not address the problem of playlist generation.

The SOM, which was first proposed by Kohonen, is one of the best known unsupervised artificial neural network models [9,10]. The SOM is a competitive learning algorithm, which looks for similar patterns, features, and relations among samples. The SOM can be used for

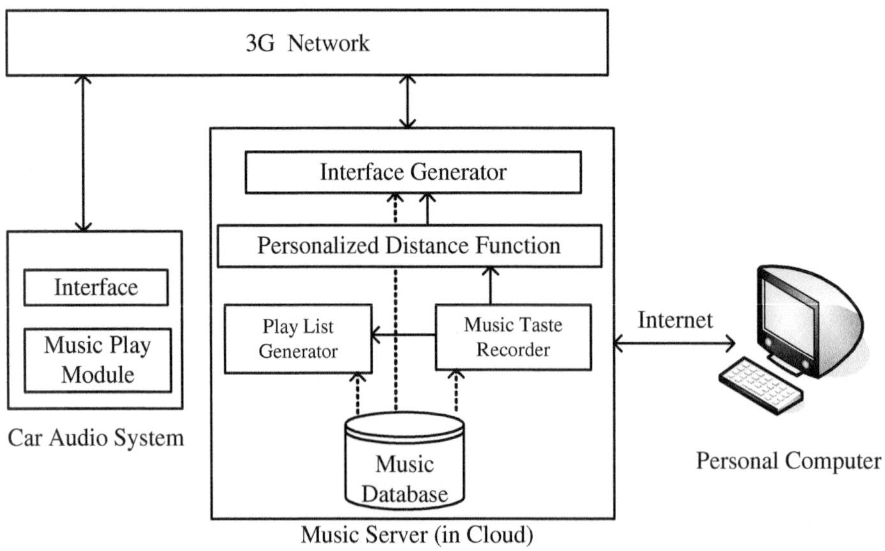

Figure 1 System architecture.

data classification or clustering. After clustering using the SOM algorithm, an arbitrary dimensional input vector can be mapped to a lower discrete grid (usually one- or two-dimensional (2D)). The SOM retains most of the distance relations which were present in the original data in this lower discrete grid. There is a large body of research on the SOM. Some studies focus on improved algorithms for faster convergence, such as the fast SOM method proposed by Su and Chang [11], who added the *k*-means algorithm to obtain an initial clustering, which was used to speed up the convergence of the SOM learning. In Su et al. [12], it was proposed to use grid partitions to replace *k*-means. A hierarchical self-organizing map (HSOM) [13] uses a multi-layer architecture that splits the original single layer neuron network into multiple layers. After the upper layer, parent neurons are trained, their child neurons are trained, etc. Due to the superior clustering performance of the SOM on high-dimensional data, it has been employed for music data clustering, recommendation, and querying. The SOM-enhanced JukeBox, proposed by Rauber et al. [8] automatically creates an organization of music archives based on audio similarity. In Dickerson [14], the SOM was used to describe the properties of music and created a query-by-content system that operates independent of the user's query style.

Concurrent recommendation systems can be categorized into content-based filtering (CBF) recommendation, collaborative filtering (CF) recommendation, and hybrid recommendation, based on the underlying algorithm. Research has focused on implementing effective music management and categorization using features such as metadata. Others proposed content-based methods [15,16] generate recommendations by identifying preferred content features and extracting items similar to previous choices. Lyric feature analysis [17] is used to find similar items, focusing on lyrics that describe race conflicts and social issues. The authors proposed a content that features analysis-based customer playlists, generated according to user profile and feedback information [18]. The CF mechanism uses rating and user profile information to identify users with similar interests. Users with common interests recommend items to each other. The CF mechanism is a widely applied recommendation technology [15,19] where user preferences are obtained from visitor records or browsing behavior. Although the CF mechanism is widely used, it has many unsolved problems, such as the cold-start problem and sparse information, which affect performance during system initialization. Several researchers have proposed recommendation methods that combine both CBF and CF approaches to build a recommendation system. The hybrid approach is able to partially solve some of the problems associated with recommendation loss. For example, collaborative user groups are constructed based on tagged labels [16,20]. Users who have tagged the same label are included into the same group. The initial grouping can reduce calculation time. Further analysis is based on music content and is aimed at providing mutual recommendations.

For playlist generation, a popular method allows the user to specify some tracks as seeds. The algorithms then find similar music and make a list [21]. A similar example [22] creates playlists with an inherent sequential order to have a smooth transition between the start and end of each song. Another approach to playlist generation populates the playlist according to a set of user-specified constraints about the desired content [23]. A third method generates a playlist according to user preferences and situations [24]. The system produces a playlist for users in similar situations using content-based retrieval on the playlists. These three methods rely on examining metadata attached to music tracks. Another method uses the listening history of the user and generates a listener model for the user's preferences. For instance, the system proposed by Andric and Haus [25] records the frequency of two music tracks being played together to find the relevance between them. A vector space is constructed where each vector represents a track. The vectors are then grouped such that each grouping of the tracks indicates different listening situations. The music playlist is then generated according to the situation indicated by the user. Another method to generate a playlist was proposed based on users' interests and time parameters, utilizing personalized neural networks in Liu and Hsieh [26].

3. Music features and personalized distance function

The results of music clustering are often affected by the distances between them. In this paper, we design a mechanism to build a personalized distance function for each user. Simple surveys are used to record user perceptions of the distances between songs in order to achieve personalization. The personalized distance function forms the basis for generating the personalized interface.

3.1. Extracted features

Several studies [27,28] have used features extracted from musical data for classification and recommendation. The features extracted from both symbolic data and audio are adopted due to the fact that they provide more useful information [29]. For each track in the database, the associated musical instrument digital interface (MIDI) file is manually created. The MIDI format can describe a range of musical forms and characteristics, which can accurately record pitch and tone. Only the MIDI data for the main melody of the track are used, which are usually monophonic. The following features are extracted from the main melody [30]:

1. Average pitch indicates whether the piece of music is of high or low pitch.
2. Pitch entropy indicates the degree of variation in the piece of music.
3. Pitch density represents the number of different pitches that occur in the piece of music.
4. Average duration describes the rhythm of the piece of music, for example fast or slow.

5. Duration entropy indicates the degree of variation in rhythms present in the piece of music. The pitch interval represents the difference between consecutive pitches. It is commonly used in key-invariant music query systems, which are not based on pitch. Extracting features from a pitch interval string is useful since two melodies with the same pitch contour may be considered identical even when they have different pitch scales.

6. Pitch interval entropy represents the degree of variation in the piece of music under key-invariant conditions. Refrains in pop music are common. They often form the basis by which people decide if they like a music track. Therefore, the same set of features is extracted from refrains. In total, 12 features are utilized from MIDI files. For audio files, the features are extracted from the frequency domain, which include formants and Mel-frequency cepstral coefficients (MFCCs). As pop music is the most common, the tone color (or timbre) of the singer's vocals is a possible feature in determining whether the user accepts the recommendations. The vocal sections are manually identified (usually in the refrain) for formants.

7. MFCC is the best-known feature in the area of speech and speaker recognition, based on human perception sensitivity in terms of audio frequencies. In this system, the input signal of the entire track is segmented into 20 ms frames with an optional overlap of 1/2 the frame size. Only the first five bins of MFCCs are extracted; this avoids high-dimensional data. The average and the standard deviation of each bin are calculated as features.

8. Formant. Vocals are usually evaluated based on formants. Formants are the frequencies that have higher energy or amplitude in the spectrum. For example, the formant of a female voice is typically higher than a male voice. To limit the number of features, only the first (F1), the second (F2), and the third (F3) formants are used because they are more useful for distinguishing between different sound sources. The mean and standard deviation of F1, F2, and F3 within a window are calculated for use as the features of the specific vocalist.

A total of 16 features are extracted from the audio file, and together with the features extracted from the corresponding MIDI file, the feature vector of a piece of music contains 28 dimensions.

Normalization is applied on each dimension to fix the values between zero and one:

$$N_d = \frac{d - D_{\min}}{D_{\max} - D_{\min}} \tag{1}$$

where d is the value to be normalized, $D_{\max}$ is the maximum value for this dimension, and $D_{\min}$ is the minimum value in the dimension.

In a music clustering system, the distance formula usually determines the performance. Therefore, a method to personalize the distance formula is proposed. This is the subject of the following subsection.

3.2. Weighted squared Euclidean distance function generated from maximum likelihood estimation

Most research on content-based music recommendation has adopted the Euclidean distance function [31] (Euclidean distance is a special case of the Minkowski distance). When using the Euclidean distance to measure music similarity, each dimension is given the same importance. However, each music feature has a different impact on the similarity measure. For example, rhythm is usually more important in similarity calculations for rock music than it is for classical music [32]. Taking the weight of each dimension into consideration in the Euclidean distance calculation gives the weighted Euclidean distance. To limit the computation, the square root is not calculated. Instead, the weighted squared Euclidean distance is used as the music distance. As every user has a different rating, the weight is separately computed for each.

Before the system can compute the weight for each dimension, it needs to obtain information from the user. To achieve this, the user is asked to rate the similarity between some tracks from 1 to 10. User-specific weights are calculated using maximum likelihood estimation (MLE). Assume that the seed track randomly selected by the system is Q; it can be deemed a point in the d-dimensional feature space. For any track X in database, the weighted squared Euclidean distance between X and Q is as follows:

$$D(X, Q) = \sum_{n=1}^{d} w_n (x_n - q_n)^2 \tag{2}$$

where x_n and q_n are the nth-dimensional values of the feature vectors of tracks X and Q, respectively.

The initial weight for each dimension, w_k, is 1. *k-NN* (k - nearest neighbor) search is used to find the k

closest tracks to Q. Suppose M is the set of k tracks closest to Q found using k-NN, $M = \{S_i \mid S_i \in k\text{-NN of } Q, i = 1,2,\ldots,k\}$. Users are then asked to rate the similarity between S_i and Q. The user assigns a rating only if they feel S_i and Q are similar to some degree. Suppose r_i is the similarity rating, where a higher r_i signifies a higher similarity between S_i and Q. Assume M_p is the set of non-empty scores, $M_p = \{m_i \mid m_i \in M \text{ and } r_i \neq \phi\}$. According to [33], m_i is an independent event. The probability for the set of events M_p can be represented by the following formula:

$$P(M_p|W,Q) = \prod_{m_i \in M_p} P(m_i|W,Q^{r_i})$$
$$W = \langle w_1, w_2, .., w_d \rangle \tag{3}$$

where r_i is the user's rating for m_i. A larger r_i has a higher impact on the probability. The weight for each dimension is adjusted in vector W such that $P(M_p|W, Q)$ is maximized:

$$W' = \arg\max_W P(M_p|W,Q) \tag{4}$$

Assume that $P(m_i|W,Q)$ follows a Gaussian distribution:

$$P(m_i|W,Q) = \frac{1}{(2\pi)^{\frac{d}{2}}\sqrt{\prod_{n=1}^{d}\frac{1}{w_n}}} \exp\left(-\frac{1}{2}D(m_i,Q)\right) \tag{5}$$

The partial derivatives of $P(M_p|W,Q)$ with respect to each w_i can be computed as:

$$\nabla\{P(M_p|W,Q),W\} \equiv \left[\frac{\partial P(M_p|W,Q)}{\partial w_1}, \frac{\partial P(M_p|W,Q)}{\partial w_2}, \ldots, \frac{\partial P(M_p|W,Q)}{\partial w_d}\right] \tag{6}$$

For W to maximize $P(M_p|W,Q)$, the partial derivative for each dimension should be zero. In order to simplify this calculation, the log of $P(M_p|W,Q)$ is taken:

$$\ln(P(M_p|W,Q)) = \sum_{m_i \in M} r_i \left[\frac{1}{2}\left(\sum_{n=1}^{d} \ln w_n - D(m_i,Q) - d \times \ln(2\pi)\right)\right] \tag{7}$$

and,

$$\frac{\partial \ln(P(M_p|W,Q))}{\partial w_n} = 0 \tag{8}$$

Thus:

$$\sum_{m_i \in M_p} r_i \frac{1 - w_n(m_{i,n} - q_n)^2}{2w_n} = 0 \tag{9}$$

where $m_{i,n}$ is the nth-dimensional value of m_i, and q_n is the nth-dimensional value of Q.

Finally, the new weight w'_n for the nth dimension is as follows:

$$w'_n = w_n = \frac{\sum_{m_i \in M_p} r_i}{\sum_{m_i \in M_p} r_i (m_{i,n} - q_n)^2} \tag{10}$$

A single question does not provide enough data to estimate the user's preferences in terms of music distance. The above process is repeated several times to compute weights that are more representative. Suppose the user provides T independent responses, then the final weight of the nth dimension is:

$$\bar{w}_n = \frac{\sum_{t=1}^{T} w'_{n,t}}{T} \tag{11}$$

where $w'_{n,t}$ is the re-weighted value of the nth dimension of W in round t.

After recording the user's perception of distances through surveys, the above formula is utilized to obtain the personalized distance function. This function forms the basis for music clustering in the next step.

4. Music selection interface using SOM index

The SOM is capable of mapping multi-dimensional data into one- or two-dimensional neurons and is hence ideal for selecting music and generating playlists. First, the personal distance function obtained in the previous section is used to select the winner neurons in the SOM and cluster all the songs according to their feature values. This two-dimensional space is then displayed on the touch screen of the car audio system. Users can press or slide their finger on the touch screen to select songs or playlists.

4.1. SOM and HSOM clustering

The competitive learning procedure of the SOM with a personalized distance function is as follows [34]:

Step 1: Initialization. Set initial synaptic weights of neurons $v \in V$ (V is the set of neurons) to small random values, in the interval [0, 1]. Assign a small positive value to the learning rate parameter α. Set j and k as the linear decay functions. R is the radius used for determining neighbors.

Step 2: Activation and similarity matching. For training X, Kohonen network is activated and find the winner neuron j_x at iteration t using the minimum personal distance (weighted squared Euclidean distance):

$$j_X = \arg\min_{v \in V} \|X - v\| \tag{12}$$

Step 3: Learning. For u in neighbor (j_x, R), update the synaptic weights as:

$$C_{iu}(p+1) = C_{iu}(p) + \Delta C_{iu}(p) \tag{13}$$

where $\Delta C_{iu}(p) = \alpha \times [x_i - C_{iu}(p) \times w_i$ is the weight correction in iteration p. w_i is the weight of dimension i in the personal distance function.

Step 4: Iteration. Update the learning rate and radius as:

$$a = k \times a \tag{14}$$

$$R = j \times R \tag{15}$$

Increase p by 1, return to Step 2 and continue until the criterion is satisfied, or no noticeable changes occur in the feature map.

Suppose there are $n \times n$ neurons in a two-dimensional SOM. After the SOM clustering is complete, the tracks in the database will be clustered based on content, into at most $n \times n$ clusters. As shown in the example of Figure 2, the SOM has 4 × 4 neurons, resulting in a division into a maximum of 16 clusters (the SOM with 4 × 4 neurons is only an example for illustration purposes, and in the actual experiments, a 10 × 10 grid was utilized). To generate the user interface, the screen area is divided into $n \times n$ regions, i.e., one for each neuron. When the user presses a particular region of the touch screen, a cluster of tracks corresponding to that neuron will be retrieved. For example, if the user selects region A, the system will play tracks from Music Set A. If region B is selected, the system will play tracks from Music Set B, etc. The system randomly chooses tracks from the selected music set to play.

There are two problems that need to be solved in the above design. (1) According to the characteristics of the SOM clustering, some neurons may not be winner neurons of any tracks. Therefore, when a user selects that region, there will be no tracks to play. (2) Some clusters may have a large number of tracks, which may have widely varying characteristics. To address these problems, two strategies are proposed as follows.

The neurons after the SOM clustering have a property, i.e., the sets corresponding to nearby neurons are usually closer than the sets corresponding to far away neurons. Taking Figure 2 as an example, the distance between Music Set A and Music Set B is smaller than the distance between Music Set A and Music Set G because their corresponding neurons are closer. Using this property, if the selected neuron has no relevant tracks, the closest neurons are checked in order until a neuron is found that has relevant tracks.

In Figure 3, an example with 5 × 5 neurons is shown. Suppose that the user pressed region T, and region T contains no tracks. To find a track to play, the neighboring regions are searched in their numbered order, until a region with tracks is found. If several regions in the same level all have tracks, the distances between the neuron vectors of these regions and the vector of neuron T are calculated, and the tracks are selected in the nearest neuron as candidates. If more than one neuron is equally close, a neuron is picked at random.

The formula for selecting the replacement region is as follows:

Assume V_n is the neuron set in level n for the selected neuron X, and the music set belonging to the neuron of V_n is not empty. The replacement region

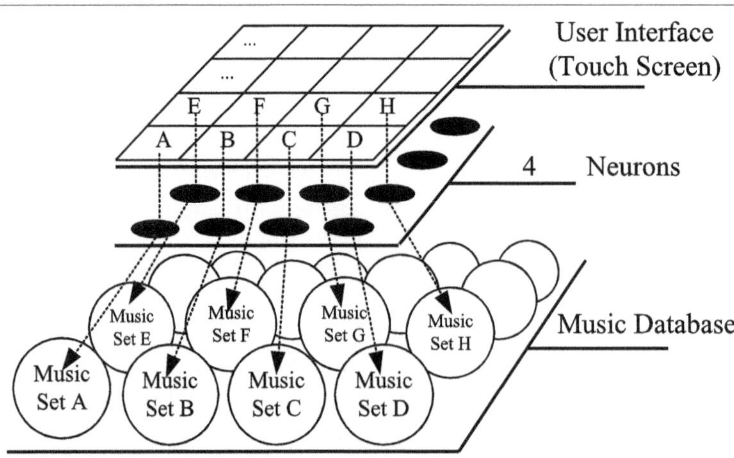

Figure 2 Sample interface of the SOM clustering.

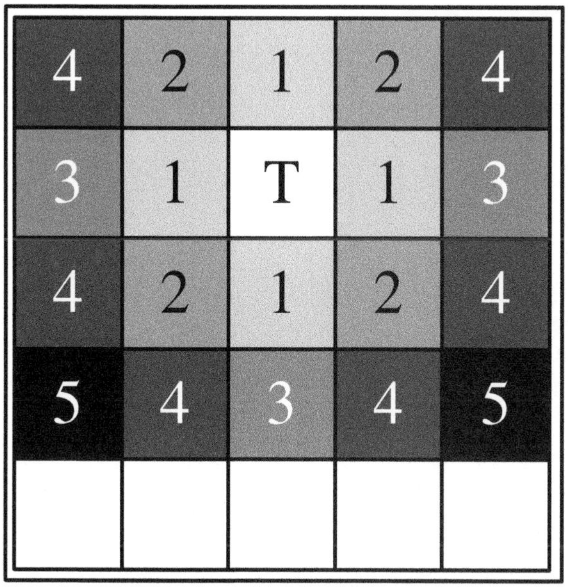

Figure 3 Example for searching songs from neighboring regions.

of X is as follows:

$$R_X = \arg\min_{v \in Vn} \|X - v\| \tag{16}$$

The weighted squared Euclidean distance function is used in the formula.

Since many tracks may be clustered to the same neuron, those tracks are clustered further in the next layer. In other words, the multi-layer clustering algorithm, HSOM, is adopted for clustering songs.

Assume that the maximum number of tracks that can be clustered to a neuron is m. If there is a cluster with size larger than m after the upper layer SOM clustering is complete, another round of SOM clustering for those tracks is performed in the next layer, as shown in Figure 4. In layer 0, the A and C clusters have more than m tracks, so they are further clustered in the next layer. In layer 1, B' and D+ also have too many tracks, and so they are also further clustered. The limit of tracks that can be in each cluster can be different for each layer.

With the music index generated by the HSOM, an easy method for selecting tracks is provided to the driver, which avoids the effort needed for browsing multiple layers of neurons. The operation is as follows: the user selects a neuron in the upper layer SOM. If the user double clicks the region of that neuron and it has a lower layer SOM clustering, then the lower layer is displayed. This process is repeated until the user does not double click any more or the bottom layer of the HSOM is reached. Finally, tracks belonging to the last neuron are treated as candidates. The selection of music from candidates can be randomized or be made using artificial intelligence. The artificial intelligence-based method is discussed in subsection C.

4.2. Playlist generator

In addition to pressing regions of the touch screen, the driver can also draw paths on the screen. The system will then use these paths to generate a playlist. Figure 5 shows an example. The user draws three paths on the screen. The system can arrange songs in the playlist according to the types forming the paths. In this example, the types are ordered as follows: A->G->H->I->E->P->L->W->S->T. If the region covered contains no tracks, the procedure is processed as discussed in the previous section.

If the user does not want to generate a playlist in the first layer, they can make more detailed selections on the lower layers.

Take Figure 6 as an example. The user first draws a path going through regions E and F on layer 0, then double clicks region C. The system will display the next layer of clustering on the screen. The user continues to draw a path going through the E+F+B+C+ sub-clusters, which all belong to cluster C. Next, the user double clicks region D+. The sub-clusters of D+ are displayed on the interface, and the user draws a path through the F#E#G# regions. Next, the user returns to layer 0 (two-finger slide input is defined as returning to the last layer) and draws a path through the G and H regions. The above operations will generate a playlist with tracks in this order: E->F-> E+–>F+–>B+–>C+–> F#-> E#->G#->G->H.

4.3. Strategies of music selection from clusters

In the previous section, how to select certain categories of tracks is discussed. When there are too many music in some categories, the SOM clustering can be applied multiple times, but it takes longer for the driver to select tracks if there are too many layers. Therefore, the number of layers in the HSOM is limited. For tracks in the same categories, two strategies are designed to select the tracks to be played. The first strategy is fast and does not require data to be collected beforehand. The second one uses artificial intelligence to select the tracks.

4.3.1. Strategy I: random selection

This method selects tracks randomly. When a type (i.e., a cluster) of tracks is selected, the probability of a track m being played is as follows:

$$P(m) = 1/(\text{size of the cluster}) \tag{17}$$

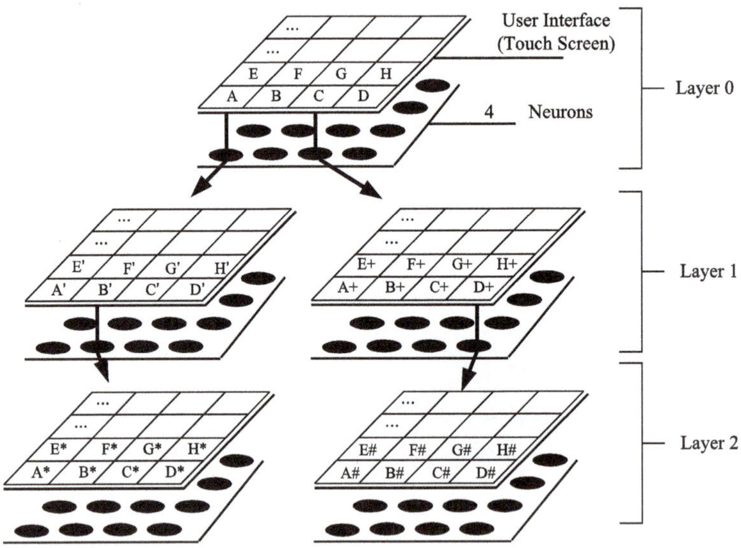

Figure 4 Example of using HSOM clustering for interface design.

4.3.2. Strategy II: selection by artificial neural networks

The above method is simple, but it does not take user preferences into consideration and may play a track that the user does not like, which may distract the driver into changing the music. To address this problem, an intelligent mechanism should be developed. This mechanism can select tracks that the user will likely want to be played based on past user preferences. Typically, this kind of intelligent mechanism requires collecting user preferences, i.e., the driver would need to rate the tracks played. However, as it is not easy to rate music when driving, the ratings are recorded when the vehicle is stationary. For example, when the driver uses a PC at home or another hand-held device to listen to music, they can rate the music. The decision about not allowing drivers to rate music while driving is made in the interest of safety. The input data for music selection engine includes the attributes of the music features. After the engine evaluates these inputs through an intelligent mechanism, it outputs the predicted value, which dictates user preference for the musical piece. This value can be regarded as the class label in the prediction process, a type of classification. Artificial neural networks (ANN) have been adopted as the core of the music selection engine in this work due to its ability to classify continuous values (i.e., the values of input features).

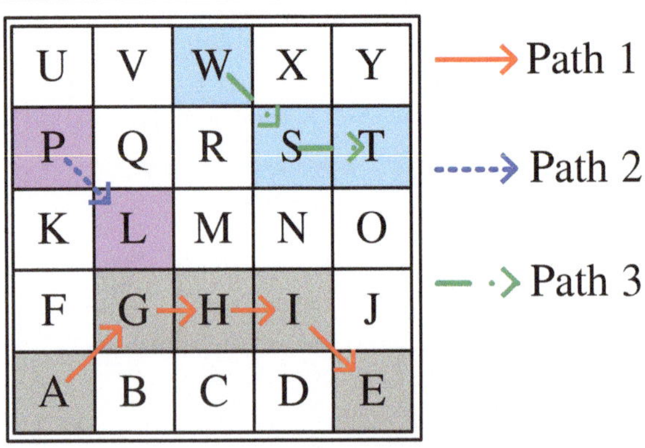

Figure 5 User interface and sample operations.

- Training data of artificial neural networks. In order to enable the neural networks to learn user preferences, the system lists scores with the associated music file name (or ID). As shown in Table 1.
- The kernel of the playlist generator: a mixed ANN. The previous method from [30] is utilized. The training time of the ANN can be higher than normal in two cases: firstly, when the training set data for an ANN have higher dimensionality or consist of a larger amount of training samples, and secondly, when the prediction does not satisfy the user requirement and hence new feedback is recorded to retrain the ANN. Retraining the system takes time if all the user data are used as the training sample. To reduce the retraining time, the structure is modified as shown by Figure 7. This structure includes two ANNs: a *short-term ANN* and *long-term ANN* to record the user's short- and long-term preferences, respectively.

Both the long-term and short-term ANNs are multi-layer feed-forward neural networks with two hidden layers. To predict a user's preference for a musical piece

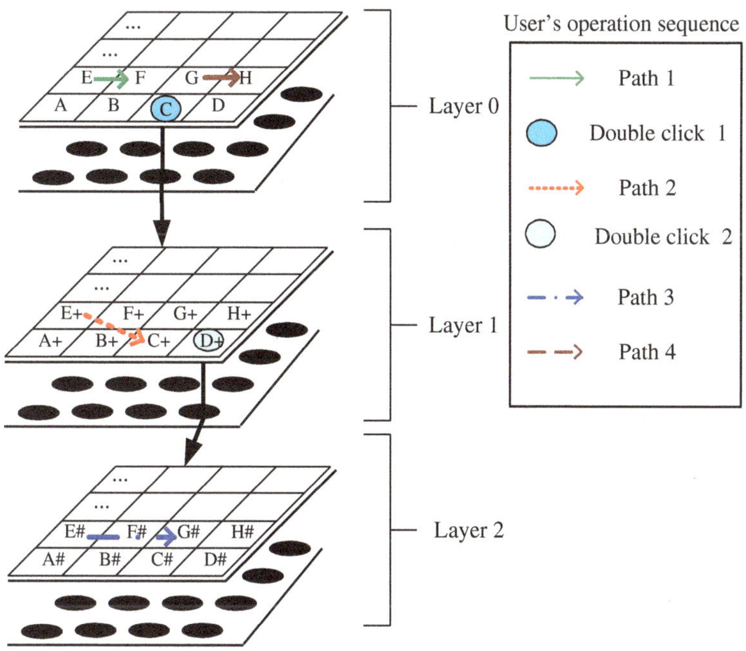

Figure 6 Sample music selection operation on the interface with HSOM clustering.

Mx, the features of *Mx* form the input vector of the input units for the two ANNs. Two output values are obtained, namely $Output_{ANN1}$ and $Output_{ANN2}$, which are combined to form a single value using a fusion function. The output of this fusion function is the predicted user score for the music. The fusion function is defined as:

$$F(\text{Output}_{\text{ANN1}}, \text{Output}_{\text{ANN2}}) = w_1 \times \text{Output}_{\text{ANN1}} + w_2 \times \text{Output}_{\text{ANN2}} \quad (18)$$

where $w_1 + w_2 = 1$.

The two weight parameters w_1 and w_2 can be adjusted by the user. If the user increases the value of w_2, preference is given to musical tracks that the user has recently heard and given higher scores to. Conversely, if the user increases the value of w_1, preference is given to tracks that the user heard in the past and gave higher scores to.

The system builds a personal mixed ANN for each user. The scores recorded in the user's profile provide the basis of the training data for the personal mixed neural networks, i.e., the long-term and short-term ANN. The score record contains the music ID, score, and playing date. Through a database search using the music ID, the system first obtains the music features that are associated with it. These features form training data input, which include the basic music features that the user has assigned the score to. It is intuitive to say that the output is the associated score value. The date of the training data for the long-term ANN will be older. The training process uses supervised learning. The bias weight adjustment uses error back-propagation.

Table 1 Score table in the user profile

Music ID	Score	Date
20	10	2011/11/4
101	8	2011/11/5
252	3	2011/11/5

If the trained personal mixed ANN does not satisfy the user, and the user decides to rebuild the bias weights, the short-term ANN is first retrained with the user-modified score table. The long-term ANN will only be retrained if the resultant short-term ANN cannot satisfy the user.

Once the user selects the music type in the SOM interface, the above mechanism to calculate the scores for the tracks in that cluster is used. To avoid playing just the highest scoring track every time, the scores are used as the probabilities of the tracks that will be played. In other words, for a Music Set C, if song *s* has a score *F*(*s*), then the probability that *s* is played is as follows:

$$P(s) = \frac{F(s)}{\sum_{s' \in C} F(s')} \quad (19)$$

5. Experiment and evaluation

To study the mobility and stability of a car audio system, a tablet is used as the client side platform for the car

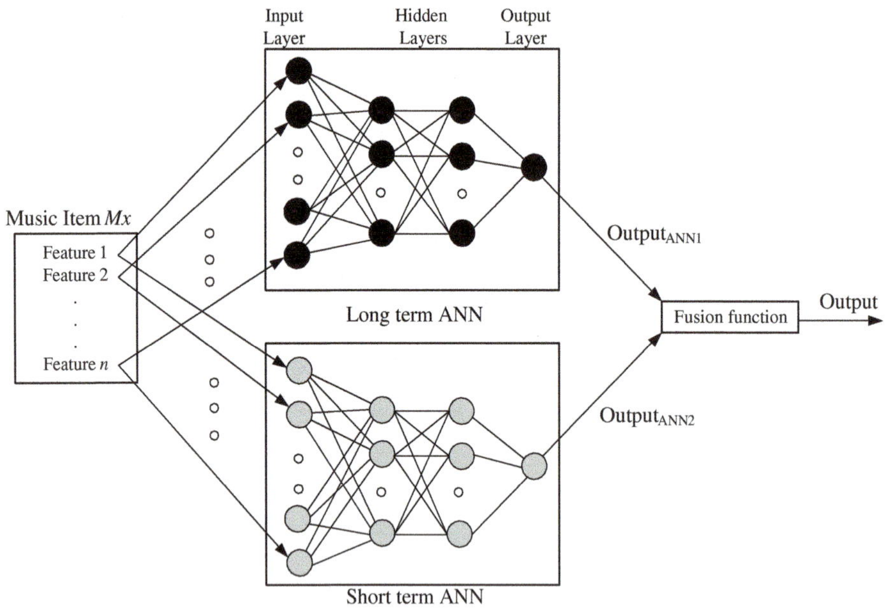

Figure 7 The structure of a personal mixed ANN.

audio system. Storing the music files, finding the personalized distance function, and the generation of the user interface and other computations all take place on the server side. The client side provides the interface to select the music cluster, handles the data transmission, and plays the music. The simulated car audio system is shown in Figure 8.

In the experiment, 560 MP3 music files and associated MIDI files are loaded into the music database. The music genres included rock and roll, blues, country, dance, heavy metal, and pop. Features with 28 dimensions are extracted from the audio and MIDI files. There are 20 enlisted undergraduate students to take part in the experiment. The age of the test subjects ranged from

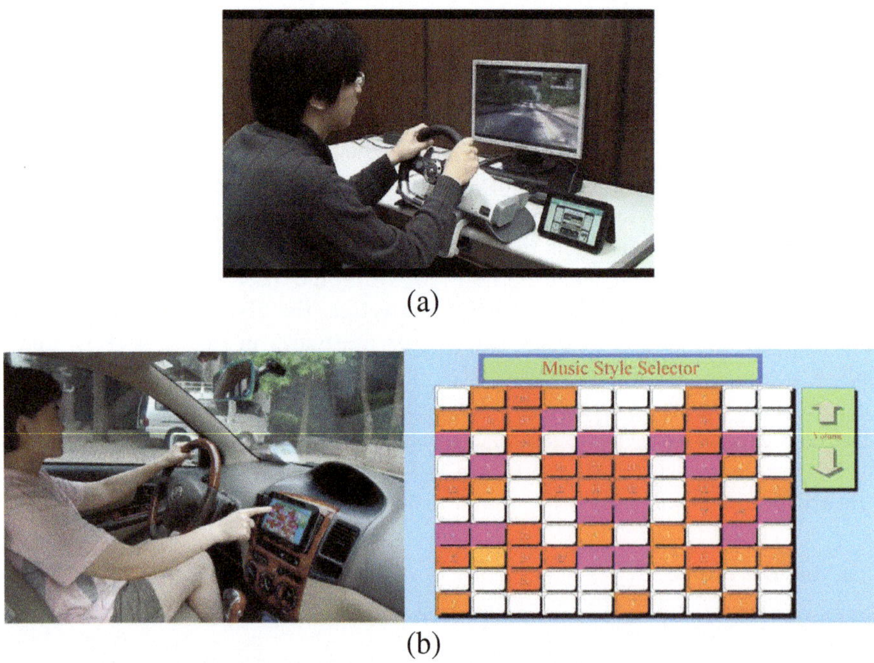

Figure 8 The simulated car audio system. (a) Simulated car audio system. **(b)** Prototype in car and user interface (the colored area means that some tracks are clustered to the associated neuron).

Table 2 Clustering performance of the MLE using all features according to users' ratings

	MLE using all features according to user's ratings (average precision)					
	$n = 3$	$n = 4$	$n = 5$	$n = 6$	$n = 7$	$n = 8$
$k = 1$	0.37	0.39	0.41	0.42	0.44	0.45
$k = 2$	0.37	0.42	0.43	0.45	0.47	0.46
$k = 3$	0.38	0.45	0.48	0.52	0.53	0.51
$k = 4$	0.41	0.49	0.53	0.61	0.67	0.62
$k = 5$	0.45	0.51	0.54	0.59	0.62	0.61

Table 4 Clustering performance of the MLE using all features according to the genre label

	MLE using all features according to the genre label (average precision)					
	$n = 3$	$n = 4$	$n = 5$	$n = 6$	$n = 7$	$n = 8$
$k = 1$	0.18	0.21	0.24	0.26	0.26	0.25
$k = 2$	0.21	0.24	0.23	0.25	0.28	0.31
$k = 3$	0.24	0.24	0.26	0.31	0.33	0.29
$k = 4$	0.26	0.29	0.32	0.37	0.38	0.38
$k = 5$	0.32	0.31	0.34	0.36	0.36	0.35

19 to 23. They had not been trained in music theory and listened to all types of music. The server side is based on an SQL server that records user profiles and rating results. The number of neurons in the SOM is 100, i.e., 10 × 10. The maximum number of songs in the first layer of the HSOM is 50. It halves in the second layer to 25, and so on.

Due to some of the features in our system that require manual annotation, it is not easy to be implemented in a large music database. Therefore, in the following experiment, we compare the effectiveness of using all the features, and only those features that can be extracted automatically (i.e., MFCCs). Moreover, the first experiment is also used for the selection of n and k for the weighted squared Euclidean distance. In the proposed method, users are asked to rate the similarity between n tracks from the database and k candidate tracks for each of those n tracks. The weighted squared Euclidean distance is then calculated for each user. The system employs the personalized distance function to cluster tracks using the SOM. In the experiment, two tracks are randomly selected in the same cluster and then played to the user. The user decides whether it is appropriate to put them in the same cluster. The number of randomly selected pairs is 20, which means the subject has to listen to 40 tracks for this experiment. In addition to the user determining whether songs are related, another method of judging whether songs are similar is to see if they belong to the same genre. The performance of the clustering system is defined as:

$$\text{precision} = \frac{N_a}{N} \tag{20}$$

where N is the number of pairs selected, and N_a is the number of pairs with positive ratings (for human judgment) or with the same genre label (for system judgment).

The experiment shows how different values of n and k used in the survey affect the clustering performance when using different feature sets. Users have a web page where they listen to music and answer the survey questions. The method is called MLE. Experimental results are shown from Tables 2, 3, 4, and 5. The average precision is calculated across the 20 users. The average precision using the traditional squared Euclidean distance as the music distance measure for recommendation is 0.35 using all of the features. Moreover, the average precision is 0.23 only using MFCCs features.

The proposed method outperforms the simple squared Euclidean distance as long as there are a sufficient number of ratings. This suggests that it is unreasonable to treat each dimension equally when calculating music similarities. Initially, the precision of the proposed method improved as n and k increased. However, after a threshold, the precision increase is insignificant and may even decrease. When users are presented with too many questions, they may lose patience and not give objective

Table 3 Clustering performance of the MLE using MFCC features according to users' ratings

	MLE using MFCC features according to users' ratings (average precision)					
	$n = 3$	$n = 4$	$n = 5$	$n = 6$	$n = 7$	$n = 8$
$k = 1$	0.24	0.26	0.31	0.34	0.36	0.35
$k = 2$	0.27	0.29	0.35	0.38	0.39	0.39
$k = 3$	0.31	0.33	0.36	0.41	0.43	0.42
$k = 4$	0.35	0.38	0.39	0.40	0.41	0.39
$k = 5$	0.36	0.39	0.41	0.42	0.40	0.41

Table 5 Clustering performance of the MLE using the MFCC features according to the genre label

	MLE using the MFCC features according to the genre label (average precision)					
	$n = 3$	$n = 4$	$n = 5$	$n = 6$	$n = 7$	$n = 8$
$k = 1$	0.15	0.15	0.14	0.16	0.21	0.19
$k = 2$	0.18	0.21	0.23	0.22	0.24	0.24
$k = 3$	0.19	0.23	0.24	0.28	0.31	0.29
$k = 4$	0.20	0.23	0.27	0.32	0.32	0.31
$k = 5$	0.24	0.26	0.25	0.31	0.32	0.29

Table 6 The clustering performances of different maximum layers using HSOM

Layers of the HSOM	1	2	3	4	5
Average precision	0.47	0.53	0.61	0.64	0.68

ratings. In addition, many candidate tracks have to be rated for each one, and it becomes harder to give clear ratings. Unclear ratings negatively affect the recommendation performance.

Comparing the precision of different feature sets, the precision obtained when only using MFCCs is inferior to that obtained when using all features. This result indicates that human input is required to improve the accuracy of clustering, even though it takes more time to process the retrieved information.

In addition, the use of different methods to measure precision will lead to different results. If we calculate precision according to whether the music belongs to the same genre, we find that the average precision is lower than when the user decides if the music belongs to the same cluster. From this, we find that most music that is classified as belonging to the same genre is actually perceived very differently by the user.

Aside from the above experiments, we carry out a similar experiment on whether there is a common distance function that all users are able to use. This experiment is different from the first experiment as it uses the estimated distance provided by all users to calculate a weighted squared Euclidean distance function. However, the average precisions obtained from the distance function are less than 0.3. In addition, the variation in the multi-dimension parameters for each user's distance function is quite large, indicating the necessity of personalizing the distance function.

The second experiment measured the effect of the number of HSOM layers on precision. In the experiment, the maximum number of layers for HSOM is established. When the maximum number of layers is established, even if too many songs are clustered in the same layer, we do not add another layer. In addition, when the experiment selects a pair, it is always chosen from the lowest cluster. According to the previous tables, the values of n and k used for calculating the personalized distance function are 7 and 4, respectively. All of the features are used in this experiment. The experimental result is shown in Table 6.

Table 6 shows that a higher maximum number of layers lead to more effective segmentation. However, when considering the ease of use of the user interface, too many layers may lead to users spending excessive time controlling the car audio system. However, if we only use a single-layer SOM, the similar characteristics of the songs may lead to their grouping in certain clusters, making it difficult for the user to select the type of song they would like to hear. Therefore, in subsequent experiments on the operation of the car audio system, the number of HSOM layers is determined by the total number of songs. In other words, when too many songs are grouped in one cluster, we created a new level.

As the personal mixed ANN has several parameters that are not easy for the user to determine, including the time period of training data for the short term as well as the weighing of short-term and long-term ANN (i.e., w_1 and w_2), we tested, under different parameters, the proportion of the songs recommended by the personal mixed ANN using different clusters that the user could accept.

The performance of the recommendations generated by the personal mixed ANN is defined as:

$$R\_\text{precision} = \frac{P_a}{P} \tag{21}$$

where P is the number of tracks recommended by the system and P_a is the number accepted by the user.

In the experiment, ten songs were recommended to each user under different parameters before the average *R*_precision value was calculated. The results are shown in Table 7.

From Table 7, it is clear that the best results are obtained when the time period of training data for the short-term ANN is set to 1 week and $w_1 = 0.4$, $w_2 = 0.6$. The results show that users are more interested in songs that they have recently awarded high scores.

The next experiment investigates the time the driver spent operating the audio system while driving. A conventional car audio system was used. Some of the common options offered included selecting a music file using folder selection or file selection. The music files

Table 7 Recommendation performance of different parameters values in the personal mixed ANN

	$w_1 = 0.1$	$w_1 = 0.2$	$w_1 = 0.3$	$w_1 = 0.4$	$w_1 = 0.5$	$w_1 = 0.6$	$w_1 = 0.7$	$w_1 = 0.8$	$w_1 = 0.9$
1 week	0.53	0.62	0.76	*0.87*	0.81	0.74	0.67	0.59	0.57
2 weeks	0.51	0.59	0.69	0.77	0.71	0.66	0.64	0.59	0.54
3 weeks	0.52	0.54	0.64	0.72	0.68	0.67	0.58	0.57	0.53
4 weeks	0.49	0.57	0.61	0.73	0.67	0.63	0.59	0.52	0.51

Note that $w_2 = 1 - w_1$ in Table 7.

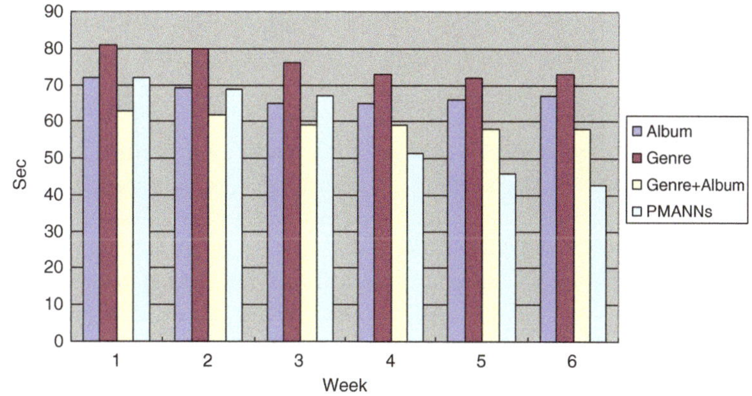

Figure 9 Average time spent in operating car audio system in 1 h over 6 weeks.

were manually categorized prior to the experiment to provide hierarchical menu lists. One choice for the menu list type is designed to categorize the music files by album. The user can choose a specific album or songs to play via the user interface. Another choice of menu list is designed to categorize music files based on the genres, including classical, folk, opera, jazz, pop, blues, rock, R and B. The user can choose music files within a category or outside a category. The user can choose whether to activate the repeat or random options when selecting the category. The third type of menu list combines the features in the previous two. The user can categorize the music files based on the album and then sort the album based on the genres. The user can have a menu list that contains three layers.

In our approach (personal mixed ANNs named as PMANNs), the values of n and k used for calculating the personalized distance function are 7 and 4, respectively. The weights for long-term and short-term ANN are set as 0.4 and 0.6, respectively, and the time period of training data for the short-term ANN is set to 1 week. The user also controls the timing for retraining the personal mixed ANNs. Due to a limited number of devices, the experiment was not performed on real cars, and instead, a simulator was used. The experiment procedure is as follows: the user listens to tracks on the simulated car audio system 1 h every day. To allow a fair comparison of the operation time, each song is cropped to 2 min. When the user hears a song that is not desired, or when a song has finished playing, the user needs to operate the car audio system to select the next song to play. We then perform statistical analysis of the average operation time for each interface for the car audio system within 1 h.

Figure 9 shows the experiment conducted over 6 weeks.

It can be seen from the experimental data that the time spent on each method has reduced as the number of weeks increased. The users gradually became familiar with each interface, which helped to speed up the time taken to locate the desired music files. The method we proposed had poor performance in the first 3 weeks; however, from week 4 onwards, the time the user spent on operation was significantly less than that of the other

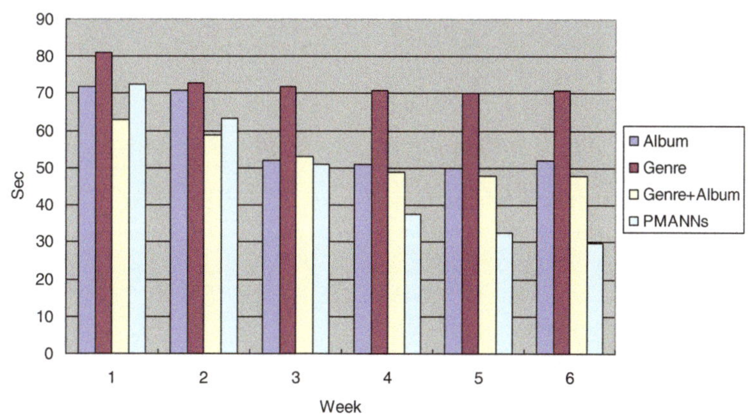

Figure 10 Average time spent looking at the monitor in 1 h over 6 weeks.

methods. The reason is that after several weeks of data collection, the PMANNs are seen to be able to infer user preferences more accurately. The time spent on operation using conventional interfaces stabilized after 3 weeks. The method using both the Genre + Album required less time. Visual ability is not required when operating the audio system, since the user would have already memorized the interface. Our third experiment uses a camera to measure the time the user takes their eyes off the road to look at the car audio. Figure 10 shows the experiment result.

It can be seen from the experiment data that during the first week, the user needs to look at the audio interface in order to operate it. However, starting from the second week, the user can reduce the time spent on the monitor based on the memories of previous operations. Our method allows the user to select music files using 2D images. The user can rapidly complete the operation by touching specific locations on the touch panel. In contrast, for a conventional operating interface, the user would need to spend an extended period of time interacting with the monitor in order locate a specific song from a long list of the file names, since each genre category contains a large number of songs. Observation of the results from the two experiments described above demonstrates that our proposed method requires the least operation time and requires the least time for the user to look at the monitor. The reduced time allows the driver to concentrate on the road while driving, which reduces the risk of an accident.

6. Conclusions

Along with the rapid development of telecommunications technology and tablet PCs, next-generation car audio systems are being investigated. Making future car audio systems more useful is an important research topic. A new method for music clustering is proposed, and a user interface is designed for such a system, which reduces the interaction needed, while still providing the required music output. First, a personalized music distance function is estimated for each user based on their preferences. This function is then used to perform personalized SOM clustering and generate a personalized user interface by utilizing the ability of the SOM to map data from high-dimensional space onto a two-dimensional plane. The experiments demonstrated that this new interface reduced the average operating time. In addition, ANNs are used to learn user preferences for music and automatically select tracks according to those preferences. This artificial intelligence-based method can reduce the risk of accidents occurring from drivers operating the audio system while in motion. There is still much room for improvement in this car audio system, for example, improving the music distance functions. Further work is also needed to develop a better user interface, in contrast to using a simple grid.

In this system, the user needs to complete a questionnaire on similar music tracks, which may create an unreasonable burden. In the future, we plan to increase the number of experimental subjects to establish whether information of the distance function obtained from certain users can be used for the music clustering of other users. For drivers, being able to listen to the music they want to hear without having to operate the car stereo system is our ideal scenario. In the future, we will also carry out research on recommendation systems, for example, the use of collaborative filtering methods to create music streaming schedules.

Competing interests

The author declares that he has no competing interests.

Acknowledgements

This work was supported in part by the NSC in Taiwan under the contact numbers NSC102-2218-E-020-002 and NSC101-2221-E-020-025.

References

1. A Paya, DC Marinescu, A cloud service for adaptive digital music streaming, (Proceedings of the 8th International Conference on Signal Image Technology and Internet Based Systems, Naples, 25-29 Nov 2012), pp. 910–917
2. D Kern, A Schmidt, J Arnsmann, T Appelmann, N Pararasasegaran, B Piepiera, Writing to your car: handwritten text input while driving, in *Proceedings of the 27th International Conference Extended Abstracts on Human Factors in Computing Systems* (Boston, 4-9 Apr 2009), pp. 4705–4710
3. M Pfeiffer, D Kern, J Schöning, T Döring, A Krüger, A Schmidt, A multi-touch enabled steering wheel: exploring the design space, in *Proceedings of the 28th International Conference Extended Abstracts on Human Factors in Computing Systems*. Atlanta **28**, 3355–3360 (Apr 2010)
4. FE Sandnes, YP Huang, YM Huang, An eyes-free in-car user interface interaction style based on visual and textual mnemonics, chording and speech, in *Proceedings of the International Conference on Multimedia and Ubiquitous Engineering of ACM*. Busan **24–26**, 342–347 (Apr 2008)
5. D Schnitzer, T Pohle, P Knees, G Widmer, One-touch access to music on mobile devices, in *Proceedings of the 6th International Conference on Mobile and Ubiquitous Multimedia of ACM*. New York **12–14**, 103–109 (Dec 2007)
6. D Prescher, G Weber, M Spindler, A tactile windowing system for blind users, in *Proceedings of the 12th International ACM SIGACCESS Conference on Computers and Accessibility*. New York **25–27**, 91–98 (Oct 2010)
7. F Alonso, S Frutos, C Montes, RJ Navajo, A generic blind user interface model. Syst Man Cybern **2**, 1133–1138 (1998)
8. A Rauber, E Pampalk, D Merkl, The SOM-enhanced JukeBox: organization and visualization of music collections based on perceptual models. J New Music Res **32**(2), 193–210 (2003)
9. T Kohonen, Self-organized formation of topologically correct feature maps. Biol Cybern **43**, 59–69 (1982)
10. T Kohonen, *Self-organizing maps. Volume 30 of Springer Series in Information Sciences*, 3rd edn. (Springer, Berlin, 2001)
11. MC Su, HT Chang, Fast self-organizing feature map algorithm. IEEE Trans. Neural Networks **11**(3), 721–733 (2000)
12. MC Su, TK Liu, HT Chang, Improving the self-organizing feature map algorithm using an efficient initialization scheme. Tamkang Journal of Sci Eng **5**(1), 35–48 (2002)
13. A Duarte, D Neto, JAF Costa, MLA Netto, Hierarchical SOM applied to image compression, in *Proceedings of the International Joint Conference on Neural Networks*. Washington, DC **15–19**, 442–447 (Jul 2001)
14. K Dickerson, A SOM-based multimodal system for musical query-by-content, (Proceedings of the International Joint Conference on Neural Networks, San Jose, 31 Jul-5 Aug 2011), pp. 291–298

15. O Celma, P Lamere, If you like radiohead, you might like this article. AI Mag **32**(3), 57–66 (2011)
16. H Wang, W Wang, J Yang, PS Yu, Clustering by pattern similarity in large data sets, in *Proceedings of ACM Special Interest Group on Management of Data*. Madison **3–6**, 394–405 (Jun 2002)
17. D McEnnis, SJ Cunningham, Sociology and music recommendation systems, (Proceedings of the International Conference on Music Information Retrieval, Vienna, Sept 2007), pp. 185–186
18. S Pauws, S Wijdeven, User evaluation of a new interactive playlist generation concept, (Proceedings of the International Conference on Music Information Retrieval, London, 11-15 Sept 2005), pp. 638–643
19. JL Herlocker, JA Konstan, Content-independent task focused recommendation. IEEE Internet Comput **5**(6), 40–47 (2001)
20. CS Firan, W Nejdl, R Paiu, The benefit of using tag-based profiles, (Proceedings of the Latin American Web Conference, Washington DC, 31 Oct-2 Nov 2007), pp. 32–41
21. X Zhu, YY Shi, HG Kim, KW Eom, An integrated music recommendation system. IEEE Trans. Consum Electron **52**(3), 917–924 (2006)
22. A Flexer, D Schnitzer, M Gasser, G Widmer, Playlist generation using start and end songs, (Proceedings of the 9th International Conference on Music Information Retrieval, Philadelphia, 14-18 Sept 2008), pp. 173–178
23. JJ Aucouturier, F Pachet, Scaling up music playlist generation, in *Proceedings of IEEE International Conference on Multimedia and Expo*. New York **26–29**, 105–108 (Aug 2002)
24. K Kaji, K Hirata, K Nagao, A music recommendation system based on annotations about listeners' preferences and situations, (Proceedings of the First International Conference on Automated Production of Cross Media Content for Multi-Channel Distribution, 30 Nov-2 Dec 2005), pp. 231–234
25. A Andric, G Haus, Automatic playlist generation based on tracking user's listening habits. Multimedia Tools and Appl **29**(2), 127–151 (2006)
26. NH Liu, SJ Hsieh, Intelligent music playlist recommendation based on user daily behavior and music content. Lect Notes Comput Sci **5879**, 671–683 (2009)
27. J Foote, S Uchihashi, The beat spectrum: a new approach to rhythm analysis, in *Proceedings of the International Conference on Multimedia Expo*. Tokyo **22**, 881–884 (Aug 2001)
28. G Tzanetakis, P Cook, Music genre classification of audio signals. IEEE Trans. Speech Audio Process **10**, 293–302 (2002)
29. NH Liu, HY Kung, JoMP: a mobile music player agent for joggers based on user interest and pace. IEEE Trans. Consum Electron **55**(4), 2225–2233 (2009)
30. NH Liu, SJ Shieh, CF Tsai, An intelligent music playlist generator based on the time parameter with artificial neural networks. Expert Syst Appl **37**(4), 2815–2825 (2010)
31. J Foote, M Cooper, U Nam, Audio retrieval by rhythmic similarity, (Proceedings of the International Society for Music Information Retrieval Conference, Paris, 23-30 Sept 2002), pp. 265–266cbrs
32. B Shao, DD Wang, T Li, M Ogihara, Music recommendation based on acoustic features and user access patterns. IEEE Trans Audio Speech Lang Process **17**(8), 1602–1611 (2009)
33. CC Chiang, JY Wu, MT Yang, WK Tai, Independent query refinement and feature re-weighting using positive and negative examples for content-based image retrieval. Multimedia Tools and Appl **41**(1), 27–53 (2009)
34. M Negnevitsky, *Artificial Intelligence: A Guide to Intelligent Systems*, 2nd edn. (Addison-Wesley, New York, 2002)

17

Homogenous ensemble phonotactic language recognition based on SVM supervector reconstruction

Wei-Wei Liu[1,2], Wei-Qiang Zhang[1*], Michael T Johnson[3] and Jia Liu[1]

Abstract

Currently, acoustic spoken language recognition (SLR) and phonotactic SLR systems are widely used language recognition systems. To achieve better performance, researchers combine multiple subsystems with the results often much better than a single SLR system. Phonotactic SLR subsystems may vary in the acoustic features vectors or include multiple language-specific phone recognizers and different acoustic models. These methods achieve good performance but usually compute at high computational cost. In this paper, a new diversification for phonotactic language recognition systems is proposed using vector space models by support vector machine (SVM) supervector reconstruction (SSR). In this architecture, the subsystems share the same feature extraction, decoding, and N-gram counting preprocessing steps, but model in a different vector space by using the SSR algorithm without significant additional computation. We term this a homogeneous ensemble phonotactic language recognition (HEPLR) system. The system integrates three different SVM supervector reconstruction algorithms, including relative SVM supervector reconstruction, functional SVM supervector reconstruction, and perturbing SVM supervector reconstruction. All of the algorithms are incorporated using a linear discriminant analysis-maximum mutual information (LDA-MMI) backend for improving language recognition evaluation (LRE) accuracy. Evaluated on the National Institute of Standards and Technology (NIST) LRE 2009 task, the proposed HEPLR system achieves better performance than a baseline phone recognition-vector space modeling (PR-VSM) system with minimal extra computational cost. The performance of the HEPLR system yields 1.39%, 3.63%, and 14.79% equal error rate (EER), representing 6.06%, 10.15%, and 10.53% relative improvements over the baseline system, respectively, for the 30-, 10-, and 3-s test conditions.

Keywords: Phonotactic language recognition; Support vector machine (SVM) supervector reconstruction; Phone recognition-vector space modeling (PR-VSM)

1 Introduction

Spoken language recognition (SLR) refers to the task of automatic determination of language identity. It is estimated that there are about 6,000 spoken languages in the world [1]. An increasing number of multilingual speech processing applications require spoken language recognition as a frontend, with the result that SLR continues to grow in importance. Spoken language recognition is an enabling technology for a wide range of intelligence and security applications for information distillation, such as spoken document retrieval, multilingual speech recognition, and spoken language translation [2].

Language cues can be categorized according to their level of knowledge abstraction as acoustic (spectrum, phone inventory), prosodic (duration, pitch, intonation), phonotactic (sequence of sounds), lexical (vocabulary, morphology), and syntax (phrases, grammar) [3,4]. Language recognition systems are usually identified by the features they employ, e.g., acoustic systems, phonotactic systems, prosodic systems, and lexical systems. Currently, acoustic language recognition (LR) systems [5] and phonotactic LR systems [3] are both widely used.

Generally, the performance of SLR systems can be improved in two ways: (1) longitudinally, through the development of new techniques to perform the SLR

*Correspondence: wqzhang@tsinghua.edu.cn
[1] Tsinghua National Laboratory for Information Science and Technology, Department of Electronic Engineering, Tsinghua University, Beijing 100084, China
Full list of author information is available at the end of the article

tasks more precisely, e.g., i-vector [6-8], JFA [9], discriminative training [10] methods, N-gram modeling methods [3], and support vector machines (SVMs) [11]; (2) transversely, by adding variety to the SLR subsystems, which extracts and integrates more information from the utterances. State-of-the-art language recognition systems fuse multiple subsystems in parallel via a post-processing backend [12]. In the National Institute of Standards and Technology (NIST) language recognition evaluation (LRE) tasks, teams from all over the world compete to build the best SLR system and have shown that better results can be obtained by combining more subsystems, creating larger and larger SLR systems. In NIST LRE 2011, all submitted language recognition systems were stacked ensembles of at least five language recognition subsystems [13-15]. Much effort goes into trying different variations of subsystems. Generally, the phonotactic LR subsystems can be varied in three ways: (a) extracting various acoustic features to provide feature diversification, for example, Mel-Frequency Cepstral Coefficients (MFCC) [16], Perceptual Linear Predictive (PLP) [17], and Temporal Patterns Neural Network (TRAPs/NN) [18]; (b) training phone recognizers on multiple language-specific speech data to provide phonetic diversification [3], e.g., the Russian, Hungarian, Czech, and English phone recognizers developed by Brno University of Technology (BUT) [18] or universal phone recognizer (UPR) [19]; and (c) training phone recognizers on the same language-specific speech data but using different acoustic models to provide acoustic diversification [20], such as the Artificial Neural Network-Hidden Markov Model (ANN-HMM) [21], Gaussian Mixture Model-Hidden Markov Model (GMM-HMM) [22], and Deep Neural Network-Hidden Markov Model (DNN-HMM) [23]. Certainly, in phonotactic language recognition systems, the subsystems must undergo different process of feature extracting, decoding, N-gram counting, and vector space modeling, which means an added computational cost of N times than single subsystem, where N is the number of the subsystems.

This paper demonstrates an architecture to provide a new diversification for phonotactic language recognition systems. The underlying motivation of these algorithms is to provide richer language identifying information without significant additional computation. The subsystems are verified using SVM supervector reconstruction (SSR) algorithms to provide vector space modeling diversification. In this architecture, the subsystems share the same preprocessing of feature extracting, decoding, and expected counting, but models in different vector space, so we call it homogeneous ensemble phonotactic language recognition (HEPLR) system. The HEPLR subsystems increase the variety of the SVM supervector, decrease the computational cost, and improve the SLR accuracy. There are many SVM supervector reconstruction algorithms such as recurrent neuron network (RNN) SVM supervector reconstruction [24]. In this paper, we present three SVM supervector reconstruction algorithms including relative SVM reconstruction [25], functional SVM reconstruction, and perturbative SVM reconstruction.

The remainder of the paper is organized as follows. Section 2 presents the lattice-based phonotactic language recognition system used as a baseline in this paper. Section 3 describes the proposed approaches, includes relative, functional, and perturbative SVM supervector reconstruction. Section 4 demonstrates the architecture for the homogeneous ensemble phonotactic language recognition system. The experimental setup to evaluate our proposed method is described in Section 5. Results obtained in language recognition experiments on the NIST LRE 2009 database are presented and discussed in Section 6. Finally, conclusions and future work are outlined in Section 7.

2 Baseline phonotactic SLR system

In this work, the phone recognition-vector space modeling (PR-VSM) [26] phonotactic language recognition system is employed as a baseline system. The motivation behind phonotactic language recognition approach is the belief that a spoken language can be characterized by special probabilities of its lexical-phonological constraints. An N-gram vector space model (VSM) is built for the language recognition task using phone transcriptions, which is a stochastic model describing the probabilities of phoneme strings. In the PR-VSM system, each N-gram VSM of produces a likelihood score by SVM classifier to a given utterance. The languages used for training the phone recognizers need not be the same with any of those recognized.

The traditional PR-VSM language recognition system works by mapping the input utterances from data space $\mathcal{X}$ into a high-dimensional feature space $\mathcal{F}$: $\Phi : \mathcal{X} \rightarrow \mathcal{F}$ and then building linear machines in the feature space to find a maximal margin separation. The vectors built in the high-dimensional feature space are SVM supervectors, which consist of N-gram counts of features representing the phonotactics of an input speech wave sample.

In PR-VSM systems, an utterance x can be mapped to the high-dimensional feature space as follows:

$$\mathbf{\Phi} : x \rightarrow \varphi(x), \tag{1}$$

where $\varphi(x)$ is the SVM supervector computed as

$$\varphi(x) = \left[p\left(d_1|\ell_x\right), p(d_2|\ell_x), \ldots, p(d_F|\ell_x)\right], \tag{2}$$

where ℓ_x is the lattice produced from data x by a phone recognizer, d_i is the N-gram phoneme string [27] $d_i = s_i \ldots s_{i+n-1}$ $(n = N)$, and F is the dimension of the SVM supervector. $p(d_i|\ell_x)$ is the probability of the phoneme sequence d_i in the lattice, which is computed as

$$p(d_i|\ell_x) = \frac{c(d_i|\ell_x)}{\sum_i c(d_i|\ell_x)},$$

where $c(d_i|\ell_x)$ denotes the N-gram occurrence of d_i given the lattice ℓ_x. This is calculated over all possible hypotheses in the lattice as follows [27]:

$$\begin{aligned} c(s_i \dots s_{i+N-1}|\ell) &= \sum_{S\in\ell} p(S|\ell)c(s_i \dots s_{i+N-1}|S) \\ &= \sum_{s_i \dots s_{i+N-1}\in\ell} \left[\alpha(s_i)\beta(s_{i+N-1}) \prod_{j=i}^{i+N-1} \xi(s_j) \right], \end{aligned}$$

where $p(S|\ell)$ denotes the probability of the sequence S in the lattice ℓ, $\alpha(s_i)$ and $\beta(s_{i+N-1})$ are the forward probability of the starting node and the backward probability of the ending node in the N-gram $s_i \dots s_{i+N-1}$, respectively. $\xi(s_j)$ is the posterior probability of the edge phoneme s_j.

The SVM supervector $\varphi(x)$ is sent to the SVM classifier and a decision is made based on the most likely hypothesis score. In a PR-VSM system, the decision is made based on the SVM output score using

$$f(\varphi(x)) = \sum_l \alpha_l K_{\text{TFLLR}}(\varphi(x), \varphi(x_l)) + d, \tag{3}$$

where $\varphi(x_l)$ are support vectors obtained from training set using the Mercer condition. The term frequency log-likelihood ratio (TFLLR) kernel K_{TFLLR} is computed as [28]:

$$K_{\text{TFLLR}}(\varphi(x_i), \varphi(x_j)) = \sum_{q=1}^{F} \frac{p(d_q|\ell_{x_i})}{\sqrt{p(d_q|\ell_{all})}} \frac{p(d_q|\ell_{x_j})}{\sqrt{p(d_q|\ell_{all})}}, \tag{4}$$

where $p(d_i|\ell_{all})$ is calculated from the observed probability of d_i across all lattices. In this paper, the training stage is always carried out with a one-versus-rest strategy between the positive set (the samples in the target language) and negative set (all other samples).

3 SVM supervector reconstruction algorithms

The motivation behind SVM supervector reconstruction is to provide vector space modeling diversification to improve the performance of the overall language recognition system. In the language recognition system employed in this paper, we focus on how a change in the input to the SVM affects the output.

Given an SVM supervector $\varphi(x)$, we define a function ϕ_{SSR} which operates on $\varphi(x)$:

$$\mathbf{\Phi}_{\text{SSR}} : x \to \phi_{\text{SSR}}(\varphi(x)). \tag{5}$$

We are interested in understanding how $\phi_{\text{SSR}}(\varphi(x))$ affects the behavior of the output scores of the SVM. The goal is to define the relationship between $\varphi(x)$ and $\phi_{\text{SSR}}(\varphi(x))$ to enhance the variety of the supervector input.

Selecting SVM supervector reconstruction methods is an open question, so here we propose some typical methods to the implementation. In this section, three SVM supervector reconstruction methods are proposed: relative SVM supervector reconstruction, functional SVM supervector reconstruction, and perturbative SVM supervector reconstruction. Relative SVM supervector reconstruction has been presented in [25], while functional and perturbative reconstructions are new methods. Relative reconstruction is a linear reconstruction, while functional and perturbative reconstructions are non-linear ones.

3.1 Relative SVM supervector reconstruction

The relative SVM supervector method uses a relative feature approach. Relative features in contrast to absolute features, which represent directly calculable information, are defined by the relationship between an utterance and a set of selected datum utterances. We have presented the concept of relative features in [25].

Calculating relative features requires a relationship measurement, such as distance on similarity. By selecting a proper relationship measurement, the identifiable characteristics can be strengthened and nuisance attributes of the utterance can be discarded. Unlike absolute features, relative features make utterances more convenient to classify by showing the relationship between the utterances and the datum database directly.

Here, we introduce a relative SVM supervector reconstruction defined using the similarity between the utterance SVM supervectors. The widely used kernel methods offer efficient similarity measurements between two SVM supervectors. In this paper, the empirical kernel [29] is introduced into language recognition and a relativized SVM supervector developed. Kernel methods have been used for face recognition [30] and handwritten digit recognition [31] and achieved higher robustness to noise [30]. Using the SVM supervectors that are already built into a language recognition system, we can easily compose a new relativized SVM supervector with only a small increase in computation.

The architecture of the relative SVM supervector reconstruction subsystem is shown in Figure 1. To construct the SVM supervector relativization map, a database $\boldsymbol{s} = [s_1, s_2, \dots s_m]$ containing m utterances is used as the datum mark of similarity. The datum database is stochastically selected from some corpus, whose language need not be the same with the target language. $\boldsymbol{s}$ is mapped into vector space:

$$\boldsymbol{s} \to \varphi(\boldsymbol{s}) = [\varphi(s_1), \varphi(s_2), \dots, \varphi(s_m)]. \tag{6}$$

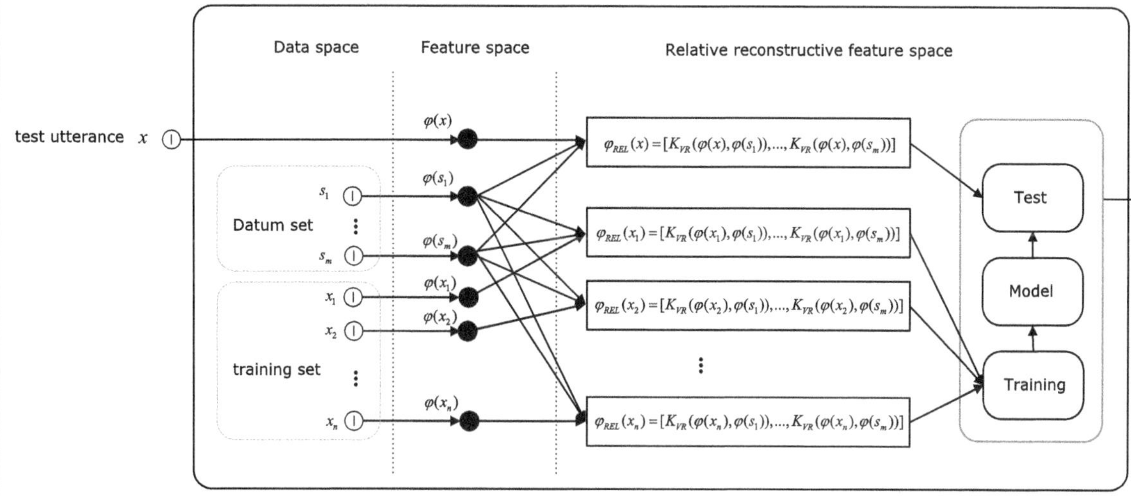

Figure 1 Architecture of relative SVM supervector reconstruction subsystem.

The vector relativizational (VR) kernel between two supervectors $\varphi(x_i)$ and $\varphi(x_j)$ is

$$K_{\mathrm{VR}}\left(\varphi(x_i), \varphi\left(x_j\right)\right) = < \varphi(x_i), \varphi(x_j) > = \sum_{q=1}^{F} \frac{p(d_q|\ell_{x_i})}{\sqrt{p(d_q|\ell_S)}} \frac{p\left(d_q|\ell_{x_j}\right)}{\sqrt{p(d_q|\ell_S)}}. \quad (7)$$

The VR kernel is similar to the TFLLR kernel, but normalized by the observed probability across all lattices of the datum dataset $p(d_i|\ell_s)$. This kernel reflects the degree of similarity between two supervectors.

The utterance x is mapped from the input data space $\mathcal{X}$ to a relativized m-dimensional Euclidean space $\mathcal{R}^m$: $\Phi_{\mathrm{REL}} : \mathcal{X} \rightarrow \mathcal{R}^m$ as follows:

$$\begin{aligned} &\mathbf{\Phi}_{\mathbf{REL}} : x \rightarrow \varphi_{\mathrm{REL}}(x) \\ &= K_{\mathrm{VR}}(\varphi(x), \varphi(\mathbf{s})) = < \varphi(x), \varphi(\mathbf{s}) > \\ &= \left[K_{\mathrm{VR}}(\varphi(x), \varphi(s_1)), \ldots, K_{\mathrm{VR}}(\varphi(x), \varphi(s_m))\right]. \end{aligned}$$

In general, $K_{\mathrm{VR}}(\varphi(x), \varphi(\boldsymbol{S}))$ defines a space in which each dimension corresponds to the similarities to a prototype. Thus, $K_{\mathrm{VR}}(\cdot, \varphi(\boldsymbol{S}))$ can be viewed as a mapping onto an m-dimensional relativized vector space.

The SVM output score is computed as

$$f'(\varphi_{\mathrm{REL}}(x)) = \sum_{l'} \alpha_{l'} K'(\varphi_{\mathrm{REL}}(x), \varphi_{\mathrm{REL}}(x_{l'})) + d', \quad (8)$$

where $\varphi_{\mathrm{REL}}(x_{l'})$ are support vectors obtained from the training set using the Mercer condition. Selecting a radial basis function (RBF) kernel, K'_{RBF} is computed as

$$K'_{\mathrm{RBF}}(\varphi_{\mathrm{REL}}(x_i), \varphi_{\mathrm{REL}}(x_j)) = \exp\left(-\frac{|\varphi_{\mathrm{REL}}(x_i) - \varphi_{\mathrm{REL}}(x_j)|^2}{D_{\mathrm{RF}}}\right), \quad (9)$$

where D_{RF} is the dimension of the relativized SVM supervector. Selecting a TFLLR kernel, K'_{TFLLR} is computed as

$$K'_{\mathrm{TFLLR}}\left(\varphi_{\mathrm{REL}}(x_i), \varphi_{\mathrm{REL}}\left(x_j\right)\right) = \sum_{q=1}^{m} \frac{K_{\mathrm{VR}}(\varphi(x_i), \varphi(s_q)) K_{\mathrm{VR}}(\varphi(x_j), \varphi(s_q))}{K_{\mathrm{VR}}(\varphi(x_{\mathrm{all}}), \varphi(s_q))}. \quad (10)$$

3.2 Functional SVM supervector reconstruction

In actual test conditions, the training and test data are variable in speakers, background noise, and channel conditions. To achieve higher robustness to variable test conditions, the widely used kernel methods offer efficient similarity measurements between two SVM supervectors in PR-VSM system [30]. The geometrical structure of the SVM vector space is completely determined by the kernel, so the selection of the kernel has a crucial impact on the performance of the language recognition systems. The functional SVM supervector reconstruction method defines a mixture between the functional and the original kernels, which can offer the robust discriminative information of the data and get robust language model.

But how to select a proper function is an open problem. There are many functions that can be used to the reconstruction, while not every function is available for the reconstruction that can reduce the equal error rate (EER). What we need to do is to find out what kind of functions can be used in feature reconstruction. The functions need to satisfy the following conditions: (1) monotonic and (2) can make the identifiable characteristics strengthened and nuisance attributes of the utterance discarded. The proposed functional SVM supervector reconstruction method does not rely on prior knowledge to select the functional to reconstruct the supervector. A development

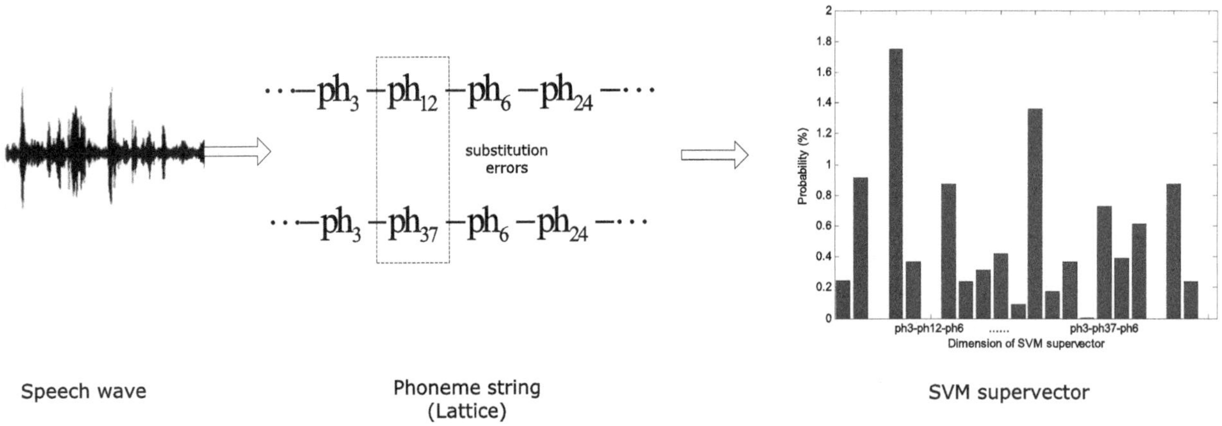

Figure 2 Effections of decoding errors on SVM supervector.

database is used for cross validation to select the function. So, here, three functions selected to be used in this paper include

(a)

$$\varphi_{\text{FUN}}(p(d_i|\ell_x)) = \sin(p(d_i|\ell_x)) + \cos(p(d_i|\ell_x)), \quad (11)$$

(b)

$$\varphi_{\text{FUN}}(p(d_i|\ell_x)) = p(d_i|\ell_x) + (p(d_i|\ell_x))^2, \quad (12)$$

(c)

$$\varphi_{\text{FUN}}(p(d_i|\ell_x)) = p(d_i|\ell_x) - (p(d_i|\ell_x))^2 + (p(d_i|\ell_x))^3. \quad (13)$$

The utterance x is mapped onto a functionalized vector space:

$$\Phi_{\text{FUN}} : x \rightarrow \varphi_{\text{FUN}}(x) = \left[\varphi_{\text{FUN}}(p(d_1|\ell_x)), \varphi_{\text{FUN}}(p(d_2|\ell_x)), \ldots, \varphi_{\text{FUN}}(p(d_F|\ell_x))\right]. \quad (14)$$

The three functions are all monotonic in the range of amplitude of the SVM supervector. Selecting a TFLLR kernel, K'_{TFLLR} is computed as

$$K'_{\text{TFLLR}}(\varphi_{\text{FUN}}(x_i), \varphi_{\text{FUN}}(x_j)) = \sum_{q=1}^{F} \frac{\varphi_{\text{FUN}}(p(d_q|\ell_{x_i}))\varphi_{\text{FUN}}(p(d_q|\ell_{x_j}))}{\varphi_{\text{FUN}}(p(d_q|\ell_{all}))}. \quad (15)$$

The SVM output score is computed as

$$f'(\varphi_{\text{FUN}}(x)) = \sum_{l'} \alpha_{l'} K'(\varphi_{\text{FUN}}(x), \varphi_{\text{FUN}}(x_{l'})) + d'. \quad (16)$$

3.3 Perturbational SVM supervector reconstruction

For spoken language recognition, the first and most essential step is to tokenize the running speech into sound units or lattices using a phone recognizer. The phoneme error rate is around 40% to 60% [32] when tokenizing an utterance. The decoding errors are deletion, insertion, and substitution errors, which are expressed as some discrete 'noise' when mapped to the high-dimensional SVM supervector space (shown in Figure 2). So, here, we introduce a perturbational denoising method for the SVM supervector. Given a supervector $\varphi(x)$ and some perturbation operator on $\varphi(x)$, we are interested in understanding how a small perturbation added to the supervector affects the behavior of the SVM [33]. This relationship can be represented using a mapping onto a perturbational vector space.

There are three purposes of proposing perturbational SVM supervector reconstruction method: first, adding perturbational noise to reduce the impact of noise in the SVM supervector introduced by the decoding errors; second, generating a more robust language model to provide input variety to the SVM classifier; and third, highlighting the most discriminative information of the SVM supervector and drowning the non-discriminative information into the perturbation (shown in Figure 3).

To accomplish the above goals, the type and strength of the perturbation must be selected carefully. How to define a proper perturbation is an open problem. There are a wide variety of perturbations,which can be categorized into multiple ways, including (1) global perturbation and local perturbation, (2) stochastic perturbation and constant perturbation according to the amplitude, (3) absolute perturbation and relative perturbation according to the relationship between the SVM supervector and the perturbation, and (4) addictive perturbation and multiplicative perturbation.

For feature supervectors in vector space, the perturbations are always discrete, maybe random in a certain range or change with the amplitude of the expected value of the supervector. So, we consider both deterministic

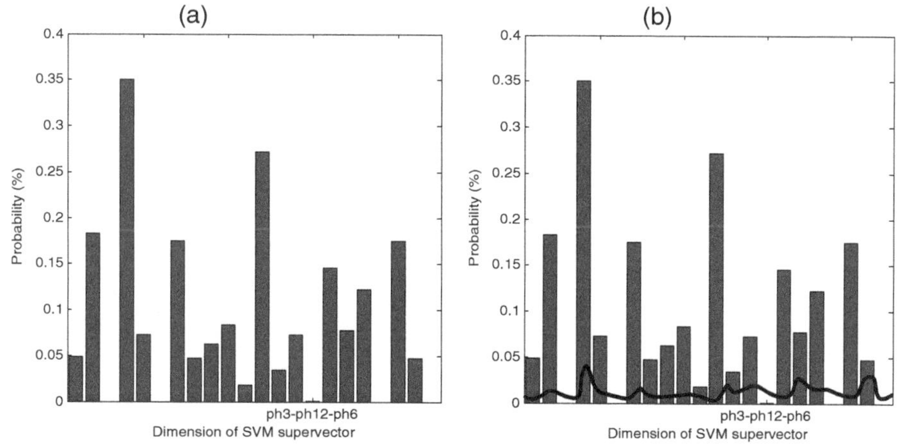

Figure 3 SVM supervector of an utterance (a) before and (b) after perturbation.

perturbation $\delta = w^{*}E_{p(d|\ell_x)}$ and stochastic perturbation $\delta = w^{*}\text{uniform}\left[0, E_{p(d|\ell_x)}\right]$, where $E_{p(d|\ell_x)}$ is the mean of the SVM supervector, uniform $\left[0, E_{p(d|\ell_x)}\right]$ is the uniform distribution between 0 and $E_{p(d|\ell_x)}$, and w^{*} is the perturbation weight. More details of the perturbation methods are discussed below.

3.3.1 Perturbational approach 1 (deterministic additive perturbation)

$$\boldsymbol{\delta} = w^{*}E_{p(d|\ell_x)}, \quad \varphi_{\text{PER}}(p(d_i|\ell_x)) = p(d_i|\ell_x) + \boldsymbol{\delta}. \tag{17}$$

This kind of perturbation represents the assumption that the expected count of every phoneme sequence is perturbed by an equivalent additive amount.

3.3.2 Perturbational approach 2 (stochastic additive perturbation)

$$\boldsymbol{\delta} = w^{*}\text{random}\left[0, E_{p(d|\ell_x)}\right], \quad \varphi_{\text{PER}}(p(d_i|\ell_x)) = p(d_i|\ell_x) + \boldsymbol{\delta}. \tag{18}$$

This perturbation represents the assumption that the expected count of every phoneme sequence is perturbed by an amount proportional to the frequency of the phoneme sequences.

3.3.3 Perturbational approach 3 (deterministic multiplicative perturbation)

$$\boldsymbol{\delta} = w^{*}E_{p(d|\ell_x)}, \quad \varphi_{\text{PER}}(p(d_i|\ell_x)) = p(d_i|\ell_x)\boldsymbol{\delta}. \tag{19}$$

This kind of perturbation represents the assumption that the expected count of every phoneme sequence perturbed by an equivalent multiple amount.

3.3.4 Perturbational approach 4 (stochastic multiplicative perturbation)

$$\boldsymbol{\delta} = w^{*}\text{random}\left[0, E_{p(d|\ell_x)}\right], \quad \varphi_{\text{PER}}(p(d_i|\ell_x)) = p(d_i|\ell_x)\boldsymbol{\delta}. \tag{20}$$

This perturbation represents the assumption that the expected count of every phoneme sequence is perturbed by a proportional to the frequency of the phoneme sequences.

From above, it can be seen that methods 1 and 2 implement absolute perturbation, and methods 3 and 4 implement relative perturbation. All are global perturbation algorithms, operating across the entire vector space. We can also investigate local perturbation using these same approaches. Local perturbation is more flexible and realistic for the noises would have effect on part of the expected counting. The proposed methods also do not rely on prior knowledge to put noising into the supervector; we use development database for cross validation to select a better perturbation.

The utterance x is mapped onto a perturbational vector space:

$$\begin{aligned}\boldsymbol{\Phi}_{\text{PER}} : x &\rightarrow \varphi_{\text{PER}}(x)\\ &= \left[\varphi_{\text{PER}}(p(d_1|\ell_x)), \varphi_{\text{PER}}(p(d_2|\ell_x)), \ldots, \varphi_{\text{PER}}(p(d_F|\ell_x))\right].\end{aligned} \tag{21}$$

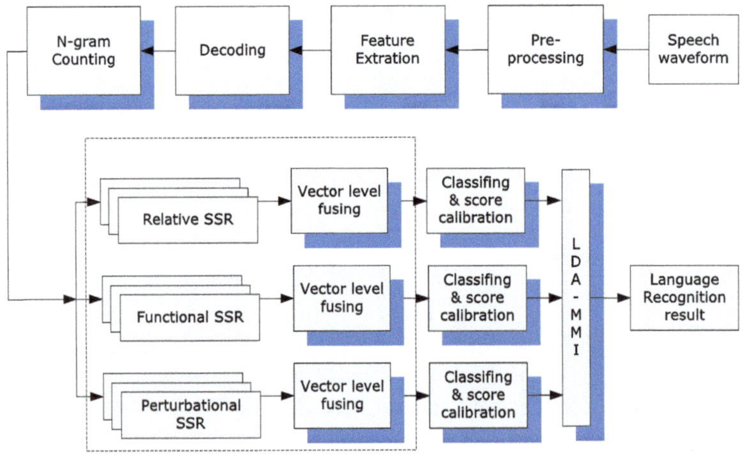

Figure 4 Architecture of the HEPLR system.

where $\varphi_{\mathrm{PER}}(x)$ is a perturbation of $\varphi(x)$. Selecting a TFLLR kernel, K'_{TFLLR} is computed as

$$K'_{\mathrm{TFLLR}}(\varphi_{\mathrm{PER}}(x_i), \varphi_{\mathrm{PER}}(x_j)) = \sum_{q=1}^{F} \frac{\varphi_{\mathrm{PER}}(p(d_q|\ell_{x_i}))\varphi_{\mathrm{PER}}(p(d_q|\ell_{x_j}))}{\varphi_{\mathrm{PER}}(p(d_q|\ell_{all}))}. \quad (22)$$

The SVM output score is computed as

$$f'(\varphi_{\mathrm{PER}}(x)) = \sum_{l'} \alpha_{l'} K'(\varphi_{\mathrm{PER}}(x), \varphi_{\mathrm{PER}}(x_{l'})) + d'. \quad (23)$$

4 Homogeneous ensemble language recognition system

The architecture of the HEPLR system is shown in Figure 4. All the SVM supervectors are reconstructed into the corresponding vector space and fused at the vector level for training and testing. In this paper, we use the classical method of vector fusion, which is to group several sets of reconstructed SVM supervectors into a large composite supervector [34]. Suppose $\varphi_{\mathrm{REL}_{N_{1_1}}}(x), \ldots, \varphi_{\mathrm{REL}_{N_{1'_m}}}(x), \varphi_{\mathrm{FUN}_{N_{2_1}}}(x), \ldots, \varphi_{\mathrm{FUN}_{N_{2'_m}}}(x)$, and $\varphi_{\mathrm{PER}_{N_{3_1}}}(x), \ldots, \varphi_{\mathrm{PER}_{N_{3'_m}}}(x)$ are the reconstructed SVM supervectors for an input utterance x. The concatenated SVM supervectors can be represented as $\varphi_{\mathrm{REL}}(x)$, $\varphi_{\mathrm{FUN}}(x)$, and $\varphi_{\mathrm{PER}}(x)$, respectively. Denoting each SVM supervector as d_m-dimensional, the concatenated SVM supervectors are $(d_1 + \ldots + d'_m)$-dimensional. The concatenated SVM supervectors are defined by

$$\varphi_{\mathrm{REL}}(x) = \left[w_{N_{1_1}} \varphi_{\mathrm{REL}_{N_{1_1}}}(x), \ldots, w_{N_{1'_m}} \varphi_{\mathrm{REL}_{N_{1'_m}}}(x)\right], \quad (24)$$

$$\varphi_{\mathrm{FUN}}(x) = \left[w_{N_{2_1}} \varphi_{\mathrm{FUN}_{N_{2_1}}}(x), \ldots, w_{N_{2'_m}} \varphi_{\mathrm{FUN}_{N_{2'_m}}}(x)\right], \quad (25)$$

$$\varphi_{\mathrm{PER}}(x) = \left[w_{N_{3_1}} \varphi_{\mathrm{PER}_{N_{3_1}}}(x), \ldots, w_{N_{3'_m}} \varphi_{\mathrm{PER}_{N_{3'_m}}}(x)\right], \quad (26)$$

where $w_{N_{j_i}} = \min_{\forall i}\left(E_{j_n}\right)/E_{j_i}$ $(j = 1, 2, 3)$ with E_{j_i} the *priori* knowledge of the EER performance of the development data of the subsystem. The logistic regression optimized weighting (LROW) method is used to optimize the reconstructed SVM supervector weighting coefficients. Since not all the SVM supervector reconstruction subsystems are effective when fused, we also extend the work by formulating quantitative measures to select the subsystems for fusion. The output score of the SVM classifier is computed as follows:

$$f^*(\varphi^*(x)) = \sum_{l^*} \alpha_{l^*} K^*(\varphi^*(x), \varphi^*(x_{l^*})) + d^*, \quad (27)$$

where the reconstruction methods are represented by means of '*', and $\varphi^*(x_{l^*})$ are support vectors obtained from the reconstructed SVM supervectors using the Mercer condition.

As mentioned previously, in the HEPLR language recognition system, the training stage is carried out between the positive set and negative set with one-versus-rest strategy.

The linear discriminant analysis-maximum mutual information (LDA-MMI) method is used to maximize the posterior probabilities of all the belief score vectors [35], using objective function [36]:

$$F_{\mathrm{MMI}}(\lambda) = \sum_{\forall i} \log \frac{p(\boldsymbol{x}_i|)\, P(g(i))}{\sum_{\forall j} p(\boldsymbol{x}_i|\lambda_j)\, P(j)}, \quad (28)$$

where $g(i)$ indicates the class label of x_i and $P(j)$ denotes the prior probability of class j. Vector fusion is implemented directly as

$$\boldsymbol{x} = \left[w'_1 f(\varphi_{\mathrm{REL}}(x)), w'_2 f(\varphi_{\mathrm{FUN}}(x)), w'_3 f(\varphi_{\mathrm{PER}}(x))\right], \quad (29)$$

The probability density function $p(\boldsymbol{x}|\lambda)$ is a Gaussian Mixture Model defined on the N-dimensional vector $\boldsymbol{x}$:

$$p(\boldsymbol{x}|\lambda) = \sum_{\forall m'} \omega'_{m'} \mathcal{N}(x; \mu_{m'}, \Sigma_{m'}), \quad (30)$$

Table 1 Performance of baseline language recognition system

	SVM supervector	30 s		10 s		3 s	
	dimension	EER	*Cavg*	EER	*Cavg*	EER	*Cavg*
HU	195112	2.44	2.39	7.54	7.38	24.23	23.98
RU	117649	2.26	2.06	6.23	6.07	20.53	20.38
CZ	74088	3.39	3.31	10.13	10.04	28.73	28.35

NIST LRE 2009 (EER and Cavg in percent).

Table 3 Performance of functional SVM supervector reconstruction subsystems

	30 s		10 s		3 s	
	EER	*Cavg*	EER	*Cavg*	EER	*Cavg*
HU	2.13	2.09	6.26	6.24	19.16	19.21
RU	2.06	1.98	5.40	5.34	17.64	17.71
CZ	2.86	2.84	8.36	8.47	22.89	23.12

NIST LRE 2009 (EER and Cavg in percent).

The proposed homogeneous ensemble language recognition system has three advantages. First, SVM supervector reconstruction provides vector space modeling diversification for richer language identification information. Second, in the HEPLR system, the subsystems share the same preprocessing steps for feature extraction, decoding, and expected counting, which minimizes additional computational cost. Third, fusing the reconstructed SVM supervector with the original supervector at the vector level means that more information can be retained than that given by score fusion.

5 Experimental setup

5.1 Baseline language recognition system setup

The TRAPs/NN phonotactic language recognizers developed by the BUT [37] based on phone lattices, N-gram counts, and SVM scoring are used as baseline systems. An energy-based voice activity detector that splits and removes long-duration non-speech segments from the signals is applied initially. Following this, the BUT decoders for Czech (CZ), Hungarian (HU), and Russian (RU) are applied to compute phone *posteriori* probabilities, as used in NIST LRE tasks by many groups [38,39]. The phone inventory is 43 for Czech, 59 for Hungarian, and 50 for Russian. *Posteriori* probabilities are put into the HVite decoder produced by HTK to produce phone lattices, which encode multiple hypotheses with acoustic likelihoods. The N-gram counts are produced by lattice-tool from SRILM (SRI International, Menlo Park, CA, USA) [40]. The LIBLINEAR tool [41] for multiclass SVMs with linear kernels is applied to give SVM scores. Finally, the LDA-MMI algorithm [42] is used for score calibration and fusion.

5.2 Training, development, and test datasets

Evaluation is carried out on the NIST LRE 2009 tasks. This data includes 41793 utterances including 30-, 10-, and 3-s nominal duration, closed condition. The NIST LRE 2009 core task recognition of is to recognize 23 languages, including Amharic, Bosnian, Cantonese, Creole, Croatian, Dari, American English, Indian English, Farsi, French, Georgian, Hausa, Hindi, Korean, Mandarin, Pashto, Portuguese, Russian, Spanish, Turkish, Ukrainian, Urdu, and Vietnamese. The evaluation involves radio broadcasts and conversational telephone speech channel conditions.

The training data comes from different sources including CallHome, CallFriend, OGI, OHSU, VOA, and the development corpora for the 2003, 2005, and 2007 NIST LRE evaluations.

About 25,000 utterances are selected randomly from VOA and 2003, 2005, and 2007 NIST LRE datasets used as development data.

5.3 Evaluation measures

In this work, the performance of language recognition systems is compared using: (1) EER and (2) average cost performance Cavg defined by NIST [43], which are obtained by one-versus-rest tragedy.

6 Experimental results and discussion

We demonstrate the effectiveness of our approaches on NIST LRE 2009 tasks under 30-, 10-, and 3-s conditions. Results are shown in Tables 1, 2, 3, 4, 5, and 6 and Figures 1, 2, 3, 4, 5, 6, 7, and 8 in the following sections. The EER and Cavg performance of individual subsystems and fusions is also shown in the tables below for reference.

Table 2 Performance of relative SVM supervector reconstruction subsystem, TFLLR and RBF kernel

	30 s		10 s		3 s	
	EER	*Cavg*	EER	*Cavg*	EER	*Cavg*
HU	2.38	2.40	7.11	7.17	20.32	20.49
RU	2.01	1.92	5.83	5.77	17.26	17.34
CZ	3.14	3.09	8.47	8.53	23.12	23.17

NIST LRE 2009 (EER and Cavg in percent).

Table 4 Performance of perturbational SVM supervector reconstruction subsystems

	30 s		10 s		3 s	
	EER	*Cavg*	EER	*Cavg*	EER	*Cavg*
HU	2.14	2.10	6.78	6.84	20.39	20.49
RU	2.11	1.96	5.91	5.85	18.67	18.55
CZ	2.95	2.87	8.76	8.71	25.35	25.24

NIST LRE 2009 (EER and Cavg in percent).

Table 5 Comparison of baseline SLR system and HEPLR system

	30 s		10 s		3 s	
	EER	Cavg	EER	Cavg	EER	Cavg
HU	2.44	2.39	7.54	7.38	24.23	23.98
RU	2.26	2.06	6.23	6.07	20.53	20.38
CZ	3.39	3.31	10.13	10.04	28.73	28.35
Fusion	1.52	1.58	4.04	4.05	16.53	16.10
HU-SSR	1.92	1.84	5.89	5.91	18.64	18.73
RU-SSR	1.82	1.77	5.21	5.14	16.76	16.51
CZ-SSR	2.86	2.84	8.36	8.47	22.89	22.93
Fusion-SSR	1.39	1.29	3.63	3.64	14.79	14.64

NIST LRE 2009 (EER and Cavg in percent).

6.1 Baseline PRVSM system

Table 1 shows EER and Cavg performance for the NIST LRE 2009 language recognition tasks using the baseline subsystems. In this work, the dimension of the possible 3-gram SVM supervector produced by the single Hungarian (HU) phone recognizer with 58 phones is $58 \times 58 \times 58 = 195112$. SVM supervector dimensions for the Russian (RU) and Czech (CZ) recognizers are 117649 and 74088, respectively.

6.2 Relative SVM supervector reconstruction

In this paper, 13,000 conversations which are randomly selected from the 40 languages of the 2003, 2005, and 2007 NIST LRE and VOA, CallHome, and CallFriend Corpora. These are used as the dataset to build the relative SVM supervector reconstructor.

Figure 5 shows the performance of the relative SVM supervector reconstruction subsystems whose SVM classifier uses the TFLLR and RBF kernel, respectively. Figure 5 also shows the performance of the relative SVM supervector reconstruction subsystems whose SVM classifier uses fusion of TFLLR and RBF kernel. Table 2 shows that the performance of the relative SVM supervector reconstruction subsystem is better than baseline and increases slowly with increasing size of the datum dataset.

Table 6 Comparison of real-time factor for language recognition systems

	Baseline	Relative SSR	Functional SSR	Perturbational SSR
Decoding	0.11	0.11	0.11	0.11
SV prod.	2.63×10^{-6}	0.06	2.67×10^{-6}	2.64×10^{-6}
Total	0.11	0.17	0.11	0.11

HU frontend, NIST LRE 2009, 30-s test. CPU: Xeon E5520@2.27 GHz, RAM: 8 GB, single thread. SV prod., super vector product.

A lower dimension relative SVM supervector can be chosen to trade off performance and computation cost. The language recognition performance for short utterances is significantly improved in the relative SSR subsystem compared to the baseline. The original supervector can not describe short utterances precisely because of insufficient phoneme data, while the relative reconstructed SVM supervectors use the relationship between a short utterance and a large set of datum utterances, which is a richer representation. The experimental results show that the relative SSR subsystem outperforms the baseline and can obtain better performance with relative feature using a low-dimension SVM supervector.

6.3 Functional SVM supervector reconstruction

The language recognition results of functional SVM supervector reconstruction subsystems are given in Figure 6 and Table 3. The results using this approach were similar to or slightly worse than the baseline system in 30-s test condition, but outperform the baseline system in the 10- and 3-s test conditions.

6.4 Perturbational SVM supervector reconstruction

Figure 7 and Table 4 describe the results of the four perturbational methods. Overall, approach 2 yielded better results (2.20%, 6.59%, 20.93% EER) than the other approaches. The perturbative SVM supervector reconstruction subsystems performed consistently better than the baseline subsystem; particularly, those based on approach 2 performed better than the others. We hypothesize that approach 2 outperforms other perturbation methods because the distribution of the perturbation better matches the distribution of the noise. The perturbation approach adds robustness to the language modeling.

6.5 SVM supervector reconstruction

From the experimental results and discussion, it can be concluded that some of the reconstruction methods (relative SSR) are better at identifying the language of the short utterance and others (functional SSR and perturbational SSR) are better at recognizing long utterances. Because these errors are not highly correlated, we can fuse these results together to harness the complementary behavior among subsystems and improve the language recognition performance.

Figure 8 shows DET curves of the baseline system versus the HEPLR system for NIST LRE 2009. Table 5 gives the corresponding performance numbers for all configurations. These results show that the SSR approaches proposed in this paper outperformed the baseline system in terms of EER and Cavg when considering complete fusions for the subsystems.

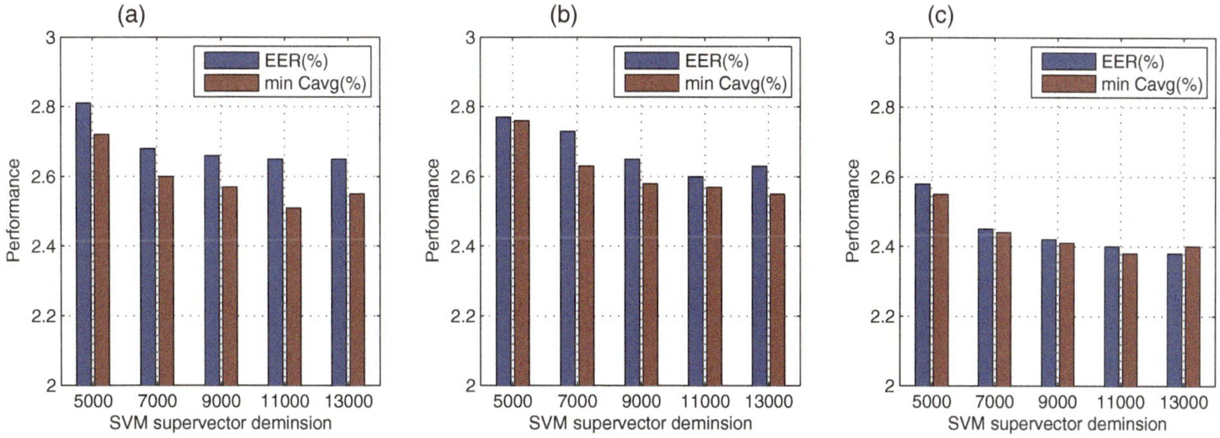

Figure 5 Performance of relative SVM supervector reconstruction subsystem versus dimension. NIST LRE 2009, 30-s, HU frontend (EER and Cavg in percent). **(a)** TFLLR kernel. **(b)** RBF kernel. **(c)** TFLLR and RBF kernel.

6.6 Real-time factors

Table 6 shows the real-time (RT) factors of each part of SSR system. From Table 6, we can see that decoding is the dominant part. Compared to PR-VSM baseline system, the computational cost increases about 1.5 times for the relative SSR and barely no increases for the functional SSR and perturbational SSR.

7 Computational cost

Let F, M, and M_{datum} denote the dimension of the phonotactic feature supervector of an utterance, the number of utterances of training dataset, and the datum dataset, respectively. And let c_{φ} denote the computation cost of the mapping from x to $\varphi(x)$, and $c_{\text{modeling}}(F, M)$ denotes the computational cost of modeling the languages, which relate to F and M. Then, the computational cost of the baseline system is

$$c_{\text{baseline}} = M \cdot c_{\varphi} + c_{\text{modeling}}(F, M) \quad (31)$$

$$c_{\varphi} = c_{\text{Pre-Processing}} + c_{\text{FeatureExtract}} + c_{\text{Decoding}} + c_{\text{N-gramCounting}}, \quad (32)$$

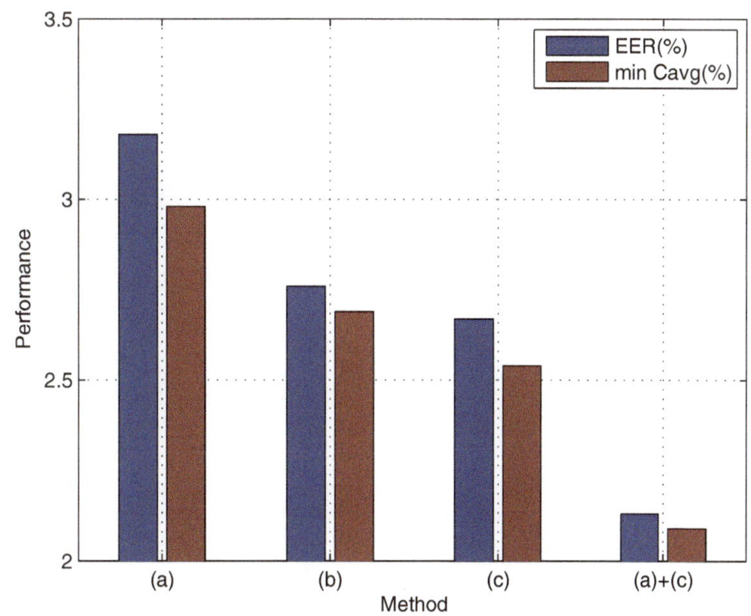

Figure 6 Performance of functional SVM supervector reconstruction subsystems. NIST LRE 2009, 30-s, HU frontend. '+' indicated fusion (EER and Cavg in percent).

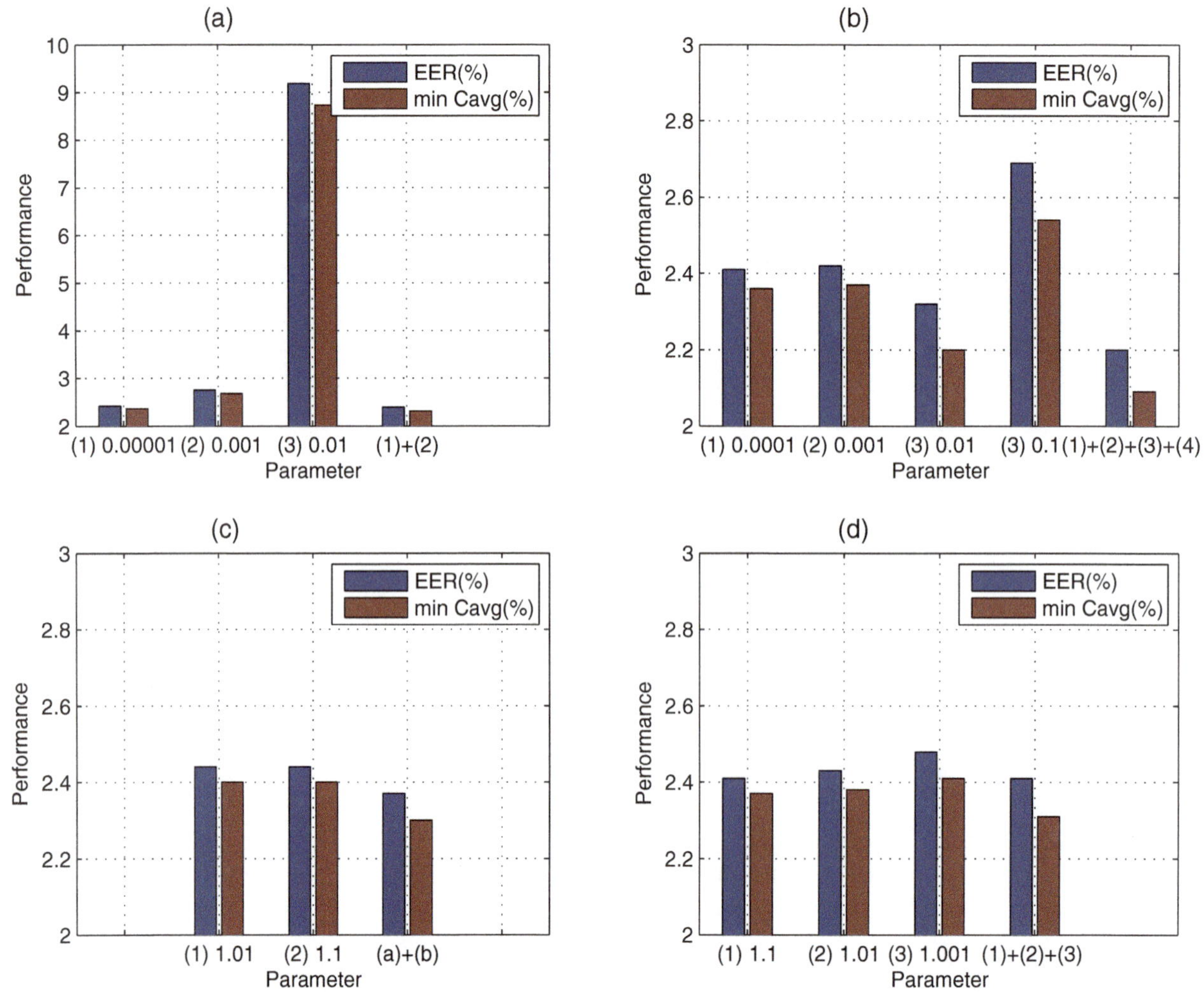

Figure 7 Performance of perturbational SVM supervector reconstruction subsystems. NIST LRE 2009, 30-s, HU frontend (EER and Cavg in percent). **(a)** Approach 1. **(b)** Approach 2. **(c)** Approach 3. **(d)** Approach 4.

where $c_{\text{Pre-Processing}}$, $c_{\text{FeatureExtract}}$, c_{Decoding}, and $c_{\text{N-gramCounting}}$ denote the computational cost of pre-processing, feature extracting, decoding, and N-gram counting, respectively.

7.1 Relative SVM supervector reconstruction

Let $c_{\text{inner}}(F)$ denote the computation cost of the inner product of the two F dimensional supervectors. Then, the computational cost of the relative SVM supervector reconstruction system is computed as

$$c_{\text{REL}} = M \cdot c_{\varphi} + M_{\text{datum}} \cdot c_{\varphi} + c_{\text{modeling}}(M_{\text{datum}}, M) + M \cdot M_{\text{datum}} \cdot c_{\text{inner}}(F) \tag{33}$$

Usually, $c_{\text{modeling}}(M_{\text{datum}}, M) < c_{\text{modeling}}(F, M) \ll M \cdot c_{\varphi}$, $M \cdot M_{\text{datum}} \cdot c_{\text{inner}}(F) \ll M \cdot c_{\varphi}$, so

$$\frac{c_{\text{REL}}}{c_{\text{baseline}}} \approx \frac{M \cdot c_{\varphi} + M_{\text{datum}} \cdot c_{\varphi}}{M \cdot c_{\varphi}} = 1 + \frac{M_{\text{datum}}}{M} \tag{34}$$

In this paper, $M = 30996$, when considering RU frontend, then $F = 117649$. When $M_{\text{datum}} = 13000$, $c_{\text{REL}}/c_{\text{baseline}} = 41.94\%$. That means that the relative SVM supervector reconstruction system takes 41.94% extra computation and achieves a 11.84%, 6.42%, and 15.92% relative improvements, respectively, for 30-, 10-, and 3-s compared to the baseline.

7.2 Functional SVM supervector reconstruction

Let $c_{\varphi_{\text{FUN}}}$ denote the computational cost of mapping $\varphi(x)$ to $\varphi_{\text{FUN}}(x)$. Then, the computational cost of the functional SVM supervector reconstruction system is computed as

$$c_{\text{FUN}} = M \cdot c_{\varphi} + M \cdot c_{\varphi_{\text{FUN}}} + c_{\text{modeling}}(F, M), \tag{35}$$

because preprocessing, feature extracting, decoding, and N-gram counting are more complex than the functional computation in this paper, so $M \cdot c_{\varphi_{\text{FUN}}} \ll M \cdot c_{\varphi}$. The computational cost of modeling the languages can be

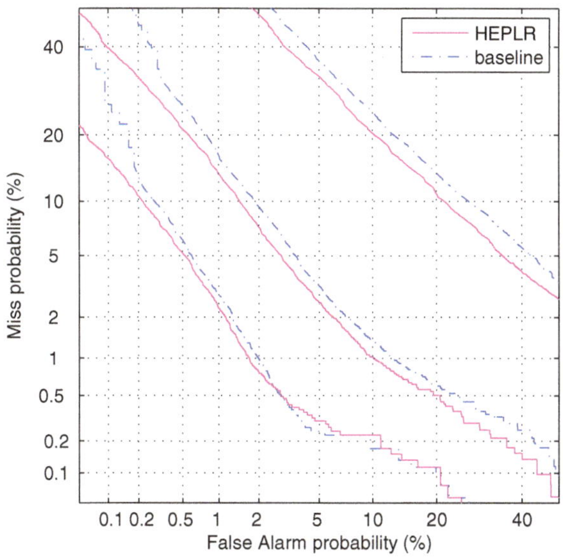

Figure 8 DET curves of baseline system and HEPLR system for NIST LRE 2009.

considered equal to the baseline. For RU frontend, the functional SVM supervector reconstruction system takes almost no extra computation and achieves 8.84%, 13.32%, and 14.76% relative improvements, respectively, for 30-, 10-, and 3-s compared to the baseline.

7.3 Perturbational SVM supervector reconstruction

Let $c_{\varphi_{\text{PER}}}$ denote the computational cost of adding perturbation to $\varphi(x)$. Then, the computational cost of the functional SVM supervector reconstruction system is computed as

$$c_{\text{PER}} = M \cdot c_{\varphi} + M \cdot c_{\varphi_{\text{PER}}} + c_{\text{modeling}}(F, M), \tag{36}$$

because preprocessing, feature extracting, decoding, and N-gram counting are more complex than the perturbational computation in this paper, so $M \cdot c_{\varphi_{\text{PER}}} \ll M \cdot c_{\varphi}$. The computational cost of modeling the languages can be considered equal to the baseline. For RU frontend, the perturbational SVM supervector reconstruction system takes almost no extra computation and achieves 6.63%, 5.13%, and 9.05% relative improvements, respectively, for 30-, 10-, and 3-s compared to the baseline.

8 Conclusions

In this article, we investigate a strategy of SVM supervector reconstruction to provide vector space modeling diversification to improve the performance and robustness of language recognition tasks with very low additional computational cost. A variety of SVM supervector reconstruction methods are employed to develop the diversified SVM supervectors. Reconstruction methods include relative SSR, perturbational SSR, and functional SSR. Relative SSR method uses the relationship of an utterance and a datum set to present the utterance.

Perturbational SSR reconstructs the SVM supervector to a slightly perturbational version and improves the language recognition performance. Functional SSR can derive effective kernel mixtures and get robust language model. The approaches do not involve significant additional computation compared to a baseline phonotactic system, but represents a way to extract more information from existing decodings.

Experimental results of the proposed HEPLR system on the NIST LRE 2009 evaluation set show better performance than the baseline system. When we fuse the three subsystems at the score level for further improvements, we achieve 1.39%, 3.63%, and 14.79% EER for the 30-, 10-, and 3-s closed-set test conditions, respectively. This corresponds to 6.06%, 10.15%, and 10.53% relative improvements.

Competing interests
The authors declare that they have no competing interests.

Acknowledgements
This project is supported by the National Natural Science Foundation of China under grant nos. 61370034, 61273268, and 61403224.

Author details
[1]Tsinghua National Laboratory for Information Science and Technology, Department of Electronic Engineering, Tsinghua University, Beijing 100084, China. [2]General Communication Station, General Logistics Department, Beijing 100036, China. [3]Department of Electrical and Computer Engineering, Marquette University, Milwaukee, WI 53233, USA.

References

1. MP Lewis, Ethnologue: languages of the world, 16th edn. (SIL International, 2009). http://www.ethnologue.com. Accessed 17 Jan 2007
2. VW Zue, JR Glass, Conversational interfaces: advances and challenges. Proc. IEEE. **88**(8), 1166–1180 (2000)
3. MA Zissman, Comparison of four approaches to automatic language identification of telephone speech. IEEE Trans. Speech Audio Proc. **4**(1), 31–44 (1996)
4. YK Muthusamy, E Barnard, RA Cole, Reviewing automatic language identification. IEEE Signal Process. Mag. **11**(4), 33–41 (1994)
5. PA Torres-Carrasquillo, E Singer, MA Kohler, RJ Greene, DA Reynolds, JR Deller, in *Proceedings of the International Conference on Spoken Language Processing (ICSLP)*, Denver, 16–20. Approaches to language identification using Gaussian mixture models and shifted delta cepstral features, (Sept 2002), pp. 33–36
6. P Matejka, O Glembek, F Castaldo, MJ Alam, O Plchot, P Kenny, L Burget, J Cernocky, in *Proceedings of the IEEE International Conference on Acoustics, Speech and Signal Processing (ICASSP)*, Prague, 22–27. Full-covariance UBM and heavy-tailed PLDA in i-vector speaker verification, (May 2011), pp. 4828–4831
7. N Dehak, PJ Kenny, R Dehak, P Dumouchel, P Ouellet, Front-end factor analysis for speaker verification. IEEE Trans. Audio Speech Language Process. **19**(4), 788–798 (2011)
8. LF Dharo Enriquez, O Glembek, O Plchot, P Matejka, M Soufifar, Rd Cordoba Herralde, J Cernocky, in *Proceedings of INTERSPEECH*, Oregon, 9–13. Phonotactic language recognition using i-vectors and phoneme posteriogram counts, (Sept 2012), pp. 42–45

9. D Garcia-Romero, C Espy-Wilson, in *Proceedings of Odyssey—The Speaker and Language Recognition Workshop*, Brno, 7–11. Joint factor analysis for speaker recognition reinterpreted as signal coding using overcomplete dictionaries, (July 2010), pp. 117–124
10. G Heigold, H Ney, R Schluter, S Wiesler, Discriminative training for automatic speech recognition: modeling, criteria, optimization, implementation, and performance. IEEE Signal Process. Mag. **29**(6), 58–69 (2012)
11. WM Campbell, JP Campbell, DA Reynolds, E Singer, PA Torres-Carrasquillo, Support vector machines for speaker and language recognition. Comput. Speech Language. **20**(2), 210–229 (2006)
12. H Li, B Ma, K-A Lee, Spoken language recognition: from fundamentals to practice. Proc. IEEE. **101**(5), 1136–1159 (2013)
13. E Singer, P Torres-Carrasquillo, D Reynolds, A McCree, F Richardson, N Dehak, D Sturim, in *Proceedings of Odyssey—The Speaker and Language Recognition Workshop*, Singapore, 25–28. The MITLL NIST LRE 2011 language recognition system, (June 2012), pp. 209–215
14. LJ Rodriguez-Fuentes, M Penagarikano, A Varona, M Diez, G Bordel, A Abad, D Martinez, J Villalba, A Ortega, E Lleida, in *Proceedings of INTERSPEECH*, Oregon, 9–13. The BLZ submission to the NIST 2011 LRE: data collection, system development and performance, (Sept 2012), pp. 38–41
15. M Penagarikano, A Varona, LJ Rodriguez-Fuentes, M Diez, G Bordel, in *Proceedings of the NIST 2011 LRE Workshop*. University of the Basque Country (EHU) systems for the 2011 NIST language recognition evaluation (Gaithersburg 6–7, Dec 2011), pp. 1–5
16. F Zheng, G Zhang, Z Song, Comparison of different implementations of MFCC. J Comput. Sci. Technol. **16**(6), 582–589 (2001)
17. A Zolnay, R Schlüter, H Ney, in *Proceedings of the IEEE International Conference on Acoustics, Speech and Signal Processing (ICASSP)*. Acoustic feature combination for robust speech recognition (Philadelphia, 18–23, March 2005), pp. 457–460
18. V Hubeika, L Burget, P Matejka, P Schwarz, in *Proceedings of INTERSPEECH*. Discriminative training and channel compensation for acoustic language recognition (Brisbane, 22–26, Sept 2008), pp. 301–304
19. H Li, B Ma, C-H Lee, A vector space modeeling approach to spoken language identification. IEEE Trans. Audio Speech Language Process. **15**(1), 271–284 (2007)
20. KC Sim, H Li, On acoustic diversification front-end for spoken language identification. IEEE Trans. Audio Speech Language Process. **16**(5), 1029–1037 (2008)
21. N Morgan, H Bourlard, in *Proceedings of the IEEE International Conference on Acoustics, Speech, and Signal Processing (ICASSP)*. Continuous speech recognition using multilayer perceptrons with hidden Markov models (Kobe, 18–22, Nov 1990), pp. 413–416
22. PA Torres-Carrasquillo, *Language identification using Gaussian mixture models*. (PhD thesis, Michigan State University, 2002)
23. W-W Liu, M Cai, H Yuan, J Xu, J Liu, W-Q Zhang, in *Proceedings of the IEEE International Symposium on Chinese Spoken Language Processing (ISCSLP)*. Phonotactic Language Recognition Based on DNN-HMM Acoustic Model (Singapore, 12–14, Sept 2014), pp. 148–152
24. W-W Liu, W-Q Zhang, Y Shi, A Ji, J Xu, J Liu, in *Proceedings of the IEEE International Conference on Acoustics, Speech and Signal Processing (ICASSP)*. Improved phonotactic language recognition based on RNN feature reconstruction (Florence, 4–9, May 2014), pp. 5359–5363
25. W-W Liu, W-Q Zhang, Z Li, J Liu, in *Proceedings of INTERSPEECH*. Parallel absolute-relative feature based phonotactic language recognition (Lyon, 25–29, Aug 2013), pp. 2474–2478
26. WM Campbell, JP Campbell, DA Reynolds, E Singer, PA Torres-Carrasquillo, Support vector machines for speaker and language recognition. Comput. Speech Language. **20**(2–3), 210–229 (2006)
27. JL Gauvain, A Messaoudi, H Schwenk, in *Proceedings of the International Conference on Spoken Language Processing (ICSLP)*. Language recognition using phone lattices (Jesu Land, 4–8, Oct 2004), pp. 1283–1286
28. WM Campbell, JP Campbell, DA Reynolds, DA Jones, TR Leek, Phonetic speaker recognition with support vector machines. Adv. Neural Inf. Process. Syst. **16**, 1377–1384 (2003)
29. B Scholkopf, J Weston, E Eskin, C Leslie, WS Noble, in *Proceedings of the European Conference on Machine Learning (ECML)*. Kernel approach for learning from almost orthogonal patterns (Helsinki, 19–23, Aug 2002), pp. 511–528
30. M Wang, S Chen, Enhanced FMAM based on empirical kernel map. IEEE Trans. Neural Networks. **16**(3), 557–564 (2005)
31. H Xiong, MNS Swamy, MO Ahmad, Optimizing the kernel in the empirical feature. IEEE Trans. Neural Networks. **16**(2), 460–474 (2005)
32. P Matejka, P Schwarz, J Cernocký, P Chytil, in *Proceedings of INTERSPEECH*. Phonotactic language identification using high quality phoneme recognition (Lisbon, 4–8, Sept 2005), pp. 2237–2240
33. P Vincent, H Larochelle, I Lajoie, Y Bengio, P-A Manzagol, Stacked denoising autoencoders: learning useful representations in a deep network with a local denoising criterion. J. Machine Learning Res. **11**, 3371–3408 (2010)
34. H Li, B Ma, C-H Lee, A vector space modeling approach to spoken language identification. IEEE Trans. Audio Speech Language Process. **15**(1), 271–284 (2007)
35. P Matejka, L Burget, O Glembek, P Schwarz, V Hubeika, M Fapso, T Mikolov, O Plchot, in *Proceedings of the NIST Language Recognition Evaluation Workshop*. BUT system description for NIST LRE 2007 (Orlando, 11–12, Dec 2007), pp. 1–5
36. D Povey, PC Woodland, in *Proceedings of the IEEE International Conference on Acoustics, Speech and Signal Processing (ICASSP)*. Improved discriminative training techniques for large vocabulary continuous speech recognition (Prague, 22–27, May 2011), pp. 45–48
37. P Schwarz, *Phoneme recognition based on long temporal context*. (PhD thesis, Brno University of Technology, 2009)
38. Z Jancik, O Plchot, N Brummer, L Burget, O Glembek, V Hubeika, M Karafiat, P Matejka, T Mikolov, A Strasheim, J Cernocky, in *Proceedings of Odyssey—The Speaker and Language Recognition Workshop*. Data selection and calibration issues in automatic language recognition-investigation with but-agnitio NIST LRE, 2009 system (Brno, 7–11, July 2010), pp. 215–221
39. PA Torres-Carrasquillo, E Singer, T Gleason, A McCree, DA Reynolds, F Richardson, D Sturim, in *Proceedings of the IEEE International Conference on Acoustics Speech and Signal Processing (ICASSP)*. The MITLL NIST LRE 2009 language recognition system (Dalas, 14–19, March 2010), pp. 4994–4997
40. A Stolcke, *SRILM - an extensible language modeling toolkit*. (SRI International, 2002. http://www.speech.sri.com/projects/srilm/. Accessed 3 April 2002
41. R Collobert, S Bengio, SVMTorch: support vector machines for large-scale regression problems. J. Machine Learning Res. **1**, 143–160 (2001)
42. W-Q Zhang, T Hou, J Liu, Discriminative score fusion for language identification. Chin. J. Electron. **19**, 124–128 (2010)
43. The 2009 NIST language recognition evaluation plan. (U.S. Department of Commerce) http://www.itl.nist.gov/iad/mig/tests/lang/2009/. Accessed 1 April 2009

18

Simulation of tremulous voices using a biomechanical model

Rubén Fraile[1*], Juan Ignacio Godino-Llorente[1] and Malte Kob[2]

Abstract

Vocal tremor has been simulated using a high-dimensional discrete vocal fold model. Specifically, respiratory, phonatory, and articulatory tremors have been modeled as instabilities in six parameters of the model. Reported results are consistent with previous knowledge in that respiratory tremor mainly causes amplitude modulation of the voice signal while laryngeal tremor causes both amplitude and frequency modulation. In turn, articulatory tremor is commonly assumed to produce only amplitude modulations but the simulation results indicate that it also produces a high-frequency modulation of the output signal. Furthermore, articulatory tremor affects the frequency response of the vocal tract and it might thus be detected by analyzing the spectral envelope of the acoustic signal.

Keywords: Vocal tremor; Biomechanical modeling; Voice production modeling and simulation

Introduction

Tremor can be defined as an *involuntary oscillatory movement of a body part* [1]. Correspondingly, vocal tremor can be defined as a *low-frequency fluctuation* (i.e. modulation) *in the amplitude or frequency (or both)* of voice [2]. Vocal tremor happens at a rate well below the fundamental frequency, between 1 and 15 Hz [2]. It differs from voice amplitude and frequency perturbations in that while these are random and fast deviations from stable underlying values of amplitude and fundamental frequency, vocal tremor involves an instability in such values [2]. Similarly, vibrato is a modulation of the acoustic voice signal too, but vocal tremor differs from it in that for vibrato, the span of typical modulation rates is narrower (between 4 and 7 Hz) [2] and the modulation is more cyclic [3]. While vocal tremor happens during speaking, vibrato is a feature of the singing voice [2] that may express emotions such as anger, happiness, fear, or sadness [4].

The presence of vocal tremor in the acoustic signal is better perceived in the phonation of sustained sounds than in running speech [5,6]. This is coherent with the nature of voice tremor, that is, instabilities in the voice production system are better noticed when a stable phonation task is performed. Acoustic analysis of the voice aimed at detecting, measuring, or modeling instabilities due to vocal tremor is usually approached by seeking for modulations in the sequence of pitch periods (e.g. [7]). The presence of such modulations by itself is not an indicator of dysphonia; instead, the presence of tremor is necessary for a natural sounding voice [2]. In fact, vocal tremor in healthy individuals (physiological tremor) has been measured, showing modulation rates below 5 Hz [8]. However, the modulation rate is not definite in discriminating between physiological and pathological voice tremors. Jiang et al. and Anand et al. showed that for pathological voice tremors, with rates between 4 and 7 Hz, the difference with physiological voice tremor was more in the modulation extent (or modulation amplitude) than in the modulation rate [9,10].

The fact that vocal tremor is more related to modulation extents than to modulation rates is consistent with results indicating that the severity of tremor is correlated with acoustic measures of amplitude and frequency perturbations [11]. This is due to perturbation measures such as *jitter* or *shimmer* not including information on the autocorrelation properties of the varying magnitude; i.e. they are perturbation functions of order 0 [12]. As a consequence, they measure the same the extent of random and rapid perturbations as the extent of lower frequency modulations.

*Correspondence: rfraile@ics.upm.es
[1] Signal Theory & Communications Department, Universidad Politécnica de Madrid, Campus Sur - Carretera de Valencia km 7, 28031 Madrid, Spain
Full list of author information is available at the end of the article

Regarding perception, frequency modulation seems to be the most relevant feature for the perceptual detection of vocal tremor [10,13] but tremor is only perceived if its intensity (i.e. modulation extent) is above a certain threshold [6].

From the point of view of the anatomy of the voice production system, vocal tremor may be produced by three different sources [14,15]:

- *Respiratory system:* Instabilities in the respiratory system can typically be detected by analyzing the voice intensity contours [16]. Some results from the computer simulation of respiratory vocal tremor also point out that this is more related to amplitude modulations than to frequency modulations [15], but more recent measurements from *in vivo* induction of respiratory tremor indicate that it also affects fundamental frequency [17].
- *Phonatory system:* To the greatest extent, tremor in the phonatory system, or laryngeal tremor, is caused by irregular tension patterns in laryngeal muscles [18], mainly the thyroarytenoid (TA), the cricothyroid (CT), and the inter-arytenoid (IA) (cf. Figure 1). The activity of the laryngeal muscles has a relevant impact on the fundamental frequency of voice [19-21], although the effect of interactions between such muscles is not fully understood yet [22] and the relation between muscles tension and fundamental frequency is complex [23]. Accordingly, the main effect of laryngeal tremor is commonly assumed to be the frequency modulation of the acoustic wave [24]. However, amplitude modulation is also expected to happen due to the interaction between harmonics and formants [25].
- *Articulatory system:* Articulatory tremor is produced by instabilities in the position of the articulators (e.g. jaw or tongue), position of the epiglottis, width of the epilarynx, etc. Similarly to respiratory tremor, articulatory tremor is thought to affect voice amplitude [24] but also to produce correlations between airflow magnitude and pressure wave amplitudes different to the case of respiratory tremor [14].

Regarding simulation of vocal tremor, only a limited number of experiments have been reported in the scientific literature up to now. Hanquinet et al. developed a synthesizer of disordered voices in which simulation of vocal tremor was possible [26]. Their simulation approach was based on modeling the acoustic voice signal, not the underlying biomechanical process that produces it. Zhang and Jiang used a two-mass model of the vocal folds to produce simulated tremulous voices with the objective of studying the nonlinear dynamics of vocal tremor [27]. Lester, Barkmeier-Kraemer, and Story utilized a kinematic model of the vocal folds to simulate tremor produced

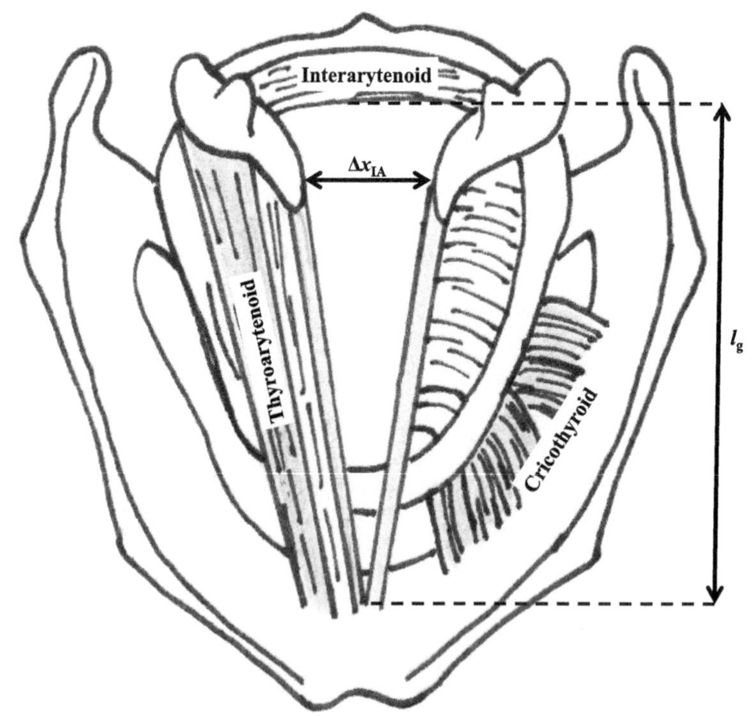

Figure 1 Larynx diagram. Sketch of the larynx with indication of intrinsic laryngeal muscles TA, CT, and IA.

by different anatomic sources [15,24]. Using a different approach, Jiang et al. [14] and Lester and Story [17] made use of *in vivo* induction of tremor to seek for relationships between acoustic features and tremor sources.

This paper reports on the results of using a high-dimensional multiple-mass vocal fold model for simulating vocal tremor. In comparison to other options, using this model allows to study the impact that the addition of tremor to specific biomechanical parameters has on the acoustic voice signal. Namely, the paper includes results from the simulation of tremulous behavior in lung pressure, vocal fold stiffness, vocal ligament stress, vocal fold length, vocal fold adduction, and jaw opening. Both respiratory [15] and phonatory tremors [24,27] had been simulated before but, to the best of authors' knowledge, this paper is the first one that reports results on the simulation of articulatory tremor. Results are presented in terms of modulation extents, both for amplitude and for frequency modulations, bearing in mind the idea that diverse anatomic sources of tremor should produce different modulation extents [24]. In the case of articulatory tremor, it is shown that amplitude and frequency modulations are not the only acoustic effects of tremor; additionally, articulatory tremor generates instabilities in the spectral envelope of the voice signal.

Voice production model

A detailed and complete description of the voice production model can be found in [28]. Here, only a brief and qualitative description is included, aimed at providing a framework for a better understanding of the tremor model. This voice production model belongs to the class of discrete, high-dimensional, biomechanical models. It is based on previous works carried out by Kob et al. [29,30], and it has been successfully employed for analyzing the pathogenesis of vocal fold nodules [31] and for simulating laryngeal disorders [28]; a resembling model was also used by Wong et al. [32] for a similar purpose.

In the biomechanical model used, each vocal fold is formed by a set of 15 contiguous elements aligned in the anterior-posterior direction. Thus, each element models the transverse section of the vocal fold in a different coronal plane. In turn, each element is composed by two masses: the upper mass that approximately models the mucosa and the lower mass that somewhat models the vocal muscle (TA muscle) and the vocal ligament. Simulated mechanical properties of the vocal-fold tissues include tissue elasticity, elasticity of the interfaces between adjacent tissues, and compression forces tending to recover form after collision. Other simulated forces are the vocal ligament stress, the pressure induced by the glottal airflow, and the subglottal pressure.

The model does not include simulation of the subglottal pressure waves and resonances, so subglottal pressure is made equal to the lung pressure. As for supraglottal structures, the vocal tract is modeled as a set of 44 concatenated cylinders having different cross-section areas. Pressure wave propagation along the vocal tract has been simulated using the Kelly-Lochbaum model plus an energy loss factor applied at each cylinder. Wave radiation at the output of the last cylinder has been simulated, assuming that no external acoustic wave arrives to the lips.

All parameters of the voice production model have been assigned the values specified in [28], except for those summarized in Table 1. These changes have been made with the purpose of increasing the naturalness on the synthesized sound waves.

Tremor model

Respiratory tremor

As mentioned before, vocal tremor may have three different anatomic sources: the respiratory system, the phonatory system, and the articulatory system. Respiratory tremor, which is caused by irregularities in the behavior of respiratory muscles, has been modeled by making the lung pressure (P_{sub} in [28]) variable instead of constant, that is:

$$P_{\text{sub}}(t) = P_0 + p(t) \tag{1}$$

where $P_0 = 700$ Pa as in [28]. Although the subglottal pressure is to some extent related to fundamental frequency, it is not a primary cause for changes in pitch [19,20]. Thus, *a priori* the main expected effect of the time variability of P_{sub} is a modulation in the amplitude of the acoustic signal [15,16].

Phonatory tremor

At laryngeal level, phonatory tremor is conjectured to be produced by instabilities in the tension patterns of intrinsic laryngeal muscles (e.g. TA and CT). Results reported by Finnengan et al. indicate that these are more related to vocal tremor than extrinsic laryngeal muscles [18]. The activity of the CT muscle (Figure 1) is crucial for

Table 1 Values of simulation model parameters that have been changed with respect to [28]

Description	Parameter	Value
Height of the upper masses	ΔMz	1.0 mm
Height of the lower masses	ΔVz	2.2 mm
Depth of the upper masses	ΔMx	1.8 mm
Depth of the lower masses	ΔVx	2.2 mm
Sum of the upper masses at each side	$\sum_{i=1}^{N} m_{Msi}$	25.7 mg
Sum of the lower masses at each side	$\sum_{i=1}^{N} m_{Vsi}$	70.5 mg
Stiffness of the link between boundary and lower mass	k_{sVB}	2.0 N/m
Stiffness of the link between upper and lower mass	k_{sMV}	1.5 N/m

the determination of the voice fundamental frequency [19-21]. Its function is twofold [23]: on the one hand, it affects the tension and the elongation of the vocal cords (i.e. vocal ligaments); on the other hand, it helps in modifying the stiffness of the vocal folds. This second function is shared with the TA muscle, which is parallel to the vocal ligament (Figure 1). Although TA activity exhibits a moderately high degree of correlation with CT activity, the TA muscle seems to play a secondary role in the control of fundamental frequency as compared with the CT muscle [21].

The activities of the TA and CT muscles can be modeled in discrete low-dimensional vocal fold simulation models as changes in the mass and stiffness of the vocal fold elements [23,33]. The use of a high-dimensional model allows to specifically model the changes in tension and elongation of the vocal ligament caused by the CT muscle. Hence, tremor induced by instabilities in the activity of TA and CT muscles is modeled by variations in the stiffness of the vocal folds and tension and length of the vocal ligament.

Changes in stiffness have been simulated by adding variability to the stiffness factors k_{sMV} and k_{sVB} in [28]. This variability has been modeled as a multiplicative factor:

$$k_s(t) = k_{s0} \cdot (1 + k(t)) \tag{2}$$

where k_{s0} has the values specified in Table 1 as k_{sVB} and k_{sMV}. Instabilities in the tension of the vocal ligament have been simulated by adding a variable term to the maximum active stress:

$$\sigma_{\mathrm{MAX}}^{\mathrm{act}}(t) = \sigma_{\mathrm{MAX0}}^{\mathrm{act}} + \sigma(t) \tag{3}$$

where $\sigma_{\mathrm{MAX0}}^{\mathrm{act}} = 60$ kPa. Last, irregularities in the elongation of the vocal cords have been modeled by multiplying the glottal length l_g (cf. Figure 1) by a variable factor, while correcting the dimensions of the vocal fold elements in order to keep their mass constant during simulations:

$$l_g(t) = l_{g0} \cdot (1 + l(t)) \tag{4}$$

with $l_{g0} = 14$ mm.

In addition to the role of TA and CT muscles in the production of vocal tremor, some research results indicate that the role of the IA muscle should not be underestimated [34]. The IA muscle, similarly to the CT, affects tension and elongation of the vocal ligaments. But it also controls the adduction of the vocal folds. Therefore, the inter-arytenoid distance, Δx_{IA} (cf. Figure 1), has been added to the variables above in order to account for the role of the IA muscle:

$$\Delta x_{\mathrm{IA}}(t) = \Delta x(t) \tag{5}$$

where it has been assumed that the default configuration is $\Delta x_{\mathrm{IA}} = 0$.

Articulatory tremor

Supraglottal structures, i.e. vocal tract, may contribute to the production of vocal tremor. This is hypothesized to be the case of patients suffering from Parkinson⊠s disease [9]. In fact, jaw and lip tremors are some of the features that help in differentiating between parkinsonian and essential tremor [35,36].

Vocal tract configuration, as in the case of the larynx, is multi-dimensional, that is, it depends on the values of several variables [37], such as the positions of jaw, hyoid, tongue body, tongue blade, lips, velum, etc. In the simulation model used, the vocal tract is mainly described by geometric parameters, specifically cross sections of concatenated elements. Therefore, articulatory tremor is modeled by making such geometric parameters variable.

Since jaw tremor is one specific feature of parkinsonian tremor, instabilities in the jaw position have been simulated. Jaw movement is modeled in [37] as an angular displacement of the lower incisors with respect to a fix point. In a similar way, jaw tremor has been simulated as an angular displacement in the boundaries of the elements forming the upper part of the vocal tract. The pharynx section, which approximately equals 1/3 of the total vocal tract length [38], has been assumed to remain unaffected by jaw position in the simulation model. Consequently, the sections corresponding to the bottom third of the vocal tract have been kept unaltered. The remaining cross sections have been made variable according to the following rule:

$$S_\delta(t) = S_{\delta 0} \cdot \left(1 + \frac{\delta - \frac{1}{3}}{\frac{2}{3}} \cdot S(t)\right) = S_{\delta 0} \cdot \left(1 + \frac{3\delta - 1}{2} \cdot S(t)\right) \tag{6}$$

where δ indicates the position along the vocal tract axis ($\delta = \frac{1}{3}$ at the top of the pharynx; $\delta = 1$ at the lips) and $S_{\delta 0}$ are the corresponding reference cross sections provided by [39]. Figure 2 gives an overview of the reference shape of the vocal tract cross section and the variability induced by jaw tremor.

Mathematical model

While tremor is often considered to be a purely oscillatory movement [1,40], some analyses indicate that tremor may not always be oscillatory [13,41]. In this paper, tremor has been simulated as random, having a certain bandwidth Δf, instead of sinusoidal. Thus, when referring to the simulation results, the term *tremor bandwidth* will be preferred to *modulation rate.* Random values for each tremor variable described above ($p(t)$, $k(t)$, $\sigma(t)$, $l(t)$, $S(t)$, and $\Delta x(t)$) have been generated every $1/\Delta f$ seconds, according to the distributions mentioned below.

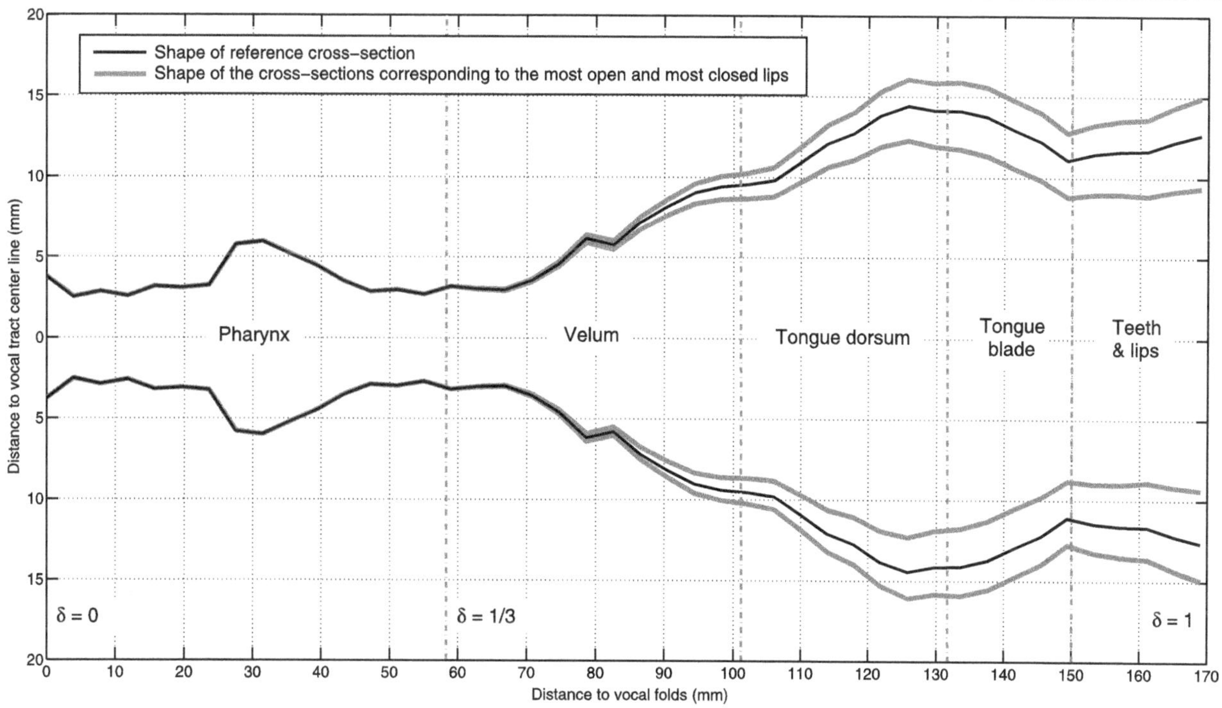

Figure 2 Vocal tract shape. Shape of the cross section of the simulated vocal tract taken from [39] and corresponding to vowel /ɑ/. The continuous black line indicates the reference shape. Continuous gray lines indicate the variation margin corresponding to one simulation run. For reference purposes, approximate indication of the vocal tract parts has been added, according to the measures in [38].

In order to calculate tremor values at intermediate time instants, a quadratic interpolation has been applied that ensures continuity of the first derivative.

Except for the case of the inter-arytenoid distance, all variable magnitudes that model tremor ($p(t)$, $k(t)$, $\sigma(t)$, $l(t)$, and $S(t)$) have been assumed to have Gaussian distributions with zero mean. In the case of the inter-arytenoid distance, it has been preferred to avoid negative values for $\Delta x(t)$ so values for it have been obtained by squaring values obtained from a Gaussian distribution, resulting in a χ^2 distribution with one degree of freedom.

Simulation parameters and acoustic analysis

The voice production and tremor models described in preceding sections have been used to generate voice signals with duration equal to 2 s and sampling rate equal to 8,000 Hz. In order to isolate the effects of different anatomical sources of tremor, time variability has only been modeled in one of the six aforementioned variables at a time. Simulations have been realized for four different bandwidths: $\Delta f\,(\text{Hz}) \in \{2.5, 5, 7.5, 10\}$ and diverse variances (i.e. modulation extents). Figure 3 shows one result of such simulations with its correspondent amplitude contour and fundamental frequency. The fundamental frequency has been estimated with a resolution of one sample, which at a sampling rate of 8,000 Hz and with voices having a mean fundamental frequency around 150 Hz corresponds to a resolution equal to approximately 3 Hz.

For each simulation, the sample normalized standard deviation of the corresponding variable has been computed as:

$$\overline{\sigma}_x = \frac{\sqrt{\mathrm{E}\left\{[x(t) - \mathrm{E}\{x(t)\}]^2\right\}}}{\mathrm{E}\{x(t)\}} \tag{7}$$

In the case of the inter-arytenoid distance, since the mean and the standard deviation are not independent for a χ^2 distribution, no normalization has been applied to the standard deviation:

$$\sigma_{\Delta x} = \sqrt{\mathrm{E}\left\{[\Delta x(t) - \mathrm{E}\{\Delta x(t)\}]^2\right\}} \tag{8}$$

Periods in the output voice signal have been identified using a simple autocorrelation-based pitch detector. Based on the periods identified, two signals have been generated: one containing information on the period peak amplitude, $A(t)$, and another one containing information about the inverse period duration, $f(t)$. $A(t)$ is the amplitude modulating signal while $f(t)$ is the frequency modulating signal. Since $A(t)$ and $f(t)$ are expected to have random values

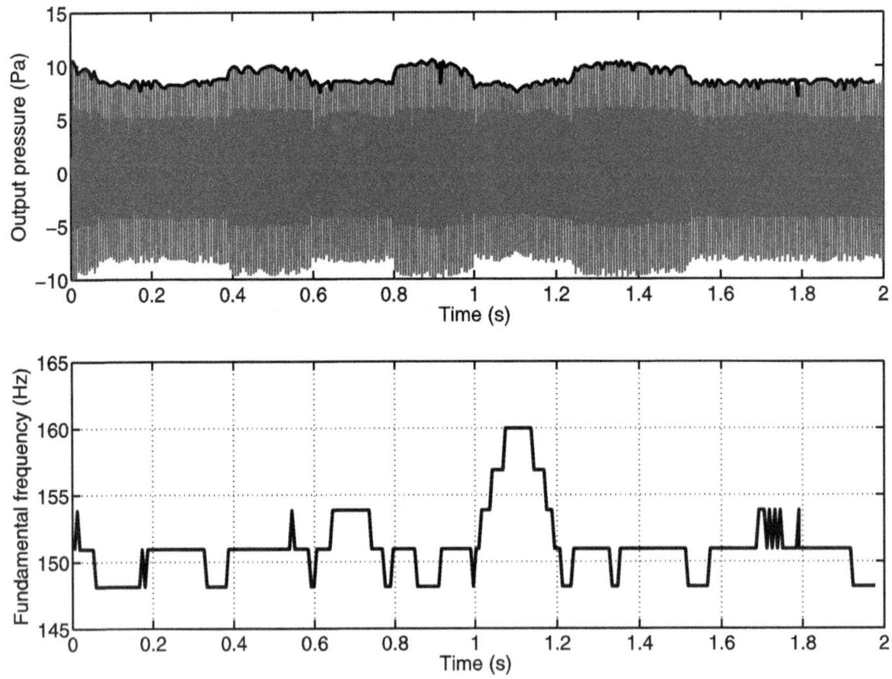

Figure 3 Simulated voice. Sample of simulated voice signals: acoustic pressure with highlighted intensity contour (top) and fundamental frequency (bottom).

obtained from unbounded distributions, the modulation extent has been defined as the normalized standard deviation of their values along a given voice signal: $\overline{\sigma}_f$ for frequency modulation and $\overline{\sigma}_A$ for amplitude modulation (Equation 7).

Results

Figure 4 shows the results of simulating respiratory tremor. For low variability in lung pressure ($\overline{\sigma}_p < 2\%$), only amplitude modulation is produced in the voice signal. For higher variabilities, frequency modulation appears but with a limited extent: $\overline{\sigma}_f < 2\%$. The extent of amplitude modulation $\overline{\sigma}_A$ varies from 3% to 20%, which means that it is from 5 to 38 times higher than the extent of frequency modulation.

Results corresponding to phonatory tremor are summarized in Figures 5, 6, 7 and 8. The impact of stiffness variations on frequency modulation is more relevant than in the case of respiratory tremor, that is, $\overline{\sigma}_A$ is only 3 to 10 times higher than $\overline{\sigma}_f$ (Figure 5). Tremor induced by instabilities in vocal fold adduction has a similar behavior (Figure 6) but with $\overline{\sigma}_f$ reaching values similar to $\overline{\sigma}_A$ in some cases. The active stress applied to the vocal ligament only has a moderate effect on vocal tremor (Figure 7), and it provides similar values for $\overline{\sigma}_f$ and $\overline{\sigma}_A$ if $\overline{\sigma}_\sigma \geq 10\%$. Last, a varying length of the vocal folds (Figure 8) seems to have a relevant effect on frequency modulation ($\overline{\sigma}_f \approx 10\%$) when the length variability goes above a certain threshold ($\overline{\sigma}_l > 1.3\%$). Below that threshold, only amplitude modulation has a certain relevance.

Finally, the effect of articulatory tremor on the modulations of the acoustic signal is depicted in Figure 9. Similarly to the case of respiratory tremor, the effect on amplitude modulation is much more relevant than the effect on frequency modulation, with $\overline{\sigma}_A$ being at least 10 times higher than $\overline{\sigma}_f$. The limited extent of frequency modulation can be noticed in the harmonics above the fifth one in the example illustrated in Figure 10. A second effect of articulatory tremor that can be measured in the acoustic signal is the instability in the frequency response of the vocal tract. This can be noticed by smoothing the spectrum using local averaging (Figure 11).

Discussion

The main effect of vocal tremor on the acoustic signal is a combination of amplitude and frequency modulation [10,13]. In contrast, other perturbation measures such as noise seem not to be related to tremor [10]. As for modulation parameters, modulation extent has been reported to be much more significant for the perception of tremor than modulation rate [9,10]. This is coherent with the fact that physiologic and pathological tremors happen within overlapped frequency ranges [1,8,9]. Thus, amplitude serves better the purpose of discriminating between them. Accordingly, the normalized standard deviations of both peak amplitude and fundamental frequency of the

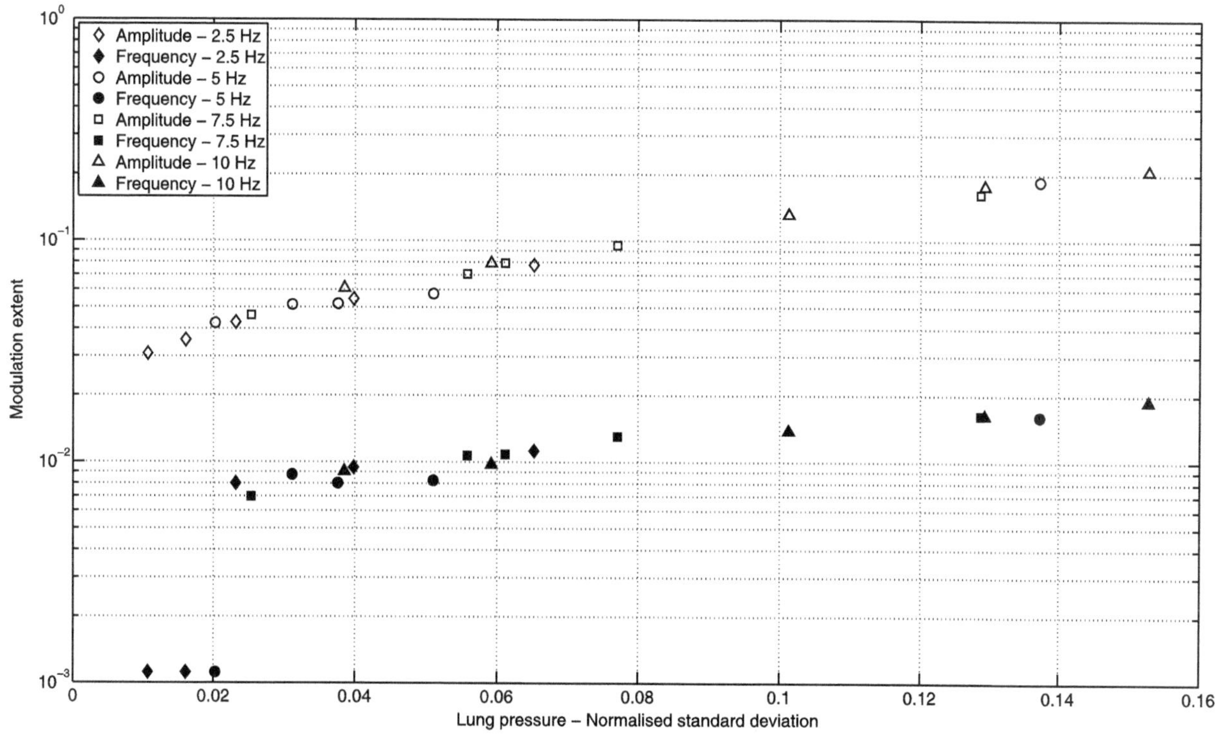

Figure 4 Respiratory tremor. Normalized amplitude and frequency modulation extents for different tremor bandwidths (2.5, 5, 7.5, and 10 Hz) as a function of the normalized standard deviation of lung pressure $\overline{\sigma}_p$.

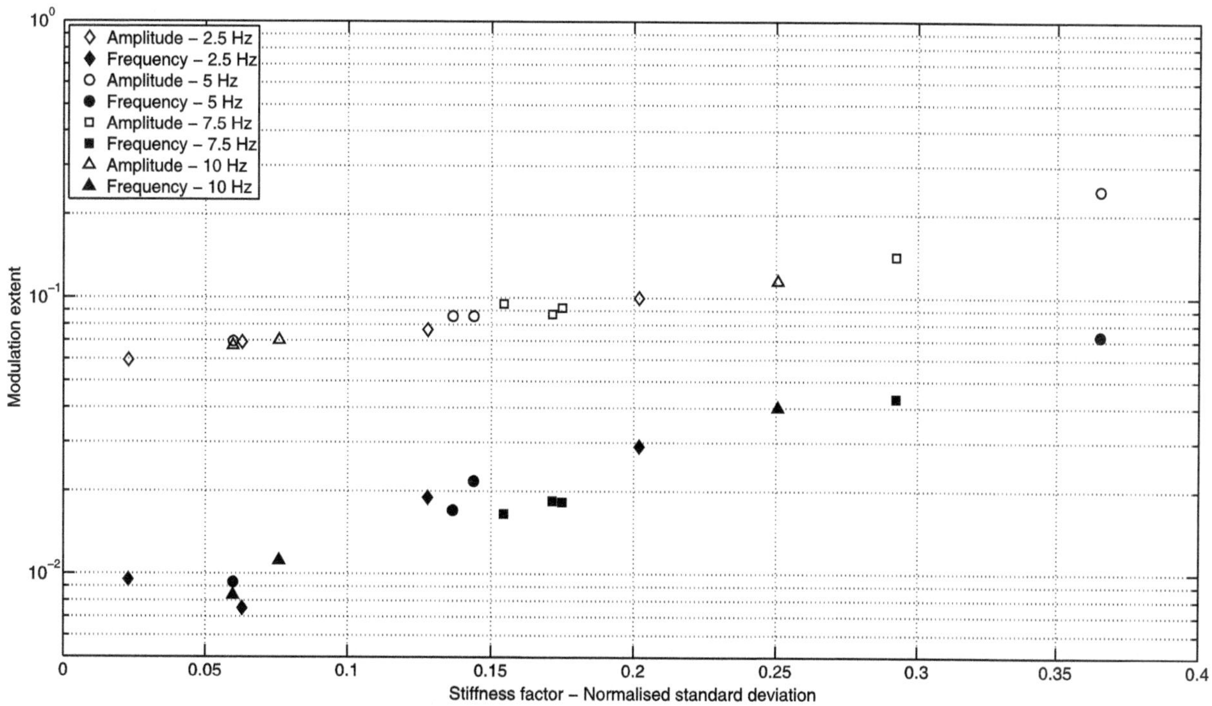

Figure 5 Phonatory tremor I. Normalized amplitude and frequency modulation extents for different tremor bandwidths (2.5, 5, 7.5, and 10 Hz) as a function of the normalized standard deviation of stiffness factor $\overline{\sigma}_k$.

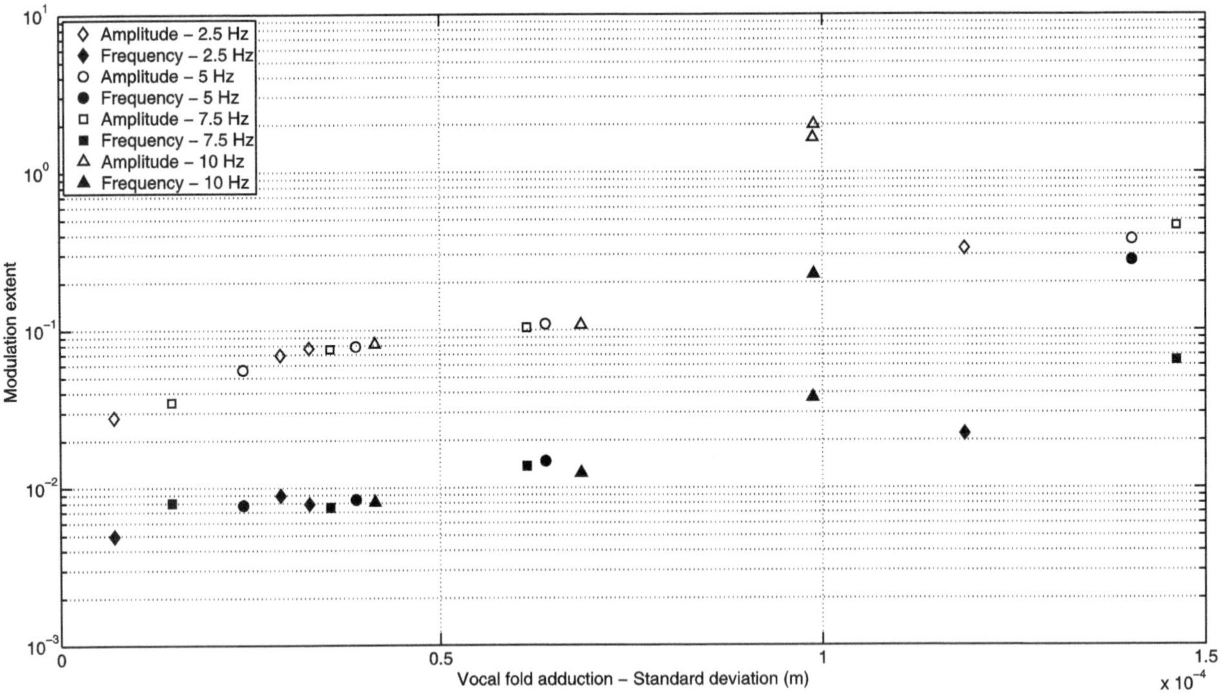

Figure 6 Phonatory tremor II. Normalized amplitude and frequency modulation extents for different tremor bandwidths (2.5, 5, 7.5, and 10 Hz) as a function of the standard deviation of inter-arytenoid distance $\sigma_{\Delta x}$.

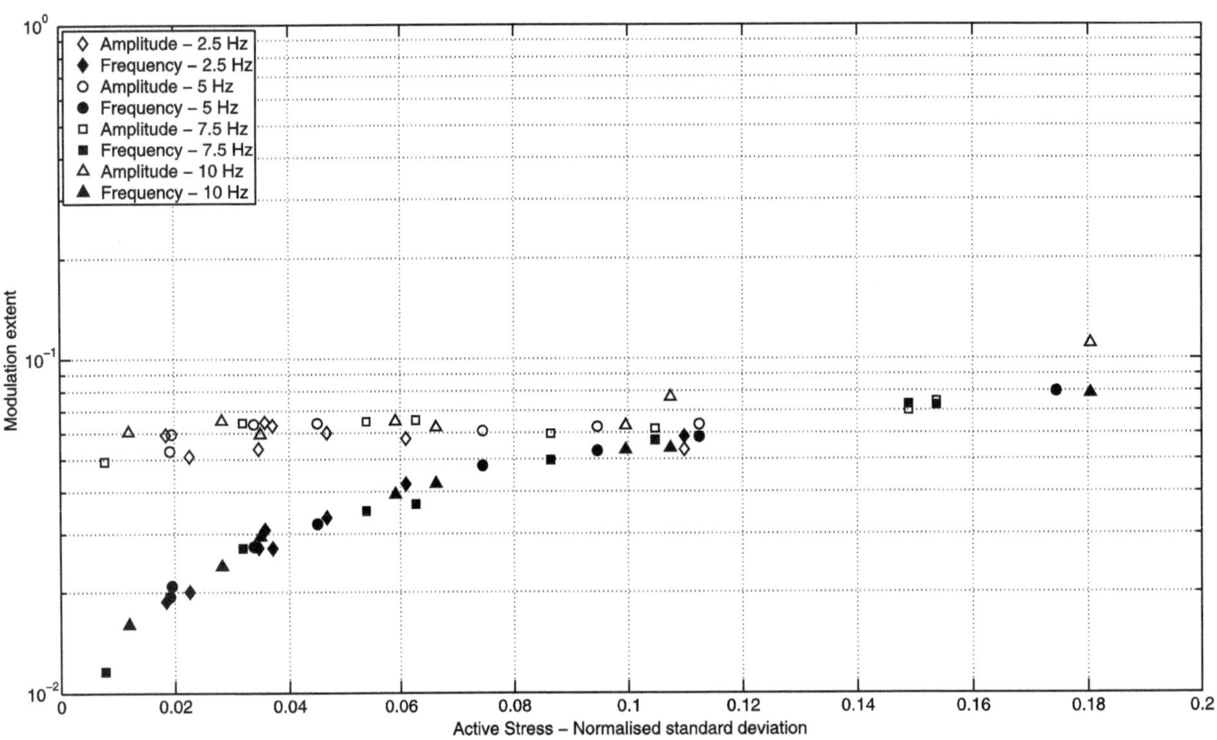

Figure 7 Phonatory tremor III. Normalized amplitude and frequency modulation extents for different tremor bandwidths (2.5, 5, 7.5, and 10 Hz) as a function of the normalized standard deviation of maximum active stress $\overline{\sigma}_{\sigma}$.

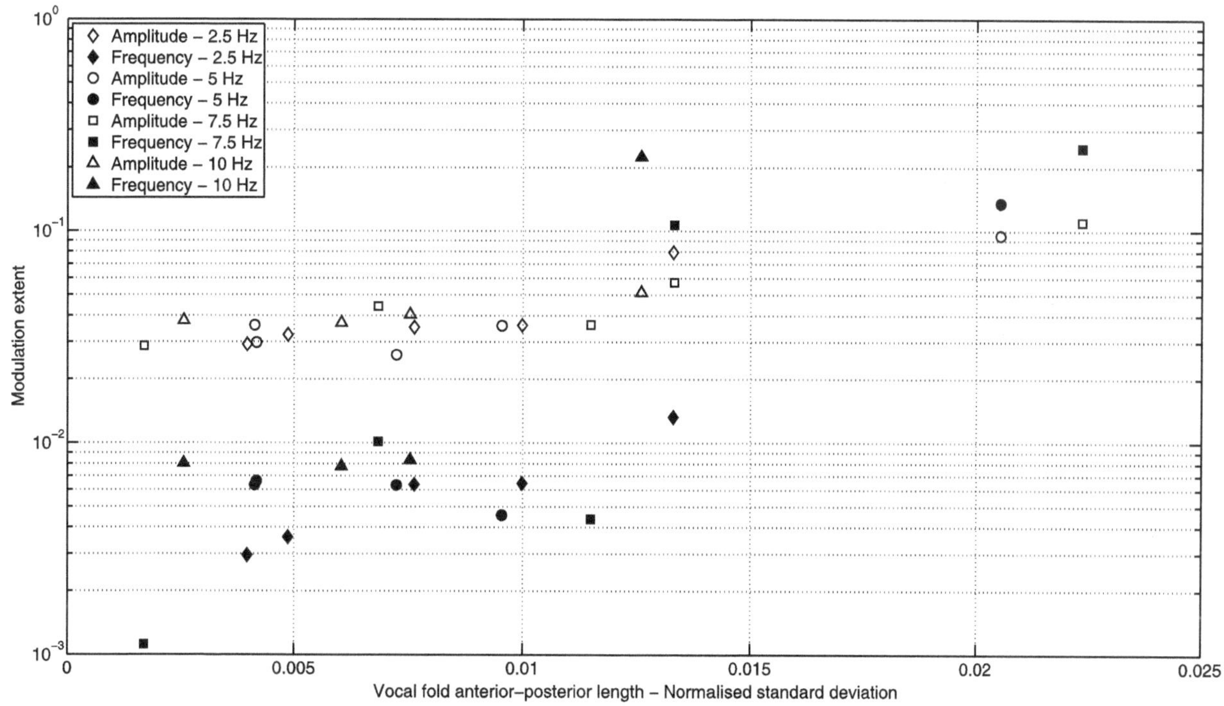

Figure 8 Phonatory tremor IV. Normalized amplitude and frequency modulation extents for different tremor bandwidths (2.5, 5, 7.5, and 10 Hz) as a function of the normalized standard deviation of vocal fold length $\overline{\sigma}_l$.

voice signal have been chosen as relevant measures of tremor for the analysis of results.

The effect of muscular tremor has been simulated independently for six physical magnitudes affecting voice production at subglottal (lung pressure), laryngeal (vocal fold stiffness, vocal fold length, vocal fold adduction, and vocal ligament stress), and supraglottal (jaw position) levels. As reported in previous studies [14,15], obtained results indicate that respiratory tremor has a relevant effect on voice amplitude but it only has a limited effect on fundamental frequency. The dominance of amplitude modulation over frequency modulation in respiratory tremor has also been recently mentioned by Lester et al. [24]. The same authors quantified the modulation extents after their simulation experiments reported in [17]: amplitude modulation extent between 5% and 50% and frequency modulation extent between 0.5% and 3.5%. Results plotted in Figure 4 are within the same range: $\overline{\sigma}_A \approx 20\%$ and $\overline{\sigma}_f \approx 2\%$ at the right of the graph.

Incidentally, the behavior of $\overline{\sigma}_f$ with respect to $\overline{\sigma}_p$ is consistent with the observation that there seems to be a threshold for the perception of respiratory tremor [6]. That is, although not formally demonstrated, it is commonly assumed that frequency modulation is more relevant for perception than amplitude modulation [13,25]. Furthermore, Kreiman et al. showed that modulation extent affects perception more than modulation rate [13]. Therefore, the jump in modulation extent that can be observed in Figure 4 near $\overline{\sigma}_p = 2\%$ is consistent with the existence of a threshold for the perception of respiratory tremor, as reported by Farinella et al. [6]. A similar threshold appears for articulatory tremor (Figure 9).

At laryngeal level, a tremulous behavior of the TA, CT, and IA muscles has a direct impact on the dynamics of the vocal folds. Such impact has been modeled by adding variability to the stiffness of the vocal folds, the stress of the vocal ligament, the length of the vocal folds, and the distance between the arytenoid cartilages on adduction. Among the four parameters, active stress in the vocal ligament seems to have the least effect on tremor (Figure 7). However, no single laryngeal parameter is likely to change independently from the rest under the joint action of laryngeal muscles. What is most relevant in the reported results is that the magnitude of frequency modulation relative to amplitude modulation is larger than for respiratory tremor, though the extents of amplitude modulations are still greater. This happens for each one of the four studied laryngeal parameters, except for the active stress, for which amplitude and frequency modulations reach similar extents when $\overline{\sigma}_\sigma \geq 10\%$. Thus, an analysis of the relative magnitudes of amplitude modulation extent and frequency modulation extent should allow to discriminate respiratory from laryngeal tremor. These results are coherent with the conclusions in [24], although

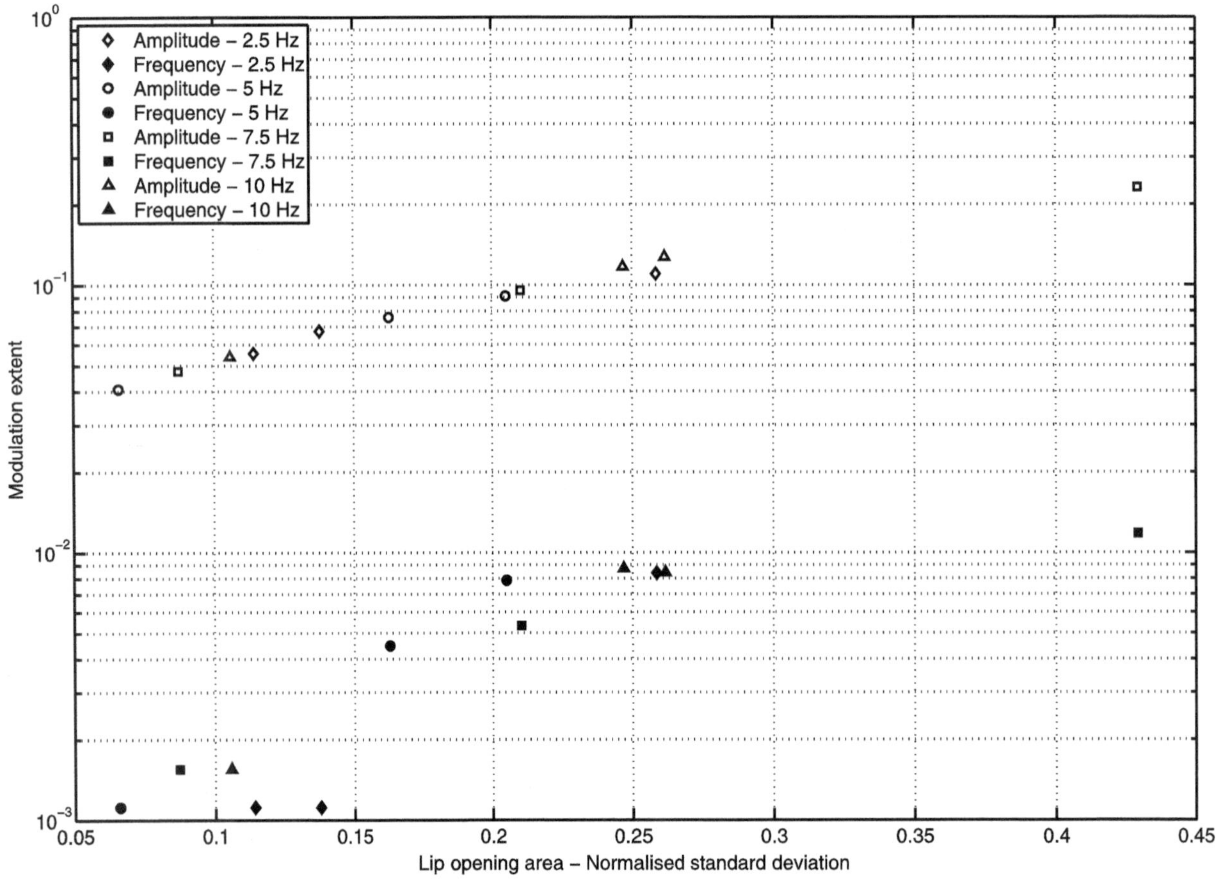

Figure 9 Articulatory tremor. Normalized amplitude and frequency modulation extents for different tremor bandwidths (2.5, 5, 7.5, and 10 Hz) as a function of the normalized standard deviation of lip opening $\overline{\sigma}_5$.

they indicate that amplitude modulations may also have a phonatory source.

Articulatory tremor may happen due to changes in shape and elastic properties at any segment of the vocal tract from the epiglottis to the lips. Among all possible models for vocal tract instability, changes in the cross section of the vocal tract segment above the pharynx have been simulated, hence modeling instabilities in the jaw position. While articulatory tremor is expected to produce only amplitude modulations of the acoustic wave [9,24], the reported results (Figure 9) indicate that frequency modulation also happens, though with a more limited extent. The classical source-filter model assumes independence between the glottal source and the vocal tract. This implies that the vocal tract cannot affect the fundamental frequency of phonation, that is, it can only amplify or attenuate signal harmonics. The expectation that articulatory tremor only causes amplitude modulations in the voice is based on this assumption. However, the existence of interaction between the glottal source and vocal tract has already been shown in several experiments [42]. As a result, vocal tract configuration can actually affect glottal flow characteristics. In the utilized simulation model, such non-linear relation is modeled to a limited extent by (Equation 23, [28]). By including this source-filter interaction in the model, it has been shown that articulatory tremor also results in frequency modulation, although the extent of this modulation is one order of magnitude smaller than the extent of the amplitude modulation (Figure 9). On the contrary, the voice production simulator does not include models for either subglottal resonances or non-linear interaction between the trachea and larynx. As a consequence, the extent of frequency modulation due to respiratory tremor may be underestimated in the results reported here.

A second effect of articulatory tremor, illustrated in Figure 11, is the instability in the frequency response of the vocal tract. As far as the authors know, this has not been used as a measure of tremor yet. However, this effect is specific from articulatory tremor and, consequently, any measure of it could reasonably be used to discriminate whether a certain tremor comes from the articulatory structures or not.

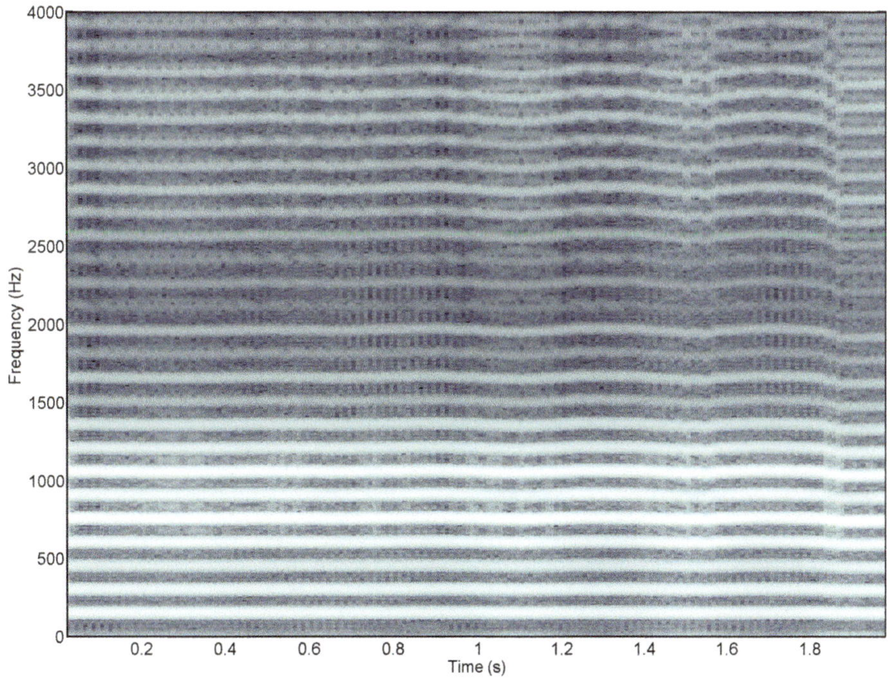

Figure 10 Articulatory tremor - spectrogram. Spectrogram of an acoustic voice signal produced as a result of articulatory tremor simulation (tremor bandwidth equal to 7.5 Hz).

Conclusions

A high-dimensional discrete voice production model has been used to simulate vocal tremor from respiratory, phonatory, and articulatory sources. Results for respiratory and phonatory tremor are consistent with the previous knowledge of respiratory tremor mainly causing amplitude modulations in the voice signal and frequency modulation coming from phonatory tremor. However, in contrast to some previous assumptions, phonatory tremor in these results causes amplitude modulations too. Apart

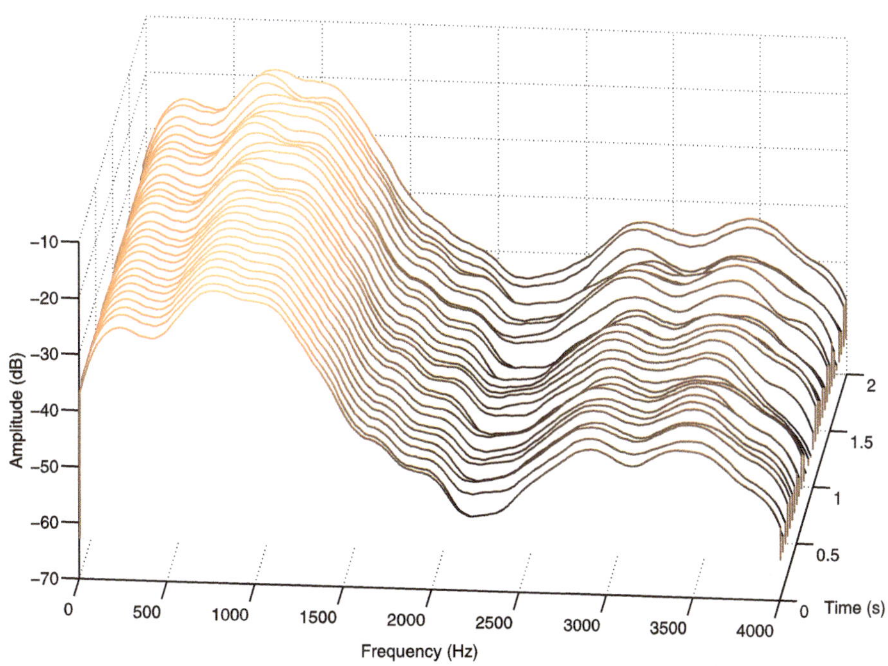

Figure 11 Articulatory tremor - smoothed spectrum. Time evolution of the smoothed spectrum of an acoustic voice signal produced as a result of articulatory tremor simulation (tremor bandwidth equal to 7.5 Hz). All plotted spectra have been normalized in energy.

from the usage of a high-dimensional discrete model, the second novelty of the paper is the simulation of articulatory tremor. Similarly to the case of respiratory tremor, its main effect is amplitude modulation of the voice signal but it also causes frequency modulation with a lower modulation extent. Another specific feature of articulatory tremor that can help in differentiating it from respiratory tremor is the instability induced in the frequency response of the vocal tract.

Competing interests

The authors declare that they have no competing interests.

Acknowledgements

This work has been carried out in the framework of project grant TEC2012-38630-C04-01, financed by the Spanish Government.

Author details

[1]Signal Theory & Communications Department, Universidad Politécnica de Madrid, Campus Sur - Carretera de Valencia km 7, 28031 Madrid, Spain. [2]Erich-Thienhaus-Institut, Hochschule für Musik Detmold, Neustadt 22, 32756 Detmold, Germany.

References

1. G Deuschl, P Bain, M Brin, Consensus statement of the Movement Disorder Society on tremor. Mov. Disord. **13**(S3), 2–23 (1998)
2. IR Titze, Workshop on acoustic voice analysis: Summary statement. Technical report, National Center for Voice and Speech (1995)
3. C Dromey, ME Smith, Vocal tremor and vibrato in the same person: acoustic and electromyographic differences. J. Voice. **22**(5), 541–545 (2008)
4. PN Juslin, P Laukka, Communication of emotions in vocal expression and music performance: Different channels, same code? Psychol. Bull. **129**(5), 770–814 (2003)
5. A Lederle, J Barkmeier-Kraemer, E Finnegan, Perception of vocal tremor during sustained phonation compared with sentence context. J. Voice. **26**(5), 668–19 (2012)
6. KA Farinella, TJ Hixon, JD Hoit, BH Story, PA Jones, Listener perception of respiratory-induced voice tremor. Am. J. Speech Lang. Pathol. **15**(1), 72–84 (2006)
7. E Yair, I Gath, On the use of pitch power spectrum in the evaluation of vocal tremor. Proc. IEEE. **76**(9), 1166–1175 (1988)
8. J Schoentgen, Modulation frequency and modulation level owing to vocal microtremor. J. Acoust. Soc. Am. **112**(2), 690–700 (2002)
9. J Jiang, E Lin, DG Hanson, Acoustic and airflow spectral analysis of voice tremor. J. Speech Lang. Hearing Res. **43**(1), 191–204 (2000)
10. S Anand, R Shrivastav, JM Wingate, NN Chheda, An acoustic-perceptual study of vocal tremor. J. Voice. **26**(6), 811–17 (2012)
11. J Gambo, FJ Jiménez-Jiménez, A Nieto, I Cobeta, A Vegas, M Ortí-Pareja, T Gasalla, JA Molina, E García-Albea, Acoustic voice analysis in patients with essential tremor. J. Voice. **12**(4), 444–452 (1998)
12. NB Pinto, IR Titze, Unification of perturbation measures in speech signals. J. Acoust. Soc. Am. **87**(3), 1278–1289 (1990)
13. J Kreiman, B Gabelman, BR Gerratt, Perception of vocal tremor. J. Speech Lang. Hear. Res. **46**(1), 203–214 (2003)
14. J Jiang, E Lin, J Wu, C Gener, DG Hanson, Effects of simulated source of tremor on acoustic and airflow voice measures. J. Voice. **14**(1), 47–57 (2000)
15. J Barkmeier-Kraemer, BH Story, Conceptual and clinical updates on vocal tremor. The ASHA Leader. **1**, 16–19 (2010)
16. H Ackermann, W Ziegler, Cerebellar voice tremor: an acoustic analysis. J. Neurol. Neurosurg. Psychiatry. **54**(1), 74–76 (1991)
17. RA Lester, BH Story, Acoustic characteristics of simulated respiratory-induced vocal tremor. Am. J. Speech Lang. Pathol. **22**(2), 205–211 (2013)
18. EM Finnegan, ES Luschei, JM Barkmeier, HT Hoffman, Synchrony of laryngeal muscle activity in persons with vocal tremor. Arch. Otolaryngology-Head Neck Surg. **129**(3), 313 (2003)
19. R Collier, Laryngeal muscle activity, subglottal air pressure, and the control of pitch in speechTechnical Report Status Report on Speech Research SR-39/40, Haskins Laboratory (1974). http://www.haskins.yale.edu/sr/SR039/SR039_09.pdf (visited May, 2013)
20. JE Atkinson, Correlation analysis of the physiological factors controlling fundamental voice frequency. J. Acoust. Soc. Am. **63**(1), 211–222 (1978)
21. T Shipp, ET Doherty, P Morrissey, Predicting vocal frequency from selected physiologic measures. J. Acoust. Soc. Am. **66**(3), 678–684 (1979)
22. DK Chhetri, J Neubauer, E Sofer, DA Berry, Influence and interactions of laryngeal adductors and cricothyroid muscles on fundamental frequency and glottal posture control. J. Acoust. Soc. Am. **135**(4), 2052–2064 (2014)
23. SY Lowell, BH Story, Simulated effects of cricothyroid and thyroarytenoid muscle activation on adult-male vocal fold vibration. J. Acoust. Soc. Am. **120**(1), 386–397 (2006)
24. RA Lester, J Barkmeier-Kraemer, BH Story, Physiologic and acoustic patterns of essential vocal tremor. J. Voice. **27**(4), 422–432 (2013)
25. J Sundberg, Acoustic and psychoacoustic aspects of vocal vibrato. Quarterly Progress and Status Report 2-3, KTH. **35**, 45–68 (1994)
26. J Hanquinet, F Grenez, J Schoentgen, in *Nonlinear Analyses and Algorithms for Speech Processing*. Synthesis of disordered voices (Heidelberg Springer, 2005), pp. 231–241
27. Y Zhang, JJ Jiang, Nonlinear dynamic mechanism of vocal tremor from voice analysis and model simulations. J. Sound Vibration. **316**(1), 248–262 (2008)
28. R Fraile, M Kob, JI Godino-Llorente, N Sáenz-Lechón, VJ Osma-Ruiz, JM Gutiérrez-Arriola, Physical simulation of laryngeal disorders using a multiple-mass vocal fold model. Biomed. Signal Process. Control. **7**(1), 65–78 (2012)
29. M Kob, Physical modeling of the singing voice. PhD thesis, Fakultät für Elektrotechnik und Informationstechnik, RWTH Aachen, Logos-Verlag (2002). http://darwin.bth.rwth-aachen.de/opus3/volltexte/2002/393/pdf/Kob_Malte.pdf
30. E Loch, S Noelle, M Kob, in *Proc. of the International Conference on Acoustics, Including the 35th German Annual Conference on Acoustics*. An approach for stable calculation of vocal fold oscillation (NAG/DAGA Rotterdam, 2009), pp. 1715–1717
31. PH Dejonckere, M Kob, Pathogenesis of vocal fold nodules: new insights from a modelling approach. Folia Phoniatrica et Logopaedica. **61**(3), 171–179 (2009)
32. D Wong, MR Ito, NB Cox, IR Titze, Observation of perturbations in a lumped-element model of the vocal folds with application to some pathological cases. J. Acoust. Soc. Am. **89**(1), 383–394 (1991)
33. IR Titze, BH Story, Rules for controlling low-dimensional vocal fold models with muscle activation. J. Acoust. Soc. Am. **112**, 1064–1076 (2002)
34. KA Kendall, RJ Leonard, Interarytenoid muscle botox injection for treatment of adductor spasmodic dysphonia with vocal tremor. J. Voice. **25**(1), 114–119 (2011)
35. N Quinn, Parkinsonism–recognition and differential diagnosis. Br. Med. J.(BMJ). **310**, 447–452 (1995)
36. J Jankovic, Parkinson?s disease: clinical features and diagnosis. J. Neurol. Neurosurg. Psychiatry. **79**(4), 368–376 (2008)
37. P Mermelstein, Articulatory model for the study of speech production. J. Acoust. Soc. Am. **53**(4), 1070–1082 (1973)
38. WT Fitch, J Giedd, Morphology and development of the human vocal tract: A study using magnetic resonance imaging. J. Acoust. Soc. Am. **106**, 1511–1522 (1999)
39. BH Story, IR Titze, Parameterization of vocal tract area functions by empirical orthogonal modes. J. Phonetics. **26**(3), 223–260 (1998)
40. AG Shaikh, K Miura, LM Optican, S Ramat, RM Tripp, DS Zee, Hypothetical membrane mechanisms in essential tremor. J. Translational Med. **6**(1), 68–78 (2008)
41. C Dromey, P Warrick, J Irish, The influence of pitch and loudness changes on the acoustics of vocal tremor. J. Speech Lang. Hear. Res. **45**(5), 879–890 (2002)
42. IR Titze, Nonlinear source-filter coupling in phonation: Theory. J. Acoust. Soc. Am. **123**(5), 2733–2749 (2008)

Linguistically motivated parameter estimation methods for a superpositional intonation model

Humberto M Torres[1]*, Jorge A Gurlekian[1], Hansjörg Mixdorff[2] and Hartmut Pfitzinger[3]

Abstract

This paper proposes two novel approaches for parameter estimation of a superpositional intonation model. These approaches present linguistic and paralinguistic assumptions for initializing a pre-existing standard method. In addition, all restrictions on the configuration of commands were eliminated. The proposed linguistic hypotheses can be based on either pitch accents or lexical stress, which give rise to two different estimation methods. These two hypotheses were validated by comparison of the estimation performance relative to two standard methods, one manual and one automatic. The results of the experiments for German, English and Spanish corpora show that the proposed methods outperform the standard ones.

Keywords: Superpositional F0 modeling; Automatic F0 contour estimation; Fujisaki model

1 Introduction

The Fujisaki model of intonation [1] has been tested for different languages [2-8], standing out for its simplicity and strong physiological basis. Currently, it is widely used in different application areas [9-13]. The model parameterizes F0 contours in an efficient way. With a small number of parameters, we can achieve a desired level of fitting accuracy. A task that has not been satisfactorily solved is the automatic model parameter extraction, that is, parameter estimation from F0 contours, since it is not directly reversable and hence there is no unique representation. As a general rule, if we desire a higher accuracy, we need to include more parameters in the model. However, if we aim at a set of parameters which will be linguistically interpretable, we invariably sacrifice accuracy in the F0 contour fitting, and estimation of parameter values must be done manually.

Several approaches have been proposed to estimate the model parameters [11,14,15]. One of the currently popular methods for parameter extraction, successfully tested for different languages, is the one proposed by Mixdorff [15] to which we will refer as the 'standard method'. Although this method is completely automatic, the author proposes a *post-hoc* manual correction to eliminate spurious commands which cannot be justified linguistically [16].

Pfitzinger and Mixdorff [17] discuss the accuracy of the current methods to estimate the model parameters solely on the basis of the extracted natural F0 contours. The authors emphasize the importance of F0 contour stylization, as a way to ensure the elimination of micro-prosody. The algorithm initialization is a critical issue, given that different sets of initial parameter values produce different model estimations with varying accuracy in fitting of the F0 contour.

Hirose et al. [18] have proposed to introduce linguistic information in the estimation process for a Japanese corpus. The model estimation is performed in two stages: First, an automatic estimation trying to fit the F0 contour, and second a correction of the parameters using *ad hoc* rules based on linguistic hypotheses. In a later work [19], they used linguistic information to obtain a first approximation of the location of the command, which is then adjusted by an iterative analysis-by-synthesis process. In this process, the linguistic information is automatically extracted using binary regression trees. These regression trees were created automatically from a portion of the corpus that was manually analyzed beforehand.

For Spanish, Torres y Gurlekian [11] proposed to consider linguistic aspects, such as the positions of pauses and syllables with lexical stress. The parameter values were

*Correspondence: hmtorres@conicet.gov.ar
[1] Laboratorio de Investigaciones Sensoriales, INIGEM, CONICET-UBA, Av. Córdoba 2351, 9 Piso Sala 2, 1120 Ciudad Autonoma de Buenos Aires, Argentina
Full list of author information is available at the end of the article

estimated by genetic algorithms (GA), achieving a performance similar to the standard method, but with too high algorithmic complexity. The ability to perform a global search in an n-dimensional solution space is the main justification of using GA, even more when the error surface has many local minima.

The purpose of this paper is to introduce linguistic information in the model estimation and also reduce the source of variability on the initialization of the parameter extraction method. Specifically, we introduce two new methods for parameter extraction, in which linguistic information is introduced to the method proposed by Mixdorff [15].

Our main idea is to initialize the estimation method with *a priori* information about the values of the model parameters. In the first case, this information is extracted from labels of tones and break indices, as proposed in the Tones and Break Index (ToBI) system [20]. In the second case, we take into account information on the locations of pauses and syllables with lexical stress. In this paper, we present the results of applying our estimation methods to German, English and Spanish speech databases.

This paper is organized as follows: In section 2, we present a brief introduction to the Fujisaki intonation model and the Mixdorff model estimation method. Our approaches to model estimation are presented in section 3. The speech databases used in this paper are introduced in section 4. The experiments and the results obtained are presented and discussed in section 5. Finally, the conclusions are discussed in section 6.

2 The Fujisaki model

This model - called superpositional - is hierarchical, additive, parametric and continuous in time.

It allows the efficient and automatic calculation of a reduced parameter set that represents real intonation contours. This model analytically describes the F0 contour in a log scale, as the superposition of three components [21]: a base frequency, accents and phrase components, as shown in Figure 1.

Phrase components are calculated as the response to a critically damped second order linear filter excited with a delta function called phrase command. Accent components result from the response to a similar filter, excited with a step function called accent command.

The F0 contour can be expressed by

$$\ln(F0) = \ln(Fb) + \sum_{i=1}^{N_f} Ap_i Gp_i(t - T0_i) + \sum_{j=1}^{N_a} Aa_j \left\{ Ga_j(t - T1_j) - Ga_j(t - T2_j) \right\}$$

$$Gp_i(t) = \begin{cases} \alpha_i^2 t \exp -\alpha_i t; & t \geq 0 \\ 0; & t < 0 \end{cases}$$

$$Ga_j(t) = \begin{cases} \min \left\{ 1 - (1 + \beta_j t) \exp -\beta_j t, \gamma_j \right\}; & t \geq 0 \\ 0; & t < 0 \end{cases}$$

where

- Fb: baseline value of fundamental frequency,
- Gp_i: impulse response of the ith phrase control mechanism,
- Ap_i: magnitude of the ith phrase command,
- $T0_i$: timing of the ith phrase command,
- Ga_j: step response of the jth accent control mechanism.
- Aa_j: amplitude of the jth accent command,
- $T1_j$: onset of the jth accent command,
- $T2_j$: end of the jth accent command,
- α_i: is the eigenvalue of the ith phrase control mechanism,
- β_j: is the eigenvalue of the jth accent control mechanism,
- γ_j: is the maximum value of the jth accent component.

The parameters α and β characterize the dynamic properties of the laryngeal mechanisms of phrase and accent

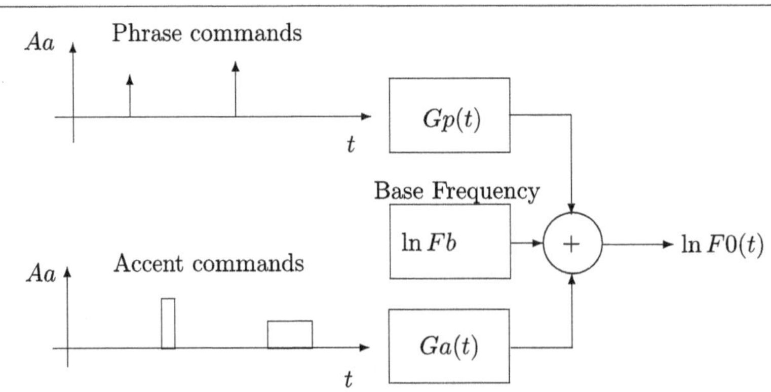

Figure 1 A scheme of the superpositional intonation model (Adapted from [21]).

control. Together with γ, they can be considered practically constant for all speakers. *Fb* must be estimated for each utterance, but is assumed to be constant for each speaker [21].

An example of Fujisaki model-based F0 contour parameterization is shown in Figure 2. We can see the slow F0 declination due to the phrase command, and local F0 movements triggered by the accent commands.

2.1 Mixdorff estimation method

Mixdorff presented an automated parameter extraction approach based on F0 measurements [15]. We will call this method *A-ME*, and it represents our baseline system. Below, we briefly present its main features.

After F0 contour interpolation and smoothing using Momel [22], the resulting spline contour is passed through a high-pass filter with a stop frequency at 0.5 Hz, similar to [23]. The output of the high-pass filter (henceforth referred to as 'high-frequency contour' or HFC) is subtracted from the spline contour yielding a 'low-frequency contour' (LFC), containing the sum of phrase components and *Fb*. The latter is initially set to the overall minimum of the LFC. Consecutive minima are detected in the HFC delimiting potential accent commands whose *Aa* is initialized to reach the maximum of *F*0 between the two minima. Since the onset of a new phrase command is characterized by a local minimum in the phrase component the LFC searches for local minima, applying a minimum distance threshold of 1 s between consecutive phrase commands. In order to initialize the magnitude value *Ap* assigned to each phrase command, the next local maximum is detected in the part of the LFC after the potential onset time *T*0. *Ap* is then calculated in proportion to F0 at that relative point while also considering contributions of preceding commands. The analysis-by-synthesis procedure is performed in three steps that are designed to optimize the initial parameter set iteratively by applying a hill-climb search to reduce the overall mean square error in the log frequency domain. At the first step, phrase and accent components are optimized separately, using respectively LFC and HFC as the targets. Next, phrase component, accent component and *Fb* are optimized together, with the spline contour as the target. In the final step, the parameters are fine-tuned by making use of a weighted representation of the extracted original *F*0 contour. The weighting factor applied is the product of degree of voicing and frame energy for every *F*0 value, hence favoring 'reliable' portions of the contour in vowel nuclei, for instance.

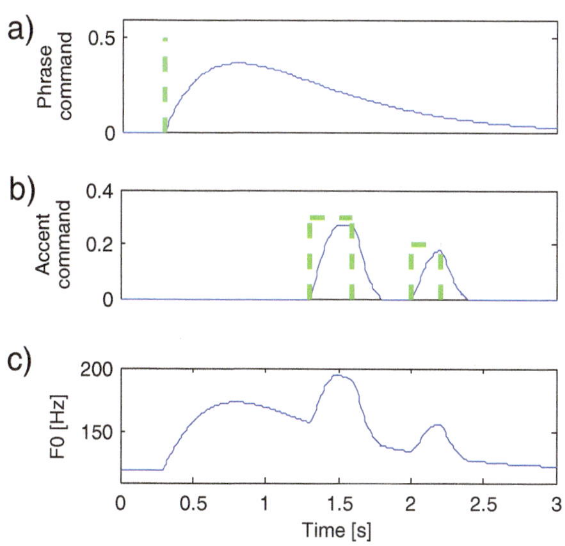

Figure 2 Example of Fujisaki model. (a) In green dashed lines, phrase command with $T0 = 0.3$, $Ap = 0.5$ and $\alpha = 2$, and its respective response in blue; **(b)** In green dashed lines, two accent command with $Aa = [0.3\ 0.2]$, $T1 = [1.3\ 2]$, $T2 = [1.6\ 2.2]$, $\beta = 20$, and $\gamma = 0.9$, and their response superimposed in blue; and **(c)** *F*0 model output, with $Fb = 120$ Hz.

In A-ME, the parameters β and γ are constants for each speaker, and their values are fixed *a priori*. *Fb* and α can be varied initially but will eventually be kept constant for a given speaker, for instance, by finding the median value of these parameters for the entire corpus. It should be stressed, however, that the Fujisaki model does not preclude optimizing these parameters for each command and each utterance in the corpus. We will call this method *A-ME+*.

Later, the automatic parameters can be manually corrected [16], in order to reduce inevitable misdetections and enforce linguistic criteria. We will call *M-ME* at this method with the hand-corrected parameters.

3 Linguistically motivated parameter estimation

In German, English and Spanish, as in many other languages, we can find two linguistic features: pitch accents and lexical stress. The pitch accents are associated with movements in the intonation contours and used to mark contrasts between different parts of a sentence [20,24]. Lexical stress are in contrast an intrinsic property of words. In general, it holds that the high tones in pitch accents are associated with lexical stress, but not vice versa.

In this work, we propose two modifications of Mixdorff's automatic method of parameter model extraction. One modification is related to the construction of a prototype for initialization, and the second modification is related to eliminate restrictions in model parameter values. The prototypes will be built from lexical stress or pitch accent information. They will be herein called lexically motivated or *L-ME* method, and tonally motivated or *T-ME* method.

The idea is to reduce parameters which are already considered speaker-dependent as much as possible and then to put the effort to estimate the remaining parameters. Those can be fixed in advance or limited to a certain range according to our linguistic hypotheses. Others will depend on higher level information, such as phrase type, intention, speaker mood, etc. Since this information is not available in advance, we will suppose that the values are only influenced by the text.

In addition, we removed some restrictions from A-ME, such as minimum or maximum values of command amplitudes, as well as minimum durations and distances between commands.

In summary, our hypothesis is that the model parameters will only depend on the text and that the speaker characteristics will remain invariant.

An outline of the proposed estimation methods are shown in Figure 3.

3.1 Lexically motivated estimation method

For the L-ME method, we made the following assumptions: First, the position of accent commands will be close to the location of syllable with lexical stress. We only considered stressed syllables in content words. Second, it is reasonable to expect accent commands occurring at or near the end of intermediate intonational phrases; these are model approximations to 'boundary tones' in some linguistic transcription methods [25]. Third, phrase commands will be near intonational phrase beginnings, as has been reported in previous studies [11,24].

To introduce these assumptions, we propose a *prototype* of initial model parameters, as follows:

- Fb, α, β and γ parameters are fixed for each corpus. These values are obtained from A-ME or M-ME.
- One phrase command per intonational phrase. The phrase command position is measured from the beginning of each intonational phrase, and we called it $T0_r$.
- One accent command for each syllable with lexical stress in content words. The position of this accent command is measured relative to location of the syllables with lexical stress. We called it $T1_r$.
- One accent command for each intermediate intonational phrase. The position of this accent command is measured relative to ending of intermediate intonational phrase. We called it $T1_r$.

3.2 Tonally motivated estimation method

For the T-ME method, we made the following assumptions: First, the position of accent commands will be close to the location of syllable with high-tone pitch accent. One way of labelling these events is through the ToBI system [20,24]. Second, it is reasonable to expect accent commands occurring at or near the end of intermediate intonational phrases, just as in the L-ME method. Third, phrase commands will be near intonational phrase beginnings, just as in the L-ME method.

For the T-ME method, we propose a *prototype* of initial model parameters, as follows:

- Fb, α, β and γ parameters are fixed for each corpus. These values are obtained from A-ME or M-ME.
- One phrase command per intonational phrase. The phase command position is measured from the beginning of each intonational phrase, and we called it $T0_r$.
- One accent command for each high tone in pitch accent. The position of this accent command is measured relative to location of syllables with high tone in pitch accent. We called it $T1_r$.
- One accent command for each intermediate intonational phrase. The position of this accent command is measured relative to ending of intermediate intonational phrase. We called it $T1_r$.

4 Speech material

The model prototypes were used as initial parameters of linguistic motivated methods L-ME and T-ME. The

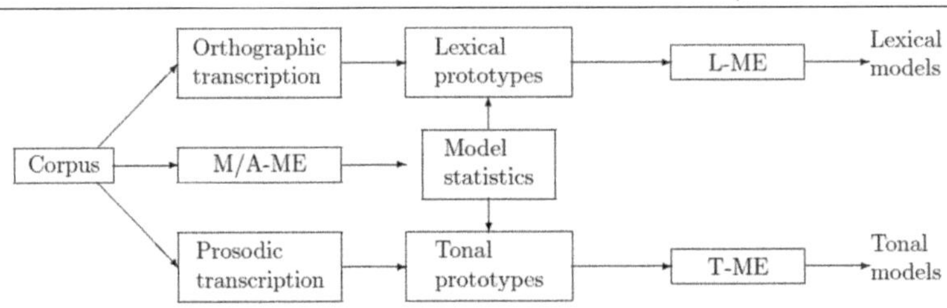

Figure 3 Outline of the two proposed methods. The *corpus* comprises wave files, transcription and phonetic F0 contours. A/M-ME means *Automatic or Manual Mixdorff Estimation method for parameter extraction*.The *orthographic transcription* (temporal alignment of pauses, part-of-speech and stressed syllables) is combined with the information extracted of models estimated by A/M-ME (mean value of Ap, $T0_r$, Aa, $T1_r$, $T2-T1$, α, β, γ and Fb), to obtain the *prototypes* for model initialization. With these prototypes we initialize the L-ME and we re-estimated the parameters of *lexical model*. In the same way, *prosodic transcription* (temporal alignment of pauses and syllables with tonal accents) is used to estimate the *tonal model*.

estimated model, phrase and accent command amplitudes and positions, as well as *Fb*, α, β and γ, were used to resynthesize the F0 values by means of the Fujisaki model. The semitone scale was used to evaluate the resulting contours versus the real F0 contour.

We tested our method in three different languages: German, English and Spanish. Below is a brief description of each corpus used in our experiments.

4.1 German database

For German, we used the IMS Radio News Corpus [26] which consists of German news texts read by professional speakers. The data were automatically segmented into phonemes according to the German SAMPA [27] inventory followed by manual corrections. Prosody of the data was manually labelled following the G-ToBI [28] conventions. The syllables with lexical stress also were manually labelled.

The reference data for the Fujisaki model were extracted automatically [15] and manually corrected following linguistic criteria [16] and using the interactive FujiParaEditor [29]. Although raw F0 data are provided with the corpus extracted in 10 ms steps *via* get F0 of ESPS waves, a substantial correction was necessary. Our data selection comprises 73 news articles read by one male speaker.

4.2 English database

For English, we used the CSTR US KED Timit database [30] which contains 453 phonetically balanced utterances spoken by a US male speaker. The database was hand-labelled in phonemes, syllables and words, and carefully corrected. The syllables with lexical stress were also manually labelled.

For this corpus, we have neither labelling of prosody nor manually extracted intonation model parameters. Therefore, we only used information on lexical stress for estimating the parameter values.

4.3 Spanish databases

We use two databases in Spanish to test our models. Both were recorded by two professional female announcers, natives of Buenos Aires.

The first one, which we call DB1, was created with the aim to study prosody [24]. The corpus sentences had marked natural inflections and with different number of intonational phrases. Its text corpus consists of 741 declarative sentences extracted from Argentine newspapers published in Buenos Aires. The sentences contain 97% of all Spanish syllables, in both stress and unstressed conditions, and all possible syllabic positions within the word.

The second database was created to be used in a text-to-speech system, and we call it DB2. DB2 text corpus was based on DB1 supplemented with new sentences, in order to reach a broad coverage of diphones [31]. Also, we included 235 interrogative sentences. The corpus contains 1,826 sentences.

Recordings were made in a sound-proof chamber, with an AKG dynamic microphone and 16 Khz/16 bit sampling rate conversion. The speakers were instructed to read the sentences with natural tonal variations. The speech material collected was of approximately 40 min for DB1 and 140 min for DB2.

For the two databases, each sound file was manually labelled twice, by musically trained speech therapists who distinguished prosodic occurrences as intonational groups and accents [32]. The files were labelled on different tiers: phonetic according to Argentinian SAMPA [33], orthographic, break levels between words, and tonal marks according to an extended ToBI method for Argentine Spanish [24]. Part-of-speech and syntactic layers were also indicated.

Table 1 Parameter values used to build the prototypes for all models

Language	German	German	German	English	Spanish	Spanish	Spanish
Method	T-ME	T-ME	L-ME	L-ME	T-ME	L-ME	L-ME
Reference	M-ME	A-ME	A-ME	A-ME	A-ME	A-ME	A-ME
Corpus	IMS	IMS	IMS	KDE	DB1	DB1	DB2
α	0.95	2	2	2	2	2	2
β	20.3	20	20	20	20	20	20
Fb	50	50	50	60	100	100	100
Ap	1.13	0.75	0.75	1.0	0.8	0.8	0.75
$T0_r$	−0.1	−0.5	−0.5	−0.5	−0.5	−0.5	−0.5
Aa	0.3	0.4	0.4	0.34	0.4	0.4	0.33
$T1_r$	−0.045	−0.1	−0.1	−0.1	−0.15	−0.15	−0.1
$T2$-$T1$	0.24	0.3	0.3	0.3	0.25	0.3	0.25

Table 2 Results for manual and two automatic estimation approaches for German

	RMSE	*Ap* rates	*Aa* rates
M-ME	1.48 ± 0.19	0.44 ± 0.05	1.11 ± 0.09
A-ME	1.33 ± 0.14	0.44 ± 0.06	1.69 ± 0.1
A-ME+	1.33 ± 0.21	0.44 ± 0.05	1.10 ± 0.08

The RMSE is given in ST and the rates in commands per second. Standard deviation is included as scattering measure.

5 Results

5.1 Prototypes

Prototypes are built from the analysis of the reference models and linguistic restrictions explained in section 3. First, we have to obtain the reference models, M-ME or A-ME. Second, the mean values of the parameters are employed to build the prototype. Mean values are estimated over all parameter instances and for each analyzed database.

In Table 1, values used to build all the models analyzed in this work are shown. $T2$-$T1$ is the duration of accent commands. For both methods, the positions of commands, $T0_r$ and $T1_r$, were extracted from manual labelling, but there are alternatives for automatic labelling [34,35].

Finally, with these benchmarks, the prototypes for each sentence to be processed are built.

5.2 German

5.2.1 Estimation approaches for M-ME, A-ME and A-ME+

For German, we have M-ME- and A-ME- estimated models. The root mean square error (RMSE) in semitones (ST) and the average command density per second of the different experiments are shown in Table 2. We have also included the standard deviation as measured dispersion values. The two automatic models have the same performance, with an RMSE slightly lower than that of the manual. But A-ME+ models have meaningful lower numbers of accent commands than A-ME. We can assume that a higher number of free parameters in A-ME+ reduces accent command density.

The histograms of parameter values are shown in Figure 4. The relative position of the accent commands, $T0_r$ and $T1_r$, are measured as described in section 5.1. The histograms of M-ME and A-ME+ parameters are similar, with the exception of the *Ap* parameter, where the manual models presents higher value amplitudes. The A-ME models have accent commands with greater amplitude and duration than the others. Moreover, the phrase command amplitudes in the A-ME have less dispersion on their values. The temporal location of the phrase commands is similar for the three models, but the A-ME models have an extra peak in the histogram of the accent command positions, approximately at 0.25 s. This indicates the presence of accent commands that appear after the pitch accent.

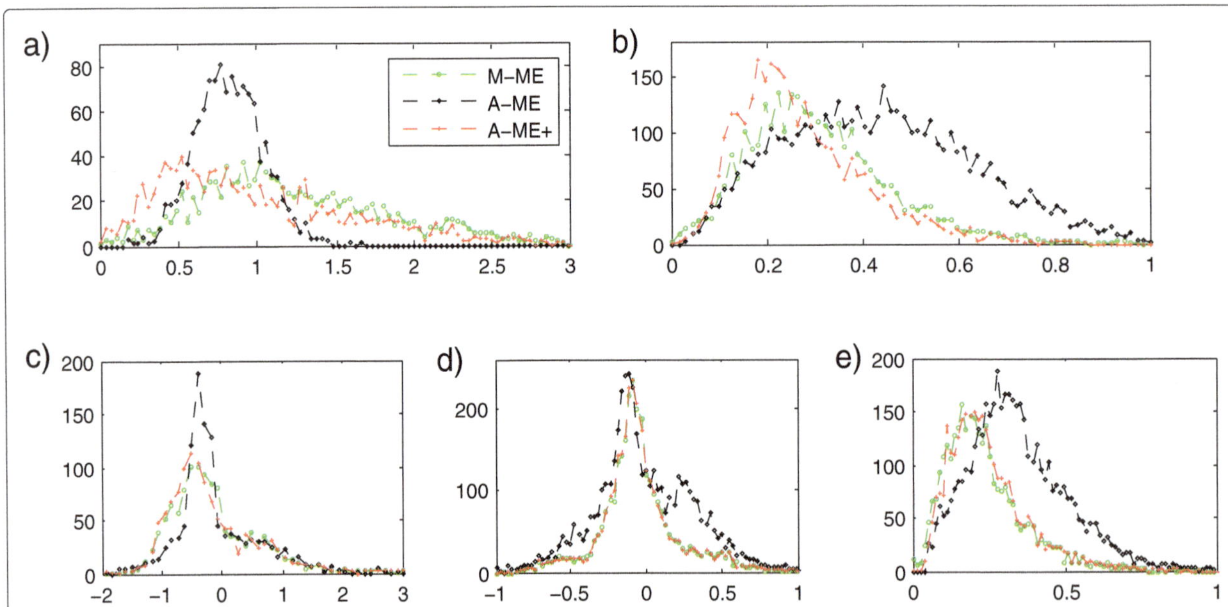

Figure 4 Histogram of amplitudes, relative positions and duration values of accent and phrase commands for German models. The methods were Mixdorff's manual (M-ME), automatic (A-ME) with global parameter optimization, and automatic+ (A-ME+) with local parameter optimization. **(a)** Phrase command amplitude, **(b)** accent command amplitude, **(c)** phrase command relative position, **(d)** accent command relative position, and **(e)** accent command duration.

Table 3 Results for the tonal method with two initialization methods for German

	RMSE	*Ap* rates	*Aa* rates
M-ME	1.48 ± 0.19	0.44 ± 0.05	1.11 ± 0.09
A-ME	1.33 ± 0.14	0.44 ± 0.06	1.69 ± 0.1
T-ME (by M-ME)	1.46 ± 0.27	0.49 ± 0.06	1.27 ± 0.13
T-ME (by A-ME)	1.50 ± 0.21	0.49 ± 0.06	1.35 ± 0.14
T-ME Prototype		0.49 ± 0.06	1.53 ± 0.16

The RMSE is given in ST and the rates in commands per second. Standard deviation is included as scattering measure.

In what follows, we will only use the A-ME, where the values of *Fb*, α, β and γ shall be considered constants for all sentences of each speaker. Keeping these values constant enables us to compare the distributions of the amplitudes, location and duration of the model commands.

5.2.2 T-ME

As mentioned above, there are two possible ways of creating prototypes to initialize the automatic method: (1) from parameters estimated automatically (A-ME) or (2) corrected manually (M-ME). We explore the two alternatives for the tonally based method, without appreciable differences in results.

Results are shown in Table 3. Information from the prototypes generated to initialize the methods are also added. Results for two initialization approaches are comparable to the manual method, and with a performance slightly lower than the automatic estimation method.

As can be seen, the density of phrase commands remains unchanged. Moreover, the accent command density is similar for both alternatives of initialization, but they are lower than that of the prototypes and the automatic approach, and are higher than those obtained from the manual. Since the models obtained with the proposed method have a lower density of accent commands than the prototypes, the optimization algorithm removes those commands which are considered unnecessary. This is a very interesting property of the method, as it reduces model redundancy.

The histograms of parameter values are shown in Figure 5. Something interesting to note is that the two new tonal model histograms apparently match the histogram of its prototype. Moreover, both peaks correspond to the values assigned by the respective prototypes. We can assume that the method of parameter estimation determined that these values are a good approximation, and does not modify their values in the process of parameter setting.

In Figure 5e, we can note the presence of accent commands of very short duration. In the original version, there is a restriction of minimum accent duration which is allowed, that we have eliminated here.

In view of the results obtained, we create initialization prototypes taking as references the models from the automatic estimation.

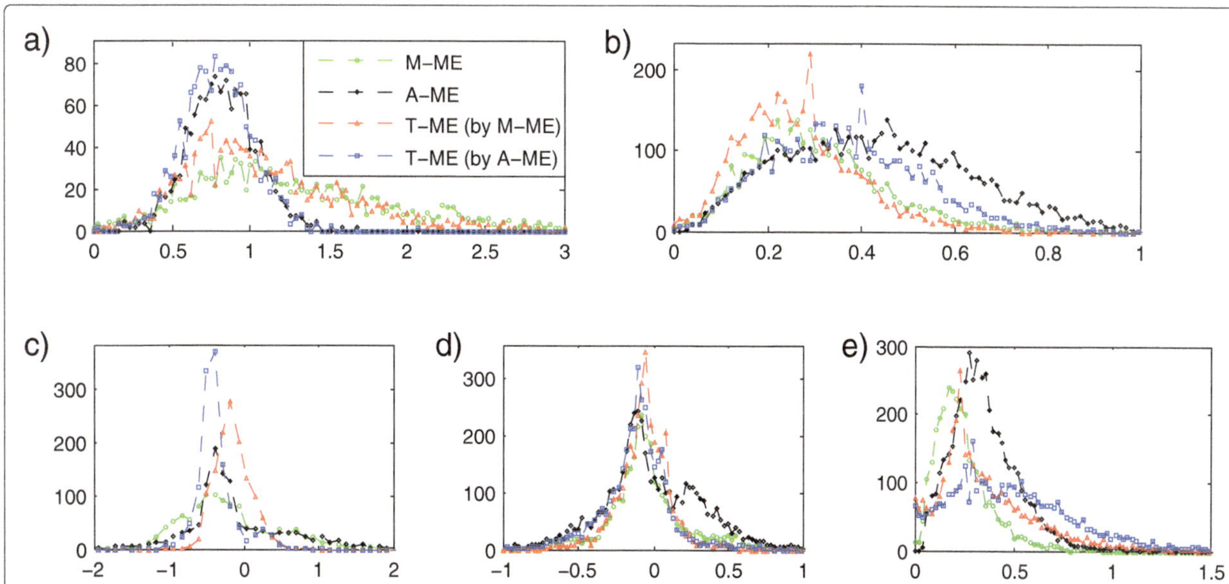

Figure 5 Histogram of amplitudes, relative positions and duration values of accent and phrase commands for tonal German models. The methods were Mixdorff's manual (M-ME), automatic (A-ME) and tonal with Mixdorff's manual (T-ME by M-ME) and automatic (T-ME by A-ME) initialization. **(a)** Phrase command amplitude, **(b)** accent command amplitude, **(c)** phrase command relative position, **(d)** accent command relative position, and **(e)** accent command duration.

Table 4 Results for the lexical method on a reduced data set of German corpus

	RMSE	*Ap* rates	*Aa* rates
M-ME	1.81 ± 0.12	0.48 ± 0.05	1.12 ± 0.06
A-ME	1.48 ± 0.15	0.38 ± 0.05	1.65 ± 0.11
T-ME	1.75 ± 0.24	0.50 ± 0.05	1.42 ± 0.13
L-ME	1.59 ± 0.23	0.50 ± 0.05	1.63 ± 0.17
L-ME Prototype		0.50 ± 0.05	1.72 ± 0.16

The RMSE is given in ST and the rates in commands per second. Standard deviation is included as scattering measure.

5.2.3 L-ME

The use of lexical stress was another alternative to create the prototype initialization of the parameter estimation method. The occurrence of an accent command is associated with a lexical stress, as described in section 5.1.

The results of the experiments with this approach are shown in Table 4. When we gathered these results, only a small number of sentences labelled with lexical stress was available. Therefore, our results are restricted to this subset. For this reason, we include the results of the other estimation methods on the same subset of sentences in this table. The lexical method gives better results than the manual and the tonal, but worse than the automatic. The density of the accent commands is similar to the automatic method, but much higher than the manual. Again, we can also observe how the proposed method significantly reduces the number of original accent commands, by removing the unnecessary commands from the prototypes.

In Figure 6, the histograms of command parameter values for this approach are shown. Its behavior is similar to that observed in Figure 5 for tonal models.

Figure 7a shows a sentence example of fundamental frequency generated from the Mixdorff methods, manual and automatic, and the methods presented in this paper, tonal and lexical. We have also included the syllabic phonetic labelling and the original F0 contour for comparison. In Figure 7b,c,d,e, we can see the accent and phrase commands inserted by the four methods. Also, the RMSE in ST of the whole sentence for each method is shown. It is interesting to note that different settings of commands generate an acceptable parametrization of the original F0 contour. We can also observe that the different models take into account some movements of the F0 contour, while smoothing out others.

5.3 English

For the English corpus, we explored the lexical method. In this case, we fixed *a priori* the *Fb*, α, β and γ parameter values for automatic model estimation. The manual labelling of phonemes in content words and syllables with lexical stress were available in the database to build the prototype of intonation model, in a similar fashion to the previous case of the lexical German models. Here, we do not have index pauses between words, so we assigned a phrase command after each pause. This is a very rough first approximation, given that many sentences of this

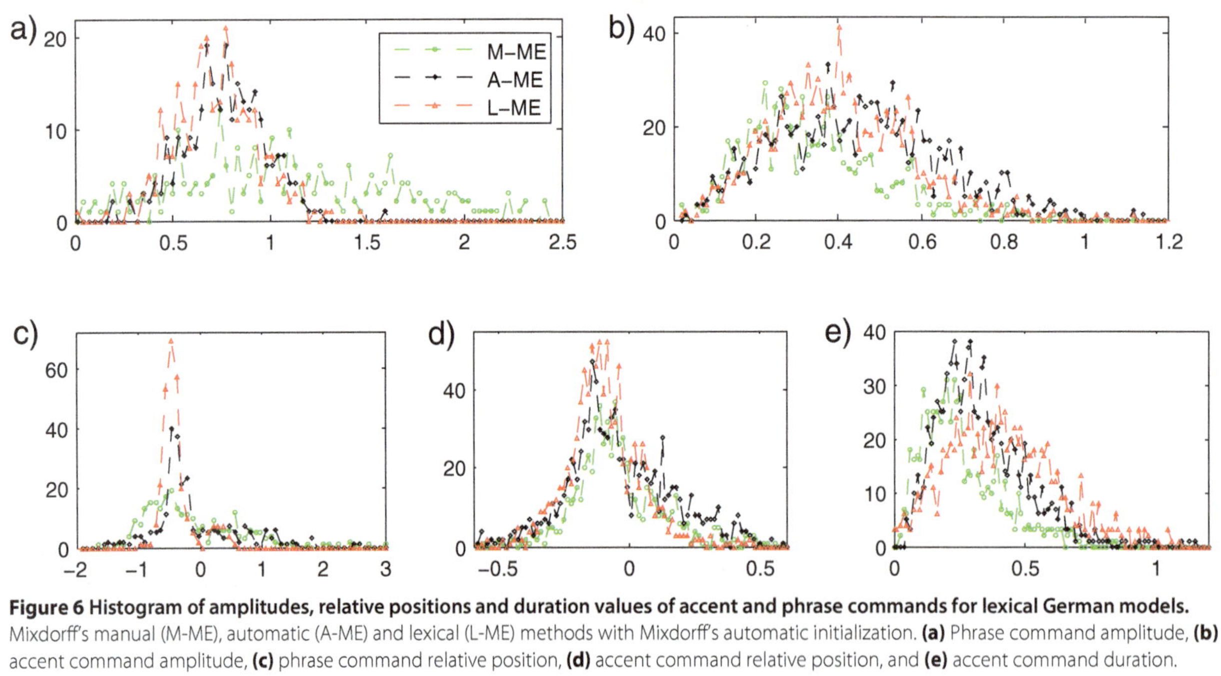

Figure 6 Histogram of amplitudes, relative positions and duration values of accent and phrase commands for lexical German models. Mixdorff's manual (M-ME), automatic (A-ME) and lexical (L-ME) methods with Mixdorff's automatic initialization. **(a)** Phrase command amplitude, **(b)** accent command amplitude, **(c)** phrase command relative position, **(d)** accent command relative position, and **(e)** accent command duration.

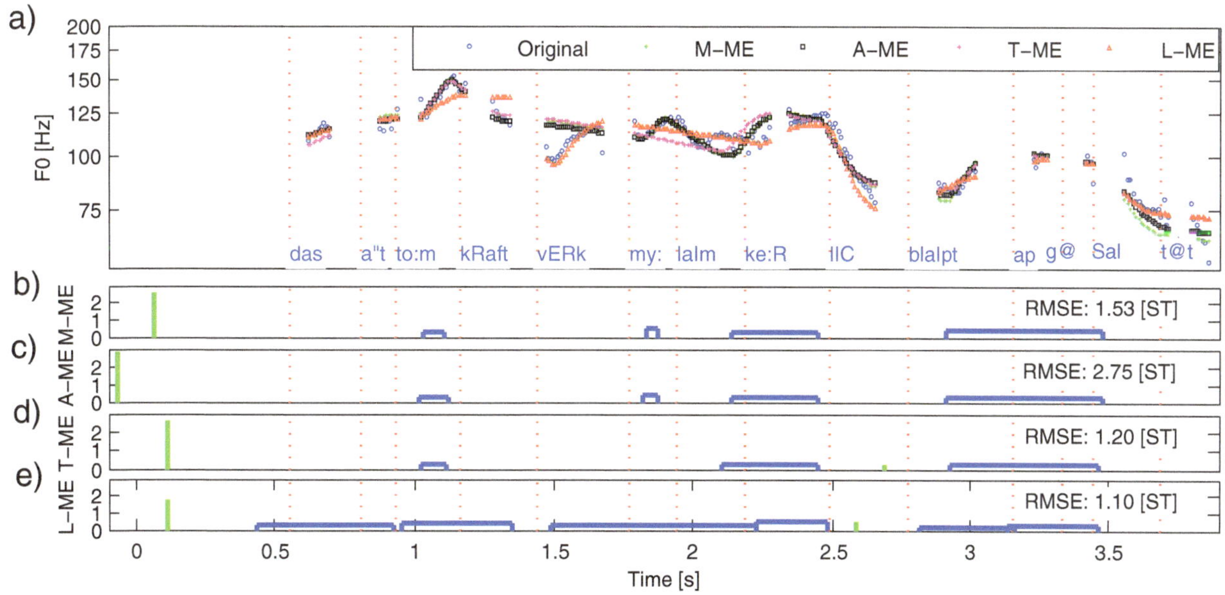

Figure 7 Example of F0 contour estimation of the four methods for German. (a) F0's and phonetic syllable transcription. **(b** and **c)** Commands for Mixdorff's manual (M-ME) and automatic (A-ME) methods. **(d** and **e)** Commands for tonal (T-ME) and lexical (L-ME) methods of estimation. RMS error in ST for a whole sentence were included. The sentence corresponds to a portion of a long sentence, and for the text *'Das Atomkraftwerk Mühlheim-Kehrlich bleibt abgeschaltet'* (*'The nuclear power plant Mühlheim-Kehrlich stays shut down'*).

corpus, between pause and pause have at least two intonation phrases.

The results are shown in Table 5. The results on the English corpus are better than for German. The lexical approach has an error slightly lower than the automatic model at the expense of a higher density of accent commands. As stated above, a high density of both phrase and accent commands improves the fitting of the F0 contour. Moreover, in most cases, English content words are mainly shorter, monosyllabic and disyllabic, than German and Spanish, where content words have two or three syllables on average [36,37]. Unlike the German case, the estimation method does not eliminate many accent commands from the English prototypes. Five percent of the accent commands from the prototypes are removed in lexical German models, and only 1% are removed in the English models.

In Figure 8, the histograms of command parameter values for this approach are shown. Histograms of the automatic and lexical parameters are similar. We can see the low dispersion of the positions of the command phrase which suggests taking a fixed value for this parameter.

Table 5 Results for the lexical method for English

	RMSE	*Ap* rates	*Aa* rates
A-ME	1.02 ± 0.49	0.68 ± 0.26	1.91 ± 0.53
L-ME	0.88 ± 0.34	0.57 ± 0.15	3.02 ± 0.64
L-ME Prototype		0.57 ± 0.15	3.05 ± 0.61

The RMSE is given in ST and the rates in commands per second. Standard deviation is included as scattering measure.

Figure 9 shows an example of fundamental frequency generation. We have included the results of command prototypes and A-ME commands for comparison. In this figure, we can see how the search algorithm optimizes the initial command prototypes to obtain a better parametrization of the F0 contour.

5.4 Spanish

For Spanish, we built the tonal models with the DB1 database and lexical models with the DB1 and DB2 databases. The prototypes for the DB1 corpus were constructed similarly to the German corpus, and prototypes for DB2 were built in a similar way as the English corpus. For both cases, we also built the reference A-ME models.

The experiment results are shown in Table 6. For DB1, the tonal model has a similar performance to the automatic model and the lexical approach has a slightly better performance of about 6% than T-ME and A-ME. The phrase command density is the lowest for the automatic model, approximately 40% lower than the proposed models. The tonal model has the lowest density of accent commands, and the lexical model has the highest density. A high number of commands improve the approximation of the models to the original F0 contours. We must also emphasize that the methods proposed removes approximately 15% of accent command prototypes.

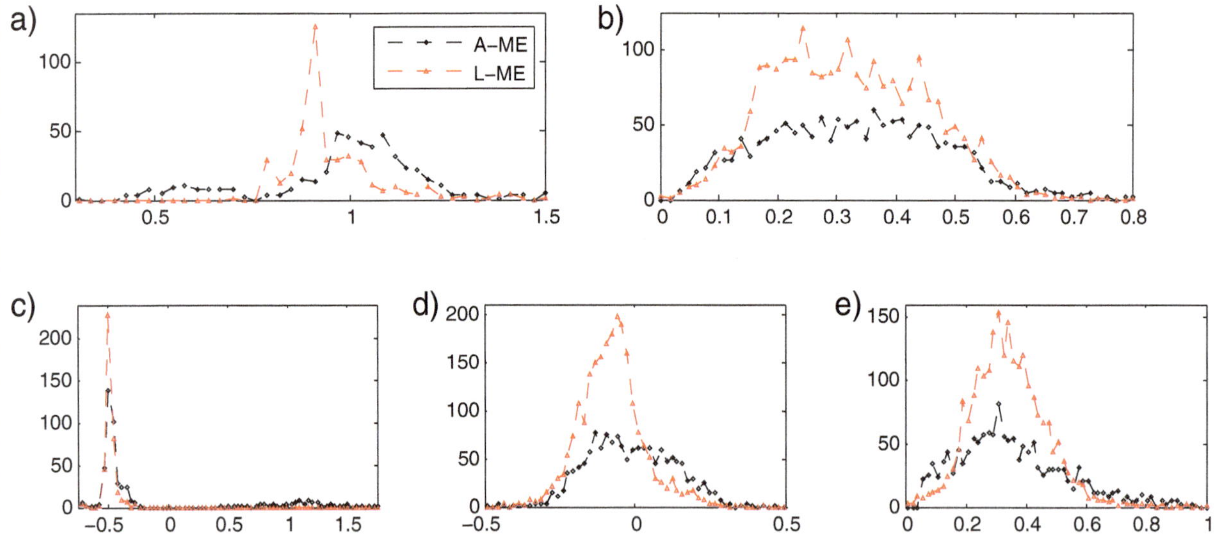

Figure 8 Histogram of amplitudes, relative positions and duration values of accent and phrase commands for English models. Mixdorff's automatic (A-ME) and lexical (L-ME) methods. **(a)** Phrase command amplitude, **(b)** accent command amplitude, **(c)** phrase command relative position, **(d)** accent command relative position, and **(e)** accent command duration.

For the DB2 database, lexical models have a performance lower than automatic approach. Unlike other corpora analyzed in this paper, however, the lexical models have a density of commands lower than the automatic models: 40% fewer phrase commands and 14% fewer accent commands.

The large difference in the density of phrase commands between the two corpora is explained by the fact that, in DB1, we assign one command for each intonation phrase, and in DB2, we assign one command for each pause.

In Figure 10, the histograms of command parameter values are shown. For DB2, the phrase command

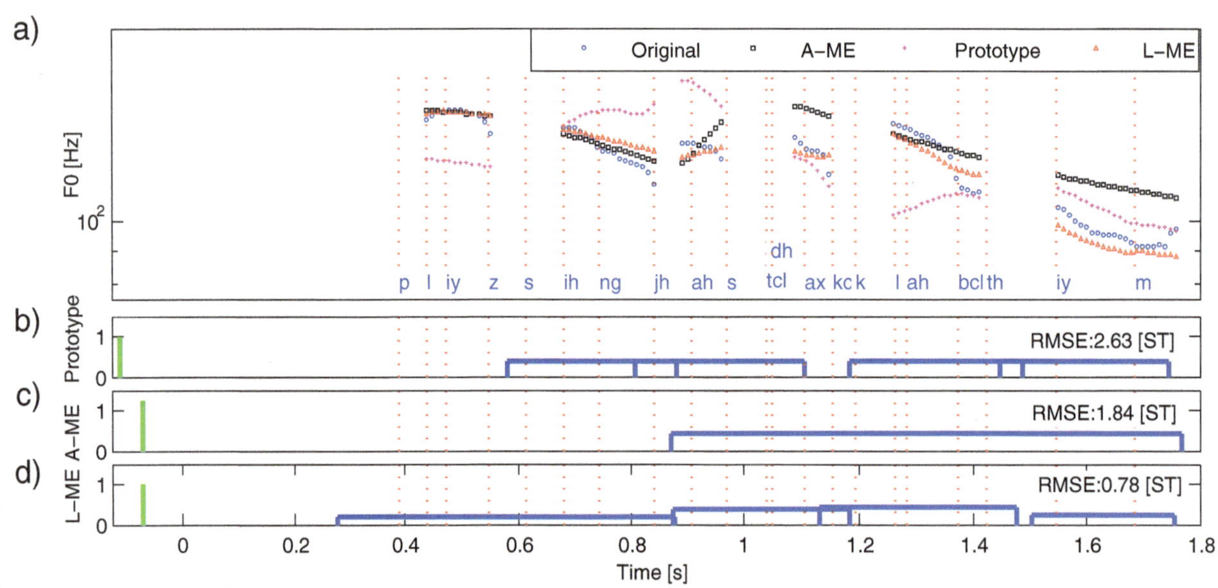

Figure 9 Example of F0 contour estimation for English. (a) F0's and phonetic syllable transcription. **(b)** Prototype commands, **(c)** commands for Mixdorff's automatic (A-ME) method, and **(d)** commands for the lexical (L-ME) method of estimation. RMS error in ST for a whole sentence was included. The text sentence is *'Please sing just the club theme'*.

Table 6 Results for DB1 and DB2 Spanish databases

		RMSE	*Ap* rates	*Aa* rates
DB1	A-ME	1.36 ± 0.47	0.61 ± 0.21	1.84 ± 0.35
	T-ME	1.32 ± 0.45	1.04 ± 0.20	1.57 ± 0.40
	L-ME	1.28 ± 0.47	1.04 ± 0.20	1.99 ± 0.51
	T-ME Prototype		1.04 ± 0.20	1.89 ± 0.36
	L-ME Prototype		1.04 ± 0.20	2.35 ± 0.43
DB2	A-ME	1.15 ± 0.41	0.74 ± 0.25	1.91 ± 0.41
	L-ME	1.49 ± 0.43	0.45 ± 0.22	1.65 ± 0.53
	L-ME Prototype		0.45 ± 0.22	1.76 ± 0.53

The RMSE is given in ST and the rates in commands per second. Standard deviation is included as scattering measure.

amplitudes are slightly higher than in the DB1. Furthermore, for the DB2 database, the accent command durations of lexical models is slightly higher than those of automatic models. The distributions for other parameters are similar for all models. Again, Figure 6e reveals accent commands of short duration.

Figure 11 shows an example of fundamental frequency generated. This sentence was taken from DB2. In this figure, we can see how the automatic method inserts greater amount of phrase commands than the lexical method. We can also see that in the second part of the sentence, approximately from 2.7 to 3.8 s, the automatic method introduces an accent command of long duration, which simplifies two tonal movements in the F0 contour.

6 Discussion and conclusions

The main contribution of the methods proposed in this paper is a successful link between *tonal movements* and both *pitch accent* and *lexical stress* through *accent commands*. This mutual relationship could be used in the design of an automatic tonal tagger.

The results from this study confirm our hypotheses and our assumptions about the location of the commands. Furthermore, this approach has proved to work in three different languages, with different word lengths and different syntactic structures.

The use of initial prototypes permits a better estimation of F0 contours than that obtained in a fully free approach. Prototypes can be seen as initial conditions that delimit the number of commands. Our methods assign a command for each linguistic event and also take care of removing those commands that are not necessary. Preliminary studies show that this is due to a large percentage of low boundary tones for which there is no need to insert an accent command. Another reason is the proximity between prototype accents that may lead to fusion or elimination of one of them [38]. More work should be carried out to elucidate which command accent prototypes are erased, in order to omit them in the initial prototype of our algorithm.

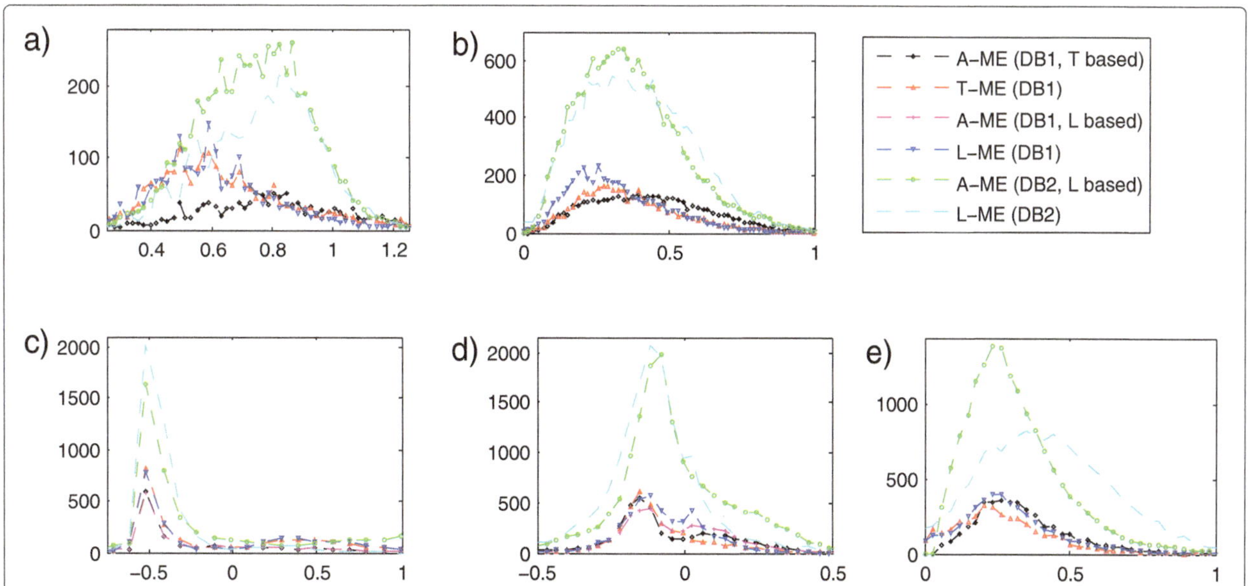

Figure 10 Histogram of amplitudes, relative positions and duration values of accent and phrase commands for Spanish models. Mixdorff's automatic method based on pitch accent over DB1 corpus (A-ME, DB1, T based), tonal method over DB1 corpus (T-ME, DB1), Mixdorff's automatic method based on lexical stress over DB1 corpus (A-ME, DB1, L based), lexical method over DB1 corpus (L-ME, DB1), Mixdorff's automatic method based on lexical stress over DB2 corpus (A-ME, DB2, L based), lexical method over DB2 corpus (L-ME, DB2). **(a)** Phrase command amplitude, **(b)** accent command amplitude, **(c)** phrase command relative position, **(d)** accent command relative position, and **(e)** accent command duration.

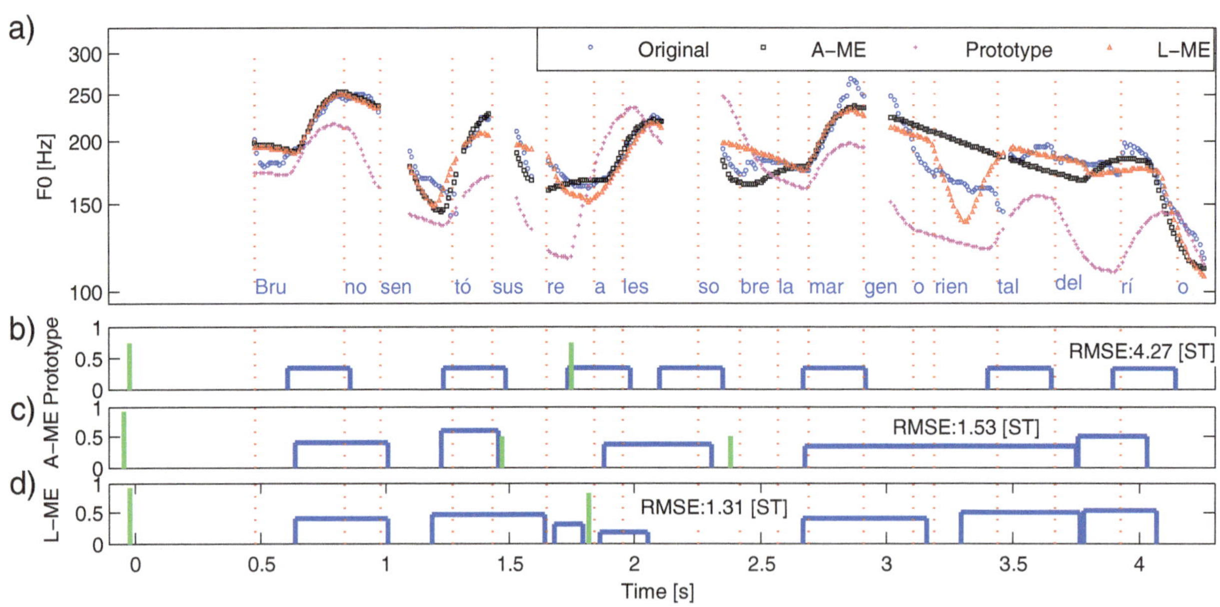

Figure 11 Example of F0 estimation for Spanish. (a) F0's and phonetic syllable transcription. **(b)** Prototype commands, **(c)** commands for Mixdorff's automatic method, and **(d)** commands for lexical method of estimation. RMS error in ST for a whole sentence were included. The text sentence is *'Bruno sentó sus reales sobre la margen oriental del río'* (*'Bruno sat on his royals over the eastern bank of the river'*).

When we compare the histograms of the accent command durations, our approaches versus the standard methods, we can see a significantly increased amount of accent commands of very short duration. These short accents are manifested as tiny movements of F0 contour, with an amplitude lower than 1.5 semitones, a value taken as minimum difference in order to be perceived as a prominence [39,40]. In the future, we should consider the role that these small commands play in the Fujisaki model, and eventually modify our algorithm to suppress accent commands that are modeling micro-prosody.

We have lifted restrictions on the amplitudes, overlaps, minimum/maximum parameter values of commands that were imposed by the original Mixdorff algorithm. We believe that physiological studies should be conducted in order to determine the existence of such restrictions and what values they take in case they do occur.

The histograms of the phrase command relative positions have a very sharp peak at −0.5 s, for both methods and for all corpora analyzed. This T0 value generates a peak in the F0 contour at the intonation phrase onset. In future studies, we will evaluate the possibility of maintaining this parameter as a constant.

The performance of the methods to fit the F0 contour in each of the languages analyzed cannot be compared, due to differences in the sentence structures of each corpus. In all cases, however, their performances are equal to or better than those obtained with the reference models.

Competing interests

The authors declare that they have no competing interests.

Acknowledgements

This research has been carried out with the support of Ministerio de Ciencia y Tecnología and Consejo Nacional de Investigaciones Científicas y Técnicas, Argentina.

Author details

[1]Laboratorio de Investigaciones Sensoriales, INIGEM, CONICET-UBA, Av. Córdoba 2351, 9 Piso Sala 2, 1120 Ciudad Autonoma de Buenos Aires, Argentina. [2]Department of Computer Sciences and Media, Beuth University Berlin, Berlin, Germany. [3]Pfitzinger Voice Design, Lübeck, Germany.

References

1. H Fujisaki, K Hirose, Analysis of voice fundamental frequency contours for declarative sentences of Japanese. J. Acoust. Soc. **5**(4), 233–242 (1984)
2. PS Rossi, F Palmieri, F Cutugno, ed. by B Bel, I Marlien, A method for automatic extraction of Fujisaki-model parameters, in *Proceedings of Speech Prosody 2002* (Aix-en-Provence, 11–13 April 2002), pp. 615–618
3. B Uslu, HG Ilk, Fujisaki intonation model in Turkish text-to-speech synthesis, in *The IEEE 17th Signal Processing and Communications Applications Conference, 2009. SIU 2009* (Antalya, Turkey, 844). doi:10.1109/SIU.2009.5136528
4. E Navas, I Hernadez, Modelado de la entonación en euskera utilizando el modelo de fujisaki y árboles de regresión binarios, in *Resumenes de las I Jornadas de Tecnologías del Habla* (Sevilla, Spain, 2000)
5. H Fujisaki, S Narusawa, S Ohno, D Freitas, Analysis and modeling of F0 contours of Portuguese utterances based on the command-response model, in *Proceedings of Eurospeech 2003*, vol. 3 (Geneva, Switzerland, 1–4 September 2003), pp. 2317–2320
6. H Fujisaki, Proceedings of Joint International Conference of SNLP and Oriental COCOSDA (Hua-Hin, Thailand, 9–11 May 2002), pp. 1–10

7. H Fujisaki, C Wang, S Ohno, W Gu, Analysis and synthesis of fundamental frequency contours of standard Chinese using the command-response model. Speech Commun. **47**(1–2), 59–70 (2005)
8. H Fujisaki, S Ohono, K-i Nakamura, M Guirao, J Gurlekian, Computational modeling of accent and intonation in declarative sentences in Spanish. J. Acoust. Soc. Am. **95**(5), 2949 (1994)
9. K Hirose, H Fujisaki, H Kawai, Generation of prosodic symbols for rule-synthesis of connected speech of Japanese, in *Proceedings of the IEEE International Conference on Acoustics, Speech, and Signal Processing (ICASSP86)*, vol. 4 (Tokyo, Japan, 7–11 April 1986), pp. 2415–2418
10. HM Torres, *Generación automática de la prosodia para un sistema de conversión de texto a habla*, PhD thesis, (Universidad de Buenos Aires, Buenos Aires, Argentina, Agosto 2008)
11. HM Torres, JA Gurlekian, Parameter estimation and prediction from text for a superpositional intonation model, in *Proceedings of the 20 Konferenz Elektronische Sprachsignalverarbeitung* (TUDpress Verlag der Wissenschaften, 2009), pp. 238–247
12. M O'Reilly, AN Chasaide, ed. by R Cowie, F de Rosis, Analysis of intonation contours in portrayed emotions using the Fujisaki model, in *Proceedings of the The Second International Conference on Affective Computing and Intelligent Interaction* (Lisbon, Portugal, 12–14 September 2007), pp. 102–109
13. P Zervas, IMN Fakotakis, G Kokkinakis, ed. by A Grigoris, G Potamias, C Spyropoulos, and D Plexousakis, Employing Fujisaki's intonation model parameters for emotion recognition, in *Advances in Artificial Intelligence.* Lecture Notes in Computer Science, vol. 3955 (Springer Berlin, 2006), pp. 443–453
14. S Narusawa, N Minematsu, K Hirose, H Fujisaki, A method for automatic extraction of model parameters from fundamental frequency contours of speech, in *Proceedings of the IEEE International Conference on Acoustics, Speech, and Signal Processing (ICASSP '02)*, vol. 1 (Orlando, Florida, 13–17 May 2002), pp. I-509–I-512
15. H Mixdorff, A novel approach to the fully automatic extraction of Fujisaki model parameters, in *Proceedings of the 2000 IEEE International Conference on Acoustics, Speech, and Signal Processing (ICASSP '00)*, vol. 3 (Istanbul, Turkey, 5–9 June 2000), pp. 1281–1284
16. H Mixdorff, *An integrated approach to modeling German prosody.* Phd thesis, (Universitatsverlag, Dresden, Germany, 2002)
17. H Pfitzinger, H Mixdorff, Valuation of F0 stylisation methods and Fujisaki-model extractors, in *Proceedings of the 20 Konferenz Elektronische Sprachsignalverarbeitung* (Dresden, Germany, 21–23 September 2009), pp. 228–237
18. K Hirose, Y Furuyama, S Narusawa, NMH Fujisaki, Use of linguistic information for automatic extraction of F0 contour generation process model parameters, in *Proceedings of the 8th European Conference on Speech Communication and Technology (EUROSPEECH 2003)* (Geneva, Switzerland, 1–4 September 2003), pp. 141–144
19. K Hirose, Y Furuyama, N Minematsu, Corpus-based extraction of F0 contour generation process model parameters, in *Proceedings of the INTERSPEECH 2005 - Eurospeech, 9th European Conference on Speech Communication and Technology* (Lisbon, Portugal, 4–8 September 2005), pp. 3257–3260
20. M Beckman, G Ayers, Guidelines for ToBI labelling, (Ohio State University, Version 3.0, 1997). http://www.ling.ohio-state.edu/~tobi/ame_tobi/labelling_guide_v3.pdf. Accessed 1 July 2013
21. H Fujisaki, Prosody, information and modelling with emphasis on tonal features of speech, in *Proceedings of Workshop on SLP* (Mumbai, India, 9–13 December 2003), pp. 5–14
22. D Hirst, R Espesser, Automatic modelling of fundamental frequency using a quadratic spline function. Travaux de l'Institut de Phonétique d'Aix. **15**, 75–85 (1993)
23. V Strom, Detection of accents, phrase boundaries and sentence modality in German with prosodic features, in *Proceedings of the European Conference on Speech Communication and Technology* (Madrid, Spain, 18–21 September 1995), pp. 2039–2041
24. JA Gurlekian, H Rodriguez, L Colantoni, HM Torres, ed. by BBird Liberman, Development of a prosodic database for an Argentine Spanish text to speech system, in *Proceedings of the IRCS Workshop on Linguistic Databases* (SIAM, University of Pennsylvania Philadelphia, USA, 2001), pp. 99–104
25. P Prieto, *Las Teorías Lingüísticas de La entonación*, (Ariel, Barcelona, 2003)
26. S Rapp, *Automatisierte erstellung von korpora für die prosodieforschung*, PhD thesis, (Univ. Stuttgart, 1998). PhD Dissertation
27. D Gibbon, German SAMPA. (Dokumentation V1.0. Bielefeld, 1995). http://coral.lili.uni-bielefeld.de/Documents/sampa-d-vmlex.html. Accessed 1 April 2013
28. M Grice, S Baumann, Deutsche intonation und gtobi. Linguistische Ber. **191**, 267–298 (2002)
29. H Mixdorff, FujiParaEditor (2009). http://public.beuth-hochschule.de/~mixdorff/thesis/fujisaki.html. Accessed 1 March 2013
30. Cstr us ked timit database (2001). http://www.festvox.org/dbs/dbs_kdt.html. Accessed 1 December 2013
31. H Torres, Creación de un corpus de texto para la construcción de un sistema tts. Informe técnico ISSN 0325-2043, Laboratorio de Investigaciones Sensoriales, UBA-CONICET, (Buenos Aires, Argentina, Diciembre 2012)
32. L Colantoni, J Gurlekian, Convergence and intonation: historical evidence from Buenos Aires Spanish. Bilingualism: Lang. Cogn. **7**(2), 107–119 (2004)
33. JA Gurlekian, L Colantoni, HM Torres, El alfabeto fonético SAMPA y el diseño de córpora fonéticamente balanceados. Fonoaudiológica. **47**(3), 58–70 (2001)
34. A Rosenberg, J Hirschberg, Detecting pitch accent using pitch-corrected energy-based predictors, in *Proceedings of Interspeech* (Antwerp, Belgium, 27–31 August 2007), pp. 2777–2800
35. H Torres, J Gurlekian, ed. by B Bel, I Marlien, Automatic determination of phrase breaks for Argentine Spanish, in *Proceedings of the Speech Prosody 2004 (SP-2004)*, (Nara, Japan, 23–26 March 2004), pp. 553–556
36. M Guirao, MAG Jurado, *Estudio Estadístico del Español.* (CONICET, Buenos Aires, 1993)
37. KJ Kohler, *Einführung in die Phonetik des Deutschen.* (Erich Schmidt Verlag, Berlin, 1977)
38. G Toledo, J Gurlekian, Choque de acentos tonales frente al fraseo. Revista Philologica Romanica. **11**, 43–66 (2011)
39. ACM Rietveld, C Gussenhoven, On the relation between pitch excursion size and prominence. J. Phon. **13**, 299–308 (1985)
40. AP Bertrán, AMF Planas, EM Celdrán, AO Escandell, MCA Céspedes, ed. by JD García, Umbrales tonales en español peninsular, in *Actas del II Congreso de Fonética Experimental* (Sevilla, 2002), pp. 272–278

Permissions

All chapters in this book were first published in EURASIP-JASMP, by Springer; hereby published with permission under the Creative Commons Attribution License or equivalent. Every chapter published in this book has been scrutinized by our experts. Their significance has been extensively debated. The topics covered herein carry significant findings which will fuel the growth of the discipline. They may even be implemented as practical applications or may be referred to as a beginning point for another development.

The contributors of this book come from diverse backgrounds, making this book a truly international effort. This book will bring forth new frontiers with its revolutionizing research information and detailed analysis of the nascent developments around the world.

We would like to thank all the contributing authors for lending their expertise to make the book truly unique. They have played a crucial role in the development of this book. Without their invaluable contributions this book wouldn't have been possible. They have made vital efforts to compile up to date information on the varied aspects of this subject to make this book a valuable addition to the collection of many professionals and students.

This book was conceptualized with the vision of imparting up-to-date information and advanced data in this field. To ensure the same, a matchless editorial board was set up. Every individual on the board went through rigorous rounds of assessment to prove their worth. After which they invested a large part of their time researching and compiling the most relevant data for our readers.

The editorial board has been involved in producing this book since its inception. They have spent rigorous hours researching and exploring the diverse topics which have resulted in the successful publishing of this book. They have passed on their knowledge of decades through this book. To expedite this challenging task, the publisher supported the team at every step. A small team of assistant editors was also appointed to further simplify the editing procedure and attain best results for the readers.

Apart from the editorial board, the designing team has also invested a significant amount of their time in understanding the subject and creating the most relevant covers. They scrutinized every image to scout for the most suitable representation of the subject and create an appropriate cover for the book.

The publishing team has been an ardent support to the editorial, designing and production team. Their endless efforts to recruit the best for this project, has resulted in the accomplishment of this book. They are a veteran in the field of academics and their pool of knowledge is as vast as their experience in printing. Their expertise and guidance has proved useful at every step. Their uncompromising quality standards have made this book an exceptional effort. Their encouragement from time to time has been an inspiration for everyone.

The publisher and the editorial board hope that this book will prove to be a valuable piece of knowledge for researchers, students, practitioners and scholars across the globe.

List of Contributors

Jen-Tzung Chien and Hsin-Lung Hsieh
Department of Electrical and Computer Engineering, National Chiao Tung University, Taiwan 30010, Republic of China

Ali Pourmohammad and Seyed Mohammad Ahadi
Electrical Engineering Department, Amirkabir University of Technology, 424 Hafez Ave., Tehran 15914, Iran

Sumxin Jiang, Rendong Yin and Peilin Liu
Department of Electronic Engineering, Shanghai Jiao Tong University, Shanghai 200241, China

Yanna Ma and Akinori Nishihara
Department of Communications and Integrated Systems, Tokyo Institute of Technology, Tokyo 152-8552, Japan

Santitham Prom-on
Department of Computer Engineering, Faculty of Engineering, King Mongkut's University of Technology Thonburi, Bangkok 10140, Thailand
Department of Speech, Hearing and Phonetic Sciences, University College London, London WC1N 1PF, UK

Peter Birkholz
Department for Phoniatrics, Pedaudiology and Communication Disorders, University Hospital Aachen and RWTH Aachen University, Aachen 52074, Germany

Yi Xu
Department of Speech, Hearing and Phonetic Sciences, University College London, London WC1N 1PF, UK

Caleb Rascon, Gibran Fuentes and Ivan Meza
Instituto de Investigaciones en Matematicas Aplicadas y en Sistemas, Universidad Nacional Autonoma de Mexico, Circuito Escolar S/N, 04510 Mexico, Mexico

Jack Leegaard, Jan Østergaard and Søren Holdt Jensen
Department of Electronic Systems, Aalborg University, Niels Jernes Vej 12, Aalborg 9220, Denmark

Ram Zamir
Department of Electrical Engineering-Systems, Tel Aviv University, Ramat Aviv, Tel Aviv 69978, Israel

Mohammadali Sebghati and Hamidreza Amindavar
Electrical Engineering Department, Amirkabir University of Technology, Tehran 15914, Iran

James A Ritcey
Electrical Engineering Department, University of Washington, Seattle, WA 98195, USA

Toru Nakashika
Graduate School of System Informatics, Kobe University, 1-1 Rokkodai, Nada-ku, 657-8501 Kobe Japan

Tetsuya Takiguchi and Yasuo Ariki
Organization of Advanced Science and Technology, Kobe University, 1-1 Rokkodai, Nada-ku, 657-8501 Kobe, Japan

Wen-Lin Zhang, DanQu and Bi-Cheng Li
Zhengzhou Information Science and Technology Institute, Zhengzhou 450000, China

Wei-Qiang Zhang
Tsinghua National Laboratory for Information Science and Technology, Department of Electronic Engineering, Tsinghua University, Beijing 100084, China

Gen-Fang Chen
Department of Computer Science and Technology, Hangzhou Normal University, Hangzhou, Zhejiang 310036, China

Jia-Shing Sheu
Department of Computer Science, National Taipei University of Education, Taipei 10671, Taiwan

Xin Liu and Chang-Chun Bao
Speech and Audio Signal Processing Lab, School of Electronic Information and Control Engineering, Beijing University of Technology, 100124 Beijing, China

Furong Yan, Yanbin Zhang and Jiachang Yan
Beijing University of Posts and Telecommunications, No.10 Xitucheng Road, Beijing 100876, China

Ali Al-Haj
Department of Computer Engineering, King Abdullah II Faculty of Engineering, Princess Sumaya University for Technology, Al-Jubeiha, PO Box 1438, Amman 11941, Jordan

Diego Castán, Alfonso Ortega and Antonio Miguel
Departamento Ingeniería Electrónica y Comunicaciones, Universidad de Zaragoza, María de Luna, 1, Zaragoza 50018, Spain

Eduardo Lleida
Departamento Ingeniería Electrónica y Comunicaciones, Universidad de Zaragoza, María de Luna, 1, Zaragoza 50018, Spain

Ning-Han Liu
Department of Management Information Systems, National Pingtung University of Science and Technology, 1, Shuefu Road, Neipu, Pingtung 912, Taiwan

Wei-Wei Liu
Tsinghua National Laboratory for Information Science and Technology, Department of Electronic Engineering, Tsinghua University, Beijing 100084, China
General Communication Station, General Logistics Department, Beijing 100036, China

Wei-Qiang Zhang and Jia Liu
Tsinghua National Laboratory for Information Science and Technology, Department of Electronic Engineering, Tsinghua University, Beijing 100084, China

Michael T Johnson
Department of Electrical and Computer Engineering, Marquette University, Milwaukee, WI 53233, USA

Rubén Fraile and Juan Ignacio Godino-Llorente
Signal Theory & Communications Department, Universidad Politécnica de Madrid, Campus Sur - Carretera de Valencia km 7, 28031 Madrid, Spain

Malte Kob
Erich-Thienhaus-Institut, Hochschule für Musik Detmold, Neustadt 22, 32756 Detmold, Germany

Humberto M Torres
Laboratorio de Investigaciones Sensoriales, INIGEM, CONICET-UBA, Av. Córdoba 2351, 9 Piso Sala 2, 1120 Ciudad Autonoma de Buenos Aires, Argentina

Jorge A Gurlekian
Laboratorio de Investigaciones Sensoriales, INIGEM, CONICET-UBA, Av. Córdoba 2351, 9 Piso Sala 2, 1120 Ciudad Autonoma de Buenos Aires, Argentina

Hansjörg Mixdorff
Department of Computer Sciences and Media, Beuth University Berlin, Berlin, Germany

Hartmut Pfitzinger
Pfitzinger Voice Design, Lübeck, Germany